2016 IEEE 22nd International Symposium for Design and Technology in Electronic Packaging (SIITME 2016)

Oradea, Romania
20-23 October 2016

IEEE Catalog Number: CFP1607I-POD
ISBN: 978-1-5090-4446-7

Copyright © 2016 by the Institute of Electrical and Electronics Engineers, Inc
All Rights Reserved

Copyright and Reprint Permissions: Abstracting is permitted with credit to the source. Libraries are permitted to photocopy beyond the limit of U.S. copyright law for private use of patrons those articles in this volume that carry a code at the bottom of the first page, provided the per-copy fee indicated in the code is paid through Copyright Clearance Center, 222 Rosewood Drive, Danvers, MA 01923.

For other copying, reprint or republication permission, write to IEEE Copyrights Manager, IEEE Service Center, 445 Hoes Lane, Piscataway, NJ 08854. All rights reserved.

***This publication is a representation of what appears in the IEEE Digital Libraries. Some format issues inherent in the e-media version may also appear in this print version.**

IEEE Catalog Number: CFP1607I-POD
ISBN (Print-On-Demand):   978-1-5090-4446-7
ISBN (Online):   978-1-5090-4445-0

Additional Copies of This Publication Are Available From:

Curran Associates, Inc
57 Morehouse Lane
Red Hook, NY 12571 USA
Phone: (845) 758-0400
Fax: (845) 758-2633
E-mail: curran@proceedings.com
Web: www.proceedings.com

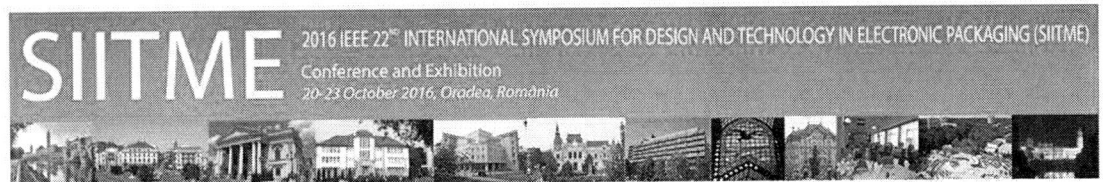

2016 IEEE 22nd International Symposium for Design and Technology in Electronic Packaging (SIITME).

20th–23rd of October 2016, Oradea, Romania

"Politehnica" University of Bucharest
Technological Electronics and Interconnections
Techniques Research Center,
UPB-CETTI
www.cetti.ro

The University of Oradea
www.uoradea.ro

Contact: Gabriel Chindris, Publication Chair
Tel: +40 264-401469 Fax: +40 264-594806 E-mail: gabriel.chindris@ael.utcluj.ro

978-1-5090-4446-7/16 $31.00 © 2016 IEEE

Foreword

Welcome to the 22nd edition of the International Symposium for Design and Technology in Electronic Packaging!

SIITME, now at the 22nd edition, has become during its long existence a meaningful event focused on highlighting topics that are relevant to the electronic industry in our region. One of the main objectives of SIITME is to foster a strategic partnership between the academia and the electronics industry. This is a relevant goal especially since the electronics industry has become an important economic player in many parts of our region. In this context of unprecedented development of the electronics industry, in a globalized economic environment, it is natural to ask ourselves what course of action to take. Some time ago, Prof. Rao Tummala, President of the IEEE CPMT Society at that time, and current Director of the prestigious Packaging Research Center at Georgia Institute of Technology, said:" **The global electronic industry acts as the engine for science, technology, advanced manufacturing, and the overall economy of the countries that participate in it."** As such, it is reasonable to be interested in the fact that countries in our region take part in the evolution of the electronics industry, which is why a strategic partnership between the industry and the educational environment involved in training human resource is a "sine qua non" condition. An ecosystem should be created to ensure positive effects on the prosperity of those involved. It is extremely important to establish dialogue, open communication between those who train the human resource and the economic environment. The resulting synergy leads to benefits for those involved, academia and electronics industry both. In the context of providing human resource able to be involved in the electronics industry, it is useful to note Professor James Morris's assertion during a communication at ECTC (Electronic Component Technology Conference), one of the most prominent manifestations of the electronics industry worldwide that is held annually in the United States. Among others he said: **"Electronics packaging has become recognized as a critical technology for the continued growth of the nation's electronics industry."** This claim is consistent with the current educational environment necessary to be developed in our region to assure the electronic industry with proper human resource. If we want to remain active players in the evolution of the electronics industry, we have to keep them in sight. We are aware that in harmonizing, sometimes diverging, interests of the "actors" involved, we are still at the beginning. However, it is extremely important that we have established a dialogue between the parties involved, with all "cards on the table", trying to identify mutual beneficial solutions. Towards this goal, this edition brings a continued direct and massive involvement of representatives of the electronics industry. Representatives of important companies participate in the Industry Exhibition. Some of them also hold keynote speeches or presentations. We hope that this trend continues in future editions and that SIITME will become a bridge between academia, research and industry.

Finally, to those who are present in the very nice city of Oradea at the 22nd edition of SIITME, we wish a pleasant stay, fruitful discussions and professional satisfaction.

Paul SVASTA,
SIITME General Chair

Delia UNGUR,
General Industrial Co-Chair

Dan Pitică,
General Academic
Co-Chair

Dear participants and guests,

I'm pleased and honoured, on behalf of the local organizing committee, to welcome you all at Băile Felix, Oradea. Romania, to the 2016 IEEE 22nd International Symposium for Design and Technology in Electronic Packaging (SIITME).

Oradea is a north-west border Romanian city, a multicultural town which is highly appreciated by the different ethnic groups living on the banks of the Crisul Repede River. Because the Romanian industry must cope with a fierce global competition, Oradea City Hall developed the Euro Business Industrial Park containing some important companies involved in electronic industry and manufacturing.

Băile Felix spa is located in the north-west of Romania, Bihor county, at a distance of about 9 km from Oradea and 22 km from the border crossing point Bors. Located in the Crisuri Plain, the resort has a moderate continental climate with mild winters and temperate summers, with Mediterranean influence, ideal for spa tourism. The whole year, Băile Felix spa offers relaxing conditions, vacation and recreation recovery.

University of Oradea is aware that a very important target in near future is to develop a network that promotes the development of human resources for innovation. So, the Electronics and Telecommunication Department is deeply involved in developing a strong connection between industry and technical academic courses and has a growing interest for electronic packaging.

Our local organizing committee is confident that SIITME 2016 will be a chance for companies involved in electronics industry and Romanian technical academic schools to unite their interests and activities.

We wish all SIITME 2016 participants a very pleasant and successful attendance! Welcome to Băile Felix!

Prof. Cornelia GORDAN, Ph.D.
SIITME 2016 Conference Chair

University of Oradea, Romania

CALL FOR PAPERS - SIITME2016

Dear Colleagues,
The organising committee of SIITME 2016 kindly invites you to submit an abstract/paper to the **2016 IEEE 22nd International Symposium for Design and Technology in Electronic Packaging (SIITME).** The scientific event will take place in Oradea, on **October 20th–23rd, 2016. Web page:** Web page: http://www.siitme.ro

SUBMISSION OF ABSTRACTS AND PAPERS:

Interested authors are invited to submit a two-page abstract according to the template, as MS-Word document, Version 97 or later, and as PDF document, to the SIITME 2016 Conference Management System.

Papers meeting the quality criteria will be included in the IEEE Xplore Digital Library.

Abstracts and papers will be reviewed by the international Scientific Committee.

1. Each abstract will be reviewed on:
- suitability for one of the topics of the conference
- scientific content and level, and the relevance of presented results;
- correspondence with the abstract template, English usage and grammar.
Authors of accepted abstracts will be invited to submit a full-length conference paper according to the conference paper template.

2. Papers meeting the following criteria will be published in the Conference Proceedings and will be available through IEEE Xplore:
- the comments of the reviewers have been taken into consideration;
- originality of the paper is given*;
- it corresponds with the paper template, English usage and grammar are correct,
- signed IEEE copyright form has been submitted.

TOPICS
A. Emerging Technologies & Trends in Advanced Packaging, Microsystems, Heterointegration, Printed Electronics, Smart Textiles, Healthcare
B. Components, Assembling, and Manufacturing Technology
C. Design of Electronic Circuits and Systems
D. Electronics Simulation & Modelling
E. Electronics Applications: Optoelectronics, Advanced Communication, Automotive, Aerospace and Power Electronics
F. Applied Reliability
G. Challenges in Global Education

CHAIRS:
General Chair: **Paul SVASTA**, "Politehnica" University of Bucharest, Ro
General Academic Co-Chair: **Dan PITICĂ**, Technical University of Cluj-Napoca, Ro
General Industrial Co-Chair: **Delia UNGUR**, Euro Business Park, Oradea, Ro
Conference Chair: **Cornelia GORDAN**, University of Oradea, Ro
Technical Program Chair: **Detlef BONFERT**, Fraunhofer EMFT, Munich, De
Technical Program Co-Chair: **Norocel CODREANU**, "Politehnica" University of Bucharest, Ro

Awards Committee Chair: **Heinz WOHLRABE**, Dresden University of Technology, De
Scientific Committee Chair: **Balázs ILLÉS**, Budapest University of Technology and Economics, Hu
Scientific Committee Co-Chair: **Heinz WOHLRABE**, Dresden University of Technology, De
Scientific Committee Co-Chair: **Ciprian IONESCU**, "Politehnica" University of Bucharest, Ro
Publication Committee Chair: **Gabriel CHINDRIȘ**, Technical University of Cluj- Napoca, Ro
Publication Committee Co-Chair: **Bogdan MIHĂILESCU**, "Politehnica" University of Bucharest, Ro
International Publication Advisor: **Zsolt ILLYEFALVI-VITÉZ**, Budapest University of Technology and Economics, Hu
Human Resource Education and Training Committee Chair: **Aurelia FLOREA,** MIELE Tehnica, Brașov, Ro
Human Resource Education and Training Committee Co-Chair: **Cornelia GORDAN,** University of Oradea, Ro
Human Resource Education and Training Committee Co-Chair:**Delia UNGUR**, Euro Business Park Oradea, Ro

IMPORTANT DATES:
Submission of abstracts: September 1st, 2016
Notification of abstract acceptance: September 19th, 2016
Submission of camera-ready papers and copyright forms: October 8th, 2016
Confirmation of registration and submission of e-copyright form: October 13th, 2016

*The originality of papers will be checked by the IEEE CrossCheck plagiarism detection and prevention software.

Contents

TOPIC A – EMERGING TECHNOLOGIES & TRENDS IN ADVANCED PACKAGING, MICROSYSTEMS, HETEROINTEGRATION, PRINTED ELECTRONICS, SMART TEXTILES, HEALTHCARE

PP. 25 - 31:

CHARACTERIZATION OF THE SHAPE OF GOLD NANOPARTICLES PREPARED BY THERMAL ANNEALING

A. Bonyár, J. Kámán, I. Csarnovics

PP. 32- 35:

A STUDY OF SW150 CONDUCTIVE PASTE AS A POSSIBLE USE IN SOLDERLESS ASSEMBLY FOR ELECTRONICS

G. Varzaru, C. Marghescu, Al. Vasile, P. Svasta and B. Buta

TOPIC B – COMPONENTS, ASSEMBLING AND MANUFACTURING TECHNOLOGY

PP. 36 - 39:

INVESTIGATING THE EFFECT OF SOLDER PASTE VISCOSITY CHANGE ON THE PRESSURE DURING STENCIL PRINTING

O. Krammer, B. Varga, D. Busek

PP. 40 - 43:

OPTIMIZATION OF PCB ASSEMBLY PROCESS

L. Tarba , P. Mach

PP. 44 - 47:

THERMOPHYSICAL PROPERTIES OF SOME LOW TEMPERATURE LEAD-FREE SOLDER PASTES DEDICATED TO AUTOMOTIVE APPLICATIONS

M. Branzei, I. Plotog, G. Varzaru, T. Cucu

PP. 48 - 54:

INVESTIGATIONS ON HEAT TRANSFER WITH DIFFERENT PCB SUBSTRATES DURING VAPOUR PHASE SOLDERING

L. Fazekas, D. Nagy, D. Busek, A. Géczy

PP. 55 - 58:

SENSITIVITY OF RESISTANCE, NOISE AND NONLINEARITY OF CONDUCTIVE ADHESIVE JOINTS TO CHANGES IN ADHESIVE

S. Barto, P. Mach, A. Duraj;

TOPIC C – DESIGN OF ELECTRONIC CIRCUITS AND SYSTEMS

PP. 59 - 62:

IMPLEMENTATION OF A MICROWAVE ELLIPTICAL LOW PASS FILTER WITH RADIAL STUBS

M. Al. Ilie, D. Pitica, L. Viman

PP. 63 - 65:

THREE OMEGA PROBE WITH AUTO-ZEROING

I. Ates, L. Cetin, A. Turgut, M. Chirtoc

PP. 66 - 69:

REVISION OF THE SAMPLING THEOREM

M. Bujor

PP. 70 - 73:

DEVICE FOR INTERCEPTING AND DISRUPTING THE HIDDEN HEADSETS

C. Cherciu, and D-I. Nastac

PP. 74 - 77:

ASPECTS OF USING LOW LAYER COUNT PCBS FOR EMBEDDED SYSTEMS WITH FPGA DEVICES IN BGA PACKAGES

A. Drumea, M. Pantazica

PP. 78 - 81:

HOST EMULATOR FOR NEXT GENERATION BATTERY CHARGERS

C. Grecu, M. Pantazica, C. Iordache

PP. 82 - 85:

DEVELOPMENT OF UNDERWATER SENSOR UNIT FOR STUDYING MARINE LIFE

M. Hnatiuc, I. Lazar, A. Ghilezan

PP. 86 - 89:

REAL TIME SYSTEM FOR EXTRACTION AND PLAYBACK OF AN INSTRUMENTAL SOUND

L. M. Ionescu, I. Lita

PP. 90 - 93:

REMOTE COMMUNICATION INTERFACE FOR SOUND AND VIBRATION SENSORS

D. Al. Visan, L. M. Ionescu and A. I. Lita

PP. 94 - 97:

FPGA-ENABLED HARDWARE MULTITASKING APPLICATIONS IN ENERGY HARVESTING LABORATORIES

O. Machidon, P. A. Cotfas, D. T. Cotfas

PP. 98 - 101:

PERSONALIZED RING OSCILLATOR-BASED TRUE RANDOM NUMBER GENERATOR ANALYSIS USING NON-INVASIVE ATTACKS

A. Marghescu, D.-C. Vasile, E. Simion, P. Svasta

PP. 102 - 105:

WIRELESS DIAGNOSIS AND MONITORING SYSTEM OF SENSOR NETWORK FROM CIVIL STRUCTURES

S. Pop, V. Bande

PP. 106 - 109:

CONTINUOUS RESPIRATORY MONITORING DEVICE FOR DETECTION OF SLEEP APNEA EPISODES

C. Rotariu, R. Bozomitu, Al. Pasarica, C. Cristea, and D. Arotaritei

PP. 110 - 113:

DESIGN AND SETUP OF POWER ANALYSIS ATTACKS

M. Safta, A. Marghescu, M. Dima, P. Svasta

PP. 114 - 117:

MACHINE-TO-MACHINE COMMUNICATIONS FOR CLOUD-BASED ENERGY MANAGEMENT SYSTEMS WITHIN SMES

G. Suciu, Al. Vulpe, O. Fratu, L. Necula, A. Pasat, V. Suciu

PP. 118 - 121:

MOUSE AND DISPLAY DRIVER ON A SINGLE MICROCHIP TESTED ON FPGA AND BUILT FOR AN ASIC

R. Szabo, A. Gontean

PP. 122 - 125:

CREATION OF A FIGHT GAME IN BORLAND PASCAL WITH THE POSSIBILITY TO BE PORTED ON AN FPGA

R. Szabo, A. Gontean

PP. 126 - 129:

DSP BASED INTERCONNECTION CIRCUIT OF THE RENEWABLE ENERGY SOURCES TO A SMART GRID

N. D. Trip, O. Neamtu

PP. 130 - 133:

IMPROVED TAMPER DETECTION CIRCUIT BASED ON LINEAR-FEEDBACK SHIFT REGISTER

D.-C. Vasile, A. Marghescu, P. Svasta

PP. 134 - 138:

HIGH RELIABILITY WIRELESS SENSOR NODE FOR BEE HIVE MONITORING

M. Vidrascu, P. Svasta, M. Vladescu

TOPIC D – ELECTRONICS SIMULATION & MODELLING

PP. 139 - 141:

COUPLED SURFACE PLASMON RESONANCE ON GOLD NANOCUBES - INVESTIGATION BY SIMULATION

A. Bonyár, G. Szántó, and I. Csarnovics

PP. 142 - 145:

ENHANCING THERMAL CAPABILITIES OF COMPONENT PACKAGING

A. Fodor, G. Chindris, and D. Pitica

PP. 146 - 150:

ANALYSIS OF CROSSTALK EFFECTS ON SINGLE ENDED SIGNAL LINES CROSSING SPLIT REFERENCE PLANES

M. Manofu, R. Vladu?a and C. Negrea

PP. 151 - 154:

ANALYSING OF HALF-BRIDGE INVERTER USING THE SIMULINK PLATFORM

I. Baciu, S. Pop, V. Bande

PP. 155 - 158:

ANALYSIS OF LEDS THERMAL PROPERTIES

N. Badalan, P. Svasta

PP. 159 - 164:

STABILITY EVALUATION METHOD USING PHASE RESPONSE MEASUREMENTS

R. Belea, S. Epure

PP. 165 - 170:

SIMULATION & MODELLING OF A TUNGSTEN FILAMENT WITH COMSOL FOR ELECTROTHERMAL PROCESS

S. Cadar, R. Etz, T. Patarau, D. Petreus, S. M. Fonou

PP. 171 - 174:

AN IMPROVED METHOD FOR THE ELECTRICAL PARAMETERS IDENTIFICATION OF A SIMPLIFIED PSPICE SUPERCAPACITOR MODEL

I. Ciocan, C. Farcas, and A. Tulbure

PP. 175 - 178:

DEVELOPING A MULTI SENSORS SYSTEM TO DETECT SLEEPINESS TO DRIVERS FROM TRANSPORT SYSTEMS

I. Costea, C. Dumitrescu, F. Nemtanu, I. Badescu, A. Banica

PP. 179 - 183:

MODELLING AND PSPICE SIMULATION OF A PHOTOVOLTAIC/THERMOELECTRIC SYSTEM

P. A. Cotfas, D. T. Cotfas, O. Machidon

PP. 184 - 187:

STATISTICAL METHODS FOR DETERMINING COMPONENTS' NON-LINIARITIES, FROM THERMOLUMINESCENT DEVICES

M. Dima, D. David-Rus, C. Ionescu

PP. 188 - 191:

BOND-GRAPH MODELLING OF THE EQUIVALENT CIRCUIT OF AN ON-CHIP SPIRAL INDUCTOR

A. Grava, C- Grava

PP. 192 - 195:

AN PHOTOVOLTAIC SYSTEM TESTER WITH THREE-PHASE OFF-GRID SUPPLY

M. Neamtu, N. D. Trip

PP. 196 - 199:

COMPARISON BETWEEN ZUBIETA MODEL OF SUPERCAPACITORS AND THEIR REAL BEHAVIOR

R. Negroiu, P. Svasta, Al. Vasile, C. Ionescu, C. Marghescu

PP. 200 - 204:

THE FRAMEWORK OF USING MODELS FOR COMPARATIVE ASSESSMENT OF TRAFFIC SENSORS

F. Nemtanu, I. Costea

PP. 205 - 208:

REAL-TIME 3D NEAR-FIELD VISUALIZATION USING LED FIELD SENSORS

A. Petrariu and E. Coca

PP. 209 - 212:

ELECTRO-THERMAL SIMULATION STUDY OF DIFFERENT CORE SHAPE PLANAR TRANSFORMER

C. Ropoteanu, P.Svasta and C. Ionescu

TOPIC E – ELECTRONICS APPLICATIONS: OPTOELECTRONICS, ADVANCED COMMUNICATION, AUTOMOTIVE, AEROSPACE AND POWER ELECTRONICS

PP. 213 - 216:

EYE BLINKING DETECTION TO PERFORM SELECTION FOR AN EYE TRACKING SYSTEM USED IN ASSISTIVE TECHNOLOGY

A. Pasarica, R. G. Bozomitu, V. Cehan and C. Rotariu

PP. 217 - 222:

VIBROMOD - AN ELECTRONIC EQUIPMENT FOR DATA VIBRATION MEASUREMENT AND ANALYSIS

I. Nacu, L. Luca, N. Roman, D. Aiordachioaie

PP. 223 - 226:

THERMAL SIMULATION OF TRAFFIC LIGHTS IN EXTREME WEATHER CONDITIONS

N. Badalan, P. Svasta, C. Marghescu

PP. 227 - 231:

ONE GLASS SOLUTION TOUCH PANEL PERFORMANCE VARIATION OVER TEMPERATURE EXPOSURE

H.-T. Cutlac, P. M. Svasta, S. Calea

PP. 232 - 235:

ROBUST AUDIO FORENSIC SOFTWARE FOR RECOVERING SPEECH SIGNALS DROWNED IN LOUD MUSIC

R. Dobre, C. Elisei-Iliescu, C. Paleologu, C. Negrescu, D. Stanomir

PP. 236 - 239:

FORMULA STUDENT SINGLE USER RACE CAR -ELECTRONIC CONTROL

V. Lupu, C. Gerigan, and P. L. Ogrutan

PP. 240 - 243:

WIRELESS SENSOR NETWORKS AS PART OF EMERGENCY SITUATIONS MANAGEMENT SYSTEM

C. Lung, A. Buchman, S. Sabou

PP. 244 - 247:

INVESTIGATIONS ON AVAILABLE BANDWIDTH IN VISIBLE-LIGHT COMMUNICATIONS

A. Marcu, R. Dobre, M. Vladescu

PP. 248 - 251:

E-BIKE ELECTRONIC CONTROL UNIT

F. Dumitrache, M. C. Carp, and Gh. Pana

PP. 252 - 255:

LABVIEW SIMULATOR FOR TERRESTRIAL MIMO COMMUNICATIONS

L. Perisoara, D. Sacaleanu, R. Stoian

PP. 256 - 261:

SINGLE-PHASE INVERTER FOR SOLAR ENERGY CONVERSION CONTROLLED WITH DSPACE DS1104

D. Petreus, T. Patarau, R. Truta, R. Etz

PP. 262 - 265:

OBSTACLE AVOIDANCE ALGORITHM

S. Sabou, C. Lung

PP. 266 - 269:

WIRELESS SENSOR NODE FOR FRUIT GROWING MONITORING

D. Sacaleanu, L. Perisoara, R. Stoian, L. Sucu

PP. 270 - 273:

TESTING IMMUNITY TO PORTABLE TRANSMITTERS WITH HELICAL ANTENNAS: KEY CONCEPTS

A. M. Silaghi, A. De Sabata, Al. M. Silaghi

TOPIC F –APPLIED RELIABILITY

PP. 274 - 278:

ELECTROCHEMICAL MIGRATION OF SN AND AG IN NACL ENVIRONMENT

B. Medgyes, D. Szivós, S. Ádám, L. Tar, P. Tamási, L. Gál, R. Berényi, G. Harsányi

PP. 279 - 282:

A LOW-COST PAVEMENT IMAGE ACQUISITION SYSTEM

C. Chiculita, L. Frangu

PP. 283 - 289:

FFT BASED INVESTIGATIONS ON LIGHT FLICKER IN NEW LIGHTING SYSTEMS

C. Ionescu, M. Dima, D. Bonfert

PP. 290 - 293:

ON SPECTRAL COMPONENT ESTIMATION USING NEURAL NETWORKS FOR ROLLING BEARING FAULT DIAGNOSIS

V. Nicolau, M. Andrei

PP. 294 - 299:

CHARACTERIZATION OF TIN PEST BY ELECTRICAL RESISTANCE MEASUREMENT

A. Skwarek, B. Illés, A. Géczy

TOPIC G – CHALLENGES IN GLOBAL EDUCATION

PP. 300 - 303:

ADAPTIVE USER INTERFACE FOR HIGHER EDUCATION BASED ON WEBTECHNOLOGY

M. Ciolacu, R. Beer

PP. 304 - 307:

EXPERIMENTAL MODULE FOR ASSISTIVE TECHNOLOGIES APPLICATIONS

I. Lita, D. Al. Visan and A. Gh. Mazare

Impressum

General Conference e-mails and website:
siitme@cetti.ro; siitme@siitme.ro; http://www.siitme.ro/

Location:
Hotel Poienita, Calea Beiușului, nr. 26, Băile Felix, Oradea, Romania | http://www.hotel-poienita.ro/

Organizer:

"Politehnica" University of Bucharest (UPB-CETTI) (http://www.cetti.ro)
Bd. Iuliu Maniu, nr. 1-3 Complex LEU, Corp B, cam. 301
Tel.: +40 21 3169633
Fax: +40 21 3169634
e-mail: cetti@cetti.ro

Committees:

General Chair:
Paul SVASTA, "Politehnica" University of Bucharest, Romania

General Academic Co-Chair:
Dan PITICĂ, Technical University of Cluj-Napoca, Romania
General Industrial Co-Chair:
Delia UNGUR, Euro Business Park Oradea, Romania

Conference Chair:
Cornelia GORDAN, University of Oradea, Romania
Conference Co-Chair:
Daniel Nistor TRIP, University of Oradea, Romania

Technical Program Chair:
Detlef BONFERT, Fraunhofer EMFT, München, Germany
Technical Program Co-Chair:
Norocel CODREANU, "Politehnica" University of Bucharest, Romania

Awards Committee Chair:
Heinz WOHLRABE, Dresden University of Technology, Dresden, Germany

Scientific Committee Chair:
Balázs ILLÉS, Budapest University of Technology and Economics, Hungary
Scientific Co-Chairs:

Heinz WOHLRABE, Dresden University of Technology, Germany
Ciprian IONESCU, "Politehnica" University of Bucharest, Romania

Human Resource Education and Training Committee Chair:
Aurelia FLOREA, MIELE Tehnica, Braşov, Romania

Human Resource Education and Training Committee Co-Chair:
Cornelia GORDAN, University of Oradea
Human Resource Education and Training Committee Co-Chair:
Delia UNGUR, Euro Business Park Oradea

Publication Chair:
Gabriel CHINDRIŞ, Technical University of Cluj- Napoca, Romania
Publication Co-Chair:
Bogdan MIHĂILESCU, "Politehnica" University of Bucharest, Romania
International Publication Advisor:
Zsolt ILLYEFALVI-VITÉZ, Budapest University of Technology and Economics, Hungary

International Steering Committee
Chaired by the General Chair:
Paul SVASTA, "Politehnica" University of Bucharest, Romania

Tiberiu ABRAHAM, ROEL, Romania
Dorel AIORDĂCHIOAIE "Dunărea de Jos" University of Galaţi, Romania
Karlheinz BOCK, Dresden University of Technology, Germany
Detlef BONFERT, Fraunhofer EMFT, Munich, Germany
Atilla BONYÁR, Budapest University of Technology and Economics, Hungary
Alexandru BORCEA, ARIES, Romania
Paul Nicolae BORZA, "Transilvania" University of Braşov, Romania
Mihai BRÂNZEI, "Politehnica" University of Bucharest, Romania
Attila BUCHMAN, North University, Center of Baia Mare, Romania
Ioan BUCIU, University of Oradea, CCTIEA Research Center President, Romania
Vlad CEHAN, "Gheorghe Asachi" Technical University of Iaşi, Romania
Emilian CEUCĂ, "1 Decembrie 1918" University of Alba Iulia, Romania
Norocel CODREANU, "Politchnica" University of Bucharest, Romania
Liviu COSEREANU, ACTTM Bucharest; AFCEA Bucharest chapter, Romania
Ilona COSTEA, "Politehnica" University of Bucharest, Romania
Sorin CURILA, University of Oradea, Romania
Gabriel DIMA, "Politehnica" University of Bucharest, Romania
Mihai Octavian Dima, Institute for Nuclear Physics and Engineering - Horia Hulubei, Romania
Florin DRAGHICI, "Politehnica" University of Bucharest, Romania
Petrin DRUMEA, Research Institute for Hydraulic & Pneumatics, Romania
Carmen GERIGAN, "Transilvania" University of Braşov, Romania
Aurel-Ştefan GONTEAN, "Politehnica" University of Timişoara, Romania
Cornelia GORDAN, University of Oradea, Romania
Adrian GRAUR, "Ştefan cel Mare" University of Suceava, Romania
Patrick HASPEL, Cadence Design Systems GmbH, Germany
Mihaela HNATIUC, Maritime University of Constanta, Romania

Hartmut HOHAUS, MIELE Tehnica, Braşov, Romania
Balázs ILLÉS, Budapest University of Technology and Economics, Hungary
Zsolt ILLYEFALVI-VITÉZ, Budapest University of Technology and Economics, Hungary
Ciprian IONESCU, "Politehnica" University of Bucharest, Romania
Radu IONESCU, Radioconsult srl, Bucharest, Romania
Olivér KRAMMER, Budapest University of Technology and Economics, Hungary
Ioan LIŢĂ, University of Piteşti, Romania
Pavel MACH, Technical University of Prague, Czech Republic
Alexandru MARIN, "Politehnica" University of Bucharest, Romania
Bálint MEDGYES, Budapest University of Technology and Economics, Hungary
Ioan P. MIHU, "Lucian Blaga" University of Sibiu, Romania
Cosmin MOISE, Continental Automotive, Timisoara, Romania
James E. MORRIS, Portland State University, OR, USA
Johann NICOLICS, Vienna University of Technology, Austria
Petre OGRUŢAN, "Transilvania" University of Braşov, Romania
Gheorghe PANĂ, "Transilvania" University of Braşov, Romania
Alena PIETRIKOVA, Technical University of Kosice, Slovakia
Dorin PETREUŞ, Technical University of Cluj-Napoca, Romania
Dan PITICĂ, Technical University of Cluj- Napoca, Romania
Dan POPA, Maritime University of Constanţa, Romania
Valentin POPA, "Ştefan cel Mare" University of Suceava, Romania
Jerzy POTENCKI, University of Technology, Rzesow, Poland
Marius RANGU, "Politehnica" University of Timişoara, Romania
Wilfried SAUER, Technical University of Dresden, Germany
Paul ŞCHIOPU, "Politehnica" University of Bucharest, Romania
Paul SVASTA, "Politehnica" University of Bucharest, Romania
Iuliu SZEKELY, „Sapientia" University of Târgu Mureş, Romania
Marius TOADER, ALFA TEST, Timişoara, Romania
Daniel TRIP, University of Oradea, Romania
Adrian TULBURE, „1 Decembrie 1918" University of Alba Iulia, Romania
Carmen TURCU, INTRAROM Bucharest, Romania
Slavka TZANOVA, Technical University of Sofia, Bulgaria
Daniel VIŞAN, University of Piteşti, Romania
Gabriel VLĂDUŢ, ARIES Oltenia, Romania
Dan VUZA, Institute of Mathematics of the Romanian Academy, Romania
Heinz WOHLRABE, Dresden University of Technology, Germany
Klaus WOLTER, Dresden University of Technology, Germany

Scientific Committee
Chair: Balázs ILLÉS, Budapest University of Technology and Economics, Hungary
Co-Chairs: Heinz WOHLRABE, Dresden University of Technology, Dresden, Germany
Ciprian IONESCU, "Politehnica" University of Bucharest, Romania

Dorel AIORDACHIOAIE, "Dunărea de Jos" University of Galaţi , Romania
Detlef BONFERT, Fraunhofer EMFT, Munich, Germany
Radu BOZOMITU, "Gheorghe Asachi" Technical University of Iaşi, Romania

Mihai BRÂNZEI, "Politehnica" University of Bucharest, Romania
Ştefan CASTRAVETE, CAE – LYNX Europe, Craiova, Romania
Vlad CEHAN, "Gheorghe Asachi" Technical University of Iaşi, Romania
Gabriel CHINDRIŞ, Technical University of Cluj- Napoca, Romania
Eugen COCA, "Ştefan cel Mare" University of Suceava, Romania
Norocel CODREANU, "Politehnica" University of Bucharest, Romania
Gabriel DIMA, "Politehnica" University of Bucharest, Romania
Mihai Octavian Dima, Institute for Nuclear Physics and Engineering - Horia Hulubei, Romania
Florin DRAGHICI, "Politehnica" University of Bucharest, Romania
Andrei DRUMEA, "Politehnica" University of Bucharest, Romania
Cristian FĂRCAŞ, Technical University of Cluj-Napoca, Romania
Laurenţiu FRANGU, "Dunărea de Jos" University of Galaţi , Romania
Alexandru GACSADI, University of Oradea, Romania
Carmen GERIGAN, "Transilvania" University of Braşov, Romania
Aurel-Ştefan GONTEAN, "Politehnica" University of Timişoara, Romania
Cristian GRAVA, University of Oradea, Romania
Mihaela HNATIUC, Maritime University of Constanţa, Romania
Zsolt ILLYEFALVI-VITÉZ, Budapest University of Technology and Economics, Hungary
Olivér KRAMMER, Budapest University of Technology and Economics, Hungary
Emil LAZARCIUC, Continental Automotive Romania
Ioan LITA, University of Piteşti, Romania
Pavel MACH, Technical University of Prague, Czech Republic
Alexandru MARIN, "Politehnica" University of Bucharest, Romania
James E. MORRIS, Portland State University, OR, USA
Cătălin NEGREA, Continental Automotive Romania
Florin Codruţ NEMŢANU, "Politehnica" University of Bucharest, Romania
Johan NICOLICS, Vienna University of Technology, Austria
Dorin PETREUŞ, Technical University of Cluj Napoca, Romania
Alena PIETRIKOVA, Technical University of Kosice, Slovakia
Dan PITICĂ, Technical University of Cluj-Napoca, Romania
Ioan PLOTOG, "Politehnica" University of Bucharest, Romania
Marius RANGU, Politehnica University of Timişoara, Romania
Paul ŞCHIOPU ,"Politehnica" University of Bucharest, Romania
Paul SVASTA ,"Politehnica" University of Bucharest, Romania
Daniel VIŞAN, University of Piteşti, Romania
Marian VLĂDESCU, "Politehnica" University of Bucharest, Romania
Dan VUZA, Institute of Mathematics of the Romanian Academy, Bucharest, Romania

Human Resource Education and Training Committee
Chair: Aurelia FLOREA, MIELE Tehnica, Braşov, Romania
Academic Co-Chair: Cornelia GORDAN, University of Oradea
Industrial Co-Chair: Delia UNGUR, Euro Business Park Oradea

Codruta Bala, Celestica, Oradea Romania
Andrei Kecseg-Fonce, Conect Group Oradea, Romania
Maria Marcovici, Continental Automotive, Timisoara, Romania
Cosmin Moisa, Continental Automotive Timisoara, Romania
Diana Nastase, Plexus, Oradea Romania

Publication Committee
Chair: Gabriel CHINDRIŞ, Technical University of Cluj- Napoca, Romania
Co-Chair: Bogdan MIHĂILESCU, "Politehnica" University of Bucharest, Romania
International Publication Advisor:
Zsolt ILLYEFALVI-VITÉZ, Budapest University of Technology and Economics, Hungary

Radu BOZOMITU, "Gheorghe Asachi" Technical University of Iaşi, Romania
Eugen COCA, „Ştefan cel Mare" University of Suceava, Romania
Radu COSĂCEANU, Cavallioti Publishers, Bucharest, Romania
Andrei DRUMEA, Politehnica University of Bucharest, Romania
Cristian FĂRCAŞ, Technical University of Cluj-Napoca, Romania
Olivér KRAMMER, Budapest University of Technology and Economics, Hungary
Cristina MARGHESCU , Politehnica University of Bucharest, Romania

Technical Program Committee
Chair: Detlef BONFERT, Fraunhofer EMFT, Münich, Germany, Chair
Co-Chair: Norocel CODREANU, "Politehnica" University of Bucharest, Romania

Bogdan ANTON, "Politehnica" University of Bucharest, Romania
Réka BÁTORFI, Budapest University of Technology and Economics, Hungary
Alexandru BORCEA, ARIES, Romania
Gabriel CHINDRIŞ, Technical University of Cluj Napoca, Romania
Andrei DRUMEA, "Politehnica" University of Bucharest, Romania
Bogdan MARGINEAN, MIELE Tehnica, Braşov, Romania
Rajmond JÁNÓ, Technical University of Cluj Napoca, Romania
Cristina MARGHESCU, "Politehnica" University of Bucharest, Romania
Ovidiu NEAMTU, University of Oradea, Romania
Ioan PLOTOG, "Politehnica" University of Bucharest, Romania
Adrian SCHIOP, University of Oradea, Romania
Alexandru VASILE, "Politehnica" University of Bucharest, Romania
Daniel VIŞAN, University of Piteşti, Romania

Local Organising Committee
University of Oradea, Romania
Chair: Cornelia GORDAN, University of Oradea, Romania
Co-Chair: Daniel Nistor TRIP, University of Oradea, Romania

Razvan ALBU, University of Oradea, Romania
Cristian BOLOVAN, Rohde&Schwarz Romania
Adrian BURCA, University of Oradea, Romania
Simona CASTRASE, University of Oradea, Romania
Nicolae DRAGHICIU, University of Oradea, Romania
Ioan GAVRILUT, University of Oradea, Romania
Mircea GORDAN, University of Oradea, Romania
Octavian MALAN, Connectronics Oradea, Romania
Liviu MOLDOVAN, University of Oradea, Romania
Lucian MORGOS, University of Oradea, Romania
Florin MURESAN, Celestica Oradea, Romania

Andrei NICORAS, Plexus Oradea, Romania
Sorin POPA, University of Oradea, Romania
Romulus REIZ, University of Oradea, Romania
Marin TOMSE, University of Oradea, Romania
Laviniu TEPELEA, University of Oradea, Romania

Technical Secretariat
Delia LEPĂDATU, "Politehnica" University of Bucharest, Romania
Liliane MARGHESCU, APTE - Association for Promoting Electronic Technology
Rodica NEGROIU, "Politehnica" University of Bucharest, Romania
Mariana PĂTULEANU, "Politehnica" University of Bucharest, Romania
Florentina STĂLINESCU, APTE, Bucharest, Romania

2016 IEEE 22ⁿᵈ International Symposium for Design and Technology in Electronic Packaging (SIITME)

Authors

Author	Page
Aiordachioaie, D.	217
Alexandru, I. M.	59
Andrei, M.	290
Arotaritei, D.	106
Ates, I.	63
Baciu, I. H.	102, 151
Bădălan (Drăghici), N.	155, 223
Badescu, I.	175, 200
Balázs, L.	25
Bande, V.	102, 151
Banica, A.	175
Barto, S.	40
Beer, R.	300
Belea, R.	159
Berényi, R.	274
Bonyár, A.	25, 139
Bozomitu, R. G.	106, 213
Branzei, M.	44
Buchman, A.	240
Bujor, M.	66
Bušek, D.	36, 48
Buta, B.	32
Cadar, S.	165
Carp, M. C.	248
Cetin, L.	63
Cherciu, C.	70
Chiculiţă, C.	279
Chindris, G.	142
Chirtoc, M.	63
Ciocan, I.	171
Ciolacu, M.	300
Coca, E.	205
Costea, I. M.	175, 200

Costiuc, M. N.	110
Cotfas, D. T.	94, 179
Cotfas, P. A.	94, 179
Cristea, C.	106
Csarnovics, I.	25, 139
Cucu, T. C.	44
Cutlac, H.	227
De Sabata, A.	270
Dima, M.	110, 184
Dobre, A. R.	244
Dobre, R. A.	232
Drumea, A.	74
Dumitrache, F.	248
Dumitrescu, C.	175
Elisei-Iliescu, C.	232
Epure, S.	159
Etz, R.	165, 256
Fărcaş, C.	171
Fazekas, L.	48
Fodor, A.	142
Frangu, L.	279
Fratu, O.	114
Gál, L.	274
Géczy, A.	48, 294
Gerigan, C.	236
Ghilezan, A.	82
Gontean, A.	118, 122
Grama, A.	171
Grava, A.	188
Grava, C.	188
Grecu, C.	78
Harsányi, G.	274
Hnatiuc, M.	82
Illés, B.	294
Ionescu, C.	184, 196, 209, 283
Ionescu, L. M.	86, 90
Iordache, C.	78
Iordache, V.	200
Kámán, J.	25

Krammer, O.	36
Krug, P.	184
Lazar, I.	82
Lita, A. I.	86, 90, 304
Luca, L.	217
Lung, C.	240, 262
Lupu, V.	236
Mach, P.	40, 55
Machidon, O. M.	94, 179
Manofu, M.	146
Marcu, A. E.	244
Marghescu, A.	98, 110, 130
Marghescu, C.	32, 196, 223
Maxime, F. S.	165
Mazare, A. G.	86, 304
Megyes, B.	274
Nacu, I.	217
Nagy, D.	48
Nastac, D. I.	70
Neamțu, M. O.	126, 192
Necula, L.	114
Negrea, C.	146
Negrescu, C.	232
Negroiu, R.	196
Nemtanu, F.	175, 200
Nicolau, V.	290
Ogrutan, P.	236
Orian, C.	256
Pajuste, E.	184
Pana, G.	248
Pantazică, M.	74, 78
Păsărică, A.	106, 213
Pasat, A.	114
Patarau, T.	165, 256
Perişoară, L. A.	252, 266
Petrariu, A. I.	205
Petreus, D.	165, 256
Pitică, D.	59, 142
Plotog, I.	44
Pop, S.	102, 151
Prikulis, J.	184

Roman, N.	217
Ropoteanu, C.	209
Rotariu, C.	106, 213
Rus, D. D.	184
Sabou, S.	240, 262
Săcăleanu, D. I.	252, 266
Safta, M.	110
Schlingensiepen, J.	200
Silaghi, A. M.	270
Silaghi, Al. M.	270
Simion, E.	98
Skwarek, A.	294
Sokovnin, S.	184
Stanomir, D.	232
Stoian, R.	252, 266
Suciu, G.	114
Suciu, V.	114
Şucu, L.	266
Svasta, P.	32, 98, 110, 130, 134, 155, 196, 209, 223, 227
Szabo, R.	118, 122
Szántó, G.	139
Szivós, D.	274
Tamási, P.	274
Tar, L.	274
Tarba, L.	55
Trip, N. D.	126, 192
Tulbure, A.	171
Turgut, A.	63
Varga, B.	36
Varzaru, G.	32, 44
Vasile, Al.	32, 196
Vasile, D. C.	98, 130
Vidrascu, M. G.	134
Viman, L.	59
Visan, D. A.	90, 304
Vladescu, M.	134, 244
Vlăduţă, R.	146

Characterization of the shape of gold nanoparticles prepared by thermal annealing

Attila Bonyár, Judit Kámán
Department of Electronics Technology
Budapest University of Technology and Economics
Budapest, Hungary
bonyar@ett.bme.hu

István Csarnovics, László Balázs
Department of Experimental Physics
University of Debrecen
Debrecen, Hungary

Abstract— Gold nanoparticles – which are intended to be used as transducers in a localized surface plasmon resonance (LSPR) sensor – were prepared by thermally annealing various layers of gold thin films deposited on glass substrate. The size and distribution of nanoparticles were investigated by atomic force microscopy (AFM). The changes in the nanoparticle shape in function of the deposition and annealing parameters are characterized. A novel parameter called localization factor was used to investigate the shape of the resulting particles. A common problem concerning the AFM imaging of nanoparticles, namely the tip convolution effect was studied, and possibilities to use the localization factor parameter to optimize surface reconstruction algorithms via tip deconvolution is demonstrated.

Keywords— *nanometrology, gold nanoislands, localization factor, imaging artifacts, AFM*

I. INTRODUCTION

Gold nanoparticles are widely used as signal amplification elements in various electrochemical and optical sensor applications [1]. This includes for example the conjugation of Au NPs with biomolecules as biological tags in biosensor application, but Au NP modified sensor transducer surfaces (e.g. thin films) are also quite frequent. The nanoparticles on the transducer surface can in one hand increase the effective electrode area in electrochemical applications, while in optical sensors the so called localized surface plasmon resonance (LSPR) phenomenon – the collective oscillations of delocalized electrons in response to an external electric field – is utilized to amplify the electric near field and thus the sensor sensitivity.

Au NPs can be synthesized in several ways [2]. Perhaps one of the simplest methods is the thermal annealing of pre-deposited gold thin films on glass or silicon surfaces. In this method the parameters of the annealing process (time, temperature) and the pre-deposited thin film thickness influence and define the resulting size and distribution of the Au NPs on the surface [3]. While the size and distribution of the resulting particles can be easily measured with AFM, the shape of the prepared nanoparticles are harder to characterize. For this reason a novel parameter called localization factor is to be used.

The localization factor parameter was introduced for the characterization of AFM images by Bonyár in 2012 [4]. This parameter can be obtained by calculating the so called

generalized localization (which is the structural entropy and spatial filling factor function pair) of an AFM image, and it is meant to characterize the typical shape of the surface structures. Since its introduction the possible application of the localization factor was demonstrated for the characterization of gold thin film microstructures [4]; for the shape change of gold grains during thermal annealing [5]; and also for the detection of tin oxidation and oxide grain formation [6]. Now we intend to demonstrate the possible use of this parameter for the characterization of the shape of gold nanoparticles prepared by thermal annealing.

A common problem during AFM imaging originates from the very principles of using a scanning probe to measure the topography of surfaces, namely from the geometrical convolution of the tip and the surface structures. As illustrated in Fig. 1, the difference between the obtained structures imaged with a sharp tip and a blunt tip (caused by either damage, wear or contamination) can be significant, especially in the case of sharp surface features. This problem is common with nanoparticles since they have both small size and high contact angle at their sides. The amount of structure dilation experienced while measuring small nanoparticles is strongly depending on the quality of the tip. There are built-in algorithms in most AFM image post-processing software to handle this issue. It is possible to estimate the geometry of the used tip based on the image, and it is also possible to try to reconstruct the image, by using deconvolution with the assessed tip geometry. The two problems with such algorithms are that 1) it is not possible to perfectly reconstruct an image due to the loss of information at the convoluted areas; and 2) they are strongly depending on the parameters used for tip assessment. In this work we aim to demonstrate, that the localization factor parameter can be used to optimize the surface reconstruction process.

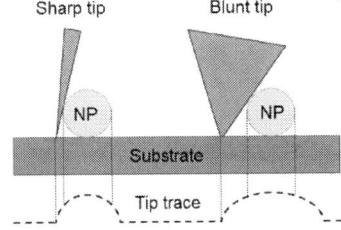

Fig. 1. Illustration of the geometrical convolution and the widening of features with sharp and blunt AFM tips.

II. THEORY

In this section we discuss the necessary mathematical background to understand the localization factor parameter. For the detailed description of this parameter see our original articles regarding its introduction for the characterization of thin film surfaces [4, 6].

To calculate the localization factor one must calculate the *generalized localization* of an AFM image. The generalized localization can be derived from the *structural entropy* (S_{str}) and the *spatial filling factor* (q) functions of the image. Since the intensity distribution (I_i) of a contact-AFM image can be normalized in a way to fulfill equations (1) and (2), the image can be well characterized with the structural entropy and the spatial filling factor.

$$Q_i \geq 0 \text{ for } i = 1,...,N \qquad (1)$$

$$\sum_{i=1}^{N} Q_i = 1 \qquad (2)$$

The spatial filling factor is derived from the *participation ratio* (or delocalization measure, D) which is a well-known quantity in density matrix analysis, introduced by Bell and Dean [7] and Pipek [8] independently. The definition modified to the case of pixel intensities has the following form:

$$D = \frac{1}{\sum_{i=1}^{N} I_i^2} \qquad (3)$$

D shows the approximate number of filled lattice sites or pixels the distribution expands to. The spatial filling factor (q) is the participation ratio divided by the total number of pixels (N):

$$q = \frac{D}{N} \qquad (4)$$

The structural entropy (S_{str}) is derived from the Shannon and Rényi entropies and was introduced by Pipek and Varga [9] to study the structure of many electron density functions. For pixel intensity distributions the Rényi entropies (the generalization of the Shannon entropy) are defined in the following way:

$$S_n = \frac{1}{1-n} \log \left(\sum_{i=1}^{N} I_i^n \right) \qquad (5)$$

It can be proven that the Shannon entropy is the first member of the Rényi entropies. The Shannon entropy measures how much the pixel intensity distribution (I_i) deviates from the uniform distribution (when all I_i have the same value, then the image is homogeneous gray). The second Rényi entropy is the entropy of a uniform distribution over D sites (also called the extension part of the Shannon entropy) and corresponds to the number of pixels with significantly high intensity. The difference between S_1 and S_2 characterizes the structure of the system, and it is called the structural entropy (S_{str}):

$$S_{str} = S_1 - S_2 = S - \log D \qquad (6)$$

The structural entropy characterizes purely the deviation of the intensity distribution from the step function (which is in fact the black and white image).

The pair of functions (q; S_{str}) of the distribution $\{I_i, i = 1,...,N\}$ is called generalized localization, and it can be used for analyzing the topology free structure of the observed distribution. Our postulated new parameter, the localization factor is the value (α) which yields the smallest square error between the generalized localization of the AFM image and the generalized localization of the $exp(-x^\alpha)$ probe function. In other words, it can be defined as the minimum of the $E(\alpha)$ error function based on (14).

$$E(\alpha) = \left[S_{str}(q)_{image} - S_{str}(q)_{\exp(-x^\alpha)} \right]^2 \qquad (7)$$

In the case of the AFM images, this exponent expresses the typical shape of hills and valleys of the topography maps. Two extremities of localization factor are $\alpha = 0$ and $\alpha = \infty$ which correspond to totally flat surface and vertical-walled features, respectively.

To calculate the localization factor of an AFM image first the image is segmented to $k \times k$ number of smaller images. The value of k should be selected taking into consideration 1) the scan size and resolution of the image, 2) the average size of the features on the image, in a way to guarantee, that the size of a segment is at least twice larger than the average size of the characteristic shapes (sampling criteria). The pair of functions (q; S_{str}) will then be calculated for all the segments, and α is determined as the minima of the $E(\alpha)$ error function calculated for this distribution. For more information please read the original articles regarding the introduction of the localization factor for AFM imaging [4, 6].

III. EXPERIMENTAL

A. AFM Instrumentation

AFM measurements were performed with a Veeco (lately Bruker) diInnova type scanning probe microscope (SPM). Contact mode measurements were done with ART D160 diamond probes. The sampling rate of the images presented in the paper is 512x512 obtained with 1 Hz scan rate. During the scans the PID values of the scanner feedback were optimized according to the Veeco User Manual to gain the best image quality. For data evaluation the Gwyddion 2.36 software was used [10].

B. Nanoisland preparation

Gold layers of a thickness between 14-16 nm were deposited onto four glass substrates via thermal evaporation at the Department of Experimental Physics, University of Debrecen. The layer thickness was measured after evaporation with an Ambios XP-1 nanoprofilometer. The subsequent thermal annealing of the four samples was done in a ceramic oven at 580 $^{\circ}$C for four different time periods in Ar atmosphere: 15 min, 30 min, 60 min and 120 min, respectively.

IV. RESULTS AND DISCUSSION

A. Tip geometry and image reconstruction

In this section we aim to demonstrate the effect of tip geometry on the image quality and the size of the nanoislands; to demonstrate the possibility to model/estimate the tip geometry and reconstruct/repair the image which was made with a blunt tip, and also, the ability of the localization factor parameter to help in the optimization of this surface reconstruction process. Fig. 2 presents a 25 μm^2 tapping-mode topography AFM image of gold nanoislands, prepared by thermal annealing the deposited 14-16 nm gold thin film for 120 min. The image was obtained with a sharp, diamond AFM tip. As can be seen in Fig. 2, the nanoislands have sharp features and appear to be 'slender'.

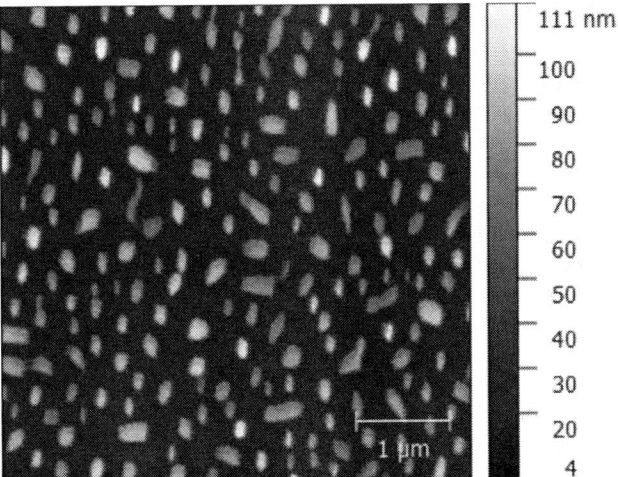

Fig. 2. Tapping-mode AFM topography image of gold nanoisland prepared by 120 min thermal annealing, measured with a sharp AFM tip. Image size: 5 μm x 5 μm

To verify the sharpness of the tip and check whether an AFM image contains any convolution artifacts which would result from a blunt/damaged tip, the Gwyddion AFM post-processing software contains useful tools. By using the 'Tip->Blind Estimation' tool, one can run an algorithm on the whole image which aims to obtain the possible geometry of the tip, which was used for the imaging. The resulting tip geometry of this estimation algorithm used on the image of Fig. 2 is presented in Fig. 3 A and B (full iteration was used on the whole image, considering a tip of 17 x 17 pixels on the 512 x 512 image).

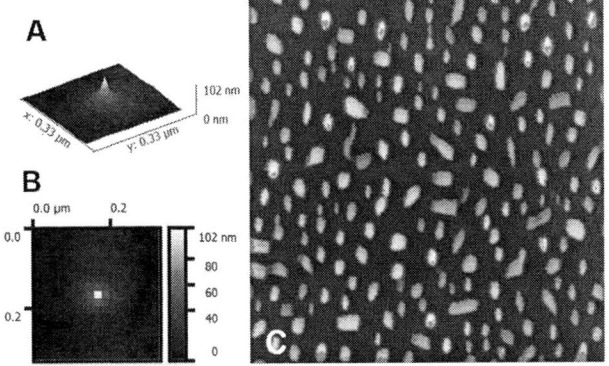

Fig. 3. A) 3D topography of the sharp tip used for the imaging. B) The same tip topography in 2D. The tip geometry was obtained by the Gwyddion 'Blind Estimation' function, using the whole image and 'Full iteration' with 17 pixels. C) 'Certainty Map' of Fig.2 using the estimated sharp tip (red areas mark the uncertain regions).

As can be seen in Fig. 3, the estimated tip appears to be sharp. Looking at the 'Certainty Map' of Fig. 3/C, which was obtained by using this estimated tip geometry on the AFM image, we can say that our image is really free of possible tip-based artifacts. In the Certainty Map the uncertain areas are marked with red mask in Fig. 3 (note that the original output of the 'Certainty Map' function was inverted here for better visualization). In these points we can suspect, that the AFM tip touched the surface in multiple points instead of only one point, which means that there is a loss of information in these points.

Fig. 4 presents a 25 μm^2 tapping-mode topography AFM image of the same nanoislands sample, but measured with a blunt AFM tip. We can instantly see that – since the scan size of the images of Fig. 2 and 4 are the same – the size of the nanoislands are significantly larger. Observing more closely we can also see, that a pattern is repeated on the surface (note the small slot at the right side of nearly every nanoisland at their middle section), which clearly means tip convolution artifact and a blunt tip, which is possibly broken or contaminated. Performing the same tip estimation and certainty map operations on the image we can see the resulting tip geometries in Fig. 3. For our investigations the assumed tip size was increased gradually from 8 pixels to 15 pixels, the result of 10 and 13 pixels are presented in Fig. 5, as an illustration (the tip was estimated on a segmented image, containing only one nanoisland for the better estimation of the possible tip geometry). It can be seen, that the estimated tip is much blunter compared to Fig. 3, and also that increasing the assumed pixel number the estimated tip diameter increases accordingly.

2016 IEEE 22nd International Symposium for Design and Technology in Electronic Packaging (SIITME)

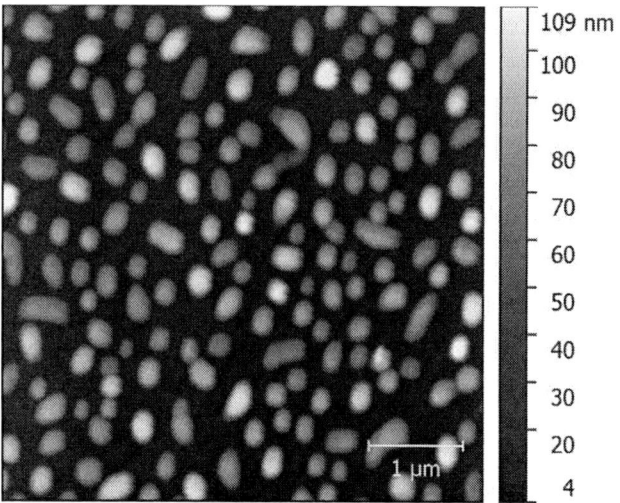

Fig. 4. Tapping-mode AFM topography image of gold nanoisland prepared by 120 min thermal annealing, measured with a blunt AFM tip.
Image size: 5 μm x 5 μm

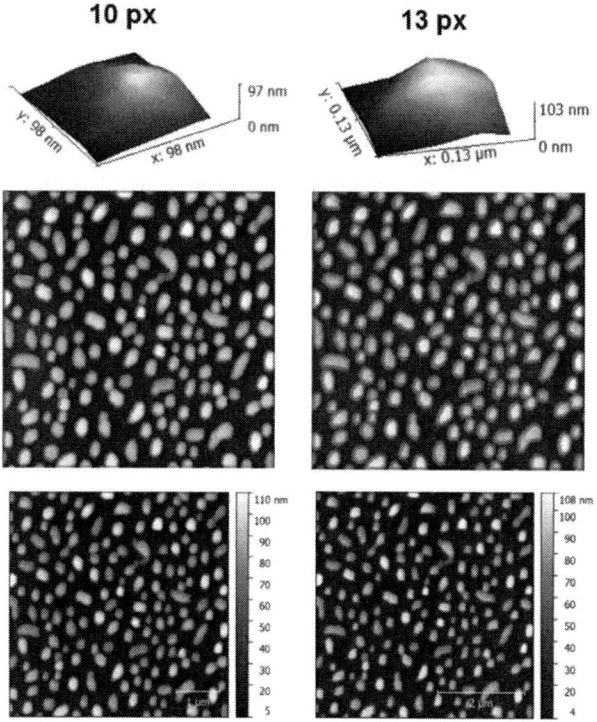

Fig. 5. Top row: 3D topography of the estimated blunt tip, using 'Blind Estimation' with 10 and 13 pixels on segmented images. Middle row: 'Certainty Maps' created by using the tips in the top row. Botton row: The topography image of Fig. 4 after 'Surface Reconstruction' using the estimated tip geometries.

The certainty maps confirm, that due to the blunt tip geometry, the side of the nanoisland are touched in multiple points, which resulted in the dilation of their size. By using the 'Tip -> Surface Reconstruction' function it is possible to try to reconstruct the image by applying a deconvolution algorithm with the estimated tip. This algorithm erodes the size of the

particles, and the larger the estimated tip is, the more is the amount of erosion, as can be seen in the bottom row of Fig. 5. Note that this surface reconstruction is only partial: the information which is missing from the points which are not touched by the tip cannot be regained by this deconvolution algorithm.

The biggest difficulty regarding the use of surface reconstruction is the precise estimation of tip geometry, since the estimation algorithm is very sensitive to input parameters, such as the assumed tip size. Depending only on the naked eye, it is impossible to precisely judge the amount of erosion which is required to properly reconstruct the surface, as can be seen on Fig. 5 for 10 x 10 and 13 x 13 pixel sizes. Is the former enough, or should we used an even bigger tip to reconstruct the image?

The function of the localization factor parameter, which was introduced in the Theory section is the quantitative characterization of the shape of the surface structures. Compared with the original image, the convoluted tip artifacts alter the shape of the surface structures, which can be detected by using this parameter. Besides, by using the segmentation algorithm for the determination of localization factor, as was introduced in our previous publications [4, 6], it will also be very sensitive to the surface ratio of the nanoparticles (total developed surface area of the nanoparticles / the scan size). Thus, the use of localization factor presents a possibility to characterize the effect of surface reconstruction, as demonstrated in Fig. 6.

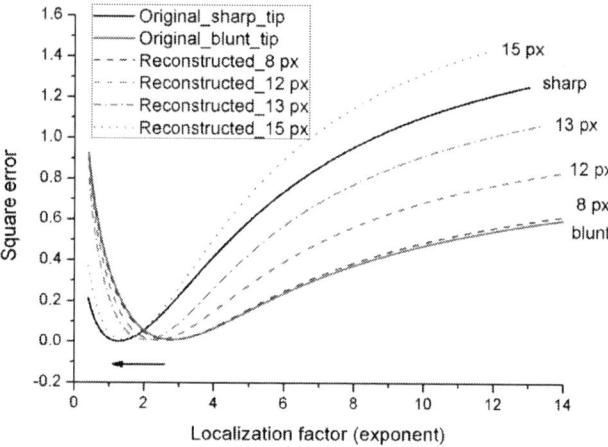

Fig. 6. Square error curves in funnction of the exponent of the probe function (localization factor). The arrow indicates the change in the localization factor (the minimum of the curves).

The curves in Fig. 6 are the error functions of Eq. (7) calculated for the AFM images and the $exp(-x^{\alpha})$ probe function. The minimum point of the curves is defined as the localization factor. It can be seen, that there is a clear difference between the curves obtained for the sharp tip (Fig. 2) and blunt tip (Fig. 4) images. The other curves were calculated on images reconstructed from the image made with the blunt tip and using estimated tip geometries with increasing assumed tip size (from 8 x 8 pixels to 15 x 15 pixels). The curves show, that by increasing the assumed tip size, and thus the amount of erosion during reconstruction, the localization factor shifts gradually in the direction of the image made with a sharp tip. Using the estimated tip geometry of 15 x 15 pixels the localization factor and the error function curve of the reconstructed image is nearly equal to the one obtained with the sharp tip, which means that this tip size is needed for the sufficient amount of erosion/reconstruction, in this example. Note, that although the two AFM images were obtained on the same sample but not on the exactly same area, the two curves would never completely fit. Also, as we previously noted it is never possible to perfectly reconstruct an AFM image due to the loss of information. Besides these notes, the localization factor clearly demonstrates the following three possibilities. If we have an AFM image from a sample which we can be sure that was measured with a sufficiently sharp tip, by obtaining the localization factor value for this image and comparing it with subsequent images from the same sample it is possible to 1) to detect, monitor and compensate tip wear/aging; 2) to decide whether other tips used to image the same sample were sufficiently sharp or not and 3) to help setting and optimizing the parameters of a surface reconstruction, as demonstrated in Fig. 6.

B. Effects of image reconstruction on the characteristic surface parameters

In this section we demonstrate the application of the described surface reconstruction method for the characterization of various nanoisland arrangements. AFM images were made on four samples, which were annealed for different time periods (15 min, 30 min, 60 min and 120 min, respectively). The resulting AFM images are presented in Fig. 7. As can be seen, there is different amount of additive tip effect convoluted on the shape of the nanoislands, so surface reconstruction was applied on the images, as discussed before.

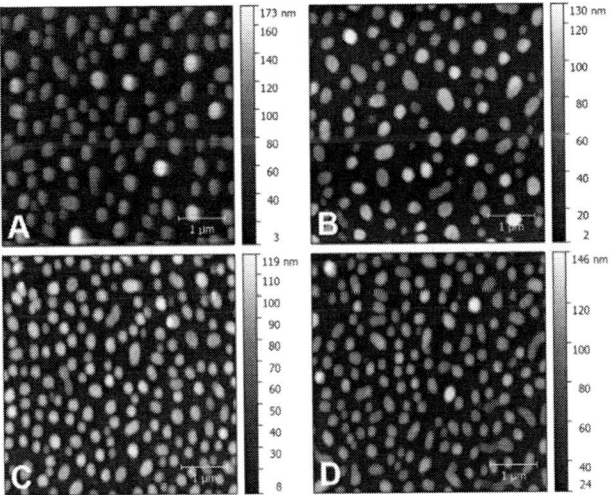

Fig. 7.   Tapping-mode AFM topography images of gold nanoisland prepared by thermal anneling for various time periods. From A-D: 15 min, 30 min, 60 min and 120 min, respectively. The blunt tip was used for the imaging. Image size: 5 μm x 5 μm

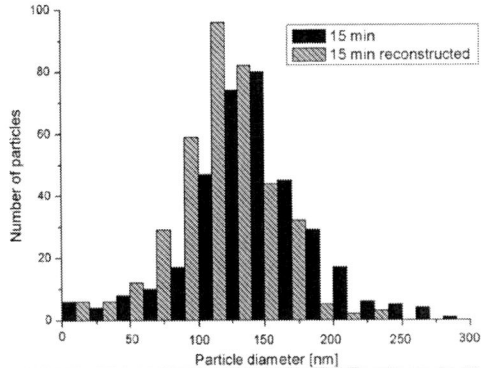

Fig. 8.   Histograms of the equivalent spherical particle diameter of the nanoislands, obtained with the blunt AFM tip. Sum of three AFM images measured on the sample after 15 min annealing, before and after surface reconstruction.

Fig. 8 presents the distribution of the equivalent disc radius of the particles for the sample with 15 min annealing time, before and after surface reconstruction (the presented histograms are the sum of three AFM images). Fig. 9 compares the histograms based on the reconstructed images from the samples after 15 min and 60 min annealing.

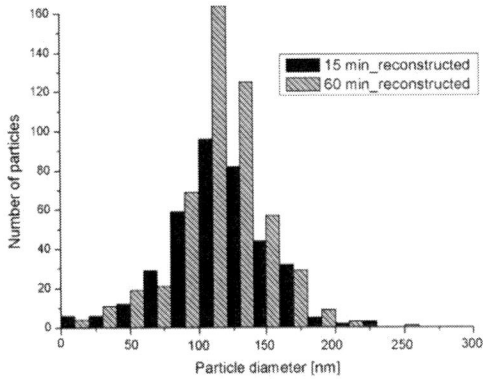

Fig. 9.   Histograms of the equivalent spherical particle diameter of the nanoislands, obtained with the blunt AFM tip. Sum of three AFM images measured on the samples after 15 min and 60 min annealing, respectively, both after surface reconstruction.

The biggest difference between the samples with increased annealing times is in the number of particles. The majority of the particles falls into the 100-150 nm diameter range, based on the reconstructed images, as can be seen in the box charts of Fig. 10. As a result of the erosion the size of the particles decreases accordingly, which is confirmed by the presented distributions.

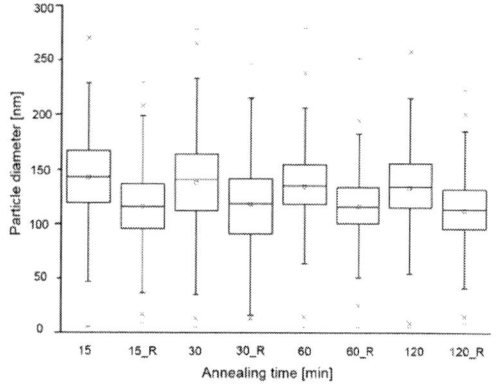

Fig. 10. Box charts illustrating the distribution of the equivalent particle diameters of the samples prepared with different annealing times. The obtained results after surface reconstruction are marked with 'R'.

Figures 11-13 present three other important surface describing parameters which are greatly influenced by the quality of the tip and thus the surface reconstruction. The nanoisland density (Fig. 11) is calculated as the number of nanoparticles per unit scan area. With a blunt tip it is possible to merge small nanoparticles which are close to each other. By using proper deconvolution it is possible te separate these partially merged particles, which will be identified as separate particles. (In our case a simple height-tresholding was applied to identify the nanoparticles.) Hence, proper surface reconstruction increases the number of the recognized nanoparticles.

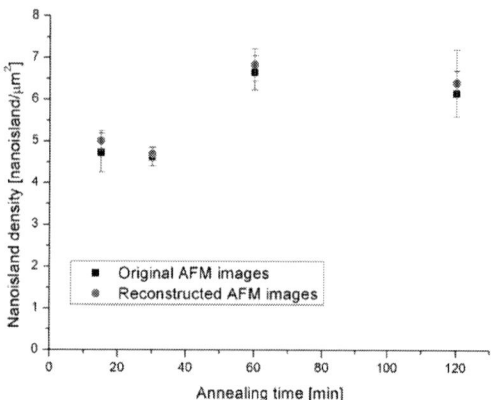

Fig. 11. Nanoisland densities calculated based on three AFM images made on the samples before and after surface reconstruction.

In the same way, as a result of the erosion, the surface ratio of the particles (total nanoparticle developed area / scan area) decreases significantly (see Fig. 12). Since the density of the nanoisland, more specifically, the size and distance between them strongly affect the coupled plasmon resonance between them, the precise estimation of these parameters are important for proper surface characterization.

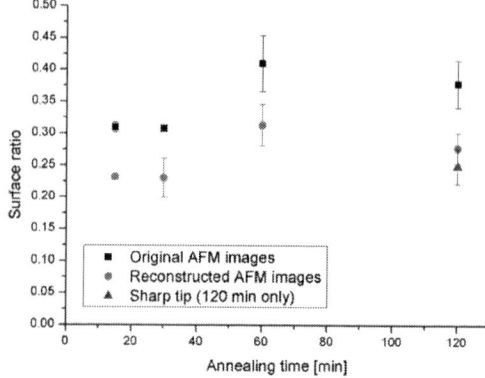

Fig. 12. Surface ratios calculated based on three AFM images made on the samples before and after surface reconstruction.

Fig. 13 presents the localization factor values measured on the AFM images with and without surface reconstruction. The deconvolution algorithm decreased the localization factor values with 0.4-0.6 depending on the sample. It can be observed, that the localization factor parameter strongly correlates with the previous parameters (the higher the surface ratio or density of the particles, the higher the localization factor). The reason for this, is that the current image segmentation method does not filter out the flat background around the particles, and the amount of these dark areas on the image strongly contribute to the obtained localization factor values. (In this work the images were segmented to 5 x 5 pieces, for the optimization of this segmentation, see our previous publication [6].) We are currently working on a new segmentation method to detect and cut only the shape of the nanoislands and use them as segments for the localization factor calculations, and thus the obtained values would only include the shape of the particles, which would be its original purpose.

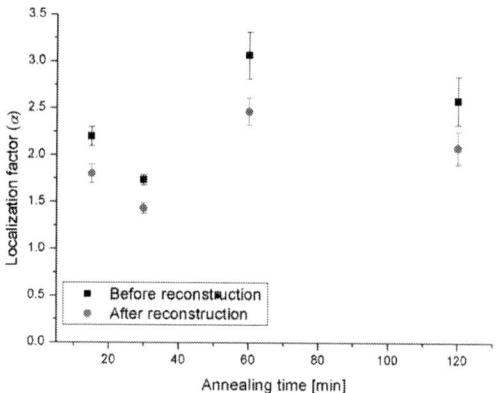

Fig. 13. Localization factor values obtained on the nanoisland arrangements, before and after surface reconstruction. (Calculated for three images per sample).

Note that the areas used for AFM imaging was 25 μm^2, which is small compared to the whole 2 cm^2 area of the samples. Although different areas on the surface were investigated, it is still not possible to completely cover the whole surface and local inhomogeneities could contribute to the observed differences.

V. CONCLUSIONS

Gold nanoisland arrangements were prepared by thermal annealing of pre-deposited gold thin films on glass substrates and were characterized with atomic force microscopy. A common problem concerning the AFM imaging of nanoparticles, namely the tip convolution effect, was studied in detail. It was demonstrated that the localization factor parameter can be used both to assess the possibility of tip effects and also to control the parameters of surface reconstruction via tip deconvolution. The effect of surface reconstruction for various nanoisland arrangements was studied and its importance for obtaining reliable characteristic surface parameters was discussed.

ACKNOWLEDGMENT

The research leading to these results has received funding from the ProProgressio foundation.

REFERENCES

[1] Xiaodong Cao, Yongkang Ye, Songqin Liu "Gold nanoparticle-based signal amplification for biosensing", Anal. Biochem., Vol. 417, pp. 1–16, 2011.

[2] Pengxiang Zhao, Na Li, Didier Astruc "State of the art in gold nanoparticle synthesis", Coord. Chem. Rev., Vol. 257, pp. 638– 665, 2013.

[3] A. Serrano, O. Rodríguez de la Fuente, M. A. García, "Extended and localized surface plasmons in annealed Au films on glass substrates", J. App. Phys., Vol. 108, pp. 074303, 2010.

[4] A. Bonyár, M. L. Molnár, G. Harsányi, " Localization factor: A new parameter for the quantitative characterization of surface structure with atomic force microscopy (AFM) ", Micron, Vol. 43, Issues 2-3, pp. 305-3010, 2012.

[5] A. Bonyár, P. Lehoczki, "An AFM study regarding the effect of annealing on the microstructure of gold thin films ", Proc. Of the 36th IEEE-ISSE International Spring Seminar on Electronics Technology, Alba Iulia, Romania, pp. 317-322, 2013.

[6] A. Bonyár, "AFM characterization of the shape of surface structures with localization factor", Micron, Vol. 87. pp. 1-9, 2016.

[7] R. J. Bell, P. Dean, "Atomic vibrations in vitreous silica", Discuss. Faraday Soc., pp. 50-55, 1970.

[8] J. Pipek, "Localization measure and maximum delocalization in molecular systems", Int. J. Quantum Chem., Vol. 36, pp. 487-501, 1989.

[9] J. Pipek, I. Varga, "Universal classification scheme for the spatial-localization properties of one-particle states in finite, d-dimensional system", Phys Rev A., Vol. 46, 3148-3163, 1992.

[10] D. Nečas, P. Klapetek, "Gwyddion: an open-source software for SPM data analysis", Cent. Eur. J. Phys. Vol. 10, Issue 1, pp. 181-188, 2012.

A Study of SW150 Conductive Paste as a possible use in Solderless Assembly for Electronics

G. Varzaru, C. Marghescu, Alex. Vasile, P. Svasta
Center for Electronic Technology and Interconnection Techniques
Politehnica University, UPB - CETTI
Bucharest, Romania
gaudentiu.varzaru@cetti.ro

B. Buta
Net Digital Service SRL
Oradea, Romania

Abstract— **Although only at the stage of proof of concept, a basic approach in Occam technology is: placement of components on a dielectric substrate, encapsulation, reversal of the substrate, drilling down to pins of components, creation of conductive traces, a.s.o. Since a reversed order of construction of a simple circuit is considered to be much cheaper in terms of required equipment, this paper presents some preliminary tests regarding conductive traces issues. As new materials [1] and new technologies [2, 3] are emerging on the market, a new conductive paste proposed to interconnect layers of a multi-layer printed circuit board through filled vias [4] was used to create traces. Printing of conductive paste and electrical measurements at the pin and trace interface were investigated.**

Keywords—conductive paste, printed circuit, solderless assembly for electronics.

I. INTRODUCTION

Lead-free technology requirements for improved reliability brought higher melting point temperature alloys which made the soldering process more aggressive and costly. The thermal stress of the entire electronic module involved by the solder joint creation, as well as all issues of soldering (whiskers, head-in-pillow, voids, a.s.o.), led to a new approach of the level 2 of electronic packaging, solder free assembling. There is already a mature technology available on the market, press-fit, however it only addresses connectors. New materials and new technologies [1, 2, 3, and 4] are emerging on the market. One innovative printed circuit board (PCB) manufacturer is proposing to interconnect internal layers of a multi-layer PCB by filling vias with conductive paste [5]. Another proposal is a conductive paste for circuit traces, SW150, which could be appropriate for building an electronic module in accordance with the concept of Occam technology. As known, this new approach proposes to build the electronic module during a single process by creating both the interconnection structure and assembling the components. A structure encapsulated in a dielectric material whose components are interconnected internally by copper traces is obtained.

The properties of the SW150 conductive paste are presented in Table I. As it can be seen, it is a mixture of silver coated copper particles and epoxy resin with a metallic ratio, by weight, comparative to common solder paste. From a technology point of view, the printability is notable: using a screen with Line/Space ratio 125μm one could expect continuous traces. The paste performs thixotropic properties, very useful for printing. Stencil printing is available too, but it requires careful design to avoid loops. The most important property is that the curing process takes place at a very low temperature compared to common lead-free solder paste (130°C vs.230-240°C for SAC alloys).

TABLE I. SW150 PROPERTIES [5]

Property	Measuring unit	Value
Filler		Silver coated Cu
		Averaged particle diameter 3~5μm
Binder		Epoxy resin/ Butyl carbitol
Fill ration of metallic	wt.%	89
Viscosity	BH type/dPa s	1800
Density	g/cm^3	3.3 … 3.5
Cure condition		130°C x 30 min
Volume resistivity	Ω cm	1.0~2.0E-04
Shelf life	0 … 5°C	>7 days (under testing)
	-10 … 0°C	>3 months (under testing)
Solder wettability (reflow)	Several type	Good
Printability	Screen printing	L/S=125/125

It should also be mentioned that the volume resistivity is two orders of magnitude higher than that of copper which could be restricting in applications dealing with small currents.

The goal was to create a simple demonstration electronic module in the manner of Occam technology in order to be able to enter in a niche market where solder joints are a major concern (aerospace, navy). The steps to follow are:

- design of the interconnection structure;

- design of the printing tool;

- printing of conductive paste;

- placement of components;
- thermal process;
- encapsulation;
- final test.

Since the interconnection structure is done with SW150 conductive paste and the components are surface mount devices (SMD) with different pin plating, several problems are arising:

- what happens to the geometry of conductive paste deposits after curing;
- what happens at the interface between the conductive trace and the pin of the component;
- how large is the electric resistance of the interface.

II. EXPERIMENTAL SETUP

A. Materials

Conductive paste SW150, year of production: 2016; two stainless steel stencils (150 μm and 178 μm thickness); different dielectric materials: polyimide films (Kapton), polytetrafluoroethylene (Teflon) substrate, and alumina substrate, different SMDs (chip components, integrated circuit, and light emitting diode). In Table II the electronic module's bill of materials is presented to point mainly the parameters related to technology.

TABLE II. BILL OF MATERIALS OF THE ELECTRONIC MODULE

Name	Description	Case	Pin plating
R1	Resistor, 100kΩ, Yageo	0603	Ni/matte Sn
R2	Resistor, 1MΩ, Yageo	0603	Ni/matte Sn
R3	Resistor, 1kΩ, Yageo	0603	Ni/matte Sn
R4	Resistor, 0Ω, Yageo	1210	Ni/matte Sn
C2, C2	Capacitor, 4.7μF, AVX	1206	Sn plating
D1	Light emitting diode, Kingbright	0805	NA (CuAg)
CI	CD4001BM, TI	SOIC14	CuNiPdAu

B. Equipment

Manual printer LT300 ZelPrint, LPKF Laser and Electronics, Programmable drying chamber 09H-300 Electronic, Townson & Mercer, Manual Pick-and-place machine SMFL3600, DIMA SMT Systems, Pulsar MV-40 handheld digital USB microscope, X-AXYS, Optical microscope MH-ZTO, DIMA SMT Systems, Stereo microscope S 6D, Leica, DM3068, Digital Multimeter, Rigol, DVM4200, True RMS Multimeter, Velleman; IT6322, Triple Output DC Power Supply, ITECH.

III. EXPERIMENTS AND RESULTS

In the attempt of constructing an electronic module we started from a design we are currently doing on FR4 substrate: an astable circuit with optical signaling. For paste deposition we decided to use stencil printing as a cheaper solution due to available equipment and materials. Two identical stainless steel stencils were designed with different thickness by laser cutting. Different patterns were cut:

- pairs of traces with different widths and spaces between them (Fig.1), 50mm long, with and without 0.2mm bridges; long traces were difficult to obtain because of the foil warpage; a small imperfection is quite visible in the smallest pair, but it is also in the stencil.

Fig. 1. Pairs of traces with same width and spacing

- a daisy chain structure for 10 0Ω chip resistors, 1206 case, specially designed for Four points measurement method for low resistance (Fig.2);

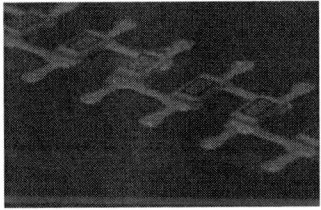

Fig.2. Daisy chain structure after curing.

- the interconnection structure of the astable circuit.

Printing itself is very critical due to the size of metal particles and the thickness of some foil substrates (Kapton, 150μm). The lack of perfect contact between the stencil and the substrate creates the reflection of the edges of the traces, like the "dog-ear" defect in solder paste printing. Print gap has to be 0 mm, warpage of flexible substrate has to be eliminated by using some adhesive.

A. Geometry investigation

After printing on two different substrates (a flexible film of Kapton, and a rigid sheet of composite material used for FR-2 PCBs) and curing, the traces were measured using a calibrated handheld microscope. The results are presented in Table III. There are differences which can be explained by processes issues. Measurements of the stencil's apertures showed some mismatches compared to the project due to the cutting process. Regarding the deposits: all the widths of the conductive traces have increased, while clearance between them has decreased

(Fig.3) dramatically from 27% to 76%. Since the paste is shrinkable these are caused by the whole printing process: stencil design, stencil manufacture, printer setup. The first stencil used for printing had a 178μm thickness which gave an excessive amount of paste. Decrease of spacing is a disadvantage because it affects the minimum distance required by dielectric strength.

TABLE III.　TRACES MEASUREMENT

Width and spacing of pairs of traces [mm]							
Project		Stencil			Polyimide/Composite		
w1,w2	s	w1	s	w2	w1	s	w2
0.321	0.18	0.107	0.322	0.186	0.234	0.127	0.420
					0.391	0.078	0.410
0.421	0.38	0.361	0.440	0.303	0.615	0.176	0.596
					0.625	0.156	0.606
0.622	0.58	0.508	0.674	0.449	0.821	0.352	0.791
					0.811	0.322	0.830
0.820	0.78	0.694	0.850	0.703	1.016	0.547	1.016
					0.977	0.586	0.977
1.022	0.98	0.909	1.036	0.938	1.221	0.752	1.202
					1.182	0.752	1.202

However, while in dry its value is 3.0 MV/m, using the encapsulation with epoxy resin the dielectric strength can increase up to 19-20 MV/m. This means that an electronic circuit having a minimum 0.05mm (2 mils) of dielectric epoxy can withstand 1000V.

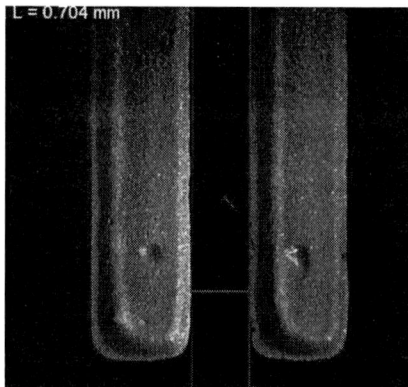

Fig. 3. Spacing measurement of printed traces

B. Electrical measurements

The goal was to investigate which is the range of the electrical resistance at the interface between the cured conductive paste and the pin of the component. A daisy-chain structure was printed with conductive paste and ten 0Ω resistors, 1206 case, were placed directly on the pads. Then the whole assembly was cured at 130°C for 30 minutes. The electric resistance was determined indirectly by measuring the voltage drop between the test pads of the Four points measurement method [6], when a fixed current flows through the circuit, as well as measuring the voltage drop on the 0Ω

resistor. The measuring stand containing a power source which can generate a constant current and a digital voltmeter is presented in Fig.4. After the first set of measurements, the polarization of the voltage is reversed and the measurements are repeated, and finally the polarization is reversed back again. Three sets of measurements were performed. The method ensures a minimization of measuring errors.

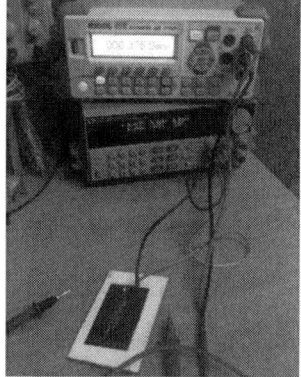

Fig. 4. The measuring stand

The voltage drop on the two interfaces between the resistor and the cured conductive paste is obtained by subtracting the voltage drop on the resistor. The electrical resistance was calculated using Ohm's Law for every measured value of the voltage drop knowing that the experiment was carried out for a fixed 0.497A value of the direct current. If R_{i1}, R_{i2}, R_{i3}, where i = 1, 2, … 10, are the sum of two interfaces resistance at both ends of the resistor i corresponding to the three sets of measurements, the resistance of one single interface resistance was finally calculated using the formula:

$$R_i = ((R_{i1}+R_{i2})/2+(R_{i2}+R_{i3})/2)/4$$

There are large differences between the resistance values. This could be due to the difference in placement position and force because the operation was done manually. The generic mean value of the electric resistance between the pin of the component and conductive trace is R = 10.05mΩ, which is greater than solder joint mean electrical resistance made by SnPb (0.85mΩ) or SAC305 (0.679) for fast cooling rate [6].

TABLE IV.　RESULT OF ELECTRICAL MEASUREMENTS

Resistance [mΩ]				
R1	R2	R3	R4	R5
6.550	7.746	8.886	8.986	11.471
R6	R7	R8	R9	R10
12.280	11.551	11.782	11.837	9.410

C. Module construction

In the new technological approach the design of the stencil does not use solder paste layer anymore, but copper layers. Since the trace loops would create islands in the stainless steel foil bridges had to be placed from place to place (Fig.5).

2016 IEEE 22nd International Symposium for Design and Technology in Electronic Packaging (SIITME)

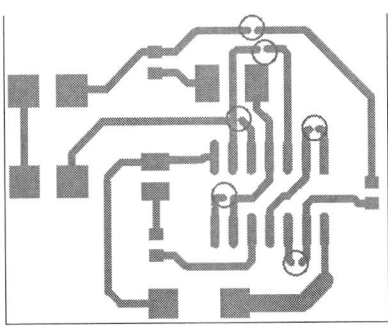

Fig.5. Original project had to be modified for making bridges on steel.

As support of the interconnection structure a sheet of alumina was used. After printing, the gaps in the traces were plugged manually. SMD components were placed except for one part, a 0Ω resistor used as bridge, whose body is over a trace. During curing the body of the component should be pressed the paste and that would have created conductive bridges. Since encapsulation with epoxy resin was not used, the part was fixed in a second curing process.

The electronic module was powered up and started functioning (Fig.6).

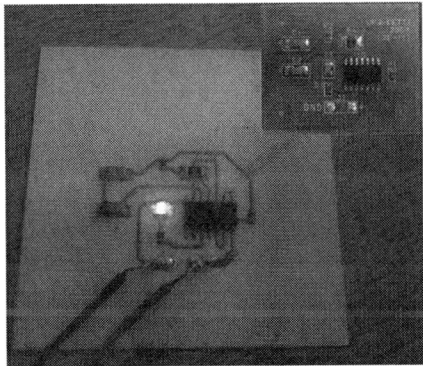

Fig.6. The electronic module made by conductive paste

IV. CONCLUSIONS

The experiments emphasized the importance of the printing process of the conductive paste for the quality of the traces, the spacing and the width. Better results could be obtained using thinner stencils (75μm, 100μm) or a screen as recommended by the paste manufacturer and a more controlled printing process that automated equipment could offer. Discontinuing traces were corrected by a new curing cycle. However, investigation on the interface between two traces obtained in two successive curing processes should be done to avoid any suspicion of discontinuity.

Electric resistance of the interconnection between the components and the trace is higher than those of solder joints. The dispersion of the resistance value leads to the conclusion that parts have to be pushed harder in the conductive paste during the pick-and-place process. This will offer a larger contact area. Again, an automated process with controlled parameters will be benefic. The influence of the pin plating should be investigated.

The study of the SW150 conductive paste was not performed on enough samples to draw a definite conclusion. Thermal diffusivity measurements will be conducted in order to characterize the thermal functionality. Also, a microstructure analysis at the interface conductive paste – pin will be performed. The shear force test could be used for evaluation but since it is supposed that the module has to be encapsulated its value is not so important.

ACKNOWLEDGMENT

This paper was published under the frame of the "Partnerships in priority areas" (PN II) Romanian Research program, developed and supported by MEN-UEFISCDI, BLCPL project, PN-II-PT-PCCA-2013-4-1546, no. 58/2014. The authors are grateful to Mike Sakaguchi from Tatsuta Electric Wire & Cables for supplying the conductive paste, to Traian Maduta from Net Digital Systems for supplying the stencil, to Joe Fjelstad from Verdant Electronics for continuous and effective support.

REFERENCES

[1] J. Fjelstad, "New material and process solutions for electronic interconnection industry", The PCB Magazine, February 2016; pp.12-22;

[2] J. Fjelstad, "Occam process for components having variations in part dimensions", Patent, US20110315302 A1, Application number US 12/228,826, 29 December 2011;

[3] J. Fjelstad, "Electronic assemblies without solder and methods for their manufacture", Patent, US20090008140 A1,Application number US 12/163,870, 8 January 2009;

[4] S. Stagon, A. Knapp, P.R. Elliott, Hanchen Huang, "Metallic glue for ambient environments making strides", Advanced Materials & Processes, vol. 174, No.1, January 2016, pp. 22 – 25;

[5] ***, "Element technologies of TATSUTA", www.tatsuta.com

[6] M. Branzei, F. Miculescu, A. Bibis, I. Cristea, I.Plotog, G. Varzaru, B. Mihailescu, "Lead/Lead Free Solder Joints Comparative Electrical Tests as Function of Microstructure and Soldering Thermal Profile", 2013 IEEE 19th International Symposium for Design and Technology in Electronic Packaging, SIITME 2013, 24-27 October 2013, Galati, Romania, ISBN: 978-1-4799-1555-2, pp. 255-258.

Investigating the Effect of Solder Paste Viscosity Change on the Pressure during Stencil Printing

Oliver Krammer, Bertalan Varga
Department of Electronics Technology
Budapest University of Technology and Economics
Budapest, Hungary
krammer@ett.bme.hu

Oliver Krammer, David Bušek
Department of Electrotechnology
Czech Technical University in Prague
Prague, Czech Republic

Abstract—The effect of viscosity change of solder pastes during stencil printing on the pressure distribution along the stencil line and on the results of numerical modelling is investigated. Rheological properties of Type 4 SAC305 (Sn96.5Ag3Cu0.5) solder paste were measured for several cycles in our previous work, to investigate the viscosity change of pastes during stencil printing. Parameters of the Cross model were fitted to the results then. In this work, the parameters of the Carreau-Yasuda model were calculated from Cross model and they are included in a finite volume model of stencil printing. The model includes the stencil, the squeegee with 55° attack angle and the solder paste as the domain of interest. The pressure distribution and the shear rates are calculated by utilising the rheological properties of the solder paste in different states (fresh, and stabilised after the 9th printing cycle). The error of numerical results is analysed for the case if viscosity parameters of the fresh paste are applied instead of that of stabilised one.

Keywords—*stencil printing, numerical modelling, solder paste viscosity, pressure on stencil line*

I. INTRODUCTION

The most common assembling method to connect electronic components to printed circuit boards (PCB) is reflow soldering technology, where the most advancing heat transfer method is vapour phase soldering [1,2]. Nevertheless, depositing solder paste by stencil printing [3] onto the soldering pads of the PCB is one of the most crucial steps of this technology [4]; almost 60% of the soldering failures can be traced back to the printing process [5]. The printing process has become even more critical by the spread of ultra-fine-pitch components, like QFNs (Quad-Flat-No-Lead) and μBGAs (Micro Ball Grid Array) [6], as they require smaller and smaller aperture sizes [7]. Investigating this process, even with numerical modelling, is therefore absolutely necessary to reach zero-defect manufacturing. One of the most crucial parameters in numerical modelling to obtain valid results is the material properties. In the modelling of stencil printing, this parameter is the viscosity of solder paste. The pressure along the stencil line in numerical modelling is proportional to the viscosity according to the Riemer's model (1)[8,9]:

$$P = \frac{1}{x}\left(\frac{2\sin^2\theta}{\theta^2 - \sin^2\theta}\right)\eta(\vec{r})v \qquad (1)$$

where: P is the hydraulic pressure, η is the apparent viscosity depending on the shear rate in a given cell, $\vec{r}$ is a pointing vector, v is the squeegee speed, θ is the squeegee angle, and x is the distance from the squeegee tip. That is the reason, why correct rheological parameters of solder pastes are required for valid numerical modelling of stencil printing.

Rheological properties of various assembly pastes were already studied before. Durairaj et al. characterised different lead-free solder pastes with viscosity measurements to apply the paste properties into numerical calculations [10,11]. The viscosity of solder pastes changes during stencil printing, from the initial, first print; and the pastes need some print strokes to stabilise. Consequently, a significant downside of these researches was the fresh sample used for every measurement. It was not possible to obtain adequate information about the time-dependent viscosity change of pastes, which particular information would be necessary for correct numerical modelling of stencil printing. Time-dependent behaviour of solder pastes was studied too, but only the viscosity change over continuous time at constant shear rates was examined [12,13,14], the time-gap between stencil printing cycles was not considered, and thixotropic behaviour of solder pastes was also neglected. Only one research dealt with the thixotropic characteristics of solder pastes [15] using non-fresh samples. In that study, a narrow shear rate sweep (0.01–0.2 1/s) was applied, and there were no rest periods between the measurements, i.e. the time-gap was 0 s. So, we performed a research to investigate the viscosity change of solder pastes during stencil printing [16]; and our aim in this paper is to analyse, how that viscosity change affects the pressure distribution along the stencil line.

II. MODELLING OF STENCIL PRINTING

A. Geometry and Boundary Conditions of the Model

The geometry of the finite volume model of the stencil printing consists of the stencil, the blade and the solder paste as the domain of the interest. The mechanical properties of the squeegee as well as the printing force can be taken in numerical calculation into account by applying the loaded squeegee angle instead of the unloaded one. The loaded angle was determined as 55° in our previous work. It should be noted that this value is valid for the following conditions: material of the squeegee blade is stainless steel; the blade thickness is 200 μm; the blade height is 15 mm; the unloaded angle is 60°; and the specific printing force is 0.3 N/mm. The boundary conditions for the modelling were as follows: the movement is represented in the frame of reference for an observer travelling with the squeegee; so the printing blade was rigid and fixed, whereas the rigid stencil was moving opposite to the blade with a chosen speed (20 mm/s and 70 mm/s) of the stencil printing process. No-slip condition was set to both of the rigid walls. The paste-air interface had a prescribed shape and free-slip condition was set to it. The model is isothermal since the temperature inside stencil printers is maintained constantly during manufacturing. The optimal mesh type and element size was chosen based on our previous work. The "hybrid mesh" – consisting tetrahedron and polar type elements – with an element size of 70 μm was applied (Fig. 1).

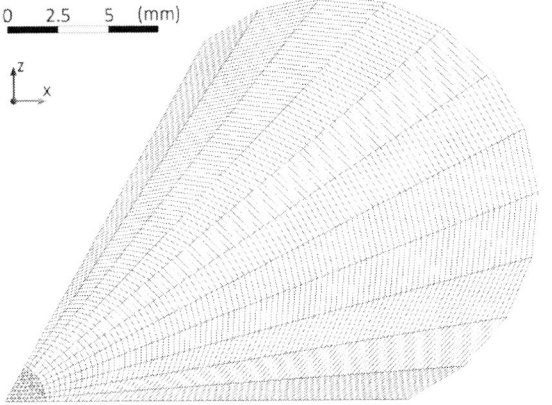

Fig. 1. Hybrid mesh type for FVM calculations.

B. Material properties of the solder paste

In our previous research [16], the viscosity change of a Type 4 SAC305 (Sn96.5Ag3Cu0.5) solder paste during stencil printing was investigated. Basically, the viscosity of solder pastes decreases during stencil printing, but increases in a given extent during the time-gaps between the printing cycles. The degree of viscosity increment between the printing cycles depends on the length of time-gap, thus the overall viscosity decrease also depends on it. In that research, three different time-gap durations were investigated: 15 s, 30 s, 60 s. The results, the viscosity change were quite similar in the cases of time-gaps 15 s and 30 s. Therefore, the time-gap 30 s was omitted and viscosity change for the 15 s (Fig. 2) and the 60 s time-gaps were investigated by the calculations in this research.

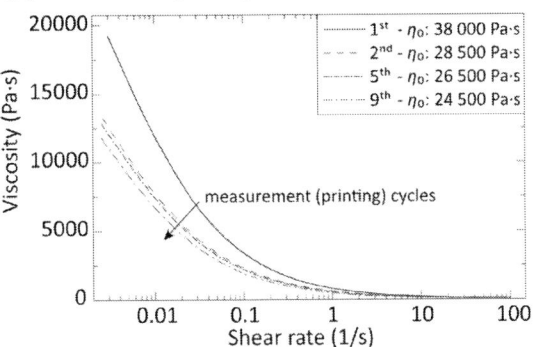

Fig. 2. The viscosity change in the case of time-gap 15 s.

It can be seen in Fig. 2, that the viscosity decrease is significant from the first printing cycle (fresh paste state) to the ninth one. Hence, the viscosity parameters for these two cycles were included in our model. The relative viscosity change as a function of the shear rate in the cases of 15 s and 60 s is illustrated in Fig. 3.

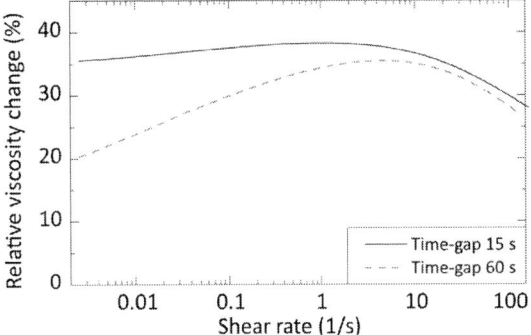

Fig. 3. Relative viscosity change in the cases of time-gap 15 s and 60 s.

It can be observed in Fig. 3, that the relative viscosity change is larger in higher shear rates than in lower shear rates; and it is almost the same in the two cases at higher shear rates (~30%). Since the shear rates during stencil printing lie in higher ranges, this suggests that the change in pressure results of numerical modelling will be higher in the case of the faster printing speed (70 mm/s) than in the case of lower printing speed (20 mm/s).

To describe the viscosity properties of a non-Newtonian fluid, the Cross or the Carreau-Yasuda model (2) is fitted to the measurement results generally:

$$\eta_a = \eta_\infty + \frac{\eta_0 - \eta_\infty}{\left(1 + \left(\lambda\dot{\gamma}\right)^a\right)^{\frac{1-n}{a}}} \qquad (2)$$

where η_a is the apparent viscosity, η_0 and η_∞ are the viscosity values at zero- and infinite shear rates respectively, $\dot{\gamma}$ is the shear rate, λ is a time constant, a is the dimensionless Yasuda exponent, and n is a power law index. The parameters for the different states of the solder paste are collected in Table 1.

TABLE I.   NON-NEWTONIAN VISCOSITY PARAMETERS FOR MODELLING

Parameter	Different states of the solder paste		
	Fresh state	*After 9th cycle – time-gap 15 s*	*After 9th cycle – time-gap 60 s*
η_0 [Pa·s]	38 000	24 500	31 500
η_∞ [Pa·s]	33	28	30
λ [s]	445	445	495
a – Yasuda exponent	0.64	0.65	0.67
n – power law index	0.36	0.35	0.33

After including the material properties in the finite volume model, the pressure distribution along the stencil line is calculated, and the difference between the cases is analysed.

III. RESULTS

The numerical calculations were performed by applying printing speeds of 20 mm/s and 70 mm/s. The shear rate for printing speed 20 mm/s is illustrated in Fig. 4.

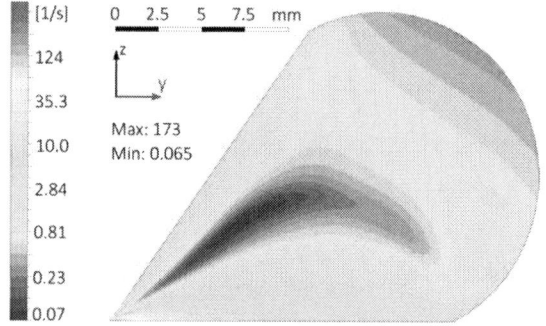

Fig. 4. Shear rate inside the solder paste for printing speed 20 mm/s.

It can be seen, that the shear rate is much higher closer to the squeegee blade and decreases exponentially at higher distances from the blade. The maximum and minimum shear rates inside the solder paste are collected in Table 2.

TABLE II.   SHEAR RATES INSIDE THE SOLDER PASTE

Shear rate	Printing speed	
	20 mm/s	*70 mm/s*
Min. global [1/s]	0.065	0.32
Max. global [1/s]	173	607
Min. by the stencil line [1/s]	3.38	11.2

The calculated pressure for fresh paste and for the printing speeds of 20 mm/s and 70 mm/s are illustrated in Fig. 5 and Fig. 6 respectively.

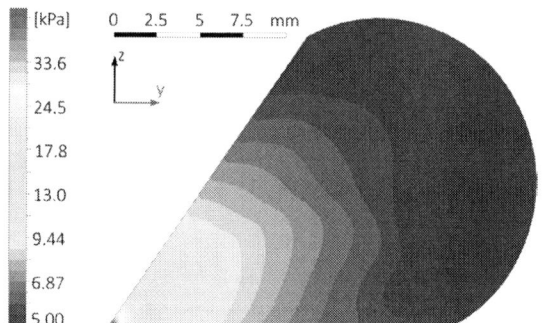

Fig. 5. Pressure – fresh paste state – printing speed 20 mm/s.

The pressure range for the printing speed 70 mm/s is remarkably higher as expected from (1). The pressure profile along the stencil line is illustrated in Fig. 7. By analyzing the pressure profile for the fresh paste state (and by applying non-Newtonian material parameters), it can be stated that the change in pressure between the printing speeds is not linear, i.e. does not follow (1), and it depends on the distance from the squeegee blade. The explanation can be that the shear stress during printing does not depends linearly on the shear strain due to the non-Newtonian fluid properties.

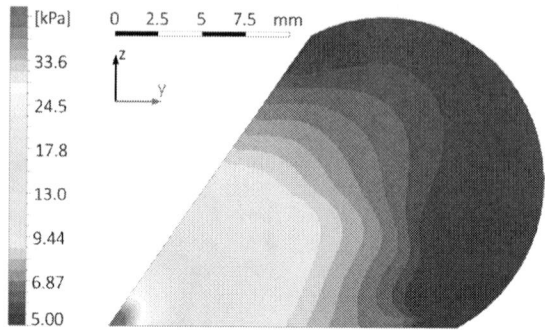

Fig. 6. Pressure – fresh paste state – printing speed 70 mm/s.

Regarding the time-gaps 15 s and 60 s, the pressure profiles are quite similar for a given printing speeds. The reason for the quite similar profiles can be that the shear rate by the stencil line (either for 20 mm/s or for 70 mm/s) lies in the range between n·1 and n·100 1/s. If we analyse Fig. 3, it can be seen that the relative viscosity changes above shear rates n·1 1/s are quite similar for the time-gaps 15 s and 60 s.

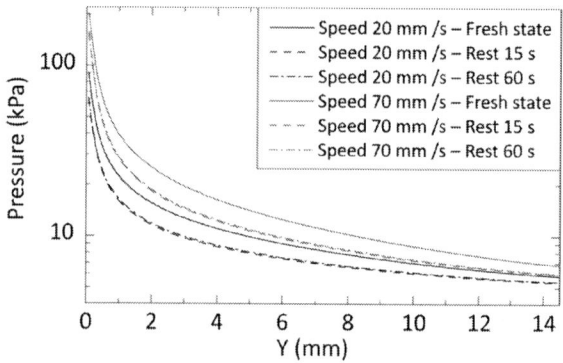

Fig. 7. Calculated pressure profile along the stencil line.

After calculating the pressure profiles, the pressure difference between the different states of the paste was calculated (Fig. 8). This difference equals to the error of calculation, if the viscosity parameters of a fresh paste are used instead of that of a stabilised one (e.g. after the 9th cycle). The change in pressure, i.e. the error of calculation is in the range between 10–30%; and it is higher closer to the squeegee blade where the higher pressure values are arisen. Besides, the average difference is slightly larger for printing speed 70 mm/s.

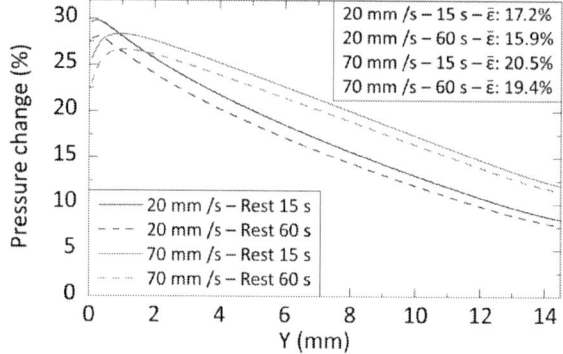

Fig. 8. Pressure difference.

IV. CONCLUSION

It can be concluded based on the research that a large calculation error can be eliminated, if correct material parameters are used. If using viscosity parameters of a paste in fresh state, the error of Finite Volume results can be as large as 30%. The order of magnitude of the error does not depend significantly on the period of resting, i.e. the time-gaps between the printing cycles. If a numerical method is used for optimising a given stencil printing process, the thixotropic properties of the local paste should be investigated to be able to apply the correct viscosity parameters in the modelling.

ACKNOWLEDGMENT

The authors would like to express their thanks for the ProProgressio foundation.

REFERENCES

[1] B. Illés, A. Géczy, A. Skwarek, D. Busek, "Effects of substrate thermal properties on the heat transfer coefficient of vapour phase soldering", Int. J. Heat Mass Tran., vol. 101, pp. 69–75, 2016.

[2] B. Illés, A. Géczy, "Investigating the heat transfer on the top side of inclined printed circuit boards during vapour phase soldering", Appl. Therm. Eng., vol. 103, pp. 1398–1407, 2016.

[3] C.S. Lau, C.Y. Khor, D. Soares, J.C. Teixeira, M.Z. Abdullah, "Thermo-mechanical challenges of reflowed lead-free solder joints in surface mount components: a review", Solder. Surf. Mount Technol., vol. 28, No. 2, pp. 41–62, 2016.

[4] A. Skwarek, B. Synkiewicz, J. Kulawik, P. Guzdek, K. Witek, J. Tarasiuk, "High temperature thermogenerators made on DBC substrate using vapour phase soldering", Solder. Surf. Mount Technol., vol. 27, No. 3, pp. 125–128, 2015.

[5] T.N. Tsai, "Modeling and optimization of stencil printing operations: A comparison study", Comput. Ind. Eng., vol. 54, no. 3, pp. 374–389, 2008.

[6] J. Pan, G.L. Tonkay, R.H. Storer, R.M. Sallade, D.J. Leandri, "Critical Variables of Solder Paste Stencil Printing for Micro-BGA and Fine-Pitch QFP", IEEE T. Electron. Pa. M., vol. 27, no. 2, pp. 125–132, 2004.

[7] D. Manessis, R. Patzelt, A. Ostmann, R. Aschenbrenner, H. Reichl, "Technical challenges of stencil printing technology for ultra fine pitch flip chip bumping", Microelectron. Reliab., vol. 44, pp. 797–803, 2004.

[8] D. Riemer, "Analytical model of the screen printing process: Part 1.", Solid State Technol., August, pp. 107–111, 1988.

[9] D. Riemer, "Analytical model of the screen printing process: Part 2.", Solid State Technol., September, pp. 85–90, 1988.

[10] G.P. Glinski, C. Bailey, K.A. Pericleous, "A non-Newtonian computational fluid dynamics study of the stencil printing process", P. I. Mech. Eng. C-J. Mec., vol. 215, no. 4, pp. 437–446, 2001.

[11] R. Durairaj, G.J. Jackson, N.N. Ekere, G. Glinski, C. Bailey, "Correlation of solder paste rheology with computational simulations of the stencil printing process", Solder. Surf. Mount Technol., vol. 14, no. 1, pp. 11–17, 2002.

[12] S. Mallik, N.N. Ekere, A.E. Marks, A. Seman, R. Durairaj, "Modelling of the Time-dependent Flow Behaviour of Lead-Free Solder Pastes used for Flip-Chip Assembly Applications", 2nd ESTC conf. Greenwich, pp. 1219–1224, 2008.

[13] A. Pietriková, M. Kravčík, "Boundary Value of Rheological Properties of Solder Paste", 34th ISSE conf. Slovakia, pp. 94–97, 2011.

[14] A. Pietriková, M. Kravčík, "Investigation of Rheology Behavior of Solder Paste", 35th ISSE conf. Austria, pp. 138–143, 2012.

[15] N.N. Ekere, D. He, L. Cai, "The Influence of Wall Slip in the Measurement of Solder Paste Viscosity", IEEE T. Compon. Pack. T., vol. 24, no. 3, pp. 468–473, 2001.

[16] O. Krammer, B. Gyarmati, A. Szilágyi, R. Storcz, L. Jakab, B. Illés, A. Géczy, K. Dušek, "Investigating the thixotropic behaviour of Type 4 solder paste during stencil printing", 40th International Microelectronics and Packaging (IMAPS Poland) Conference, unpublished, 2016.

Optimization of PCB Assembly Process

Larisa Tarba
Department of Electrotechnology
Faculty of Electrical Engineering
Czech Technical University, Prague, Czech Republic
tarbalar@fel.cvut.cz

Pavel Mach
Department of Electrotechnology
Faculty of Electrical Engineering
Czech Technical University, Prague, Czech Republic
mach@fel.cvut.cz

Abstract— **The paper is focused on different types of tools for process improvement - the synergy effect of using different tools and approaches together to drive the process to non-defective and faster execution. The goal of optimization is to decrease the total time of assembly and number of failures on assembled PCBs and to increase quality and reliability of the boards. The methodology is based on following steps: identifying the criteria for improvement, implementing the innovation technologies with modelling the criteria for quality optimization, calculating the mathematical model and finding the optimal setup for the equipment in order to minimize the overall lead time and to maximize quality of assembly. The tools used for the mentioned methodology are six sigma, lean six sigma, theory of constrains and application of fuzzy logic on lean six sigma technics.**

Keywords—PCB, Six Sigma, Quality management, Lean Six Sigma, Improvement tools, Process improvement, Fuzzy Logic, Printed Circuit Board assembly

I. INTRODUCTION

Nowadays successful functioning of the manufacturing company is determined by application of effective measures in order to organize the manufacturing process as well as dynamic researches in innovative technologies and new developing directions. Despite the constant growth in the area of electronics manufacturing the focus is aimed mostly on renewal of the existing equipment rather on analyzing all the aspects of the whole process and finding the weak points in order to not only increase the quality of the manufactured products, but also develop a strategy how to effectively manage the manufacturing process and have a model for quantitative assessment.

One of the most important criteria to evaluate the effectiveness of manufacturing processes within the company is lead time, which is dependent on complexity of the output product and on effectiveness of applied technologies and methods of organization of production system. Reduction of the lead time of the manufacturing process is one of the actual interests for each of the current manufacturing companies. Application of the process approach allows reducing execution time of any manufacturing process by optimizing its main operations, which leads as well to reduction of the lead time.

Six Sigma is a disciplined, data-driven approach and methodology for eliminating defects (driving toward six standard deviations between the mean and the nearest specification limit) in any process - from manufacturing to transactional and from product to service. Particular phases of process improvement based on quality methods were analyzed.

Lean Six Sigma approach as combination of concepts would focus on reducing cycle time in processes, thus it achieves the fastest rate of improvement in cost and quality, process speed and invested capital [1]. Fuzzy approach takes the role when computational control is needed, it will help to reduce the variability in a process and plan the sequence of production processes in the best manner.

II. METHODOLOGY FOR OPTIMIZATION OF PCB ASSEMBLY PROCESS

A. Preconditions for improvement of PCB assembly process

The prerequisites for improvement of manufacturing processes in electronics field are constant growth of different technologies as well as growth in requirements for quality control and quality of the output products. The growth of technologies and complexity of design leads to increase of usage of more different components thus makes the assembly process more expensive.

Optimization of PCB assembly process is realized with the help of implementation of innovation technologies as well as analysis based on modern methods of quality control such as six sigma, lean production and mathematical modelling using fuzzy technologies.

In order to correctly determine the acceptance and effectiveness criteria for some particular manufacturing process, it is important to identify the goal of the research. In case if the goal is the increase in productivity, then the criteria are to increase the capacity and output of existing equipment. This paper is describing the possible ways how to improve the quality of assembly processes with the help of innovation technologies, different methods of statistical tools and fuzzy logics. Therefore one of the main criteria is effectiveness factor of innovation technologies implementation, which can be determined by several sub factors.

In order to choose the correct criteria, it is important that these criteria will reach the expected goal. Taken into consideration this fact, the overall effectiveness factor ($F_{overall}$) can be divided in the following sub factors:

- Factor for minimizing the lead time of the assembly process (Fm);

- Defect factor (Fd);

- Product manufacturing quality factor (Fq);

- Factor of economic efficiency (Fe);

- Factor of staff skill qualification (Fs);

- Factor of automatization (Fa).

For each of these factors there are several parameters which are determining them.

For quantitative assessment for the individual factors it is necessary to develop mathematical model for evaluating the effectiveness of the implementation innovation process for PCB assembly. This will allow identifying the effectiveness of the whole technological system. The mathematical representation of the overall effectiveness factor is:

$$F_{overall} = \sum_{i=1}^{n} Fi \ ,$$

where Foverall is the overall factor to identify effectiveness of implementing innovation technologies; Fi is the individual factor and i - is the amount of individual factors.

For the electronics manufacturing the most important and technically complex process is the PCB assembly process. Taken it as an example, it is possible to subdivide it into individual operations and define complex metrics which can help minimizing the duration of the overall lead time of PCB assembly process.

B. Modelling the criteria for quality optimization for PCB assembly

For the whole assembly process can be applied process approach, which divides the process into several operations. Hence, PCB assembly process can be split into following operations (P1 – P5):

- Input control of PCB blanks (P1);

- Application of the lead solder paste (P2);

- Mounting of chip-components (P3);

- Mass soldering (P4);

- Quality control of the assembly process (P5).

Thus the overall process is possible to represent as following:

$$P = \{P1, P2, P3, P4, P5\}$$

Identification of constraints for reduction of the lead time of the assembly process is realized by the maximum time, spent for executing the individual operation ($t_{i\,max}$). According to the theory of Goldratt [2], the bottleneck will be the operation, which has the maximum execution time ($P_i(t_{i\,max})$). Reduction of the lead time is the main objective for optimization of the process:

$$F(t_{i\,max}) \rightarrow \min\{Ti\}$$

Minimizing criterion of the lead time of the PCB assembly process can be determined as follows:

$$F_m = \sum_{1}^{5} F_{mi} * F_{weight\_i}$$

Distribution of the weighting factors of the effectiveness criteria for minimization of the lead time of the assembly process can be done with the help of theory of constraints (TOC). The biggest value of the weighting factor is being assigned to the constraint activity, the bottleneck process so

called, in this case it is assumed as soldering process (e.g see the distribution table below).

Distribution of weighting factors for minimizing the lead time of assembly process		
Parameter	Mathematical expression	Weighting factor
Time for input control	Fm1	0.15
Time for applying the lead solder paste	Fm2	0.25
Time for chip-component mounting	Fm3	0.2
Time for soldering in the oven	Fm4	0.3
Time for controlling the PCB board	Fm5	0.1

Distribution of the weighting factors is being done according to relevance of the single parameters for particularly taken individual factor. Each of individual factors is presented by single parameters, which are determined by the ratio before and after application of innovative technologies and improvement criteria.

One of the most effective instruments for increasing the competitiveness of the company along with the constraints theory is the optimization of the manufacturing processes considering the innovative technologies. This will allow in a fast and high quality manner increase the effectiveness of the manufacturing processes based on improvement of manufacturing technologies.

After analyzing the weakest point in the chain and making it stronger, there is a necessity of reanalyzing the whole process again and finding new constraint factors, thus, there is a continuous improvement cycle being applied to the assembly process.

In order to identify the constraint it is necessary to carry out the analysis of the PCB assembly process, which consists of five consequent operations:

1. On the first stage, where input control of PCB blanks (P1) is being performed, there are buffers for continuous operation of the automatic line which have been created. Input control is important in order to identify the defects as well as suitability of PCB blanks for the automatic line. Automatic process of the PCB assembly is being performed by executing in parallel the rest of the operations P2 – P5.

2. Then PCB blanks are placed into automatic line, where plating of the lead solder paste (P2) is being done with help of jet printer. At this stage the lead solder is being placed onto the pads through the special stencil which has holes, which represent the exact placement of the pads on the PCB board.

3. Mounting of chip-components (P3) is being done by automatic installer. Its main objective is to correctly place chip-components on the PCB board. Before the mounting a special system is calculating correct placement of the components and then starts the mounting.

As this operation has the biggest amount of components with different type to be placed, hence it has the biggest lead time value (P3(t3 max)). This is a consequence of the fact, that there is a constant growth in requirements for electronic equipment as well as growth in complexity of products, thus growth in the amount of installed components.

4. Mass soldering (P4) process is when after mounting of the chip components PCB board is placed into the reflow oven, where the whole product is evenly warmed up and then cooled down.

5. At the last stage PCB board is taken by the unloader and there is a process of quality control (P5) being performed, where the board is being checked within the recommended requirements of IPC-A-610C standard.

Elimination of the constraint in the operation of the mounting of chip components will allow reducing the lead time by optimizing main parameters of this operation.

C. Application of Lean Six Sigma tools to optimize individual process

Once the constraint has been identified it is possible to deepen in analysis using the concepts which are oriented on productivity improvement – Six sigma and Lean production. However, Six Sigma is focused on reducing the variation, defects and improves quality of processes, products and services, when Lean production tries to eliminate waste and reduce cycle times in processes. Together these two methods improve the quality of processes by combining their different approaches. These methods will help to analyze the further actions in order to eliminate the constraint and improve the overall lead time of the process, output product and thus improve the quality of the process. In order to map the Lean principles to the Six Sigma methodology, i.e. DMAIC (Define, Measure, Analyze, Improve, Control) process in particular, it is essential to understand how each phase of DMAIC was used to execute a full scale Lean design effort.

- *Define and Measure phases*: the first step is to define the objectives and to stay focused on these specifics goals. It is important to identify "critical to quality" (CTQ) metrics which will be the focus. For the measurement it is important to identify how the current process is performing and how many defects are in place. For the identified constraint (mounting of chip components) it is possible to determine the operations which have higher risks of failing or giving the latency to lead time and measure which are the ones to have influenced on that the most in chosen period of time.

- *Analyze and Improve*: the next step is to define the drive of the current performance and what is the root cause of the possible delays. For the improving phase the solutions which are positively affecting the performance to be analyzed. The goal of Analyze phase is to identify potential root causes for the process problem being addressed and then confirm actual root cause with data. Having completed the Measure phase, the possible problem has been identified and the circumstances under which it may occur. Once a list of potential root causes has been compiled, the next step is to organize them in a way that makes it easier to prioritize and assess them. For chip component mounting onto the PCB board to check the accuracy of instruments, confirm key performance variables and acceptable operating range and review operating procedures. Characterize specifications for mounting procedure and try to analyze if the system uses custom experiments. For improving the existing mounting process – add and replace or calibrate the instrumentation, possible add or remove the automatic installers. The statistical analysis will identify key variables that count for the majority of process variation.

- *Control*: for the control plan is important to ensure the improvement. For mounting the components it is possible to update the operating instructions, monitor key performance variables and develop procedures for more accurate approach. Calculate possible operating cost savings. It is a very critical factor for PCB assembly how well the components will sit on the board and how well they will operate as a result. If the boards are stored or stacked, the solder height determines whether copper in the components will migrate, causing the boards to become defective. To keep solder height within control limits is critical in PCB assembly.

D. Application of fuzzy logics on Lean Six Sigma technics to optimize the lead time

One of the possible improvement measures for the chip component mounting sub process is application of fuzzy theory and focusing on the way for optimization the lead time and seeking for increasing the quality of output product.

The DMADV (Define, Measure, Analyze, Design, Verify) flow is the same as DMAIC algorithm described above and can be applied to design or redesign products or processes (Design for Six Sigma or DFSS). It will help to redesign the process for chip component mounting in the most optimal way. The process of activities of DFSS consists of four stages: Identify, Design, Optimize and Validate (IDOV).

- *Identify*: the quality function deployment (QFD) method should be employed to identify the most significant indicators with the aim of finding the weak points which effect the lead time of individual operations as well as overall lead time of the manufacturing process.

- *Design*: a theoretical function in order to analyze quality of production after particular operation incorporates fuzzy set theory and provides a way of monitoring and optimizing of the whole manufacturing system. Optimization in this context means minimizing lead time values for individual operations and thus minimizing the overall lead time of the manufacturing process. Essentially vague and subjective information often found in the fuzzy manufacturing environment of variability, complexity and constraints make it a difficult environment to control and improve. Therefore modelling with fuzzy set theory is a useful performance monitoring tool to incorporate

- *Optimize*: a transfer function is a mathematical representation of the relationship between the input and output of a system or a process. It facilitates the optimization of process output by defining the true relationship between input variables and the output. The transfer function can be presented as indicator of quality of production after the specific operation and refined with the use of fuzzy logic and is defined by following three variables:

Quality of production =

F {positioning, capturing, mounting}

Fuzzy inference is the process of formulating the mapping from a given input to an output using fuzzy logic (major stages being fuzzification, rule evaluation and aggregation, and defuzzification). The mapping of defuzzified results onto the problem situation then provides a basis from which decisions can be made:

- *Fuzzification*: the fuzzification process is performed during run time and consists of assigning membership degrees between 0 and 1 to the crisp inputs of positioning, capturing and mounting.

For *Positioning* variable the precision of placement of PCB board and stencil was measured and it is being expressed in μm. For the fuzzy scale from 1 to 5, there were 5 linguistic terms identified (i.e. 1 as low with range of more than 60μm, 2 as acceptable low with range from 50 to 60μm, 3 as regular with 45 – 50μm, 4 as high with 35-45μm, 5 as very high with more than 25μm of precision).

Capturing represents the tackiness of solder paste and is considered for the further capturing of the components and placing then onto the PCB board. Tackiness expresses the ability of the solder paste to grab the component and keep it in place and can be classified into two terms, tack time and tack force. Tack time is the length of time that a solder paste can stay tacky enough to hold a component in position after printing. The goal is to determine how long the paste stays sticky enough (the assumption is the minimum adhesion required to hold a component as more than 100gf.). For the fuzzy scale from 1 to 3, there were 3 linguistic terms identified (i.e. 1 as low with range of below 100gf, 2 as medium with range from 100 to 160gf, 3 as high above 160gf of tackiness of solder paste).

Mounting is the speed of applying the solder paste and determines the last input variable in fuzzification process. The categorization is based on the amount of components placed onto the PCB board and is presented components placed per hour. For the fuzzy scale from 1 to 3, there were 3 linguistic terms identified (i.e. 1 as low with range of below 15000 components per hour, 2 as medium with range from 15000 to 25000 components per hour, 3 as high above 25000 components per hour).

- *Rule evaluation*: the rule evaluation process consists of using the fuzzy value obtained during fuzzification and evaluating them via the rule base in order to obtain a fuzzy value for the output. Basically the use of linguistic variables and fuzzy IF-THEN- rules utilize the imprecision tolerance and uncertainty as well as simulating the ability of the human mind to summarize data and focus on decision-relevant information. These rules are based on expert knowledge and experience.

For example: if tackiness of solder paste is "medium" and if precision of placement is "regular" and if speed of applying of solder paste is "low" then the quality of production system is "acceptable". The rules are determined through expert knowledge and they can be refined following real life application which will confirm them or require them to be modified. There were 45 evaluation rules described in the fuzzy system in order to replicate the defined logic.

- *Defuzzification*: the fuzzy inference system using Mamdani's fuzzy implication rule determines the appropriate fuzzy membership value. The defuzzification process consists of combining the fuzzy values obtained from the rule evaluation step and calculating the reciprocal in order to get one and only one crisp value that the output should be equal to.

The three inputs (Positioning, Capturing and Mounting) were set within the upper and lower specification limits and the output response was calculated as a score that can be translated into linguistic terms. Using the monitorization feature of Xfuzzy tool [3] is possible to see the fuzzy output solutions for equation. The figures uncover the representation of optimal values in three dimensional plateau. It is seen that there are several peak areas, which represent the possible optimal solutions for output quality of the production system which will eventually lead to optimizing the assembly process. In Fig. 1 the quality achieves high values when the positioning is within the middle range, which corresponds to regular precision and capturing is towards higher end. As the positioning is towards higher precision, then it requires more time to place the component on the board, thus there might be less time for capturing phase and this if it goes towards higher end then output quality drops as the processing lead time would be much increased.

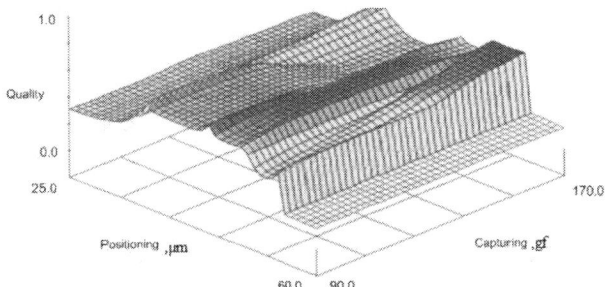

Fig.1 – Output surface of the fuzzy inference system for positioning and capturing

- *Validate* :the model is to be analyzed and validated to ensure that performance indicators provided would meet the identified requirements. The developed model should be modified and verified continuously to ensure that the model satisfies the requirements. This includes the application of fuzzy evaluation rules. The flow of IDOV activities facilitates the continuous improvement effort.

Criteria for optimizing the PCB assembly process has been analyzed and identified. The criteria are based on minimizing the lead time of the sub operations (i.e. sub operation of mounting the chip components, which has the biggest processing time because of the largest amount of components to be placed on the board) of manufacturing process which will lead to minimizing the overall lead time.

ACKNOWLEDGMENT

The work was supported by the Pre-seed project of the Ministry of Education, Youth and Sports with the title "Material research for InovaSEED", no. CZ.1.05/3.1.00/14.0301.

REFERENCES

[1] J. A. Schumpeter, "The nature and essence of economic theory". Transaction Publishers, New Jersey, 2010

[2] Goldrat, E. M., Cox, J.: The Goal: A process of Ongoing Improvement, North River Press, Atlanta, 2004

[3] XFuzzy 3.0. Centro Nacional de Microelectrnica. Instituto de Microelectrnica de Sevilla. http://www.imse.cnm.es/xfuzzy

Thermophysical properties of some low temperature lead-free solder pastes dedicated to automotive applications

M. Branzei, I. Plotog, G. Varzaru
University POLITEHNICA of Bucharest
Bucharest, Romania
mihai.branzei@upb.ro

T. C. Cucu
Alpha Assembly Solutions, Somerset NJ, USA
Traian.Cucu@alphaassembly.com

Abstract—The maximum temperature of all parts of automotive electronics equipment is given by the location and can vary between 358 K and 478 K. RoHS EU Directive in the context of the manufacturing costs reduction, necessity led automotive industry to create new Low-Temperature Lead-Free (LTLF) solder pastes family. In the paper, the thermophysical properties of the alloys from ALPHA® CVP-520 (SnBiAg) and ALPHA® OM-535 (SBX02) test samples having as a source solder pastes were compared with similar from ALPHA® OM-6106 (Sn62) solder paste. The presented work extend the theoretical studies and experiments upon heat transfer in VPSP in order to optimize the technology for soldering process (SP) of automotive electronic modules and could be extended for home and modern agriculture appliances industry.

Keywords—Thermophysical properties, Low-Temperature Lead-Free solder pastes, Vapour Phase Soldering, Soldering Process, Automotive Electronics.

I. INTRODUCTION

The control regarding solder alloys composition were practical initiated by the Directive 2002/95/EC of The European Parliament and of The Council (RoHS 1) on the restriction of the use of certain hazardous substances in electrical and electronic equipment [1], and Directive 2002/96/EC of the European Parliament and of the Council of 27 January 2003 on waste electrical and electronic equipment (WEEE) [2]. The directive was followed by RoHS 2, Recast

Directive (2011/65/EU) [3] which become law in EU from 2013. The processes and substances used on assembling line are under the restriction of Regulation (EC) No 1907/2006 of the European Parliament and of the Council concerning the Registration, Evaluation, Authorization and Restriction of Chemicals (REACH), connected to RoHS in 2011 [4].

The surface mounted technology (SMT) who is younger than wave soldering but succeeds to generate mass production on automatic assembling, without eliminating any other technology, on the contrary, coexisting with manual and wave types [5].

Consequently, the wave technologies has evolved, appearing selective wave soldering adapted to the electronic components evolution to micro and nano level with increasing number of interconnection points [6]. This trend in technology is one of actual industrial revolution, INDUSTRY 4.0, characteristics.

The automotive electronics are the result of each step from design to manufacturing of their phase life cycle, finally being the result of soldering process on assembling line. One operational condition is operating temperatures.

The minimum Temperature Range Grade (TRG) defined by Automotive Electronics Council (AEC) standard offers criteria to challenge the solder alloy from melting temperature points of view, as it shown in Table I [7].

TABLE I. THE TEMPERATURE RANGE GRADES AND TYPICAL APPLICATION FOR EACH GRADE.

Grade	Temperature Range		Passive Component Type Maximum Capability Unless Otherwise Specified and qualified	Typical/Example Application
	Min.	Max.		
0	-50 °C	150 °C	Flat chip ceramic resistors, X8R ceramic capacitors	All automotive
1	-40 °C	125 °C	Capacitor Networks, Resistors, Inductors, Transformers, Thermistors, Resonators, Crystals and Varistors, all other Ceramic and Tantalum Capacitors	Most under hood
2	-40 °C	105 °C	Aluminum Electrolytic Capacitors	Passenger compartment hotspots
3	-40 °C	85 °C	Film Capacitors, Ferrites, R/R-C Network sand Trimmer capacitors	Most passenger compartment
4	0 °C	70 °C		Non-automotive

According to the trend of Low-Temperature Lead-Free solder pastes (LTLFSP) usage in the automotive electronics production, the Vapour Phase Soldering Processes (VPSPs) request to be used under working liquid boiling point [8, 9]. The operating TRG3 at maximum 358 K (85 °C), as consequence the melting temperature of solder alloy used in soldering process could be decreased.

Transient heat transfer problems occur when the temperature distribution changes with time. the fundamental quantity that enters into heat transfer situations not at steady state is thermal diffusivity (ThD), which is related to the steady state thermal conductivity (ThC) through the equation

$$\alpha = \lambda / \rho \, C_p \qquad (1)$$

where α is the ThD, λ is the ThC, C_p is the specific heat, and ρ is the density. The ThD is the measure of how quickly a body can change its temperature, in concordance with specific heat. All three factors on the right hand of the equation are are depending on temperature.

The flash method of measuring ThD of a material is usually carried out by rapidly heating one side of a sample and measuring the temperature rise curve on the opposite side.

The time that it takes for the heat to rise on the rear face of the sample can be used to measure the ThD. According to the Parker expression [10]

$$\alpha = 0.139 \, a^2 / t_{50} \qquad (2)$$

where a is the sample thickness, and t^{50}, the half rise time (the time for the back face temperature to reach 50% of its maximum value).

The software contains an extensive comparison program that analyzes the raw data, performs corrections according to various theories [11 - 16], and determines the "goodness of fit" regression for each of them, using the theoretical thermogram as reference. In essence, it takes the measured data and sequentially applies the corrections to it, one-by-one.

Each corrected thermogram is then compared to the theoretical (calculated) one that represents the ideal case (infinitesimally narrow pulse, no heat losses, etc.). The difference within each pair is analyzed by a least-square calculation, and the one with the lowest deviation is considered to be the best.

II. MEASUREMENT SETUP

The equipment used for measuring ThD was FlashLine™ 3000 Thermal Properties Analyzer, shown in Fig. 1, according to standard ASTM E1461-13 (Standard Test Method for Thermal Diffusivity by the Flash Method).

The thermal environment for the samples to be tested is produced by an infrared (IR) heated chamber. The IR energy is concentrated (but not focused) in the central 50 mm diameter cylindrical space, where the sample holder structure is located. The space inside the liner was evacuated and then was purged with inert gas (Ar).

The sample support carrier, on which the sample support cartridge with three work station is placed. These two items are made of by ceramic to have high emissivity surfaces in the IR region. The used cartridge was equipped with three specimen placement cavities, having a diameter of 12.5 mm.

Fig. 1. FlashLine™ 3000 Thermal Properties Analyzer.

The graphite reference material NIST SRM 8425 (is a fine grain, nearly isotropic graphite) was chosen for this task and tests were conducted all in argon atmosphere at 1 bar pressure, at the same temperature mentioned as follows.

After the circular samples (disk) has been heated at desired temperatures (70 ^{0}C, 80 ^{0}C, 90 ^{0}C, 110 ^{0}C and 130 ^{0}C) and has reached the thermal equilibrium, a high intensity short duration radiant energy pulse (the average pulse width is 600 μs @ t_{50}) from a high speed Xenon discharge tube (HSXD), is applied on the bottom side of the material. In line with the wave guide, but above the sample, is the IR observation port for rear surface thermogram generation.

The data is analyzed using Clark & Taylor method. The software used in the analysis was verified by analyzing values with known value of ThD, graphite sample as reference material, certified by NIST, being used.

The equipment used for measuring thermal conductivity was Thermal Conductivity Meter QuickLine™-10C, shown in Fig. 2, according to standard ASTM E1530 (Guarded Heat Flow Meter Method).

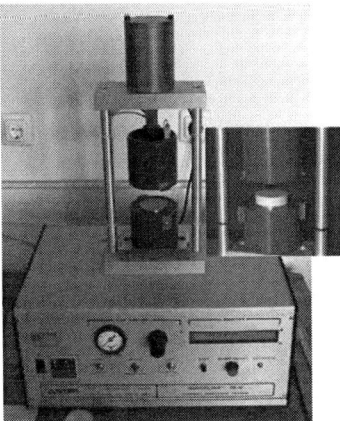

Fig. 2. QuickLine™-10C Thermal Conductivity Meter

The QuickLine™-10C is factory calibrated (has been programmed with 8 reference materials covering an approximate thermal conductivity range of 1.0-220 W/m K) using samples of known thermal resistance spanning the range of the instrument. Materials references used for calibration were Lead (35 W/m K @ 293 K) and Sapphire (1.08 W/m K @ 293 K).

The small sample of the material to be tested is held under a compressive load between two gold plated surfaces, each controlled at a different temperature. So, the contact resistance is kept small by applying a reproducible, pneumatic load to the test stack. Test duration time was set up at 30 minutes. At the end of the test duration time, the calculated thermal conductivity and sample mean temperature was displayed.

Tests were performed at ambient temperature, following that in the near future, with a new apparatus, to be performed at the same temperature with thermal diffusivity tests.

The LTLF solder pastes (ALPHA® CVP-520 and ALPHA® OM-535) alloyswere analyzed and compared with ALPHA® OM-6106 alloy as reference. The VPSPs with thermal profile characterized by two values for cooling rate, 0.5 K/s – Slow Cooling Rate (SCR), and 4 K/s - Rapid Cooling Rate (RCR), were used for experiments [3, 4]. These thermal profile are presented in Fig. 2.

In the first stage of the experiments were realized the as cast solder alloy samples having as a source the three solder pastes, ALPHA® CVP-520, ALPHA® OM-535, and ALPHA® OM-6106, VPSPs being characterized by two thermal profile presented in Fig. 2.

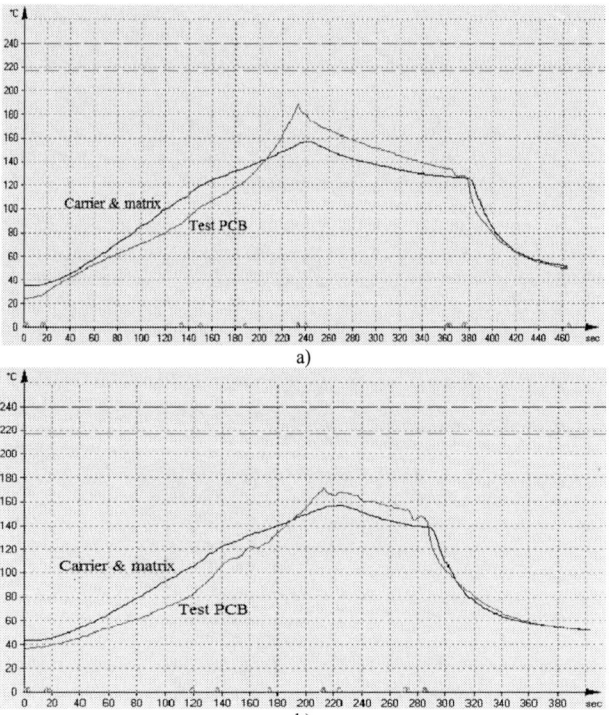

Fig. 3. VPSPs thermal profile: a) - SCR (0.5K/s); b) - RCR (4K/s).

The samples were melted in an Aluminum alloy specially designed crucible in order to have the diameter of 12.7 mm and the thickness of about 2 mm, with plane parallel surfaces for the diffusivity test. Conductivity test samples were cast at 25 mm in diameter and 30 mm height.

III. Measurements Results and Discussions

Thermal diffusivity measurements for the three samples having as a source solder pastes, ALPHA® CVP-520, ALPHA® OM-535, and ALPHA® OM-6106 as reference, at different temperatures, are presented in Table II

TABLE II. THERMAL DIFFUSIVITY MEASUREMENTS FOR THE THREE SAMPLES, AT DIFFERENT COOLING RATES AND TEMPERATURES.

	SCR		RCR	
	T[°C]	α [cm²/s]	T [°C]	α [cm²/s]
OM-6106	70	0.3396	70	0.3125
	80	0.3349	80	0.3141
	90	0.3350	90	0.3007
	110	0.3109	110	0.2956
	130	0.3087	130	0.2854
CVP-520	70	0.1326	70	0.1272
	80	0.1320	80	0.1271
	90	0.1306	90	0.1220
	110	0.1271	110	0.1116
	130	0.1198	130	0.1092
OM-535	70	0.1509	70	0.1604
	80	0.1501	80	0.1540
	90	0.1464	90	0.1431
	110	0.1320	110	0.1335
	130	0.1284	130	0.1285

It is easy to see the variation values of ThD, mainly from an alloy to another, meaning that if we consider the alloy OM-6106 as reference alloy (0.3396 cm²/s), CVP-520 one shows a drastic reduction (0.1326 @ 70⁰C) of the values, while the values for OM-535 one are reduced by about 50 percent. This variation is repeated for both samples obtained from the two-cooling rates.

For the same alloy, values did not differ significantly from one cooling rate to another, but is significantly higher at SCR. The general trend of ThD decreasing values, within the same cooling rate, with temperature increasing can be observed in all alloys type.

In this regard, a suggestive histogram is shown in Fig. 4.

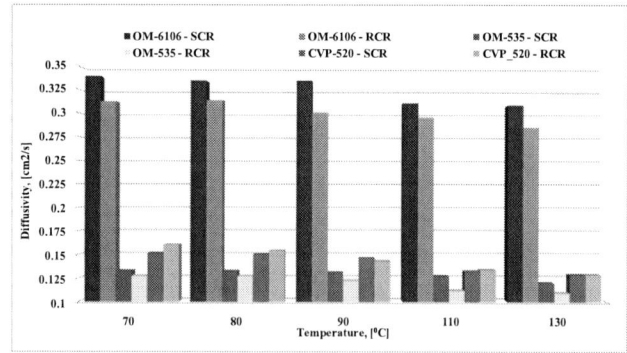

Fig. 4. Thermal diffusivity histogram for the three samples, at different cooling rates and temperatures.

However increasing the thermal properties depends not only on the diffusivity and conductivity of the added elements but also on how they affect the alloy microstructure (i.e. Intermetallics) and the final heat conduction path. For example, an addition of 0.4 wt% Ag to the binary Sn-Bi alloy, has a significant effect on the thermal conductivity, which increases by about 10 to 15 percents. Hence the difference value of the thermal conductivity of alloys presented in Table III.

TABLE III. THERMAL CONDUCTIVITY VALUES AT NEAR AMBIENT TEMPERATURES FOR THE TWO COOLING RATES.

Sample	Thickness, [mm]		Time, [min]	λ, [W/mK] / Temp., [^{0}C]	
	SCR	RCR		SCR	RCR
OM-6106	30.07	29.94	30	54,90 / 20 ^{0}C	54.20 / 20 ^{0}C
	30.07	29.94	30	55.12 / 22 ^{0}C	54.58 / 21 ^{0}C
	30.07	29.94	30	54.75 /20 ^{0}C	54.30 / 20 ^{0}C
CVP-520	29.95	30.05	30	24.43 / 20 ^{0}C	24.13 / 20 ^{0}C
	29.95	30.05	30	24.61 / 21 ^{0}C	24.33 / 20 ^{0}C
	29.95	30.05	30	24.49 / 20 ^{0}C	24.68 / 22 ^{0}C
OM-535	30.11	30.08	30	27.13 / 21 ^{0}C	26.54 / 20 ^{0}C
	30.11	30.08	30	26.94 / 20 ^{0}C	26.01 / 20 ^{0}C
	30.11	30.08	30	26.89 / 20 ^{0}C	25.96 / 20 ^{0}C

ThC measurement values follow the same variation as ThD, which shows that the materials studied have major structural variations in the soldering process.

IV. CONCLUSIONS

In conclusion, based the experiments results, the LTLF solder paste alloys ALPHA® CVP-520 and ALPHA® OM-535 registered both for the ThD and ThC comparable values, which are considerably lower than registered for ALPHA® OM-6106 solder paste alloy.

Rrelatively low values of thermophysical properties are reflected in the thermal profile allure and mainly in its ability to achieve the maximum rapid cooling rate.

These tests are a new start up in reconsidering the thermal profiles applied in VPSPs for these new types of solder pastes, in order to reduce energy consumption in packaging process.

Coefficient of Thermal Expansion (CTE) mast be taken in designing these alloys equation, considering the fact that working temperature is closer to the melting point compared to other classes of solder alloys, considering the "Homologous Temperature" concept.

The experiments results regarding thermophysical properties values of LTLF solder joints and alloys samples offer a database useful in the new LTLF solder alloys creation processes and their qualification for AE in respect of TRG3 for operating temperature range and according to RoHS 2 EU Directive. Also, this database could be a useful tool for accomplishing "zero defects" yield on assembling line.

ACKNOWLEDGMENT

This work has been performed in the frame of the "Partnerships in priority areas" Romanian Research program, developed and supported by MEN-UEFISCDI, SIOPTEF PN-II-PT-PCCA-2011-3.2-899, no. 121/2012 and BLCPL PN-II-PT-PCCA-2013-4-1546, no. 58/2014 projects. The authors are very grateful to the leading staff of IBL-Löttechnik GmbH, (which assures VPS technological support), Solder Paste Department of Alpha Assembly Solutions, a MacDermid Company, (which offered the solder pastes for experiments), and to the both teams of Technological and Business Incubator UPB CETTI-ITA and Thermophysical Testing Laboratory teams, University POLITEHNICA of Bucharest for continuous support and collaboration to fulfill the experiments presented in the paper.

REFERENCE

[1] ***, Directive 2002/95/EC of The European Parliament and of The Council (RoHS 1) on the restriction of the use of certain hazardous substances in electrical and electronic equipment, (2003).

[2] ***, Directive 2002/96/EC of the European Parliament and of the Council on waste electrical and electronic equipment (WEEE), (2003).

[3] ***, Directive 2011/65/EU of The European Parliament and of The Council (RoHS 2) on the restriction of the use of certain hazardous substances in electrical and electronic equipment (recast), (2011).

[4] ***, Regulation (EC) No 1907/2006 of the European Parliament and of the Council concerning the Registration, Evaluation, Authorization and Restriction of Chemicals (REACH), (2006).

[5] R. Strauss, "SMT Soldering Handbook", Butterworth-Heinemann Linacre House, Oxford, 1998,ISBN 0 7506 35894

[6] J. K. Puttlitz, K. A. Stalter, "Handbook of Lead-Free Solder Technology for Microelectronic Assemblies", Marcel Dekker, Inc.,2004, New York, USA, ISBN: 0-8247-4870-0.

[7] "AEC-Q200: Stress Test Qualification For Passive Components, Temperature Range Grades", 2013.

[8] T. Cucu, "Low-Temperature Soldering in Automotive", IEEE 20th International Symposium for Design and Technology of Electronics Packaging (SIITME), 23–26 Oct. 2014Bucharest, Romania.

[9] A. Geczy, P. Szoke, Z. Illyefalvi-Vitez, M. Ruszinko, R. Bunea, Soldering profile optimization for vapour phase reflow technology, IEEE-17th International Symposium for Design and Technology in ElectronicPackaging SIITME 2011, 20-23 October, Timisoara, Romania.

[10] W. J. Parker, R. J. Jenkins, C. P. Butler, and G. L. Abbot, "A Flash Method of Determining Thermal Diffusivity, Heat Capacity and Thermal Conductivity", J. Appl. Phys. 32, 1961, pp. 1679-1684.

[11] L.M.III Clark, and R. E. Taylor, "Radiation Loss in the Flash Method for Thermal Diffusivity" J. Appl. Phys. 46, 1975, pp. 714.

[12] R. E. Taylor, and J. A. Cape, "Finite Pulse-Time Effects in the Flash Diffusivity Technique", Appl.Phys. Lett. 5 (10),1964, pp. 210.

[13] R. D. Cowan, "Pulse Method of Measuring Thermal Diffusivity at High Temperatures", J. Appl.Phys. 34, 1963, pp. 926.

[14] J. A. Cape, and G. W. Lehman, "Temperature and Finite Pulse-Time Effects in the Flash Method for Measuring Thermal Diffusivity", J. Appl. Phys 34, 1963, pp. 1909.

[15] T. Azumi, and Y. Takahashi, "Novel Finite Pulse-Width Correction in Flash Thermal Diffusivity Measurement", Rev. Sci. Instrum. 52 (9), 1981, pp. 1411.

[16] A. Degiovanni, "Correction de longueur d'impulsion pour la mesure de la diffusivité thermique par la methode flash", Int. J. Heat Mass Transfer, 31 (3), 1988, pp. 2199.

Investigations on Heat Transfer with Different PCB Substrates during Vapour Phase Soldering

László Fazekas, Dániel Nagy, Attila Géczy
Dept of Electronics Technology
Budapest University of Technology and Economics
Budapest, Hungary
ifj.fazekaslaszlo@gmail.com

David Busek
Department of Electrotechnology
Czech Technical University in Prague, Faculty of Electrical
Engineering
Prague, Czech Republic

Abstract—The purpose of this article is to present our experiments with Vapour Phase Soldering, where the temperature distribution and heat transfer mechanisms were investigated on four test-boards with different substrates. We developed a method to measure the temperature gradients on-board inside an experimental vapour phase soldering oven. VP soldering process is suitable for consistent and rapid heating on boards from small scale to large ones with power components showing considerable thermal capacities. The measurements were performed on FR4, ceramic, alumina composite and flexible substrates, where the results show significant differences between the different materials, and point to investigations with thermal diffusivity for future research.

Keywords: vapour phase soldering, heat capacity, temperature gradient

I. Introduction (*Heading 1*)

Nowadays, the Vapour Phase Soldering is an up and coming reflow soldering method, based on the surface mount technology. Reflow soldering method is very common procedure in the industry of electronics, widely using forced convection and infrared heating. The VPS is usually mentioned as one of the simplest soldering process from the aspect of use, but it is complex from the aspect of heating and heat transfer investigations.

During the process, a special liquid is heated up to its boiling point. (As shown in Figure 1.) At this moment, a vapour layer appears over the surface of liquid. Continuing that boiling process, the vapour layer becomes higher and homogeneous. When a PCB assembly having room temperature is put into the vapour area, the film condensation immediately starts, liquid drops appear on the surface of merged substrate and during a fraction of a second, thin fluid film layer is created on it. The process ends, when the temperature of the substrate reaches temperature of the vapour layer. That time, the condensed film changes its phase to vapour again, because of the maxed temperature of the PCB substrate (which is practically the boiling point temperature). In this way, in the last stage, the PCB can leave the VPS oven in a dry manner [1].

This process is useable for creating soldered joins. During the industrial application, the merged PCB has positioned components and solder paste underneath their leads. When merging is finished, the temperature of solder paste grows, until the temperature arrives to the melting point. This time, the solder paste melts; meanwhile the liquid continuously condenses on the surface. When the PCB is heated up to the maximum temperature, the whole substrate has the same temperature, the boiling point of the Galden fluid.

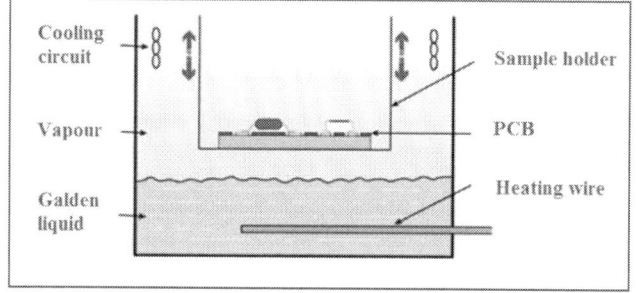

Figure 1. Basic setup of a VPS oven.

The mentioned special fluid is the Galden-type perfluoro-polyether, which is an inert fluid, used as heat transfer medium and for various high-tech applications in electrical and semiconductors technologies. It has excellent wetting properties with low viscosity; the boiling point is determined by the size distribution of molecular chains inside the liquid [2].

Heat transfer during VPS is based on the condensation theory. This phenomenon must be approached from the available models. Firstly, Nusselt investigated the issue. The calculation of heat transfer coefficients is more complicated, then in case of conduction, due to several variables. The value of the heat transfer coefficient depends on the flow of film, the formation of the flow condition, the physical properties of medium, the direction of the heat flow and the heat transfer properties of the surface. Determination of heat transfer coefficient due to the complexity and dependence on many factors can be based on experimental and modeling approaches [3] [4].

In the past few years, Leider was the first, who has summarized the knowledge of VPS. He mentioned basic context of heat transfer process, which laid on the Nusselt theory of film condensation [5]. This approach was similar to the works of Popov, Nimmo and Leppert, and Bejan who determined the first practical approach for the heat transfer coefficient in horizontal position on the top surface of sample. They also observed, that film thickness is different in the center and the corners of the horizontal sample [6]. Bejan has extended this description to the edge of the horizontal samples, where the thickness of condensate is almost zero, and the flow rate is the highest there [7]. The effect of condensate and the value of heat transfer coefficient on the bottom side of horizontal plate was investigated by Gerstmann and Griffith. They experienced, that the condensate does not form a continuous slim layer; it forms a wavy condensate [8].

The recent researches on the topic dealt with the thermal aspect of the process. Leicht figured out, that the case of non-saturated vapour proceeding can decrease the heat transfer coefficient of vapour. He revealed, the VPS ensure the best result of soldering only from 5 °C to 10 °C over the melting point of the paste, while other reflow processes need more excess heat. There are many advantage aspects of this result, as limiting stress for the components or avoiding delamination on PCB substrate [9].

That was several relevant research about thermal properties of Vapour Phase Soldering, but mostly simulations were used to find novelties in the process of condensation on PCBs. Illés and Géczy worked on a multi physics model of VPS equipment [10], and investigated the film condensation by numeric simulations [11], they also modeled and measured the heat transfer coefficient, and they examined the dynamic changes in vapour analysis [12]. Simulation studies were also performed at our department to determine the temperature and the thickness of the condensate. (Figure 2.)

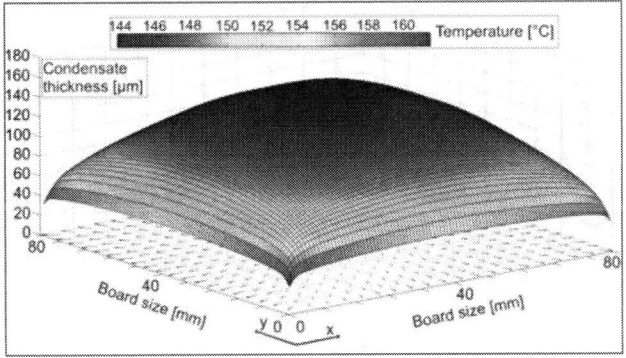

Figure 2. Condensate layer on the top side of horizontal plate (as seen in [11])

Olivér Krammer and Tamás Garami investigated the strength of solder joint, the reliability and intermetallic layer of Vapour Phase Soldering [13] (Figure 1. was also depicted from their visualization in the given paper). Hurtony et al. [15], Pietriková et al. [16], Wei et al. [17] and Rudajevova and Dusek [18] inspected the microstructure characteristics of the VPS joints, which can be highly affected by the vapour settings. Not only the structure can be influenced by different heating methods, but the intermetallic layer is also affected. Latest researches point to the practical application of the technology. Soldering power electronics is a highly focused use of VPS, as shown by Syed et al. [19] Chou and Chang [20], Hromadka et al. [21], Synkiewicz el al. [22]. Esfandyari et al. [23] also investigated energy efficiency of the process. Lately it

was shown that vapour phase soldering is a possible option for biodegradable PCB assembly [24].

II. EXPERIMENTAL

The measurements were performed at Budapest University of Economics and Technology, Department of Electronics Technology, using a Vapour Phase Soldering model equipment. It was filled with HT170 Galden (170 °C theoretical boiling point, but this temperature may be different from a few degrees in practice). Boiling point depends on the usage of the material (it can show some drift effect in boiling point), and the recorded maximum is affected by environmental effects disturbing the data acquisition units (due to cold point compensation).

Five different substrate materials were prepared, which are used in PCB technologies: FR4, aluminum-oxide Rubalit ceramic, polyimide film and FR4 with alumina layer core. The first measurement batch was prepared for boards with identical 50x50 mm surface. The second measurement batch was configured according to identical equalized thermal capacities. (This is shown later in Table 1. – "EQ" boards) Five thermocouples were attached to the substrates from the bottom side, as the Figure 3 shows.

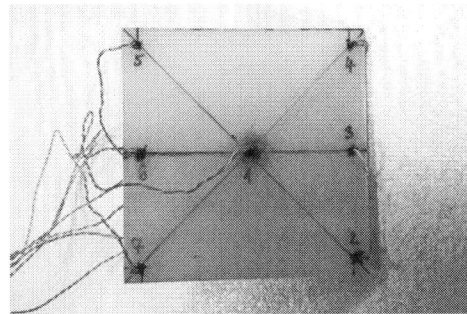

Figure 3. Attachment points of the thermocouples

0,5 mm deep blind vias were created at the measurement points, and the hot points of the T-type thermocouples were fixed into these holes with type of 3621 Loctite SMD adhesive. The cold points were connected into Novus myPCLab data collector units, which have a built-in temperature sensor, for performing the automatic cold point compensations.

Before starting the measurements, a sample holder was calibrated to zero angle horizontally. This prepared holder is shown in Figure 4 from the front.

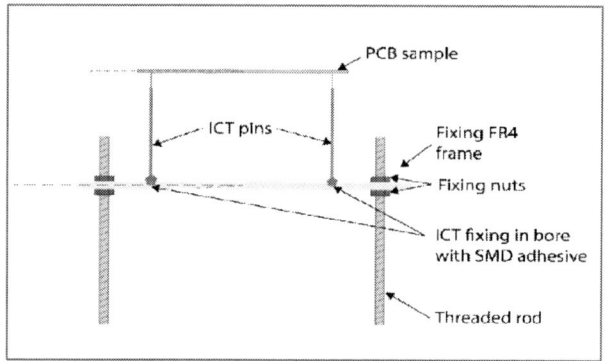

Figure 4. Prepared sample holder

The threaded rods functioned as a stand of the holder; the levelling can be performed in a simple manner because of the adjustable fixing nuts. Four pieces of in circuit tester pins were positioned to the fixing frame, which were glued to the holder with the SMD adhesive used before. It is important to note, that all of the measurements were performed in calibrated horizontal position (a clinometer device was used for this purpose), while significant inclinations to any directions are suggested to influence the heat transfer process. Table 1 presents the samples used for the investigations.

TABLE I.   SAMPLES USED FOR THE EXPERIMENTS

Calculating heat capacities of samples				
	Same size (surface)		Same heat cap. (EQ)	
Sample material	Size [mm x mm x mm]	Heat cap. [J/K]	Size [mm x mm x mm]	Heat cap. [J/K]
PCB with Al layer	50x50x1.65	7.3	29.1x29.1x1.65	~2
FR4	50x50x1.45	4.52	33.8x33.8x1.45	~2
Rubalit ceramic	50x50x0.6	4.68	33.2x33.2x0.6	~2
Polyimide PCB foil	50x50x0.1	0.36	120x120x0.1	~2

III. RESULTS

During the measurements, the prepared substrates were placed onto the sample holder in the container of Galden liquid. That time, the fluid has already reached its boiling point of 170 °C. The first measurements were performed with the plate of FR4. The goal was to investigate the film condensate of heat transfer in the created 5 measuring points. 250 ms sampling time was set in the data collector. The measurement cycles were held up as long as the temperature of PCB has reached the theoretical maximal saturation temperature. That period was approximately 60 seconds. Between each measurement cycle, the boards were cooled back to the ambient temperature by Festo air blower equipment. Totally, seven measurements were performed on the test boards, in four different orientations, as Figure 5 and Figure 6 show. The measurements were separated from the same data run, and were investigated in the same time window.

The rotation can eliminate the different variation of thermocouple fixing and accuracy. During three immersions, the temperature values were averaged for each thermocouple in a particular orientation. The repeated measurements of temperature ramps in these different measuring points is shown in Figure 6, where the highlight of different plots is omitted due to the dense visualization.

After the measuring of the individual center and corner points, five successive ramps were averaged. As it is shown in Figure 7, the measured temperature value of VPS_05 channel 1 - which is the center point of PCB – keeps under the average of corner points, as it is expected from previous literature results [11].

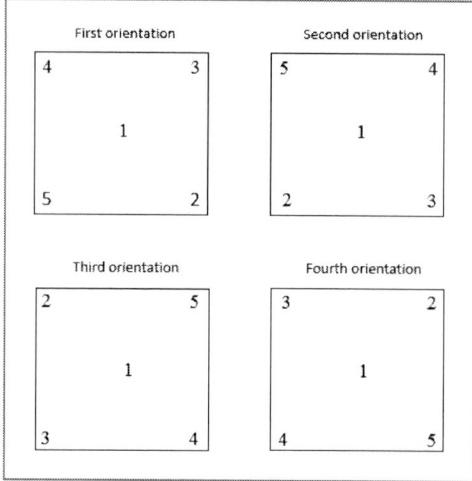

Figure 5. Definition of measuring locations

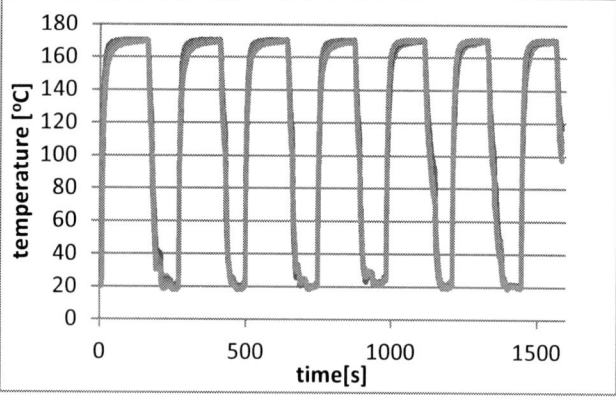

Figure 6. Repeated measurements of FR4 test board

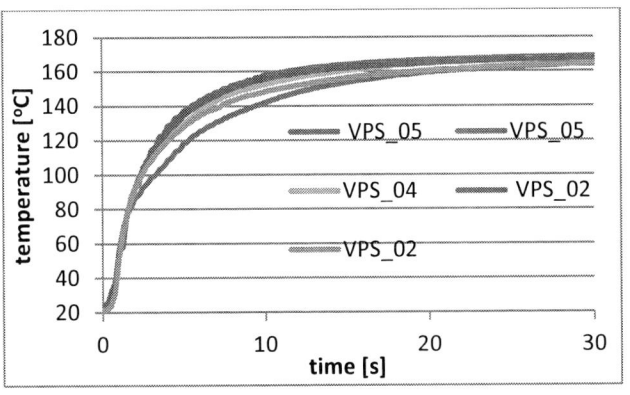

Figure 7. Temperature profile for FR4 substrate in five points

The reason of this phenomenon is the flow of condensate film during the process; the thickness of film is the thinnest at the edges, the thickest in the middle region. Due to this effect, centre positions are heating almost consistently slower, – only minor random effects present throughout the rest results in the paper. The thicker layer in the middle (practically the centre point of the flow) slows down further condensation.

Afterwards the measurements continued with the polyimide board. Figure 8 shows the results with polyimide.

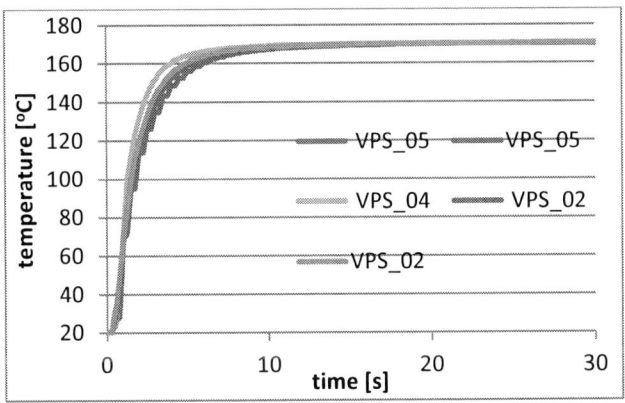

Figure 8. Temperature profile for polyimide foil in five points

According to the results, the curves run much closer to each other, so the heating process was less dependent on the position of thermocouples. The plots are practically running together (almost within the error margin of the thermocouples themselves). This also meets expectations from another aspect, where heating plots are much faster and smoother in thin polyimide film. The heat capacity of this substrate is smaller an order of magnitude, than the case of FR4 test board. Figure 8 also shows, about 15 second period was already enough time to achieve the saturation temperature, while in case of FR4 substrate, the same value was set after 30 seconds on the whole surface.

Following the investigations, Rubalit ceramic was immersed to the vapour area as in previous cases.

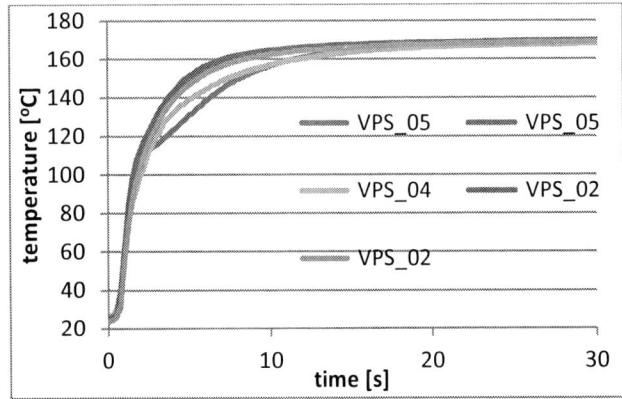

Figure 9. Temperature profile for ceramic in five points

The interpretation of the obtained curve is less evident, than the previous cases. The temperature of the point of VPS_05 channel 1 (consequent center location during our experiments) was the slowest with the point of VPS_04 channel 2, which was located in a corner. The former result meets the expectations because of the mentioned heat transfer of film condensate, latter curve may also be slow due to thermocouple inaccuracy, or simply due to the fact that ceramic substrate heats up more homogeneously than for example in the case of FR4 boards.

In the following measurement, the heat transfer process was analyzed on the PCB with aluminum layer.

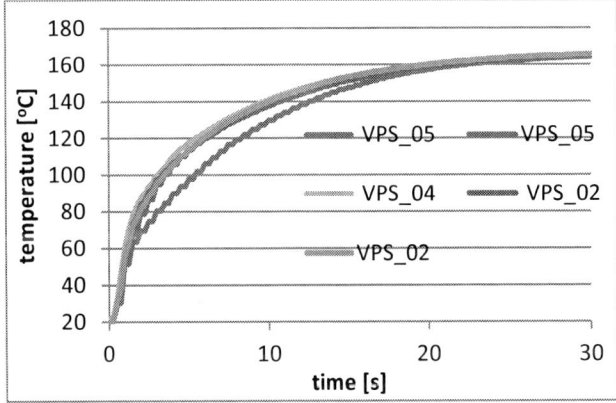

Figure 10. Temperature profile for FR4+ALU PCB in five points

Based on these results, the heat transfer process was the slowest, compared with the previous cases of PCBs, as the Figure 10 shows, after 30 seconds; the curves have not reached boiling point temperature of the Galden liquid.

At the end of the measurements, the temperature distributions of the different substrates were plotted in a summary diagram. These curves came from the averaging of temperature values for the five measurement locations.

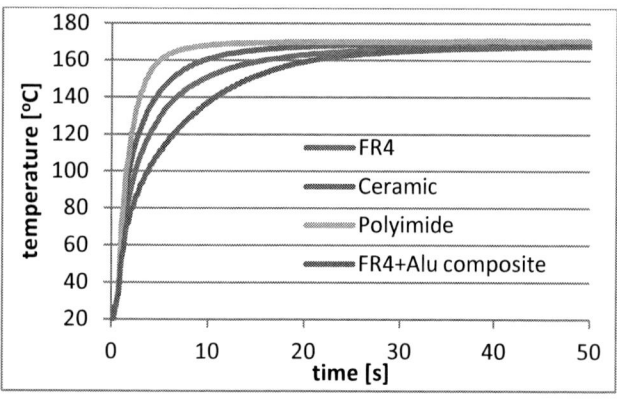

Figure 11. Temperature profile averages for all boards with identical surfaces

The results were consistent with the previous expectations, where the heat transfer of polyimide film was the fastest, then came ceramics, FR4 board followed that, and FR4+Alu was the slowest, as it can be seen in Figure 11.

The results are interesting from one further aspect. The heating curve of the ceramic board should be similar to the FR4 board, while both surfaces and the thermal capacities are similar to each other. At this point the suggestion is, that the thickness difference, and consequently the heat conduction parameter differences may cause this effect.

Further investigations were performed with the boards, where similar thermal capacities were set with different surface dimensions of the boards. Figure 12 presents temperature profile for EQ_FR4, Fig. 13 shows EQ_Polyimide, Fig. 14 shows EQ_ceramics, Fig 15 shows EQ_FR4+Alu plots, like in previous cases. It is interesting to note that overall the plots run much closer to each other. This is due to the smaller overall thermal capacitances, and smaller sizes than in the case of previous measurements with boards of identical surfaces.

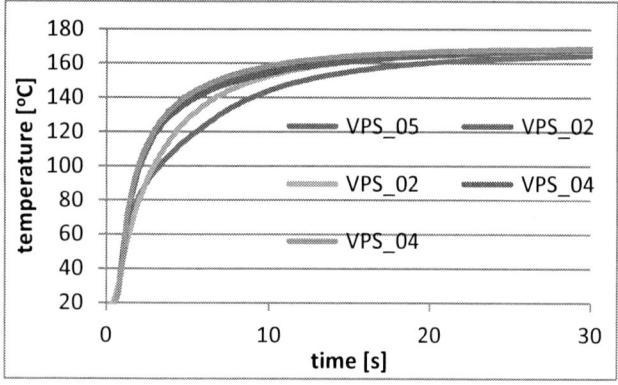

Figure 12. Temperature profile for EQ_FR4 in five points

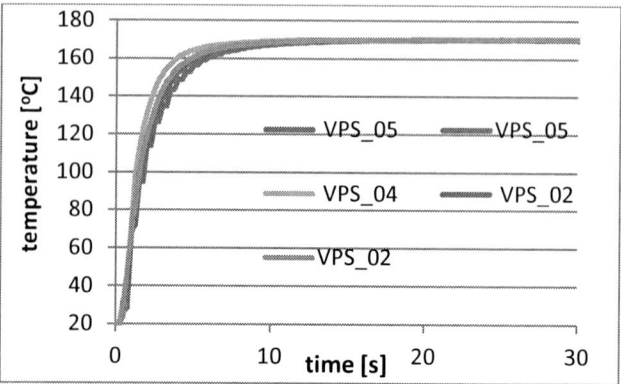

Figure 13. Temp. profile for EQ_Polyimide in five points

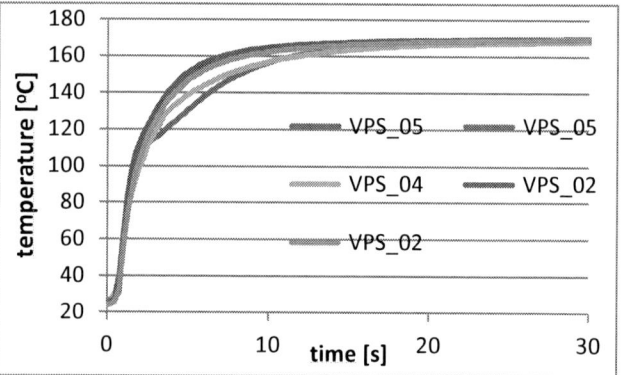

Figure 14. Temp. profile for EQ_ceramic in five points

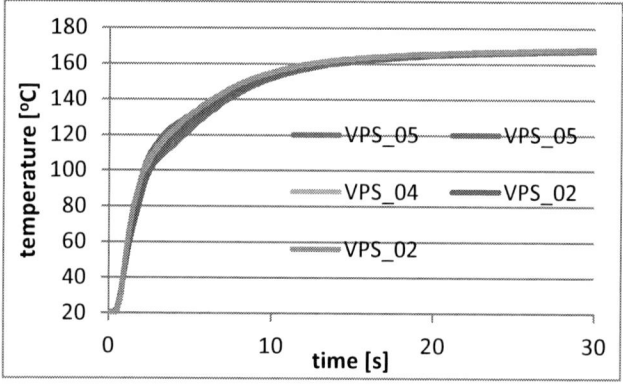

Figure 15. Temp. profile for EQ_FR4+Alu in five points

Figure 16 shows, that even with similar thermal capacities, the increasing temperature plots are not running together. The fastest heating of polyimide foil can be accounted to the fact that the surface/thickness ratio of the polyimide is the largest in this case (according to Table 1). Ceramics is also deviating from the FR4 based boards, however the area and the thickness is not different with orders of magnitudes, as in the case of polyimide. The investigation should be combined with the heat conductivity and thermal diffusivity parameter investigations for future investigations.

2016 IEEE 22nd International Symposium for Design and Technology in Electronic Packaging (SIITME)

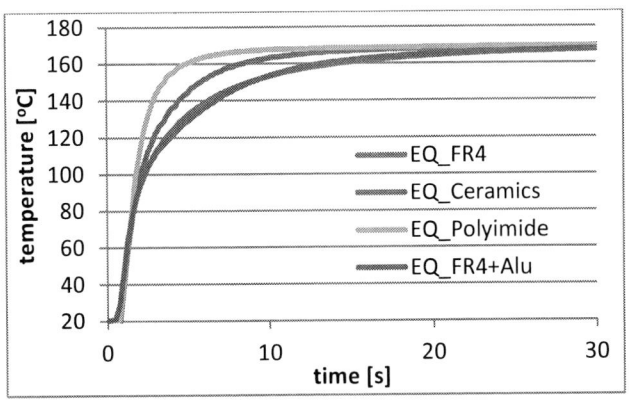

Figure 16. Overall temperature profiles for EQ boards with the same thermal capacities.

Thermal diffusivity is a composite parameter of a given material, and it can be described according to presented equation (1):

$$\alpha = \frac{k}{\rho \cdot c_p} \quad (1),$$

where α is the thermal diffusivity [m^2/s], k is the thermal conductivity [W/m·K], ρ is the density [kg/m^3] and c_p [K/kg·K] is the specific heat capacity of the given materials. Future investigations should focus on direct analysis of the thermal diffusivity values for given materials and results, to highlight the effect of this parameter on the assemblies during Vapour Phase soldering.

IV. CONCLUSION

The heat transfer process on different test boards were investigated in the horizontal setting of PCB in saturated vapour during VPS. The work focused on both identical size of surface for the different materials, and identical heat capacitances. In the former case the goal was, to justify the theoretical statement with measurements, where the heat transfer characteristics of the film condensate depend on the heat capacitances, when identical surfaces are present. Also in the latter case, the importance of board dimensions and thermal diffusivity was highlighted, while similar heat capacities may also exhibit differences in the overall heating. The basic research nature of the presented work may point to solder profile optimizations and improvements in process settings for specific PCB materials

In the future, detailed inspection is needed for the surface temperature distributions, where the differences may show interesting results according to the given materials. Also exact thermal diffusivity analysis is needed to unfold relations between the material parameters and their effects on the heating curves.

ACKNOWLEDGEMENT

The research was supported by Bolyai János Kutatási Ösztöndíj (Bolyai János Research Scholarship).

REFERENCES

[1] Claus Zabel: Condensation Reflow Soldering - The Soldering Process with Solutions for future Technological Demands, ASSCON Systemtechnik-Elektronik GmbH, Germany

[2] Galden HT170 datasheet http://www.solvayplastics.com/sites/solvayplastics/EN/sp ecialty_polymers/Fluorinated_Fluids/Pages/Galden-PFPE.aspx

[3] P.J. Marto, W.M. Rohsenow, J.P. Hartnett, Young, I. Cho, Handbook of Heat Transfer, first ed., McGraw-Hill, New York, 1998, pp. 14.1–14.27.

[4] H. Leicht, A. Thumm, Today's Vapor Phase Soldering - An Optimized Reflow Technology for Lead Free Soldering, in Proceed. of Surf. Mount Tech. Assoc. Intern. Conf., Orlando, USA, 2008, paper No. 45.

[5] W. Leider, Dampfphasenlöten, Eugen G. Leuze Verlag, Bad Saulgau, Germany, 2002

[6] B. G. Nimmo and G. Leppert: Laminar Film Condensation on a Finite Horizontal Surface, Proc.4th Int. Heat Transfer Conf., Paris, 6, Cs 2.2, 1970.

[7] A. Bejan, Film condensation on an upward facing plate with free edges, Int. J. Heat Mass Transf. 34/2 (1991) 582-587, 1991

[8] Gerstmann and P. Griffith: Laminar Film Condensation on the Underside of Horizontal and Inclined Surfaces, Int. J. Heat Mass Transfer, 10, pp. 567-580, 1967.

[9] Illés B, Géczy A (2016.) Investigating the heat transfer on the top side of inclined printed circuit boards during vapour phase soldering, Appl Therm Eng, Vol. 103 pp. 1398-1407.

[10] Balázs Illés; Attila Géczy; Multi-physics modelling of a vapour phase soldering (VPS) system; Applied Thermal Engineering; 2012 48: pp. 54-62. (2012)

[11] Balázs Illés, Attila Géczy; Numerical simulation of condensate layer formation during vapour phase soldering; Applied Thermal Engineering 70: pp. 421-429. (2014)

[12] Balázs Illés, Attila Géczy; Investigating the dynamic changes of the vapour concentration in a Vapour Phase Soldering oven by simplified condensation modelling; Applied Thermal Engineering 59:(1-2) pp. 94-100. (2013)

[13] Olivér Krammer, Tamás Garami,; Investigating the Mechanical Strength of Vapor Phase Soldering Chip Components Joints; Proceedings of IEEE Int. Symposium for Design and Technology in Electronic Packaging at Pitesti; 2010

[14] Olivér Krammer, "Comparing the reliability and intermetallic layer of solder joints prepared with infrared and vapour phase soldering," *Solder. Surf. Mt. Technol.*, vol. 26, no. 4, pp. 214–222, Aug. 2014.

[15] Hurtony T, Bonyár A, Gordon P and Harsányi G (2012.) Investigation of intermetallic compounds (IMCs) in electrochemically stripped solder joints with SEM Microelectron. Reliab. Vol. 52 pp. 1138–1142.

[16] Pietrikova A, Mach P, Durisin J and Livovsky E (2008.) Microstructure analysis and measurement of nonlinearity of vapour phase reflowed solder joints, 31th IEEE-ISSE pp. 363–366

[17] Liu W, An R, Ding Y, Wang C-Q, Tian Y-H, Shen K, (2005.) Microstructure and properties of AgCu/2 wt% Ag-added Sn–Pb solder/CuBe joints fabricated by vapor phase soldering, Rare Metals, DOI 10.1007/s12598-015-0545-y

[18] A. Rudajevova and K. Dušek, "Influence of the thermal history and composition on the melting/solidification process in Sn-Ag-Cu solders," *Kov. Mater.-Met. Mater.*, vol. 50, no. 5, pp. 295–300, 2012.

[19] Syed Khaja A, Kaestle C, Reinhardt A, Franke J (2013.) Optimized Thin-Film Diffusion Soldering for Power-Electronics Production 36th IEEE ISSE pp. 11 – 16.

[20] Chou P-C, Cheng S (2015.) Performance characterization of gallium nitride HEMT cascode switch for power conditioning applications, Mater Sci Eng B Vol. 198 pp. 43–50.

[21] Hromadka K., Reboun J, Rendl K, Wirth V, Hamacek A (2015.) Comparison of the surface properties of power electronic substrates, 38th IEEE ISSE pp. 146 – 150

[22] Synkiewicz B K, Skwarek A and Witek K (2014.) Voids investigation in solder joints performed with vapour phase soldering (VPS), Solder. Surf. Mt. Tech., Vol. 26 No. 1 pp. 8–11.

[23] Esfandyari A, Syed-Khaja A, Horvath M, Franke J (2015.) Energy Efficiency Analysis of Vapor Phase Soldering Technology through Exergy-Based Metric Appl Mech Mat Vol. 805 pp. 196-204.

[24] Bálint Kovács, Attila Géczy, Gergely Horváth, István Hajdu, László Gál, Advances in producing functional circuits on biodegradable PCBs, Period. Polytech. Elec. Eng. Comp. Sci., Vol. 60, No. 4 (2016), pp. 223-231. DOI: 10.3311/PPee.969

Sensitivity of Resistance, Noise and Nonlinearity of Conductive Adhesive Joints to Changes in Adhesive

Seba Barto

Department of Electrotechnology
Faculty of Electrical Engineering
Czech Technical University, Prague, Czech Republic
tarbalar@fel.cvut.cz

Pavel Mach

Department of Electrotechnology
Faculty of Electrical Engineering
Czech Technical University, Prague, Czech Republic
mach@fel.cvut.cz

Abstract—Conductive adhesive joints were formed by adhesive assembly of resistors with "zero" resistance on Cu pads. Adhesive with bis-phenol epoxy matrix filled by silver flakes in concentration 75 % b.w. was used for experiments. The test samples were aged at the temperature of 125 °C and the relative humidity of 64 %, in an environment with 95 % relative humidity at the temperature of 24 °C and in an environment having the temperature 85 °C and the relative humidity 85 %. Changes of basic electrical parameters of the joints, the resistance, noise and nonlinearity of the current-voltage characteristic, were examined. It was found that changes of noise and nonlinearity are significantly higher in comparison with the changes of the joint resistance. It is assumed that the reason is that the junction resistance is a result of all the mechanisms of conductivity in adhesive and in the contacts between the adhesive, component and a pad, whereas nonlinearity and noise examine the nonlinear mechanisms only, especially tunneling.

Keywords—electrically conductive adhesive; ageing; resistivity; noise; nonlinearity of current-voltage characteristic

I. INTRODUCTION

Using of electrically conductive adhesives spread since 2006, when the EU Commission released RoHS directive. This directive recommended to members of EU to avoid using some materials, which are danger for human health. One of these materials was also lead.

Soldering in electrical engineering and electronics was based on tin-lead alloys, especially on the alloy Sn50Pb48Ag2. Therefore since 2006 started rapid evolution of new types of solders, lead-free solders. Simultaneously with this process, the expanded use of electrically conductive adhesives started.

Electrically conductive adhesives are comprised of two components, of an insulating matrix and of an electrically conductive component . Wide spectrum of resins is used as the insulation matrix. As the electrically conductive component particles of different metals are most commonly used. Dimensions of these particles are between 5 – 20 μ usually and the particles of many different shapes are used. Silver flakes or silver balls are used the most often. It is also possible to meet with adhesives filled with plastic balls coated with a conductive metal layer on the market [1] – [3].

Electrical as well as mechanical, climatic and other properties of conductive adhesives do not attain properties of

solders. Also the price of adhesives is substantially higher in comparison with the price of solders.

On the other hand there are situations where this cannot be avoided. These include the assembly of components which may be damaged by the higher temperature during assembly or assembly of components with very low gap between leads. Adhesives are also used for repair of electronic devices, it is also possible to find bio-compatible adhesives on the market, and adhesives are also used in assembly of power devices and in many other applications [4].

Quality of conductive adhesives is judged according to their electrical and mechanical properties and their resistivity against climatic ageing in different types of climatic conditions [5].

Electrical properties of adhesive joints are most often characterized by measuring their electrical resistance. The electrical resistance of the joints, however, significantly depends on the tunneling between the particles of the filler in the adhesive [6]. Tunneling causes noise of the joint and also causes nonlinearity VA characteristics of the joints. It therefore seems to be logical to focus on the measurement of these parameters as well.

Measuring the resistance is simple by using a proper measuring device. Measuring of noise requests carefully screened and grounded measuring workplace, because measured voltages are very low, and the same can be said about the measurement of nonlinearity of the current-voltage characteristic.

It follows from the study of the literature that the electrical properties of adhesive joints are mostly evaluated according to their resistance. Sometimes resistivity of adhesives is also measured, but measurement of other types of electrical characteristics is omitted usually of different reasons. According to our opinion this measurement is considered insufficient for studying of mechanisms that trigger changes of electrical properties of adhesive joints during ageing.

Therefore we decided to examine sensitivity of the resistance, noise and nonlinearity of the current vs. voltage characteristic to changes in adhesives caused by different types of climatic ageing and to compare the sensitivity of these parameters to the changes in the material.

II. THEORETICAL BACKGROUND

The total resistance of adhesive joints formed of an adhesive with isotropic electrical conductivity is composed of following components:

$$R_j = R_{c1} + R_f + R_t + R_{c2} \qquad (1)$$

Where R_j ... the total resistance of the adhesive joint, R_{c1} ... the resistance of the contact between adhesive and the lead of a component, R_f ... the resistance of flakes, R_t ... the tunnel resistance between flakes, R_{c2} ... the resistance between the adhesive and the pad.

The resistance R_{c1} depends on the material and surface finish of the lead of a component. Because when creating adhesive joints, unlike solder joints, does not use flux, the basic condition for manufacturing of a high quality joint is that the surface of the lead will not be covered by oxide or another chemical compound, which prevents conductive connection of the filler of adhesive and lead of the component.

The resistance R_f represents the resistance of flakes. These particles are made of metals with high electrical conductivity such as silver, nickel, gold, palladium or some highly conducting metal alloys. Therefore, it is possible to say that this resistance does not play a significant role in the overall resistance of the adhesive joint.

Next component R_t, which influences the total resistance of the adhesive joint, is the tunnel resistance. If two electrically conductive particles of filler form a conductive contact, depending on their shape the contact is composed of two components - of a constriction resistance and of a tunnel resistance or of one component only – of a tunnel resistance. Both the resistances must be considered if a contact between two spherical particles or between a spherical particle and a plane is studied. Such the contact is formed when adhesive with the anisotropic electrical conductivity is used, because these adhesives are mostly filled by conductive balls.

In our case, when the adhesive joint formed of adhesive with isotropic electrical conductivity, which is filled by silver flakes, the constriction resistance can be omitted, because flakes can touch in more than one contact and the area of a contact is bigger in comparison with the area of contact between two balls or between a ball and a plane [2]. Therefore the tunnel resistance R_t dominates here.

The last component in the total adhesive joint resistance is the contact resistance R_{C2}, which represents the resistance between the adhesive and the pad. As no flux is used when electronic components are mounted by conductive adhesives, the surface of the pad must be carefully cleaned before the assembly, if the surface finish is not gold [8].

Because it is assumed that the contacts between the adhesive and the pad or lead of the component are almost perfect, because the surface finish has been prepared for adhesive assembly, it is possible to conclude that the nonlinearity of the current-voltage characteristic and noise are caused by the conductivity mechanisms in adhesive, it means by tunneling of electrons through the barriers between particles of filler. The tunneling mechanism has, in general, nonlinear current-voltage characteristic.

Electrical noise if generated in material of different reasons. The basic type of the noise is the thermal noise. This noise voltage depends following parameters:

$$u_{nthermal} \approx \sqrt{(T, R, B)} \qquad (2)$$

Where T ... the temperature in K, R ... the resistance of a component under test, B ... bandwidth available measuring device.

For measuring of noise of composite materials other type of noise is examined. Current noise (it is also possible to encounter naming the 1/f noise), which reflects mechanisms of conductivity in these materials is measured. This noise voltage depends dominantly on following parameters

$$u_{ncurrent} \approx (I, \frac{1}{f}) \qquad (3)$$

Where I ... DC current flowing through the component under test, f ... the frequency at which the noise voltage is measured.

It follows from the equation (3) that the noise voltage hyperbolically decreases with the frequency. Therefore it seems to be advantage to measure this voltage at low frequencies. However, the electronic components, such as capacitors and inductors used in filters for low frequencies are too large. Therefore the noise is measured in the frequency range of 0.5-20 kHz usually.

It is also under discussion where is a source of nonlinearity of a current vs. voltage characteristic and noise. Some scientists have an opinion that there is one source, which causes nonlinearity as well as noise; other scientists assume two different sources.

III. EXPERIMENTAL

A. Manufacturing of test samples

One-component electrically conductive adhesive Elpox SC 24 D (Amepox, Poland) based on bis-phenol epoxy matrix and filled by silver flakes (75 % b.w.) was used for experiments. The adhesive was cured in an air circulated oven in relative humidity of 54% in accordance with recommendation of a supplier.

Adhesive bonds were formed by adhesive assembly of resistors with "zero" resistivity (0R0 resistors) of the type 1206. The adhesive "as received" was applied by dispensing.

The test PCBs were designed to make four-point measurement of the joint resistance possible.

B. Mechanical and climatic load applied on adhesive bonds

Two types of load were used for degradation of the adhesive joints. The mechanical and subsequently the climatic load were used for loading of the joints.

The mechanical load was carried out by bonding of the test board with mounted resistors through a cylinder with a diameter of 6 mm for 10 seconds.

The climatic ageing has been carried out at the temperature of 125 °C in the relative humidity of 64 %, in an environment with 95 % RH at the temperature of 24 °C and in an environment having the temperature 85 °C and the relative humidity 85 % for 1000 hours [7].

C. Electrical measurement

Measurement of the resistance of the adhesive bonds was carried out using a precision RLC meter HP 4284A (Hewlett Packard) in four-point arrangement.

Schematic diagram of equipment for the measurement of noise of the joints is shown in Fig. 1.

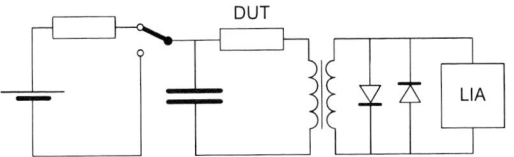

Fig. 1 Schematic diagram of equipment used for measurement of noise. DUT is device under test, LIA is lock-in amplifier

The measuring device LIA was a Lock-in amplifier PAR 124A (Princeton research). Diodes connected in parallel to the input of the LIA protect the input of the device against overload by electrical pulses.

Because the measured noise voltages are very low, the measuring equipment must be carefully screened and grounded. It is especially significant to avoid ground loops, which can contribute to the measured voltage by the noise induced in a ground loop. Therefore it is strongly recommended to connect the grounding conductors to one ground point. The best solution is using of devices with own battery powering.

Nonlinearity of the current vs. voltage characteristic was partially measured on the equipment CLT 1 (Radiometer Copenhagen) and partially using equipment based on evaluation of an intermodulation product of the third order generated by the nonlinear component when it was powered by two sinusoidal signals of different frequencies.

D. Processing of measured data

The measured data were processed as follows: at first a simple mathematic smoothing was used, the 4 maximum and 4 minimum values were deleted and at second a mathematic average of the remaining data was calculated.

IV. RESULTS AND DISCUSSION

At first changes of adhesive joints after the mechanical load were examined. The example of the changes of the joints resistance, noise and nonlinearity of the current vs. voltage characteristic is shown in Fig. 2.

It is interesting that percentage changes of electrical parameters of the joints under investigation were, after the mechanical load, not too different in contrast with changes caused by climatic stresses. This situation can be explained by the not significant changes of properties of tunnel barriers between particles of filler, which were caused by the mechanical load.

The percentage change of electrical parameters of the joints aged at the temperature of 125 °C, aged in combined ageing conditions 85 % RH / 85 °C and aged in the relative humidity of 95 % are compared in Fig. 3 – Fig. 5. It is show that whereas the resistance changes in the interval of percentage change 1 – 100, the percentage change of noise and nonlinearity is in the limits 20 – 70 000. This difference

between changes of the joint resistance and changes of these two parameters can be explained higher sensitivity of noise and nonlinearity to the properties of the tunnel barrier.

The conclusion that quality of the tunnel barrier contributes dominantly to the noise and nonlinearity can also be made with respect to the type of ageing.

Whereas the ageing at the high temperature caused changes of noise and nonlinearity in the range of 20 – 8000, ageing in high humidity and in combined climatic load high humidity-high temperature caused changes of the noise and nonlinearity in the range of 100 – 80 000, it means approximately 10 x higher. Changes of noise and nonlinearity is similar, changes of the joints resistance are lower.

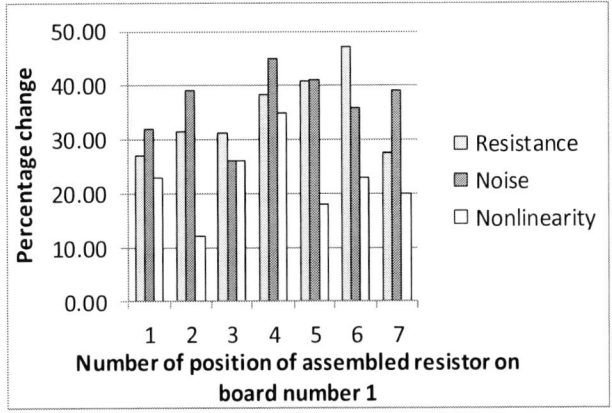

Fig. 2 Percentage change of the resistance, noise and nonlinearity of adhesive joints after application of the mechanical load

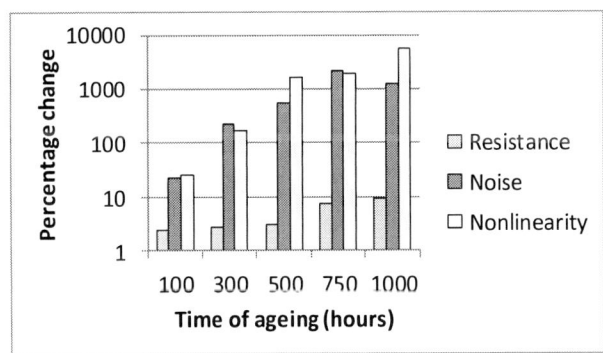

Fig. 3 Comparison of resistance, noise and nonlinearity change for the joints aged at the temperature of 125 °C

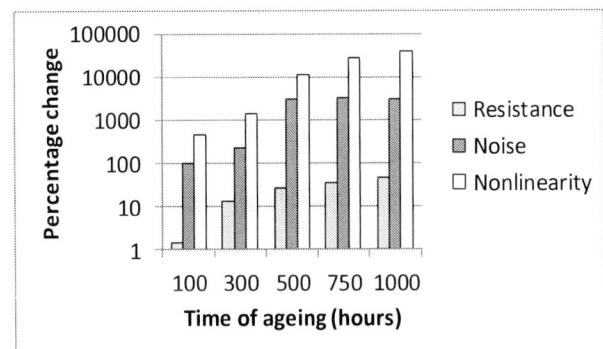

Fig. 4 Comparison of resistance, noise and nonlinearity change for the joints aged in combined ageing conditions 85% RH / 85 °C

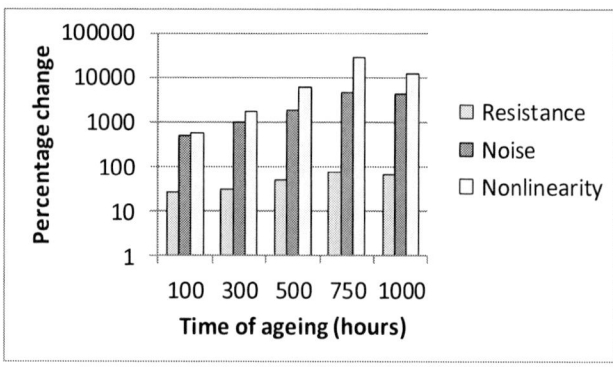

Fig. 5 Comparison of resistance, noise and nonlinearity change for the joints aged in relative humidity 95 %

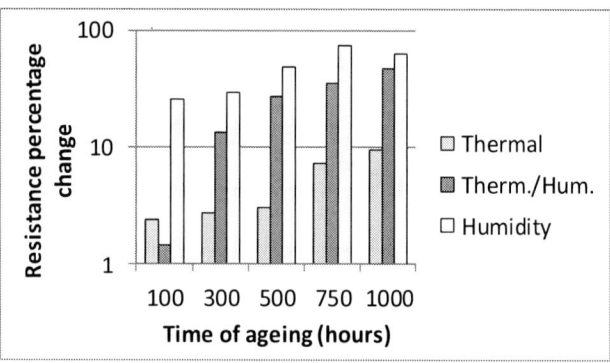

Fig. 6 Resistance change of the joints for different types of climatic load

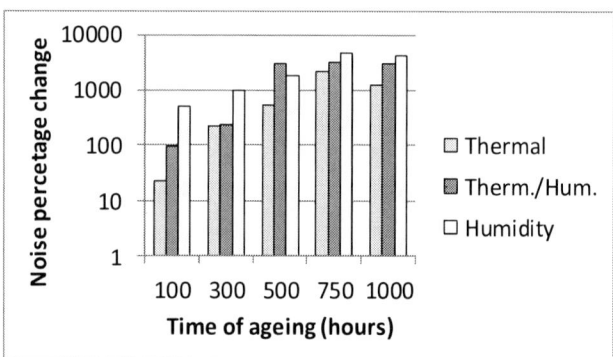

Fig. 7 Noise change of the joints for different types of climatic load

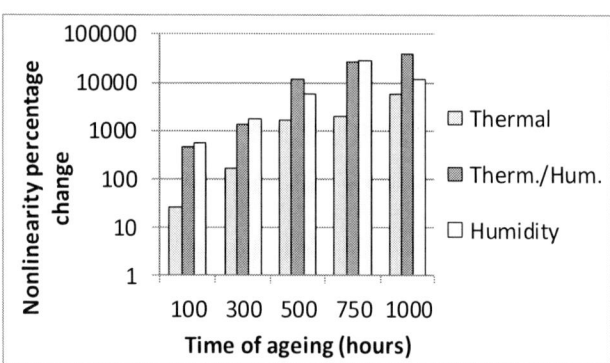

Fig. 8 Nonlinearity change of the joints for different types of climatic load

Two conclusions can be made over these two results. Significantly higher changes of noise and nonlinearity after combined climatic ageing and after ageing in high humidity in difference from ageing at the high temperature can be caused by penetration of water molecules into insulating matrix of adhesive and forming of AgOH molecule on the surface of silver flakes. Such properties of the insulating tunnel layer in the tunnel contacts between flakes changes and noise and nonlinearity that are more connected with tunnel mechanism than the resistivity change in high range.

To the same conclusions lead results presented in Fig.6 - Fig. 8. It can be concluded that for all types of ageing that were used for testing the sensitivity of the joint resistance is lower or substantially lower to the changes in material than sensitivity of the joint noise and nonlinearity.

V. CONCLUSIONS

Electrically conductive adhesive with bis-phenol matrix and silver filler with isotropic electrical conductivity was used for adhesive assembly of 0R0 resistors. The test specimens were loaded mechanically at first and then at the high temperature, in high relative humidity and in combined load high relative humidity/high temperature. The resistance of the joints, noise and nonlinearity were examined. It was found that the noise and nonlinearity are substantially more sensitive to the climatic ageing than the joint resistance, especially if the exposition was carried out in the climate with high humidity. The measured results were explained by lower sensitivity of the joint resistance to the changes in the insulating tunnel junctions between the particles.

ACKNOWLEDGMENT

The work was supported by the Pre-seed project of the Ministry of Education, Youth and Sports with the title "Material research for InovaSEED", no. CZ.1.05/3.1.00/14.0301.

REFERENCES

[1] Yi Li, Daniel Wu, C. P. Wong: "Electrical Conductive Adhesives with Nanotechnologies". Springer Science + Business Media. N.Y. 2010, pp. 166-176

[2] R. Gomatam, K. L. Mittal, "Electrically Conductive Adhesives," Koninklijke Brill NV, Leiden, Netherlands, 2008

[3] Y. Li, C. P. Wong, "Recent advances of conductive adhesives as a lead-free alternative in electronic packaging: Materials, processing, reliability and applications", Materials Science and Engineer. R 51, 2006, p. 1-35

[4] Y. Li, D. LU, C. P. Wong, "Electrical Conductive Adhesives with Nanotechnologies," Springer Science+ Business Media, 2010

[5] M. J. Yim, K W. Paik, "Review of Electrically Conductive Adhesive Technologies for Electronic Packaging," Electronic Material Letters, Vol. 2, No. 3., 2006, pp. 183-194

[6] Bin Su: "Electrical, Thermomechanical and Reliability Modeling of Electrically Conductive Adhesives." Dissertation thesis. Georgia Institute of Technology. 2006, pp. 1 – 36

[7] K. D. Kim, D. D. L. Chung, "Effect of Heating on Electrical Resistivity of Conductive Adhesive and Soldered Joint," Journal of Electronic Materials, Vol 31, No. 9, 2002, pp. 933-939

[8] W-J. Jeong, H. Nishkawa, D. Itou, T. Takemoto, "Electrical Characteristics of New Class of Conductive Adhesives," Material Transactions, Vol. 45, No. 10, 2005, pp. 2276-2281

2016 IEEE 22nd International Symposium for Design and Technology in Electronic Packaging (SIITME)

Implementation of a Microwave Elliptical low pass Filter with radial Stubs

I. M. Alexandru, L. Viman, D. Pitică

Technical University of Cluj-Napoca, Romania

alexandru.2.ilie@gmail.com

Abstract— Actual paper describes two implementation methods of an elliptical low pass filter using copper disc sectors on a PCB substrate. The main idea was to replace the series resonant branches of the filter circuit with radial stubs, which have the same equivalent circuit at high frequencies. Different approaches of the shapes and placement of the radial stubs were made during the study. Substrate material of the printed circuit board and the dielectric permittivity variation as a function of frequency were also characterized. Radial stubs are largely used for decades in microwave applications and they created a solid background for further studies or scientific research.

Keywords— Radial stubs, stepped impedance, elliptical low pass filter, transmission line.

I. INTRODUCTION

The first step is the definition of a mathematical model of the elliptic filter according to a series of design parameters: 50Ω input and output impedance, 1.6GHz cut-off frequency, tolerated pass band ripple 0.5dB and over 50dB attenuation at 2.4GHz. The elliptical filters have an extremely sharp frequency response, which makes them ideally suited for filter design cases where severe attenuations in frequencies are required [10]. The transfer function of the filter can be calculated analytical using specific equations and formulas [3], [4], or by using tables of coefficients for a certain type of filter in order to find the values for the passive filter components at a given frequency and impedance. The amplitude of ripple in each band is independently adjustable by designer. There is no other filter of equal order which can have a faster transition in gain between passband and stopband. The passive filter [7] is presented in Fig.1.

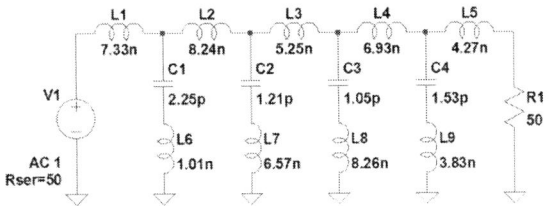

Fig. 1. 9th order elliptical low pass filter

II. FILTER DESIGN OVERVIEW

In Fig.2 can be observed the frequency response of the generated low pass filter. Simulation was done in LTSpice. There are five inductors and four series resonant branches C1-L6, C3-L7, C3-L8 and C4-L9.

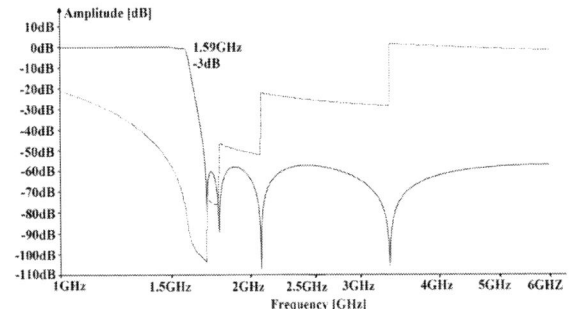

Fig. 2. Magnitude and phase plots

Resonant frequencies are: 1.716GHz, 1.785GHz, 2.07GHz and 3.34GHz. The first implementation method was based on finite impedance transmission lines, called stepped impedance. Each inductor and capacitor is equivalent with a transmission line having a certain length and characteristic impedance. Ideally, a capacitor has null impedance and an inductor has infinite impedance. Working with very small or very high values of impedance will lead to huge copper areas on the PCB for capacitors and very thin trace widths for inductors, so a tradeoff between PCB design capabilities and impedance values must exist. FR4 material has been used for this experiment. Its most important parameters are: dielectric permittivity of 4.6, 1.6mm thickness, 35um copper thickness on both layers and a loss tangent of 0.02. The bottom layer will be the ground plane and on the top layer will be printed the microstrip components. A value of 25Ω for capacitor impedance and 120Ω for inductors were chosen. Having these parameters, the length of the copper segments for microstrip capacitors and inductors can be calculated as in (1) and (2).

$$length_L[m] = \frac{L[H] * c[m/s]}{Z_L[\Omega]\sqrt{\varepsilon_{effH}}} \qquad (1)$$

$$length_C[m] = \frac{Z_C[\Omega] * c[m/s] * C[F]}{\sqrt{\varepsilon_{effL}}} \qquad (2)$$

The copper width for the microstrip capacitor and inductor is calculated as in (3) by extracting the W parameter from the equation for a certain dielectric permittivity and substrate thickness. The equation was solved for 25Ω, 120 Ω and 50Ω. The following widths were obtained: 8mm, 0.35mm and 2.95mm. 50Ω transmission lines will be used to connect the filter input and output to external the SMA board connectors.

$$\varepsilon_{eff} = \begin{cases} \dfrac{\varepsilon_r + 1}{2} + \dfrac{\varepsilon_r - 1}{2}\left[\dfrac{1}{\sqrt{1 + \dfrac{12h}{W}}} + 0.04\left(1 - \dfrac{W}{h}\right)^2\right], & \dfrac{W}{h} < 1 \\[4ex] \dfrac{\varepsilon_r + 1}{2} + \dfrac{\varepsilon_r - 1}{2}\left[\dfrac{1}{\sqrt{1 + \dfrac{12h}{W}}}\right], & \dfrac{W}{h} \geq 1 \end{cases} \quad (3)$$

There are many important parameters and characteristics of the substrate and core materials like dielectric constant, dielectric thickness, loss tangent, trace width and skin effect. They have a big impact upon power loss, working frequency range and characteristic impedance. Characteristic impedance of a transmission line with a certain core thickness, dielectric constant and trace width is calculated as in (4) [2], [11].

$$Z_0 = \begin{cases} \dfrac{60}{\sqrt{\varepsilon_{eff}}} * \ln\left(\dfrac{8h}{W} + \dfrac{W}{4h}\right), & \dfrac{W}{h} < 1 \\[3ex] \dfrac{120\pi}{\sqrt{\varepsilon_{eff}}} * \left[\dfrac{1}{\dfrac{W}{h} + 1.393 + 0.677\ln\left(\dfrac{W}{h} + 1.444\right)}\right], & \dfrac{W}{h} \geq 1 \end{cases} \quad (4)$$

$$Z_0 = 206 * \left(\frac{1}{3.2 + 0.677 * 1.18}\right) = 206 * \frac{1}{3.99} = 51.6\Omega_{w=2.9mm}$$

$$Z_C = 197 * \left(\frac{1}{6.393 + 0.677 * 1.86}\right) = 197 * \frac{1}{7.65} = 25.7\Omega_{w=8mm}$$

$$Z_L = 35 * ln\left(\frac{12.8}{0.35} + \frac{0.35}{6.4}\right) = 35 * 3.6 = 126\Omega_{w=0.35mm}$$

III. STEPPED IMPEDANCE IMPLEMENTATION

The filter has been implemented with copper areas on top layer (see Fig.3), according to the geometrical dimensions of the microstrip elements calculated before. As an experiment, the length of L7, L8 and L9 has been decreased by using meanders in order to reduce the total filter area. In Fig.4 can be observed both PCB variants.

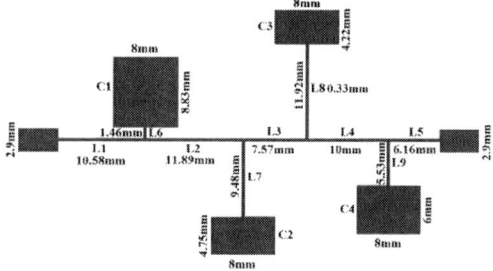

Fig. 3. Stepped impedance filter design

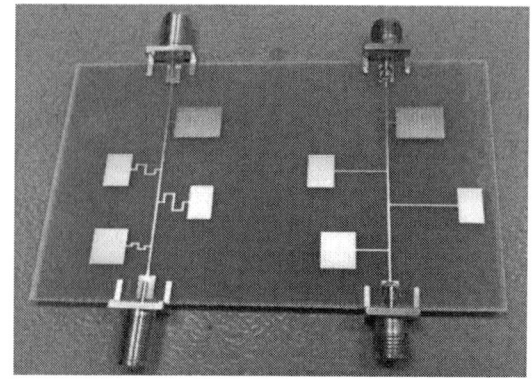

Fig. 4. Physical implementation of the filters, Top view

Frequency response of the implemented filters has been measured using a radiofrequency detector probe, a signal generator and a computer. The filters were characterized between 100MHz and 3GHz collecting 100 point per decade. It was noticed that the cutoff frequency is smaller than 1.6GHz. The first implementation had 1.212GHz cutoff frequency and the version with meanders reached 1.33GHz. In Fig.5 are represented both frequency responses of the filters and the simulated one. As a remark, the shapes are clearly indicating a low pass characteristic and the slopes are almost the same, the difference being the cutoff frequency deviation.

The main cause of this frequency difference can be the variation of the dielectric constant with the frequency. Another problem could be connected with the technique of PCB manufacturing. It is known that not only the rectangular shapes for the microstrip elements are used in microwave filters. There are resonators implemented with copper rings, spirals, discs, fractal networks etc. Another approach regarding elliptical low pass filter implementation replace the series resonant branches of the filter network with copper disc sectors. Disc sectors are modeled as a series circuit which contains an inductor and a capacitor. Disc sectors are widely used as rejection filters on power supply lines of the microwave circuits e.g. radiofrequency amplifiers. The main purpose is to reject a certain frequency, acting as a virtual short circuit to the ground plane. Fig.6 shows the reactance as function of frequency.

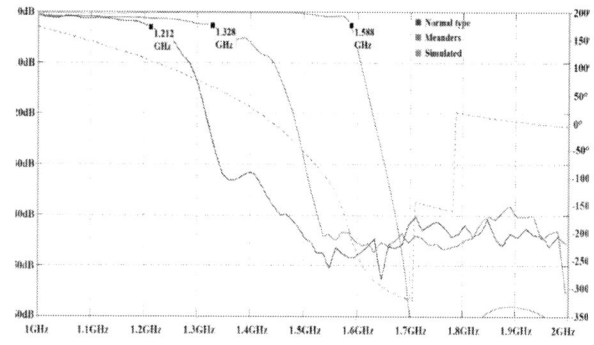

Fig. 5. Magnitude plots of the realised filters

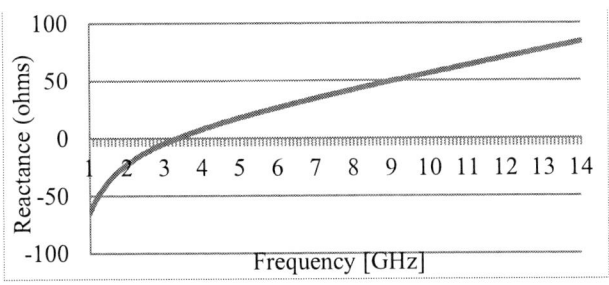

Fig. 6. Example of a figure caption. *(figure caption)*

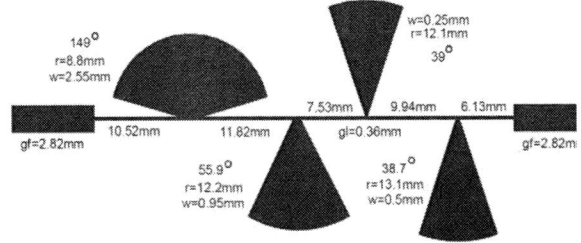

Fig. 7. Asymmetrical design with radial stubs

IV. IMPLEMENTATION WITH DISC SECTORS

Input impedance of a radial stub has been approximated by S.L. March [6] as in (5), (6) and (7).

$$Z_{in} = -j \frac{120\pi h}{r_i \theta \sqrt{\varepsilon_{eff}}} \frac{Y_0(\beta r_i)J_1(\beta r_0) - J_0(\beta r_i)Y_1(\beta r_0)}{J_1(\beta r_i)Y_1(\beta r_0) - Y_1(\beta r_i)J_1(\beta r_0)} \quad (5)$$

$$\beta = \frac{2\pi\sqrt{\varepsilon_{eff}}}{\lambda_0}; \ W_{ech} = (r_i + r_0)\sin\left(\frac{\theta}{2}\right) \quad (6)$$

$$L_{rs} = \frac{120\pi h}{\theta c}\left[\ln\left(\frac{r_0}{r_i}\right) - 0.5\right][H] \qquad C_{rs} = \frac{\theta r_0^2 \varepsilon_{eff}}{240\pi hc}[F] \quad (7)$$

$$Z_{LCser} = -j \frac{120h}{\theta}\left\{\frac{c}{fr_0^2 \varepsilon_{eff}} - \frac{2\pi^2 f}{c}\left[\ln\left(\frac{r_0}{r_i}\right) - 0.5\right]\right\}[\Omega] \quad (8)$$

$$\omega_0 = \frac{1}{\sqrt{LC}} = \frac{1}{\frac{r_0}{c}\sqrt{\frac{\varepsilon_{eff}}{2}\left[\ln\left(\frac{r_0}{r_i}\right) - 0.5\right]}}[rad/s] \quad (9)$$

A disc sector has three shape parameters: sector angle θ, outer radius, inner radius and base width. To implement a series LC branch with a radial stub is necessary to solve a system of nonlinear equation as in (10). E.g. one has to find the geometrical dimensions of a radial stub for a 3.83nH and 1.53pF series circuit. The given material constants are 4.6 for dielectric constant and 1.5mm dielectric thickness. The impedance can be written as in (8).

$$\begin{cases} \frac{180 * 120\pi h}{\theta\pi c}\left[\ln\left(\frac{r_0}{r_i}\right) - 0.5\right] = 3.83 * 10^{-9} \\ \frac{\theta\pi r_0^2 \varepsilon_{eff}}{180 * 240\pi hc} = 1.53 * 10^{-12} \end{cases} \quad (10)$$

Using numerical methods for solving the nonlinear equation system, the following solution $\theta = 55.9°, r_0 = 12.1mm$ and $W_{base} = 0.95mm$ was generated within ±0.2% tolerance. A nonlinear equation system solver for this specific purpose was made in LabVIEW. All four series resonant branches C1-L6, C3-L7, C3-L8 and C4-L9 can be replaced with a single equivalent radial stub. In Table I are shown the solutions and the resonant frequency of the elements, as in (9). The asymmetrical design is presented in Fig.7.

V. SYMMETRICAL RADIAL STUBS IMPLEMENTATION

Another way to implement the elliptical low pass filter is to separate the resonators into pairs of resonators having double inductance and half of capacity, as in (11). In this way, the filter will have a symmetrical distribution of the sectors and also the high surfaces of the radial stubs due to high capacity will be reduced. Equivalent filter schematic is shown in Fig.9.

$$Z = Z_a; \ j\left(\omega L - \frac{1}{\omega C}\right) = \frac{j}{2}\left(\omega L_a - \frac{1}{\omega C_a}\right); L_a = 2L; C_a = \frac{C}{2}; \quad (11)$$

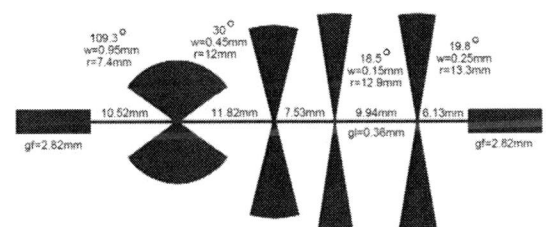

Fig. 8. Low pass filter design, symmetrical design

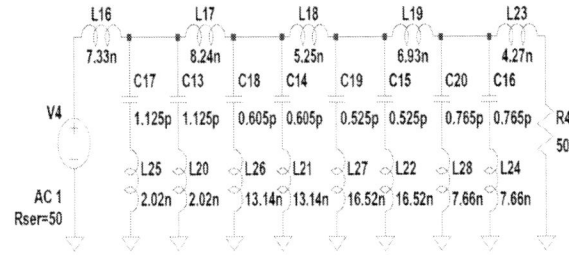

Fig. 9. Symmetrical filter design, equivalent schematic

TABLE I. GEOMETRICAL DIMENSIONS OF THE RADIAL STUBS

L	C	θ	r_0	W_{base}	f[GHz]
1.01nH	2.25pF	149°	8.8mm	2.55mm	3.336
3.83nH	1.53pF	55.9°	12.1mm	0.95mm	2.07
8.26nH	1.05pF	39°	12.2mm	0.25mm	1.716
6.57nH	1.21pF	38.7°	13.1mm	0.5mm	1.785
L*2	C/2	θ	r_0	W_{base}	f[GHz]
2.02nH	1.125pF	109°	7.4mm	0.95mm	3.336
7.66nH	0.765pF	30°	12mm	0.45mm	2.07
16.5nH	0.525pF	18.5°	12.9mm	0.15mm	1.716
13.1nH	0.605pF	19.8°	13.3mm	0.25mm	1.785

Physical implementation of the filters with radial stubs is shown in Fig.10. In this case, the frequency response is much closer to the simulated one than the first filter implementation with rectangular microstrip elements. The measured cutoff frequencies of the symmetrical and asymmetrical filter types are close to 1.87GHz as in Fig.11. Fig.5 indicates the comparison of the frequency response between filters. It can be seen that the frequency plots are following a low pass pattern and also have a sharp cutoff region.

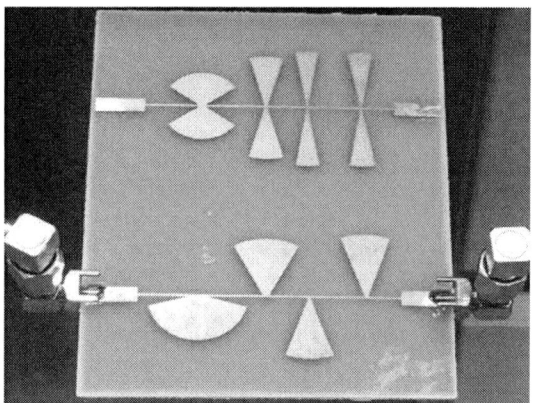

Fig. 10. Physical implementation of the filters with radial stubs

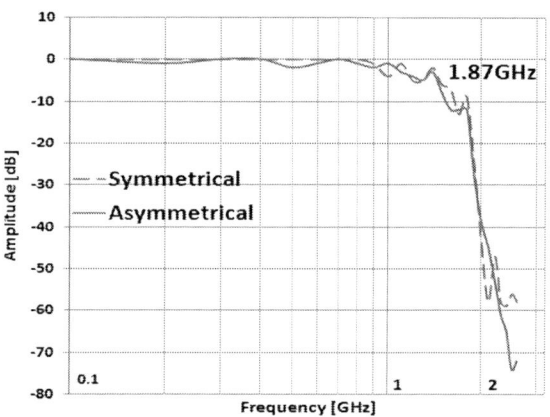

Fig. 11. Frequency response of the low pass filters realised with radial stubs

VI. CONCLUSIONS

The last filter implementation with disc sectors is much closer to the simulated filter model. Due to the fact that FR4 material is not suitable to be used above 1GHz, a couple of errors were noticed during the experiments. The biggest impact upon cutoff frequency was the variation of the dielectric constant with frequency. On the other hand, the loss tangent lies into undesired signal attenuations as the working frequency increases. Insertion loss has been calculated with (12) and (13) for FR4 substrate material. As the frequency increases, the dielectric constant decreases. The negative effect is that all the components will change their values and also the cutoff frequency will be different. Another important thing is the roughness of the copper laminate [8]. This fact is strictly connected with the skin effect. At frequencies above 1GHz, the current driven through a conductor is flowing on the surface of the material, depending on skin depth [9]. If the copper is rough, the effective conductor length increases relative to the distance over which the current would flow straight on the surface. A longer distance means undesired additional loss. This is why is important to manufacture the printed circuit board with a perfect smooth surface. For future design of the filters, ROGERS laminates can be used in order to obtain better performance. E.g: RO4350B ensures a much stable dielectric constant with frequency up to 10GHz, ten times lower tangent loss and a variety of substrate thicknesses.

$$\alpha = 2.3f[GHz] * \tan(\delta) * \sqrt{\varepsilon_{eff}} \left[\frac{dB}{inch} \right] \qquad (12)$$

$$\alpha = 2.3 * 1.6 * 0.02 * \sqrt{4.45} = \frac{0.155 dB}{inch} \qquad (13)$$

REFERENCES

[1] Microstrip, Stripline and CPW Design. http://www.qsl.net/va3iul

[2] GUPTA K.C. Ramesh Garg, Inder Bahl, Prakash Bhartia. Microstrip lines and Slotlines Second edition.

[3] Williams Fred, J.Taylor. Electronic Filter Design Handbook. 1981.

[4] KENDALL L. Su. Handbook of Tables for Elliptic Function Filter Springer US 1990.

[5] FRANK Sulak. Application of S-parameter techniques to amplifier design. 1969.

[6] Analyzing Lossy transmission lines. IEEE Transactions on Microwave theory and techniques, vol. MIT-33, NO. 3, March 1985.

[7] Dražen Jurišić and Neven Mijat. Tuning Elliptic Filters with a 'Tuning Biquad'.

[8] Barry olney. In circuit design Pty LTD. Effects of Surface Roughness on High-speed PCBs.

[9] JOHN Coonrod. Circuit Materials and High-Frequency Looses of PCBs. The PCB Magazine 2012.

[10] Experiments in Modern Electronics. Brewer and Leach, Jr., 2003. Kendall/Hunt Publishing Company. Dubuque, Iowa.

[11] A Via-Free Left-Handed Transmission Line with Radial Stubs. G.Naga Satish, K.V. Srivastava, A. Biswas, D. Kettle

Three omega Probe with Auto-zeroing

Ismet ATES
The Graduate School of Natural and Applied Sciences
Dokuz Eylul University
Izmir, Turkey

Alpaslan TURGUT
Mechanical Engineering Department
Dokuz Eylul University
Izmir, Turkey
alpaslanturgut@gmail.com

Levent CETIN
Mechatronics Engineering Department
Izmir Katip Celebi University
Izmir, Turkey

Mihai CHIRTOC
CATHERM, GRESPI
University of Reims Champagne-Ardenne
France

Abstract— The aim of this study is to build an active auto-zeroing circuit for the three omega thermal conductivity measurement method to overcome the disadvantage of the method. For this purpose an auto-zeroing algorithm is implemented which consists of two stages; simply reading the voltage drops on lower hand side resistances with high resolution and setting the resistance of the digital potentiometer to the value calculated through proposed algorithm using the difference of acquired voltages. As a result, a probe with auto-zeroing capability is developed successfully. Usage of the new probe will increase the reliability of three omega measurement results by reducing the effects of uncertainties arising from manual zeroing routines.

Keywords—Three omega method, auto-zeroing, thermal conductivity.

I. INTRODUCTION

Three omega (3ω) method has proven to be a valuable method for measuring the thermal properties in general and thermal conductivity in specific for different type of materials [1]. After the pioneering work of Cahill [2], the method is applied to wide range of materials such as thin films [3], liquids [4] and gases [5]. Basically, a sinusoidal current is applied on a linear heater (wire) at an angular frequency of ω, generating an oscillating heat source (temperature fluctuation) at 2ω which causes a temperature distribution on heater related to its thermal properties and surroundings. This oscillation in the corresponding temperature triggers the heater resistance at 2ω, leading to a voltage signal at 3ω. In other words, the term depending on 3ω is generated by nonlinear mixing of the applied AC current at ω with the electrical resistance change (due to temperature oscillation) at 2ω. By varying the configuration of the wire and its surrounding, 3ω method seen in figure 1 has been used to measure the thermal conductivity k and specific heat c of different types of materials.

II. METHOD

One of the cumbersome routine that is carried out in 3ω measurements is the zeroing of the first component. This process is necessary for avoiding negative effects due to error amplification [6]. To overcome this fallback, it is necessary for

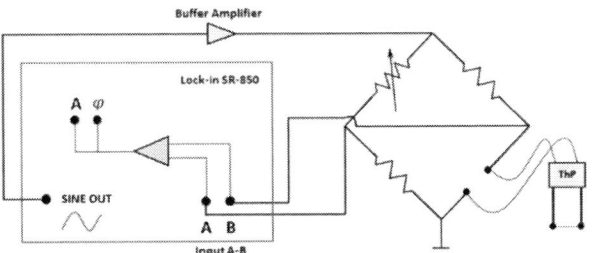

the sensing probe to be equipped with an active auto-zeroing circuit. The task of auto-zeroing can be simply described as

Fig. 1. Sensing probe for 3ω measurement

balancing the bridge circuit. In other words the ratio of resistances (or in general impedances) on each branch of the bridge should be the same. In most circuits, balancing can only be done manually via conventional potentiometers but the availability of digital programmable potentiometers may lead to autonomous design. In the light of this, the basic auto-zeroing circuit is designed to realize compensation algorithm which is implemented via digital potentiometer and a microcontroller. The realized compensation algorithm consists of two stages; simply reading the voltage drops on lower hand side resistances(at point A and B) with high resolution and setting the resistance of the digital potentiometer to the value calculated from Ohm's law by using the difference of acquired voltages. Considering the uncertain nature of the measurement process, the readings of voltages at A and B acquired during a time window and representing value is calculated via averaging. Besides, to reduce the effect of noise amplification in circuit, the A and B voltages are acquired instead of acquiring directly the bridge output.

In the prototype circuit, the AD5321 IC is utilized as 50K digital potentiometer, which can be used both with bipolar and unipolar supply voltages. It also allows usage of SPI communication protocol for setting the value for its resistance. The modified bridge circuit is given in figure 2 where it can be seen that the new bridge has new resistance values to share

2016 IEEE 22nd International Symposium for Design and Technology in Electronic Packaging (SIITME)

current avoiding a burnt out digital potentiometer. Using its communication interface digital potentiometer is controlled via

micro controller, on which the decision process is described above is coded. The designed circuit is enhanced with line low

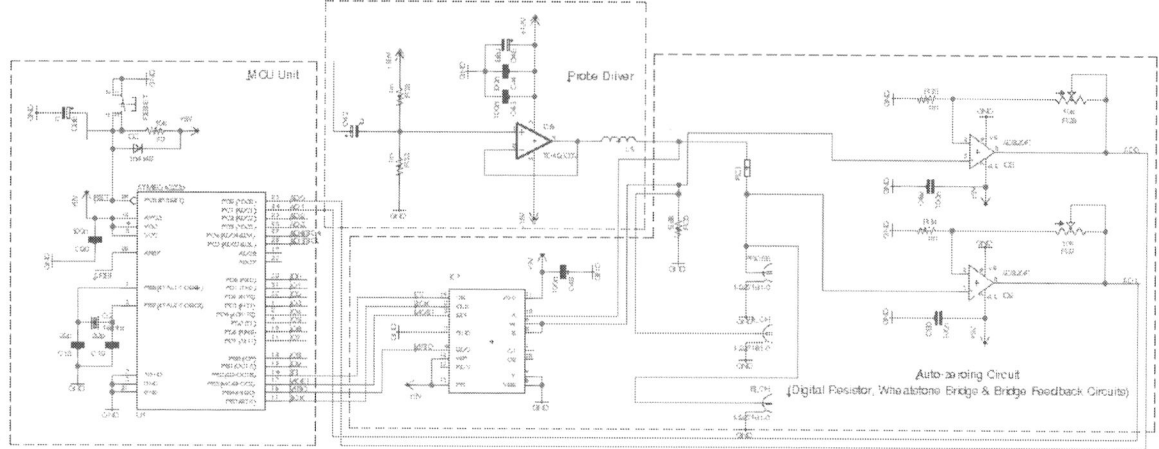

Fig. 2.

Schematic of Bridge Circuit

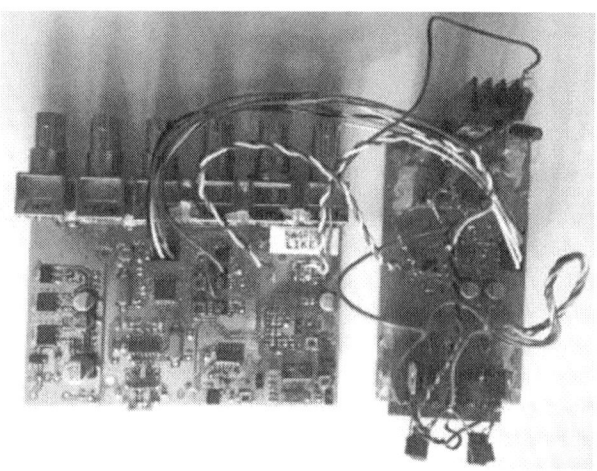

Fig. 3.   Final Prototype of Bridge Circuit

pass filters to suspend high frequency noises and a final prototype is manufactured (figure 3) .The program to control auto-zeroing is implemented on C++ and uploaded in microcontroller. The time window is taken as 2 seconds, which corresponds to 1000 samples.

III. RESULTS

The acquired signals are given figure 4 and corresponding difference voltage and resistance set value are given in table 1. The performance of algorithm in recapitulated in table 1. The first two columns show initial feedback values from bridge ports and the last two show final zeroing error and resultant digital potentiometer set.

TABLE I   Measurement Results

	Avarage of Signal A	Avarage of Signal B	Last Difference	Digital Pot Set Value

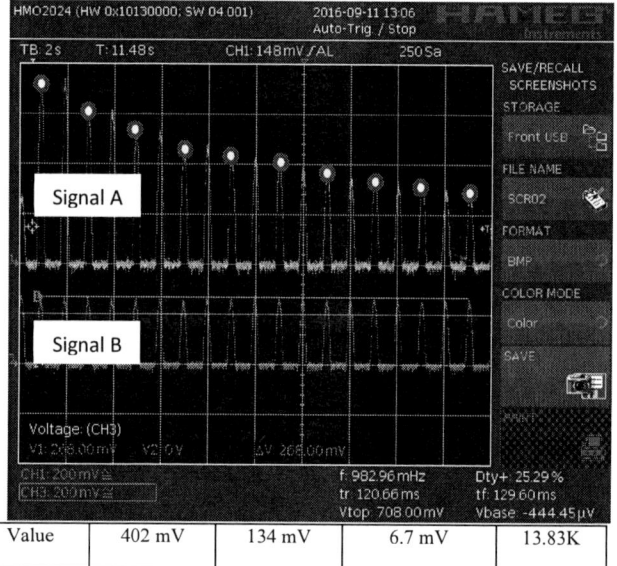

| Value | 402 mV | 134 mV | 6.7 mV | 13.83K |

Fig. 4.   A and B Signals in Bridge

IV. CONCLUSION

Paper presented a preliminary work on an automated 3ω device by considering the first step of measurement process: Auto-zeroing of the first harmonic. For this purpose an algorithm is proposed and necessary hardware modifications are presented together with experimental results which showed that the proposed method is successful in auto-zeroing applications carried out in bridge-based measurement setups.

ACKNOWLEDGMENT

This work has been supported by TUBITAK with project no 115M408.

978-1-5090-4446-7/16 $31.00 © 2016 IEEE   64   20-23 Oct 2016, Oradea, Romania

REFERENCES

[1] C. Dames and G. Chen, "1ω, 2ω, and 3ω methods for measurements of thermal properties", Review of Scientific Instruments 76, 124902. doi:10.1063/1.2130718, 2005.

[2] D. G. Cahill and D., R.O. Pohl, "Thermal conductivity of amorphous solids above the plateau" Physical review B, *35*(8), 4067, 1987.

[3] T. Tong and A. Majumdar, "Reexamining the 3-omega technique for thin film thermal characterization" Review of Scientific Instruments, 77(10), 104902, 2006.

[4] A. Turgut, C. Sauter, M. Chirtoc, J.F. Henry, S. Tavman, I. Tavman and J. Pelzl, "AC hot wire measurement of thermophysical properties of nanofluids with 3ω method", The European Physical Journal Special Topics, 153(1), 349-352, 2008.

[5] E. Yusibani, P.L Woodfield, M. Fujii, K. Shinzato, X. Zhang and y. Takata, "Application of the three-omega method to measurement of thermal conductivity and thermal diffusivity of hydrogen gas", International Journal of Thermophysics, 30(2), 397-415, 2009.

[6] J.F. Hoffmann, J.F. Henry, G. Vaitilingom, R. Olives, M. Chirtoc, D. Caron and X Py, "Temperature dependence of thermal conductivity of vegetable oils for use in concentrated solar power plants, measured by 3omega hot wire method", International Journal of Thermal Sciences, 107, 105-110, 2016.

Revision of the Sampling Theorem

Bujor Mircea
PhD student,
Polytechnic University of Timisoara
Timisoara, Romania

Abstract—**Almost every aspect of modern life benefits at some point from electronic devices. Most of them are digital, and the data they receive is digitally sampled from a sensor at some point. The Sampling Theorem stands at the core of every data acquisition system design. Recent research shows there are certain situations where information is lost, if a signal is sampled according to the Sampling Theorem. The Sampling Theorem is supposed to be true regardless of the acquisition time. Mathematically it isn't. The revision of the Sampling Theorem is about the neglected aspect of acquisition time. The acquisition time is a very limited resource in real life, and sampling can go terribly wrong if the problem is not properly addressed. In this work, the time dimension of the Sampling Theorem is covered, together with requirements for short acquisition time. A proof is provided for insufficiency of current Theorem. Another proof is provided for the revised Sampling Theorem. A discussion of what was done wrong until now and then an example from previous research is discussed. It is shown the new Sampling Theorem can help extract more information from analog to digital domain. Every digital electronic equipment used today can benefit from being fed more information.**

Keywords—sampling theorem; sampling rate; data acquisition; acquisition time

I. Introduction

There is almost no domain where digital equipment isn't used today. The information it processes is obtained by sampling analog sensors. Correct sampling is essential for data acquisition. The utility of the whole equipment depends on the quality of data fed at input.

In a recent research[1] paper, a new spectral analysis is proposed. It can independently distinguish phase and amplitude of non-orthogonal spectral components. The required sampling rate is much higher than what current Sampling Theorem recommends. In Fig. 1 is presented a sum of a 75 Hz cosine and 95Hz sine waveforms over 5 milliseconds. The signal is sampled in 64 points over that time. The resulting spectral analysis is shown in Fig. 2. The two peaks, each having a phase and an amplitude, are clearly separated by a valley. According to the current Sampling Theorem, all information within 0-100Hz bandwidth can be covered by a single sample over 5 milliseconds. This corresponds to 200 samples per second. It is obvious a single sample doesn't have a chance for the promised perfect reconstruction. It was known before that more samples produce a signal closer to the original shape. What was not known before, is the extra samples provide extra information in the lower bandwidth too. It wasn't known oversampling adds information in the low frequency spectrum too. Partly it's because the new spectral analysis wasn't available to highlight

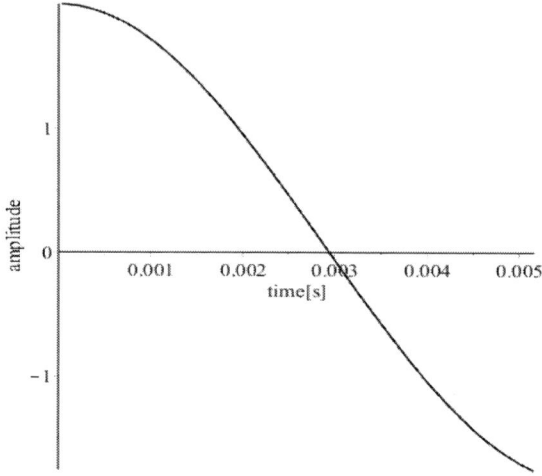

Fig. 1. A signal sum of 75Hz cosine and 95Hz sine sampled in 4 points over 5 milliseconds. Sampling a 0-100Hz bandwidth with Nyquist rate requires only one sample oer this time interval. Taken from[1].

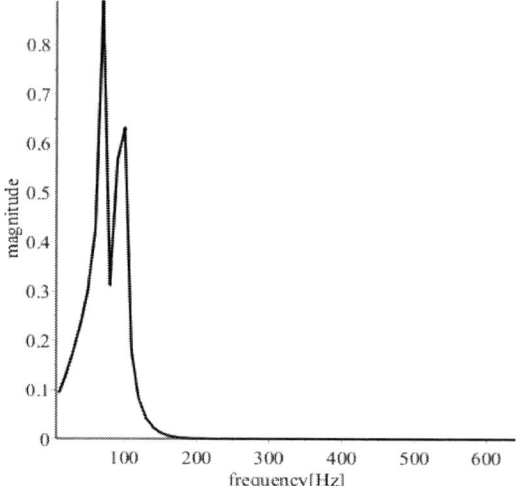

Fig. 2. The information extracted from signal in Fig. 1, shown in frequency domain. The oversampling clearly adds information within the 0-100Hz bandwidth. Taken from [1].

this point. The whole subject was looked at from a different perspective.

The mentioned spectral analysis is able to double its spectral resolution within the same bandwidth, with the doubling of the sample rate. In other words, its frequency resolution is proportional to the sampling rate. This example is a clear sign the Sampling Theorem is flawed and must be revised.

The second part is a discussion about invalidity of the current Sampling Theorem and a proof by *reductio ad absurdum*. The third part is the revised Sampling Theorem and a proof. The work ends with a discussion and a conclusion.

II. ON INSUFICCIENCY OF THE CURRENT SAMPLING THEOREM

All the data acquisition systems today are designed according to the current Sampling Theorem. The Sampling Theorem appeared in a time when digital sampling wasn't as widely available as today, and most important, the best tool for spectral analysis was the Fourier Transform. It is shown next the Fourier Transform represents the actual spectrum of a signal only under certain conditions. As a direct consequence, the current Sampling Theorem holds only under certain conditions. The theorem is wrong in it's current form, because it addresses both situations where it fits and where it doesn't. Unfortunately real life sampling is a niche where the current Sampling Theorem is not true.

A. Fourier Series and three lemmas

Let f(t) be a periodic function of time satisfying the criteria of existence of Fourier Series. That function can be expressed as in equation (1):

$$f(t) = a_0 + \sum_{n=1}^{N} b_n \cos nt + \sum_{n=1}^{N} c_n \sin nt \quad (1),$$

where indexed a, b, and c represent some coefficients.

Multiplication of equation (1) with cosine and integration gives:

$$\int_{-\pi}^{\pi} f(t) \cos(kt) dt =$$

$$= a \int_{-\pi}^{\pi} \cos(kt) dt + \sum_{n=1}^{N} b_n \int_{-\pi}^{\pi} \cos(nt) \cos(kt) dt +$$

$$+ \sum_{n=1}^{N} c_n \int_{-\pi}^{\pi} \sin(nx) \cos(kx) dx \quad (2),$$

and if orthogonality relations are applied, the equation (3) gives the formula for a b term:

$$\int_{-\pi}^{\pi} f(t) \cos(kt) dt = b_k \pi \quad (3).$$

From this introduction, three lemmas for the next steps follow:

Lemma 1: A continuous frequency spectrum corresponds to an aperiodic function, or a periodic one with infinite period.

This comes directly from equation (1), if n is made infinitesimally small. The same result is reached if the period of the signal goes to infinite.

Lemma 2: The formulas for Fourier Series coefficients are valid only if integration is performed over an integer number of periods of the function.

This is a direct consequence of equations (2), (3). Without orthogonality, equation(2) doesn't simplify into equation (3). As the Fourier Series are computed according to a wrong formula, the result is wrong.

Since real life signals have continuous spectrum, and thus infinite period, it is wrongly believed only an infinite acquisition time can give perfect, discrete spectrum. Actually the integration over one period of a sinusoid gives a discrete spectrum and is correct. If this is false, the Fourier Series is false, and we have *reductio ad absurdum*.

Lemma 3: Uniform sampling of a periodic function with a rate 2*B, where B is the bandwidth of the function in Hertz, produces only 2*B*T distinct points, where T is the period of the function.

This is illustrated in Fig. 3.

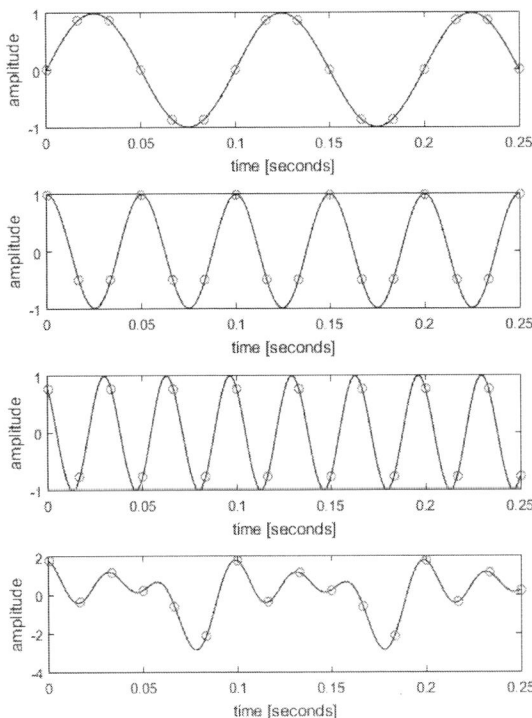

Fig. 3. Sampling points marked on a signal(bottom) and its three spectral components(the three sinusoids above).

The figure represents a periodic signal decomposed in its three components (10 Hz, 20 Hz and 30 Hz). The sampling points are marked on each of them, and are taken with a rate twice the maximum frequency. The frequencies in a Fourier Series are integer multiple of 1/T, where T is the period of the function. Let the signal have bandwidth from 0Hz to B Hz. The highest frequency is fmax = N/T=B. The Nyquist rate for sampling the bandwidth B is 2*B=2*N/T. What this rate means is 2*N samples over a period (i.e. 2N/T), uniformly spaced. In other words, the period is evenly split in 2*N points, T = 2*N. According to the properties of periodicity, the 2*N+1 sample is equal to the first sample:

$$f(2N+1) = f(T+1) = f(1) \quad (4)$$

The maximum number of distinct points sampled with twice the maximum frequency is 2*N. After that the samples start repeating.

B. Current Sampling Theorem

Original Sampling Theorem: If a function f(x) contains frequencies from 0 Hz to no higher than the B Hz, it is completely determined by giving its ordinates at a series of points spaced $T = 1/2B$ seconds apart.

For convenience, the frequency is measured in radians per second, and noted with symbol ω. The original accepted proof is:

$$f(t) = \frac{1}{2\pi} \int_{-\infty}^{+\infty} F(\omega)e^{i\omega t} d\omega = \frac{1}{2\pi} \int_{-2\pi B}^{2\pi B} F(\omega)e^{i\omega t} d\omega \quad (5)$$

if $t = n/2B$, where n is any positive or negative integer,

$$f\left(\frac{n}{2B}\right) = \frac{1}{2\pi} \int_{-2\pi B}^{2\pi B} F(\omega)e^{i\omega \frac{n}{2B}} d\omega \quad (6)$$

The time domain representation of a function was identified with a lowercase letter, and the frequency domain of the same function with the same in uppercase.

The given mathematical proof relies heavily on the Fourier Transform.

C. Proof of invalidity of Sampling Theorem in its current form

Proof by *reduction at absurdum*: let be a signal of continuous bandwidth from f1 to f2, f2-f1 = B. Sampling with Nyquist rate for an arbitrary time produces 2B*t =M samples, where t is the arbitrary time. Mathematically, a number of unknown values can be determined only if at least an equal number of values are known. Those M samples can describe the amplitude and phase of M/2 frequencies only. This contradicts the initial presumption, that there are an infinite number of sinusoids (i.e. continuous spectrum) between f1 and f2.

As real life signals are aperiodic, it follows from lemma 1 that their spectrum is continuous. Sampling a real life signal according to the current Sampling Theorem results in loss of information within the desired bandwidth. From the continuous spectrum of the bandwidth B, an infinite number of spectral components are lost and only M/2 will be preserved. The loss of information is not only in the spectrum above B.

III. REVISED SAMPLING THEOREM AND PROOF

Theorem: A signal sampled with Nyquist rate can be reconstructed only if at least one full period of the signal is sampled. If this is not possible, the quality of the signal reconstruction is proportional to the sampling rate.

Proof: A periodic signal sampled with Nyquist rate and its Fourier Series is considered. According to the lemma 3, only 2*N samples can be taken over a period of the signal. As the basic rule for a determined system of equations says the number of unknowns must be equal to the number of knowns, N frequencies of two unknowns each, phase and amplitude, can be completely described by 2*N samples. The condition is satisfied only if a full period is sampled. According to lemma 1, the period of a signal is infinite if the signal is aperiodic, which corresponds to continuous spectrum. If the signal is sampled with Nyquist rate for less than a period, it is impossible to determine all its spectral components. As the number of frequencies solved is proportional to the number of samples, the quality of the reconstructed signal is proportional to the sampling rate.

IV. DISCUSSION

A. About the proof of current Sampling Theorem, presented in II.B

Real life signals are aperiodic by default. It is known such signals need an infinite acquisition time according to lemma 2. An excuse not to do that was found: the signal outside the acquisition time is considered zero, because nothing passed through the acquisition system. The situation can be clearly seen in equation (5).

Next tool is the Fourier Transform. It is widely believed that Fourier Transform result is the same thing with the frequency spectrum of the signal fed into it. Actually this is a false belief according to lemma 2. The integration must be performed over exactly an integer number of periods of the signal. If that initial condition is not respected, we have a *reductio at absurdum*, as it was seen when lemma 2 was presented.

Going back to the real life signal in equation (5), sampled over a limited time, it is decomposed with a Fourier Transform. Even if nobody believes it, it is usually claimed a real life signal only exists when it is sampled. Let's not quibble over that, and go on with the The correct way of doing that is either zero padding the signal up to infinite, or computing every piece of the continuous spectrum, as it gives the same output. Feeding the sampled points only and computing a discrete spectrum is the usual mistake. Again we have a *reductio ad absurdum,* because Inverse Fourier Transform gives a signal that is not zero outside the acquisition time. Nevertheless, it can be seen in equation (6).

From the right or wrong frequency spectrum of the sampled signal, all components outside the targeted bandwidth are eliminated. It can be seen in equation (5), as the integration domain is replaced. Whether the bandwidth was computed continuous as it is, or discrete by mistake, is one problem. But the canceling of the high frequency components going up to infinite is again a *reductio ad absurdum*. The signal can't be limited in time without those components. It's enough to remember the sinc function. Supposing the signal is both time and frequency limited, as it is supposed in equation (5), is a *reductio ad absurdum*.

B. Windowing doesn't only add some high frequency components

According to the popular belief, leading to the proof presented in equations (5), (6), windowing only adds some high frequency components. This is way the contradiction of a signal limited both in time and frequency is accepted. Multiplying a function f(t) with a window function is not mathematically equivalent with adding some high frequency components. The window function has a spectrum going up to infinity and down to zero. Because multiplication in time domain corresponds to convolution in the frequency domain, the lower frequency components are fundamentally altered too. No filtering can undo that. Had the case been of adding some extra high frequency components only, some filtering could restore the original signal. After that filtering, a signal of infinite duration would result, containing the original bandwidth only. The problem is that the inverse of convolution is deconvolution, not filtering.

V. CONCLUSION

The sampling theorem as is used today causes the loss of a great deal of information in the process of digitization. If the revised Sampling Theorem is used, the extracted information is much richer. There aren't yet many tools able to process that information. The spectral analysis presented in [1] is an example. It decomposes the bandwidth of a signal sampled over a limited time in very fine spectral components. It can only do that if enough samples are taken. The correct form of the Sampling Theorem says other such tools must exist, even if they are not known or used today.

The use of the correct Sampling Theorem is bound to produce a significant improvement in every aspect of the digital equipment.

REFERENCES

[1] M. Bujor, "High Frequency resolution harmonic analysis," 2016 International Conference on Development and Application Systems (DAS), Suceava, 2016, pp. 203-206. doi: 10.1109/DAAS.2016.7492573

Device for intercepting and disrupting the hidden headsets

Costel Cherciu, and Dumitru-Iulian Nastac

Faculty of Electronics, Telecommunications and Information Technology
POLITEHNICA University of Bucharest, Romania
cherciu.costi@gmail.com, nastac@ieee.org

Abstract—**With the increased availability of technology and with the ability to miniaturize the electronic circuitry, more and more devices for cheating in exams are built. A cheating device is a complex system which allows a secret communication between the user and another person. However, a certain part of transmission is not encrypted in any form, and can be intercepted in a certain range. In this paper is presented one of the multiple ways to counter the devices used for cheating. The device is capable of intercepting and disrupting sounds transmitted through an inductive circuit and is hidden in a functional handheld Gameboy.**

Keywords— earpice; handheld; exam cheating.

I. INTRODUCTION

The motivation of this work is to elaborate an efficient system for discouraging the temptation of cheating during the written exams. Paradoxically this kind of unfair activity involves somehow a high level of inventivity especially for those who want to commercialize special devices for cheating. A simple but useful system was designed and built in order to easily discover most of the hidden used headsets. [1]

The starting point was to research an existing heating system in order to better understand how it works. This cheating system used a mobile phone that had replaced the speaker with an inductor. The system also had an earpiece which has its own inductor that resonates with the one inside the phone. The received signal is amplified by an active integrated audio amplifier.

After studying multiple cheating systems, we noticed that these consist of the following (basic) components: display, GSM module, Bluetooth device and headset with inductive system. For example, one of the devices has the GSM module hidden in a wristwatch, and the display is visible only with special glasses that had two polarized filters instead of lens. Other systems have the GSM module disguised as a wallet, a credit card, and they also include an inductive circuit.

The simplest device we encountered consists of a mobile phone, necklace inductive (coil) and a miniature magnet placed inside the ear canal. But even this is based on an inductive circuit.

If bidirectional communication is needed, it requires a more complex system based on a Bluetooth module that behaves as an audio device. The Bluetooth module is hidden inside a pen or another inconspicuous item (see figure 1) and is transmitting data using an integrated microphone from the user to the end-caller. It's also used to answer a call and to receive the audio signal from the phone, which is directed to an inductive circuit and received by the user's headset.

Bluetooth communication is designed in such a way that when multiple devices are operational in the same area, they don't interfere. This also makes jamming way more difficult [2]. It uses a technology named Frequency Hopping Spread Spectrum (FHSS) so the frequency is periodically changing using a pseudorandom sequence. A channel has the bandwidth of 1 MHz and the frequency domain is between 2402 and 2480 MHz.

To effectively jam a network using FHSS, an attacker must emit a similar signal (same modulation) that covers the entire bandwidth used by the network, or he has to know the hopping sequence of the target device and emit a signal on the next frequency at the right time.

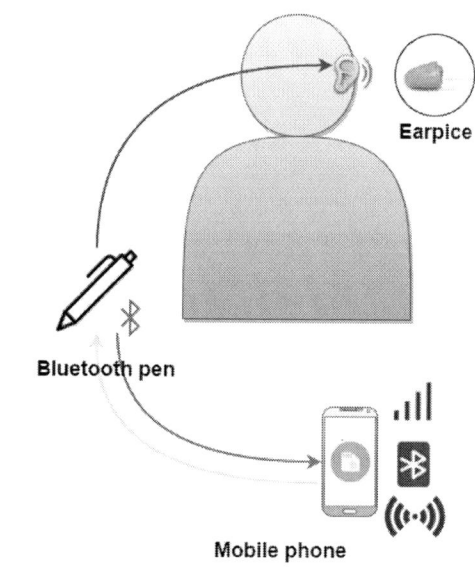

Fig. 1 Bidirectional communication of a cheating system

However, no matter how complex the system is, it still uses an inductive circuit to transmit sound to the user's in-ear device. Besides being a universal approach, this is where the system is vulnerable and easiest to attack because the data is not encrypted or modulated.

II. HARDWARE IMPLEMENTATION

The device was built starting from a functional Tetris Gameboy as a case in order to conceal the device. Additionally, this was a good choice because it allowed the use of a larger coil, thus increasing efficiency and the range of use. Inside, 2 inductors were mounted: one for one for reception of the cheater's transmission and one for transmission to the cheater (either for jamming, either for speaking directly to the cheater). For the emitting coil, a thicker wire was used to allow more current through the coil in order to increase the electromagnetic field, intensity and efficiency.

Fig. 4 First working prototype

After the demodulation stage, the cheating device will emit a raw electromagnetic audio signal wave (see figure 5). In other words, the emitted electromagnetic wave, sweeps below 20 kHz. This is where our device intercepts directly the cheater's communication and we can listen or disrupt his signal. From here on, the cheater's communication can be recorded (as a proof) or be redirected to a speaker.

However, in order to determine who is cheating, the user has to patrol along the classroom so that the device gets in the range of the cheater's transmission. Once a quiet sound is heard through the headset of our device, the user can now get closer to the target in order to hear better and identify the real culprit. Having a handheld device hidden in a functional Gameboy doesn't raise suspicion and, the user can hold the device in various angles so that its receiving coil is not perpendicular to the cheater's emitting coil.

The coils are air-core inductors, wound according to the plastic case in order to maximize the effective surface.
A simple and popular PAM8403 audio amplifier was chosen. It is a class D audio amplifier and has a power up to 3W. The low THD (total harmonic distortion) allows high-fidelity sound reproduction. Its architecture allows the device to directly drive a speaker, without any additional low-pass filter, reducing the number of external components required and circuit board size. Other useful features are the protection

---

Other existing anti-cheating system

One of the approaches is GSM signal jamming device. It attacks all mobile phones in the band's frequency. Although they are able to cover the entire examination room or even more, they require special license to use because it may block other users or emergency services. Another disadvantage is that it requires improvements as the phone network develops and uses other GSM frequency bands. They are also expensive and their cost is not justified if they are only used during the examinations.
Besides GSM dependent cheating systems, another option involves using a communication frequency programmable FM transceiver to achieve a concealed system. It can be used in an up to 1 km range and dose not uses an inductive circuit. The system is a hidden earpiece independent radio module. However, it can be easily identified with an SDR (Software Defined Radio), if its frequency of operation is known. The downside is that the receiver cannot be easily identified and located. In order to jam the (signal) communication, we built the device "Radio Sniffer" (see the figure below) which has the same operating principle as the "IRSniffer" detailed in [3].

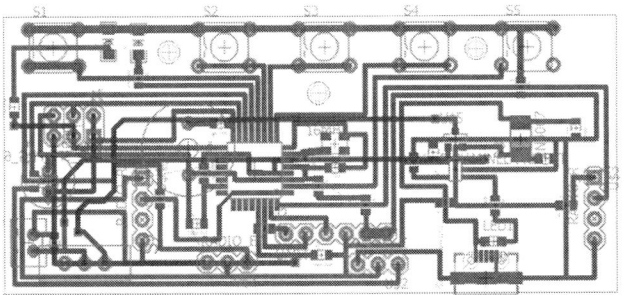

Fig. 2 PCB layout of "Radio Sniffer"

It is based on an ATMega328p microcontroller with radio modems and it can listen, decode and store modulated ASK (Amplitude Shift Keying) and OOK (On / Off Keying) radio signals, in frequency bands of 432-468 MHz and 310-318 MHz. By continuously emitting the recorded signals, the frequency band is jammed by overlapping the one the user is using for cheating.

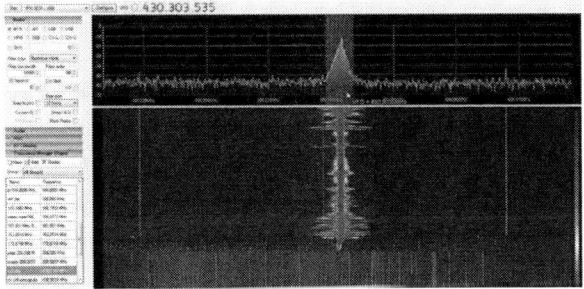

Fig. 3 FM signal captured with SDR

against short-circuit, thermal shutdown, and efficiency up to 90% which makes it well-suited for portable applications.

Receiver coil

A 0.11mm (37AWG) enameled wire was used. Starting from the size inductor wound on the case and the gauge of the wire used, the number of turns, number of turns per layer, number of layers, resistance and inductance was determined. Using an air gap inductor calculator ([4] and [5]), we could easily check the inductors parameters and adjust them accordingly.

To tune the receiver to the mobile phone's coil frequency, a small capacitor was used (forming an LC oscillator). In order to determine the capacitor value, a programmable capacitor was used in order to fine tune the LC oscillator [6]. This allowed maximum efficiency of the receiver and improved the overall performance. The capacitor was connected in parallel with the receiver coil.

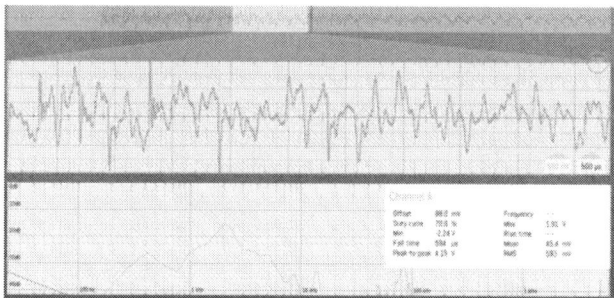

Fig. 5 Visualizing the RAW audio signal

Broadcasting Coil

Besides detecting and identifying the cheater's, we wanted to be able to talk directly to the student, to send prerecorded sounds, or to jam or disrupt the communication. In order to achieve any of these, a transmission coil was designed after an emitting coil of a typical cheating device.

This time, a capacitor can no longer be used to change the resonant frequency, so the impedance of the inductor, the number of turns and wire thickness were measured and replicated. Even if the resulted inductor has a very low impedance, this wasn't a problem for the audio amplifier. A low impedance means a higher current through the inductor, generating a stronger electromagnetic field.

For the transmitting module, a simple sound generator is used to play music files from a microSD card [7]. Here, the user controls the buttons of the Gameboy in order to select what files are played, to jam or disrupt the communication or if the received signal should be recorded, redirected to a speaker, etc. For this, an Atmel ATtiny861V AVR 8-bit microcontroller was used, which has 8kB Flash memory and high frequency PWM (Pulse Width Modulation) outputs that can provide a much clear sound than other similar AVR chips. It can also operate at 3.3V which allows direct communication with the microSD card (connected in SPI mode). The microcontroller has 20 GPIO (General-purpose input/output)

pins, which were used as an external trigger to control the music player.

The sound generator was designed to be as small as possible and to easily fit with the audio amplifier.

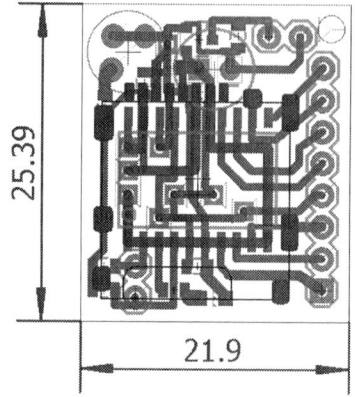

Fig. 6 PCB layout of the sound generator

Software

The supported file format for the audio files on the microSD card is WAVE coded in LPCM (Linear pulse-code modulation). These files must be 8 or 16bit encoded and must have a sampling rate between 8 and 48 kHz. Any other file type such as MP3 and AAC must be first converted to this format. The audio files must be stored in the root directory and named NNN.wav, where NNN is a 3-digit number between 001 and 255. A combination of switches will play the corresponding number in binary.

The Gameboy has 8 available buttons. Since the Gameboy's buttons are active-high, a hex inverter is used to reverse logic and drive the switches of the ATtiny861V microcontroller. Although the wav player allows 28 commands, only 8 were used for simplicity in design. Each buttons triggers one bit of the 8bit inputs. This is convenient because the Gameboy already has 8 pushbuttons.

The board also has a status LED that stays on while a music file is being played and blinks if an error occurs.

III. CONCLUSIONS AND FUTURE IMPROVEMENTS

Although it has been tested only in a few numbers of scenarios, the proposed device is a cheap alternative to other anti-cheating devices. It covers almost all popular cheating systems and for the first time, it helps to identify the cheater without interfering with GSM networks or Bluetooth communication.

The fact that the system is hidden inside a game toy makes it harder to spot, catching the students off-guard. Because it can spot the exact culprit, record the conversation as a proof, jams and disrupts in an unpleasant way, cheaters are now highly discouraged in their future attempts.

As a future improvement, 2 more coils can be added (both for emitting and transmitting side) perpendicular to the existing one in order to improve directivity and range. Also, a larger power amplifier can be used for emitting the signals, in order to extend the working radius.

IV. REFERENCES

[1] S. Tzanova and N. Codreanu, "Training microsystems technologies in an European eLearning environment," *IEEE Education Engineering (EDUCON),* p. 113 – 118, 14-16 April 2010, DOI: 10.1109/ EDUCON.2010.5493060.

[2] S. Koppe, "Bluetooth Jamming," 2012. [Online]. Available: ftp://ftp.tik.ee.ethz.ch/pub/students/2012-HS/BA-2012-16.pdf. [Accessed 02 05 2016].

[3] C. Cherciu, C. Cherciu and D.I. Nastac, "A Portable Device for Intercepting and Retransmission of Infrared Modulated Signals," in *IEEE International Symposium for Design and Technology in Electronic Packaging (SIITME)*, Braşov, 2015.

[4] A. &. S. Shipway, "CalcTool," 2008. [Online]. Available: http://www.calctool.org/CALC/eng/electronics/RLC_circuit. [Accessed 04 05 2016].

[5] "Pronine Electronics Design," 2004. [Online]. Available: http://www.pronine.ca/multind.htm. [Accessed 04 05 2016].

[6] R. Talwar, "RaptorBird Robotics," RaptorBird, 03 11 2013. [Online]. Available: https://www.kickstarter.com/projects/raptorbird/programmable-capacitor. [Accessed 10 04 2016].

[7] ChaN, "Electronic Lives Manufacturing," 15 12 2013. [Online]. Available: http://elm-chan.org/works/sd20p/report.html. [Accessed 03 05 2016].

2016 IEEE 22nd International Symposium for Design and Technology in Electronic Packaging (SIITME)

Aspects of Using Low Layer Count PCBs for Embedded Systems with FPGA Devices in BGA Packages

Andrei Drumea and Mihaela Pantazică

Department of Electronics Technology and Reliability
"Politehnica" University of Bucharest
Bucharest, Romania
andrei.drumea@cetti.ro

Abstract—**This paper shows part of the investigations conducted regarding the use of 2-layer printed circuit boards for embedded systems based on devices (microcontrollers or field programmable gate arrays) in BGA (Ball Grid Array) packages. Usually, the technical literature [1] recommends for applications involving programmable devices in BGA packages the use of multi-layer boards (4, 6 or more layers), but these boards are expensive for applications like educational kits or simple controls that runs at relatively low clock frequencies, around 1MHz. The paper analyzes the aspects of PCB design, manufacturing costs and technical limitations (maximum available I/O pins, signal integrity) of 2-layer printed circuit boards for a simple embedded system based on an FPGA device in 256-pin BGA package. Suitable PCB design techniques analyzed and layout recommendations are presented.**

Keywords—embedded systems; BGA package

I. INTRODUCTION

Embedded systems based on programmable logic devices like field programmable gate arrays (FPGAs) offer more flexibility than microprocessors or microcontroller-based solutions, especially regarding I/O lines. Such devices allow an easy implementation of an open-source soft-core processor defined in standard hardware description language that can be programmed like a normal stand-alone processor. There are many available open-source processor cores, ranging from 8-bit processors (LatticeMico8, many compatible designs for AVR, 8051, Z80, 6502 or 6800 families) to 16-bit processors (8086, openMSP430) and ending to powerful 32-bit processors like LatticeMico32 or F32C (both MIPS compatible) or RISC-V processor ([3]). Such devices can simplify the development of low-complexity embedded systems for specific applications like optics or movement detection ([4], [6], [9]).

Unfortunately, FPGA devices tend to be packaged in ball grid array (BGA) packages that require complex routing techniques in PCB design and multi-layer printed circuit boards. This paper presents aspects of the PCB design and implementation of a 2-layer PCB for an embedded system based on FPGA device in 256-pin BGA package.

II. DESIGN RESTRICTIONS

A simple development board for testing an 8-bit soft processor (LatticeMico8) was required; this board needs to be designed in order to evaluate the possibility of using 2-layer printed circuit boards for an embedded system based on a low cost FPGA device with 6864 look-up tables in 256-pin BGA package.

Different variants for PCB board were designed, with 0.8mm or 1mm ball pitch BGA devices in layout versions with top layer for I/O lines and bottom layer for power and ground (one configuration) or for I/O lines in another configuration. The pin map for the chosen FPGA device is shown in fig.1, with ground pins marked with G in dark blue, supply pins marked with V in red (core power supply, 1.2Volt) and I/O power supply pins with v in orange. Available I/O pins are represented in light blue.

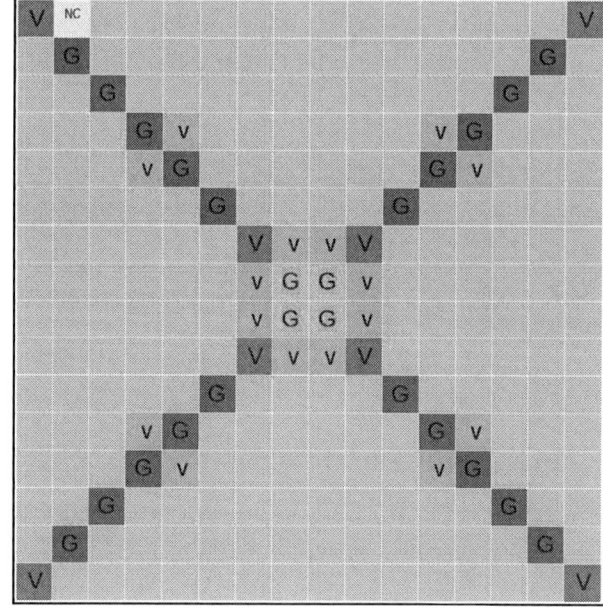

Fig. 1. Pin structure for an FPGA device in 256-pin BGA package.

978-1-5090-4446-7/16 $31.00 © 2016 IEEE   74   20-23 Oct 2016, Oradea, Romania

2016 IEEE 22nd International Symposium for Design and Technology in Electronic Packaging (SIITME)

Respecting the device manufacturer recommendations for pad sizes [1] for a 256-pin BGA with 1mm solder ball pitch allows only one escape trace between two adjacent pads (fig. 2, left side) even if specific design rules for high quality PCB manufacturing (4-mil minimum trace width, 4-mil clearance) are used for layout design. Reducing the pad diameter under manufacturer recommendations (0.45mm pad diameter instead of 0.55mm) allows routing two escape traces between two adjacent pads (fig. 2, right side), but this technique can affect the assembling of the device on the PCB and will not be considered as a viable option. If BGA devices with 0.8mm ball pitch are used, this technique cannot be used even if pad size is drastically reduced.

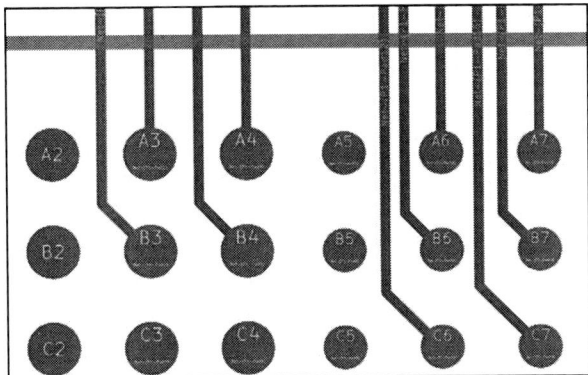

Fig. 2. BGA breakout routing on top layer.

If only one escape trace between two pads is allowed, then any standard manufacturing technology can be used for both BGA package types (0.8mm and 1mm ball pitch).

III. LAYOUT DESIGN

The first step in PCB layout design was to create virtual components (parts) in the available CAD environment. Open source software KiCAD was chosen for layout design due to its advantages compared to commercial software - no pin-count limitations, no layer restrictions, easy to use and to configure, small size and ability to run on different operating systems. An independent part was created for each BGA package type.

The FPGA device contains 6 logic banks, each with its own supply line VCCIO0...VCCIO5 for I/O pins; a core supply voltage of 1.2Volt is always required for normal device operation. The VCCIOx lines can accomodate 1.2/1.5/1.8/2.2/2.5/3.3 Volt voltages, so each bank can have its own supply voltage for its I/O lines. In our real design, all VCCIO lines are supplied with 3.3Volt, so all associated pins can be connected together and the number of decoupling capacitors can be reduced, simplifying the layout design.

Different layout configurations were designed with different numbers of available I/O pins. We designed for both BGA package types (0.8mm and 1mm ball pitch) two particular configurations, one that maximizes the number of available I/O lines and the other one that focuses on better signal integrity.

In first case, both layers (top and bottom) are used for breakout routing, each layer allowing the escape of two rows of BGA balls [2] counting them from outside to inside, as shown in fig. 3. The PCB traces on top and bottom layers are shifted a couple of mils horizontally and vertically to avoid superposition for better figure clarity.

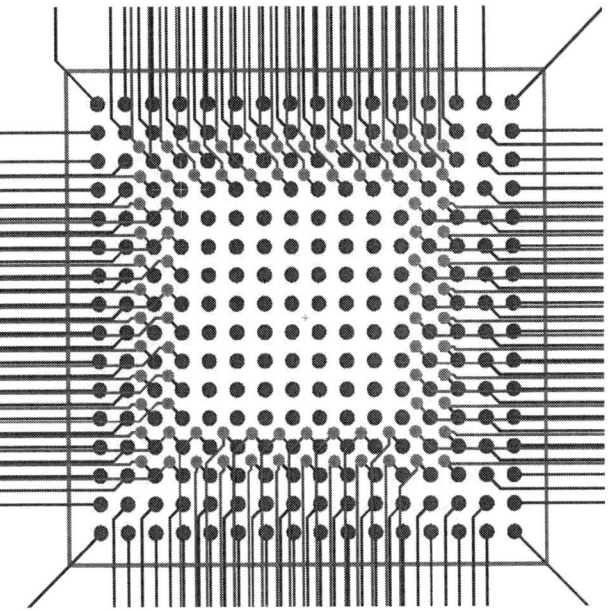

Fig. 3. BGA breakout routing on both layers.

A detail of dog-bone pattern (formed by solder ball pad, PCB trace and through hole via) used in this case for routing the inner rows of balls on bottom layer is shown in fig.4.

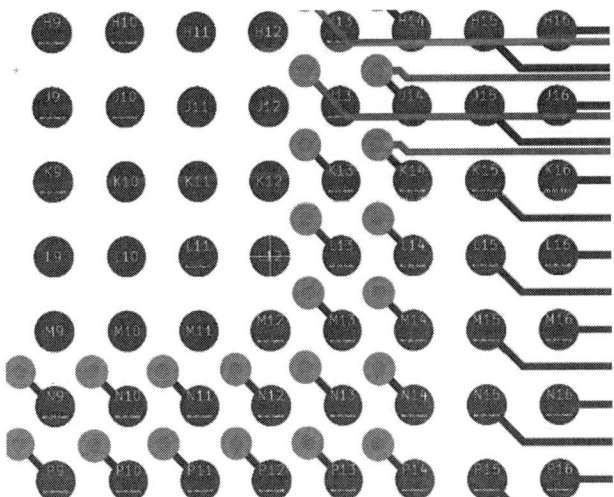

Fig. 4. Detail of BGA breakout routing and dog-bone patterns.

If we consider the BGA pattern as a NxN square decomposed in a set of empty squares with length N, N-2, N-

4 and so on, then we can route first two external squares (with N and N-2 lenghts) on top layer and the next two squares (with N-4 and N-6 lengths) on bottom layer.

A N-length square perimeter of balls yields the following number of connections:

$$Conn_N = N + N + N\text{-}2 + N\text{-}2 = 4N\text{-}4 \qquad (1)$$

Considering these aspects, the total number of available escape traces from a BGA package with N*N balls is:

$$PincountA_N = Conn_N + Conn_{N\text{-}2} + Conn_{N\text{-}4} + Conn_{N\text{-}6} \quad (2)$$

$$PincountA_N = 16N\text{-}64 \qquad (3)$$

The second design configuration uses bottom layer for ground and supply lines (for improved signal integrity), leaving no place on this layer for I/O lines, as shown in fig.5. I/O lines are routed only on top layer and no dog-bone pattern is required.

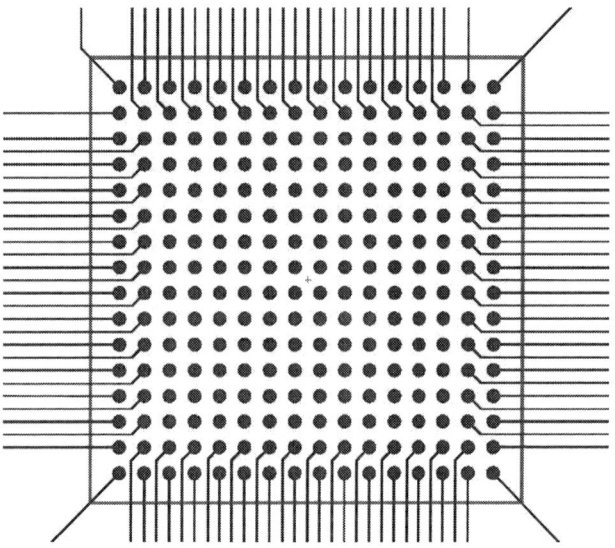

Fig. 5.   BGA breakout routing on top layer only.

This second design configuration yields the following theoretical number of I/O lines:

$$PincountB_N = Conn_N + Conn_{N\text{-}2} \qquad (4)$$

$$PincountB_N = 8N\text{-}16 \qquad (5)$$

Table 1 shows, without and with I/O lines on bottom layer, the theoretical number of available pins for the 256 pin packages (these packages have N=16) counted without taking into considering the particular pin map shown in fig.1 (first column) or considering this limitation (second column).

TABLE I.   AVAILABLE I/O LINES FOR 256-PIN BGA PACKAGE.

Layout configuration	Theoretical available I/O lines	Available I/O lines for presented package
2-layer board, top layer for I/O lines, bottom layer for power and ground	112	103
2-layer board, top and bottom layer for I/O lines	192	167

The number of available I/O lines can be further increased if a larger BGA package is used. The same FPGA device is also available in a 484-pin BGA package, so a theoretical number of 288 I/O lines can be achieved, as shown in table 2.

TABLE II.   THEORETICAL AVAILABLE I/O LINES FOR 256-PIN AND 484-PIN BGA PACKAGES.

Layout configuration	256-pin BGA	484-pin BGA
2-layer board, top layer for I/O lines, bottom layer for power and ground	112	160
2-layer board, top and bottom layer for I/O lines	192	288

The 484-pin BGA package offers more available I/O lines but its footprint is much bigger (23 x 23 mm) than the 256-pin BGA packages (14 x 14 mm for 0.8mm ball pitch version or 17 x 17 mm for 1mm ball pitch version).

IV. COST ESTIMATION AND FURTHER INVESTIGATIONS

A cost estimation was made based on prices offered by a PCB manufacturer that implements user pooling (OSH Park). This manufacturer has design rules that fit to our design requirements - 6mil minimum trace width and clearance, 13mil drill size and 7mil annular ring for two-layer boards and 5mil minimum trace width and clearance, 10mil drill size and 4mil annular ring. The price for prototype service is 5USD per square inch for 2-layer boards and 10USD per square inch for 4-layer boards, so price is double for 4-layer boards. For this price a set of 3 boards is obtained.

Our boards have aproximately 100mm length and 100mm width, so PCB manufacturing costs are 80/160USD for a set of 3 boards. Using 2-layer boards for our application has a large impact on cost reduction.

At the moment the PCB are ordered. Further investigations regarding signal integrity will be performed after receiving the boards and assembling all components

V. CONCLUSIONS

Programmable devices (FPGAs, microcontrollers) in BGA packages can be used in embedded systems manufactured with 2-layer printed circuit boards.

There are different options for routing, depending on existance of a ground plane on bottom layer; this option improves signal integrity but decreases the number of available I/O pins.

The exact number of available I/O pins depends on the specific configuration of the device pinout.

ACKNOWLEDGMENT

The presented work was supported by a grant of the Romanian National Authority for Scientific Research CCCDI - UEFISCDI, Project number: PN-II-PT-PCCA-2013-4-1673.

REFERENCES

[1] Lattice Semiconductor, "PCB Layout Recommendations for BGA Packages ", TN1074 Technical Note, pp. 1-4, September, 2016.

[2] C. Pfeil, "BGA Breakouts and Routing", Mentor Graphics, Wilsonville, Chapter 4, pp. 55-78, 2008.

[3] M. Zec, and D. Jandrevic, "FPGArduino: a cross-platform RISC-V IDE for masses", 4th RISC-V Workshop, pp. 9-11, July, 2016.

[4] C. Ionescu and R. Dobre, "Performance studies of electrochromic displays," Proc. SPIE 9258 Advanced Topics in Optoelectronics Microelectronics and Nanotechnologies VII, vol. 9258, no. 1, pp. 925833-925833-10, 2015.

[5] D.A. Visan, I. Lita, and I.B. Cioc, "Temperature Control System Based on Adaptive PID Algorithm Implemented in FPAA," Proc. of 34th International Spring Seminar on Electronics Technology (ISSE2011), pp. 501-504, May 2011.

[6] A. Drumea, and R. Dobre, "Clicks counting methods for a scope knob," Hidraulica, vol. 4, pp. 79-84, October 2013.

[7] R. Bozomitu, "A New Linearized CCII-Based Fully Differential CMOS Transconductor Used for Gm-C Active Filters Implementation," Wireless Personal Communications, vol. 74, no. 2, pp. 615-637, January 2014.

[8] M. Vidrascu, and M. Vladescu, "Programmable pulsed current driver for high power LEDs applications", 20th International Symposium for Design and Technology in Electronic Packaging (SIITME2014), pp.123-127, 2014.

[9] A. Vasile, E.L. Trifan, C. Bucsan, and C.A. Micu, " Electro-magnetic actuated vibrating platform ", 19th International Symposium for Design and Technology in Electronic Packaging (SIITME2013), pp.241-244, October 2013.

Host Emulator for next Generation Battery Chargers

Cristian GRECU, Cosmin IORDACHE
Department of Electron Devices, Circuits and Architectures,
POLITEHNICA University of Bucharest, Romania
grecucris@yahoo.com

Mihaela PANTAZICĂ
Department of Electronic Technology and Reliability,
Faculty of Electronics, Telecommunications and
Information Technology, POLITEHNICA University of
Bucharest, Romania
Mihaela.pantazica@cetti.ro

Abstract— A modern portable computer, netbook, smartphone or camera is powered by a smart battery. This is a device containing one or more Lithium cells, protection circuits and a power management circuit, interfaced by a two wire SMBus interface [1]. The need for higher efficiency and longer battery life has pushed the technology of the smart battery forward, for higher efficiency, smaller component count and more integrated functions controlled by internal registers. The internal registers are written and read via SMBus. A charger is evaluated by the test engineers and presented to the customers by the field application engineers on a demonstrator board which emulates the conditions found in a real system. The charger also needs a SMBus Master [2]. This paper presents a host emulator integrated in the demonstrator board, together with a Windows interface for accessing the internal registers. (*Abstract*)

Keywords—Smart battery, smart battery charger2

I. INTRODUCTION

A. SMBus interface

The subject of this paper has its origins back in 1995 [3]. The new standards for lower power consumption required for the portable computers lead to the development of a new low speed interface. New interface was called SMBus (from System Monitoring Bus). This interface should connect all new power management integrated circuits, such as temperature sensors, voltage regulators, fan controllers and also the smart battery and its charger. It's strongly inspired from the older I2C interface, developed in the 80s by Philips. They are both serial synchronous interfaces using two lines for communications, SDA (for data) and SCL (for clock). Both lines are connected to corresponding pins of all power management devices inside the system. The hardware layer is similar, with a supplementary signal named EVENT, used by a slave device to inform the Master that a particular event occurred and its registers must be read. [4] On the Master, it works similar with an interrupt, forcing it to interrogate the Slave which generated the event and take a decision.

SMBus protocol is oriented on registers. A slave has a specific address and at least one register, 8-bit or 16-bit wide, which can be R/W or read only. The communication may be started only by a Master.

SMBus adds timeouts. Some registers of every device must be repeated once every at 175 seconds at most. None of the data or the clock line should be kept low for more than 35ms. If any of the above intervals is not respected, then all the internal registers are erased and the integrated circuit returns to its default state.

B. Smart Battery system

First specification for smart battery control module was issued by Intel in 1995. The smart battery comes as a solution to multiple troubles generated by the rechargeable batteries in portable electronic devices. The user may never know the exact state of charge of a certain battery and operating time left. Approximations based on battery history or battery voltages were of some use in certain systems, but the data became of no use when a second battery was used. The system itself cannot know when power should be conserved. As an example, turning on a hard drive with an almost depleted battery may lead to system failure. [1]

A typical smart battery sub-system is depicted in the image below. The adapter is usually an external Flyback converter with PFC delivering a constant voltage to the system. The smart battery is a complex device containing the battery cells, protections and a power monitoring IC [5]. The power monitoring IC is interrogated by the power management host for parameters such as designated voltage, current, estimated capacity, maximum discharge current and so on. Based on this data, the Host sets the parameters for the smart battery charger.

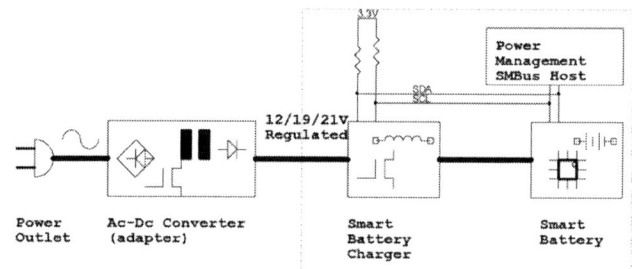

Fig. 1. The simplified schematic of a smart battery subsystem

C. Smart batteries

Usually a smart battery is composed of multiple cells connected in a series-parallel configuration together with protection and power monitoring circuitry. Most commonly used are Lithium-based batteries. They have a very large energy density in both volume and weight, superior to older nickel-hybrid metal and nickel-cadmium. Lithium batteries replaced the former in multiple applications, ranging from mobile phones to electric vehicles. The high density of energy comes with a great risk. As a recent example, the company Samsung released a recall for their latest Smartphone model, the Samsung Galaxy S7, which was prone to battery taking fire. An accidental short circuit, overcharge or over current may lead to a quick and very high energy dissipation, a rapid warm-up of one or multiple cells, and finally fire. A short circuit condition may not only be generated mechanically, but also from other circuits, such as the power management IC, thus the smart battery should be protected against this condition [6]. Another requirement of Lithium cells is balancing. In a series configuration, all cells should have the same voltage. If a cell has just slightly a higher capacity than the others, this one will be charged to a lower voltage while the others will be overcharged. The balancing circuit continuously monitors the voltage of individual cells during and after charging and discharges the cells having a higher voltage than the others [7]. Balancing was not required for Nickel cells.

The protection circuit controls two NMOS switches connected in series, in back to back configuration, with common drain or source. Those protect the battery for the unlikely events of overcharging or over discharging. Thus the protection IC permanently monitors the battery voltage and current to determine a fault condition. During normal operation both MOSFETs are ON. When the discharge MOSFET turns OFF, the battery can only be charge through its body diode, while in discharge, the body diode becomes polarized in reverse. Similarly, when the charge MOSFET turns OFF, the battery can deliver power through the body diode, but can no longer be charged.

The final safety frontier is the thermal fuse. It is a mechanical system that once overheated breaks the contact and is unable to return to its normal operating state. It usually works by means of a contact held closed using a spring. When a plastic capsule with a precise melting point melts, the spring releases the contact. Thus the protection cannot return to its initial state.

The power monitoring IC is interfaced to the SMBus host and also to the protection IC via a separate interface. It has mask ROM but also user defined Flash. The most important and complex function of this IC is capacity measurement. This parameter cannot be measured directly, so it must be computed by this circuit by integrating the measured current over the entire life of the battery. For this reason this IC is also called Gas Gauge, in comparison with the fuel measurement instrument of a car.

D. Smart Battery Charger

In the past, the smart battery chargers were built using discrete components. Such chargers needed a rather large PCB area to implement the Dc-Dc converter, control loops and protections [8]. In order to make it simple, the chargers were constant frequency, asynchronous Buck converters. The low side switch was a Schottky diode which brought some improvement to the efficiency of the converter. Power transistor was PMOS usually with totem pole drivers. PWM generator was a relaxation oscillator working at a few hunder kilohertz. Separate loops were used to control charging voltage and charging current. Last but not least, the voltage and current references were obtained using separate DACs.

Level two chargers which were used during the years 2000 had more and more integrated functions. First integrated chargers still needed external reference, but then SMBus controlled DACs were also integrated, and for greater accuracy, an integrated Bandgap voltage reference was used (ISL88731). As the processors became more efficient for lower costs, more functions had to be implemented in the charger. A modern charger for last generation of Intel processors is BQ24725.

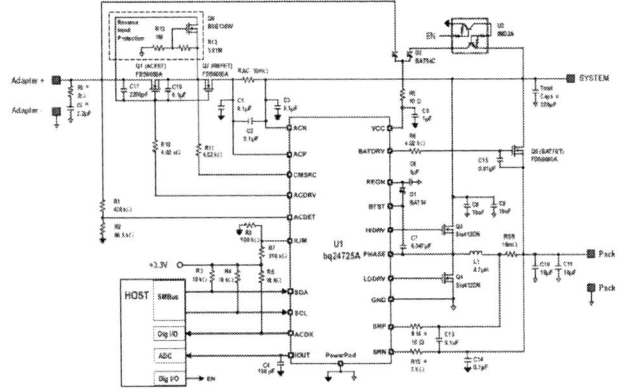

Fig. 2. Typical application schematic of BQ24725

The Dc-Dc converter is a synchronous Buck, using the transistors Q3, Q4, the inductor L1 and filter capacitors. The current is measured using the current sense resistors RAC, RSR and the internal current sense amplifiers. The system adjusts the charging parameters using three important settings: maximum charging voltage, maximum charging current and maximum adapter current.The loops are internally compensated using a type III compensation network. A few other functions are also integrated, such as the power selector, adapter presence detector or Intel Hybrid Power Boost [9] function.

The PCB layout for such a charger may be found in one of the newest notebook computers from Lenovo, the E560. The charger controller (PU101) is shown on the left side of the picture with its passive components, the power transistors are in the middle of the image and the inductor is on the right. The battery connector may be found on the lower part of the image, on the edge of the board (JBATT1). It is mandatory for the smart battery to have two or four pins for power and two pins for SMBus communication with the smart battery. Some smart batteries have other special functions pins, such as battery enabling.

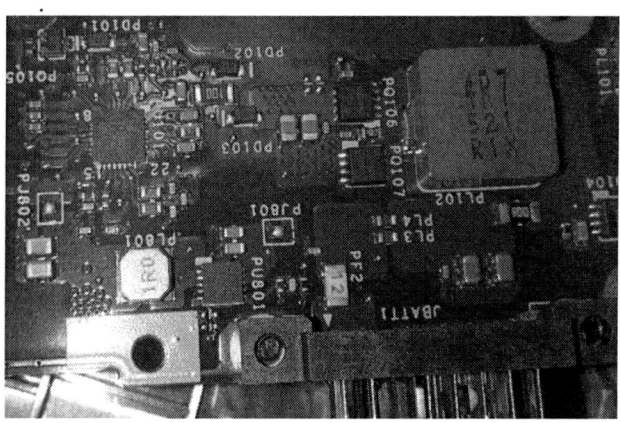

Fig. 3. Battery charger and connector of a Lenovo E560 notebook

II. THE DEMONSTRATOR BOARD

A demonstrator board is a PCB based electronic module allowing an integrated circuit to work in similar conditions as in real applications. It serves two main purposes. First it provides to the testing and application engineers an easy and accurate method for evaluation in almost any real condition. Secondly it represents a handy practical method for the field application engineer to advertise the new IC to the customers.

A. PCB Design Considerations

The demonstrator is designed according to strict design rules, so that in any case no PCB issue may disturb its functionality [10]. Multiple ground planes and screening of sensitive signals ensure low crosstalk to the power plane (DC-DC converter) [11][12]. The PCB is large enough to allow easy access to all test points, and also a proper thermal management. The emulated host should be easily connected to any computer, for quick and easy access to the registers. A photo of the Demonstrator Board is depicted below.

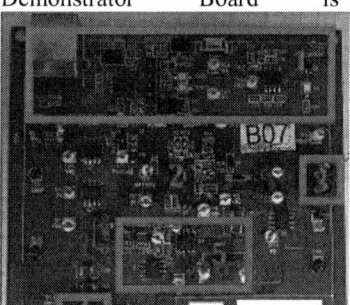

1 - Emulated SMBus host
2 - Smart Battery Charger
3 - Emulated Battery Connector
4 - Emulated System Connector
5 - Dc-Dc Converter

Fig. 4. Demonstrator Board for the new intelligent battery charger IC

B. Host Emulator

The emulated host is based on a PIC18F microcontroller which connects to a personal computer by means of an USB

type B connector. The connector is used to power the microcontroller and also for data transfer. A communication is initiated by the computer. The computer sends a data sequence to PIC, which interprets it and starts accessing the corresponding internal registers of the charger. Then another sequence is sent back to the computer containing the new values in the registers and eventual communication errors. The host is controlled from the computer by a user friendly Windows application with graphical user interface (GUI) developed in Visual Studio. The starting form of the GUI is presented in the image below.

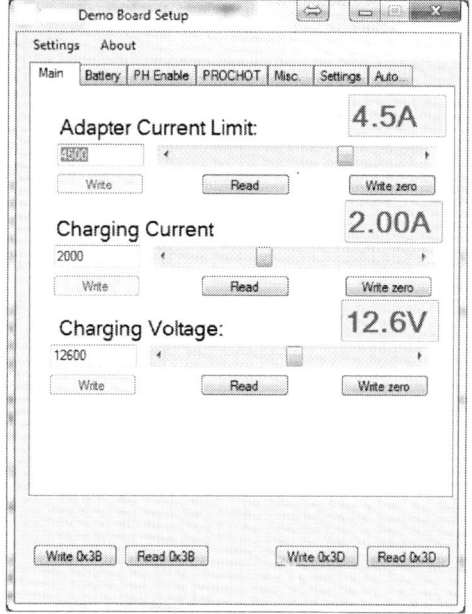

Fig. 5. Startup form of the GUI

III. CONCLUSSIONS

The firmware and Windows application were developed after the Demonstrator Board was issued. First, a general purpose I2C monitor and debugger was used. By the time the A0 version of the new charger was available, the final testing of the host was finished and the evaluation started for the newly designed charger. As an example, the main function of the charger is presented below. A SMBus sequence is sent to write the adequate registers and begin charging of the battery. SMBus data may be seen on channels 2 (SCL) and 3 (SDA) of the oscilloscope, while PWM of the Dc-Dc converter may be seen starting on chanel 1.

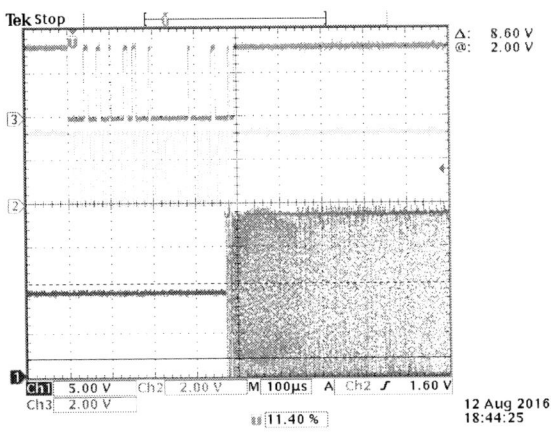

Fig. 6. Charging start upon receiving of corresponding SMBus register

The evaluation was done on the demonstrator board by the post silicon testing engineers using two types of tests. Manual tests were done using the Windows interface. This kind of tests include measurements on steady state operation of Dc-Dc converter, transients between Charge mode and Intel Hybrid Power Boost mode and behavior of the protections in extreme conditions. Automatic tests don't make use of the Windows interface. Instead a VBA script sets the equipment and the registers of the charger in real time and measures the parameters of interest [13]. The linearity of the charge current, charge voltage and adapter voltage DACs, having hundreds of possible values, may be evaluated within a few minutes. Values of voltage, current and temperature are recorded over a several charging cycles of a smart battery, while SMBus registers are continuously refreshed every few cycles. All those tests are summed up in a complex evaluation report. This is used by the design team to fix the issues, improve the behavior of the charger and provide a good, quality product for the notebook manufacturers and the end user.

REFERENCES

[1]   Duracell Inc, Intel Corporation, "Smart Battery Data Specification", rev 1.0;

[2]   Zhuang Jianhua, Diao Chao, "Implementation of SMBus Based Smart Battery System for Portable Device";

[3]   Benchmarq Microelectronics Inc., Duracell Inc., Energizer Power Systems, Intel Corporation, Linear Technology Corporation et al, "Smart Battery System Specifications, System Management Bus BIOS Interface Specification"

[4]   Intel Corporation, Microsoft Corporation, "Advanced Power Management(APM), BIOS Interface Specification"

[5]   Texas Instruments, "bq20z40EVM-001 SBS 1.1 Impedance Track™Technology Enabled Battery Management Solution EVM"

[6]   Texas Instruments, "bq29330 External Short-Circuit Protection"

[7]   Texas Instruments, "Fast Cell Balancing Using External MOSFET"

[8]   Quanta Computer Inc, "HP Omnibook 6100 Schematic Diagram"

[9]   Intel, "Intel® Turbo Boost Technology 2.0"

[10]   Henk Jan Bergveld, Wanda S. Kruijt, Peter H. L. Notten, "Smart Battery Systems", Volume 1, Chapter 2;

[11]   Norocel Codreanu, Iulian Bușu, Paul Svasta, Elena Ignat, "Advanced post layout analysis laboratory platform", The 33rd International Spring Seminar on Electronics Technology, ISSE2010, 12 – 16 May 2010, Warsaw, Poland, pp. 494 - 499, ISBN 978-1-4244-7849-1.

[12]   C.A. Tămaș, M. Pantazică, C. Marghescu, I. Marghescu, "Controller for reducing the EMI in Gas Gauge and Battery Management Units", ISSE2013, "36th International Spring Seminar on Electronics Technology – Automotive Electronics", Alba Iulia , Romania, 8-12 May 2013

[13]   C.A. Tămaș, M. Pantazică, C. Grecu, I. Marghescu, *Integrated Circuits Characterisation Platform ICCP"*, ISSE2013 36th International Spring Seminar on Electronics Technology – Automotive Electronics", Alba Iulia , Romania, 8-12 May 2013

Development of Underwater Sensor Unit for Studying Marine Life

Iulian Lazar, Alin Ghilezan, Mihaela Hnatiuc
Electronic and Telecommunication Department
Constanta Maritime University
Constanta, Romania

Contact author Mihaela Hnatiuc
mail:mihaela.hnatiuc@cmu-edu.eu

Abstract—In this paper we propose an ROV system (remotely operated vehicle) equipped with sensors and camera to monitor the marine life and offer impact estimation for the placement of oilrigs in the Black Sea reservoir. The robot is equipped with a low resolution analogue camera for real time image used by the pilot of the ROV and a HD camera for video recording of the marine ecosystem; the sensors (IMU and pressure sensor) are also used for helping the pilot maneuver the unit. All the electronic components in the robot are mounted on a specially designed frame using AutoCAD and made using a 3D printer. They will be placed in a watertight enclosure that can withstand depths up to 100 meters. The data coming from the sensors are analyzed using a LabView interface.

Keywords— underwater system, sensors placement, marine life monitoring, LabView interface

I. INTRODUCTION

The untethered underwater vehicle is called AUV, which is free from a tether and can run either a preprogrammed or logic-driven course. The difference between the AUV and the remotely operated vehicle (ROV) is the presence (or absence) of a direct hardwire (for communication and/or power) between the vehicle and the surface. However, AUVs can also be (figuratively) linked to the surface for direct communication through an acoustic modem, or (while on the surface) via an RF (radio frequency) and/or an optical link. We are concerned with the surface-directed, hard-wired (tethered) ROV.

Simplistically, an ROV is a camera mounted in a waterproof enclosure, with thrusters for maneuvering, attached to a cable to the surface over which a video signal and telemetry are transmitted (Fig. 1).

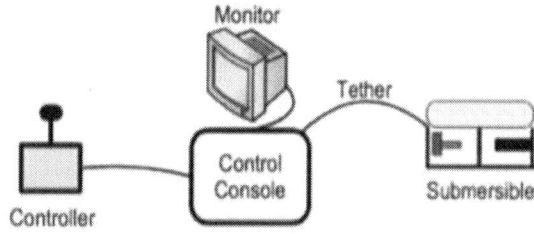

Fig. 1. The general schema of underwater system [2]

II. RELIABILITY OF ROVS IN OCEANEERING

Underwater is a hostile environment due to the high pressure, low temperature, low visibility, strong currents and even nuclear radiation and shock waves caused by explosions. Using an underwater robot for inspection operations or various operations with high-risk area is the only solution.

Some of the most important advantages of using an ROV:

- Long operating time, theoretically unlimited running water all conditions

- The ability to work in tight places, vertical walls, near the feet of oil platforms in wrecks, under the ice, nuclear facilities, etc.

- Less dependence on the environmental conditions of the area

- Bigger maneuverability due to the small size

- Lower price for the operations compared to the cost of using divers or bathyscaphs.

Here we have a table [5, 6] from o rapport made by *WTEC* in 1996, in which it was made a comparison of prices for underwater interventions made with ROVs and divers (Tabl. 1). It is about services offered by *Mobil-FSSL Diverless Intervention System* from *Heriot-Watt University*, Edinburg, UK, compared with the same services using professional divers, on a depth of 1000 meters:

- The replacement for some parts;

- Opening/ Closing a valve;

- Maintenance and tests for the pressure of a valve;

- Control of an installation for transporting raw oil trough pipes.

TABLE I.   PRICE FOR MAINTENANCE (DIVER/ ROV) [4]

Service		Diver	ROV
Unplanned Maintenance, working days	20	1.680.000 $	592.000 $
Money saved		1.088.000 $ SUA	
Unplanned		528.000 $	160.000 $

Maintenance, working days	2		
Money saved		**360.000 $**	

The first step is to identify and measure the reliability of all components integrated in its systems and after that to calculate and predict the reliability for the system. Based on this approach and having as objective a low price for production some measures are needed at the stage of system design. Some examples of these measures are: redundancy (the multiplication of vital elements in terms of increasing the reliability of the system), reliable circuit design (one example is the design of circuit which are able to function in hostile conditions like, vibration, depth limit) and component selection.

The main elements we have to examine in terms of solving some critical design problems of ROV, are as following:

- The placement of electronic modules so that they don't interfere;

- The external housing must withstand high pressures;

- The penetrators which allow cables from motors to reach the electronics must also withstand high pressures;

- Water must not infiltrate in the housing if the cable gets cut.

- The unit must check for signal loss and have a backup program if it happens.

III. METHODOLOGY

The underwater monitoring system is resulting feedback previous research presented at [1] made by our team, with impact in the field of naval electronics. Techniques and methods of building underwater systems working under outdated [2, 3]. The main problems that occur in this type of systems are how encapsulation of electronics and data and image transmission. The project idea is to provide a simple and inexpensive system to replace divers work and also to send a real information about the marine environment.

Fig. 2. The first enclosure made for testing the system

The ROV (Remotely Operated Vehicle) system presented in this article is an underwater remotely operated vehicle. ROV a tethered mobile underwater device. This is connected to a host system and a PC to send by cables the information from the depth. Monitoring the system sent on a mission into water is achieved by means of a workstation at the surface. The station at the surface receives the information required and the possible warnings from the ROV. The main information sent from the workstation are: coordinates, movement direction, acceleration, the type of the mission (Fig.2, Fig.3).

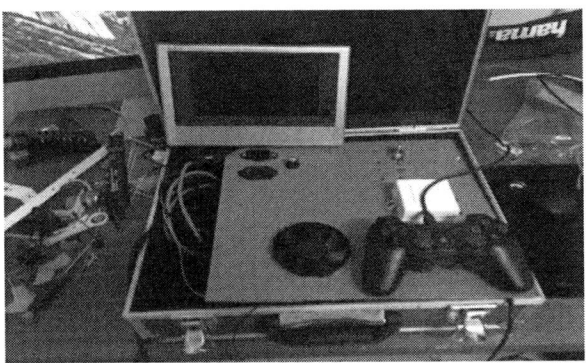

Fig. 3. The Ground Control Station

A. Abreviations

- ROV = Remotely Operated Vehicle
- GCS = Ground Control Station
- IMU = Inertial Measuring Unit
- VI = Visual Interface
- GPS = Global positioning system
- NMEA = National Marine Electronics Association, (standard for GPS coordinates)

B. The component parts of an ROV

The underwater ROV is made from a sum of basic subassemblies, with a well-defined role in the functioning and those components determine the performances and characteristics of the vehicle (Fig.4). Those are:

- The frame –compose by bytes

- Floaters/ Ballast

- Propulsion System

- Navigation and Positioning System

- Sensors

- Tools (i.e.: grippers, brush, drill)

- Ground Control Station

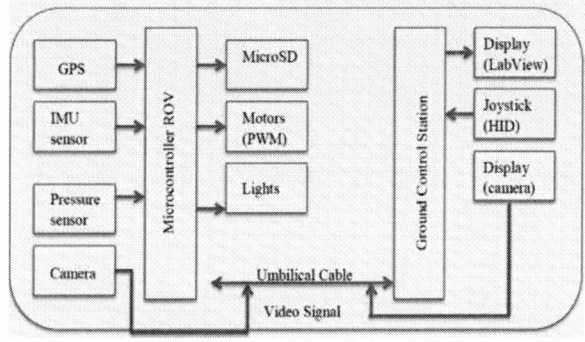

Fig. 4. The bloc diagram of underwater system described in this paper

C. Frame, floaters, enclosure:

The frame, floaters and enclosure are made from PVC with a wall thickness of 1 mm.

According to the simulations made on computer, the main body will withstand a depth of almost 10 m in sea water (Fig. 5, 6).

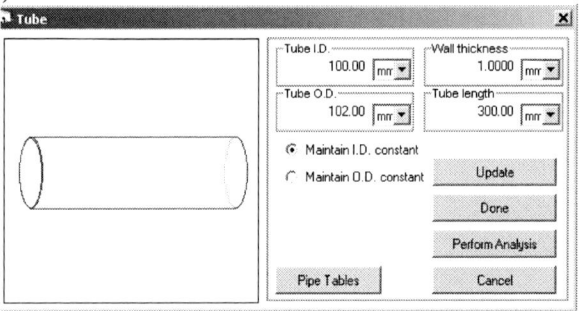

Fig. 5. The dimensions of the enclosure

D. Propulsion system

Depending on the necessity of maneuverability of the unit, the number and placement of the motors is chosen. Different placements (Fig. 6) are used for different task, a large number of motors means more degrees of freedom but also the power consumption increases.

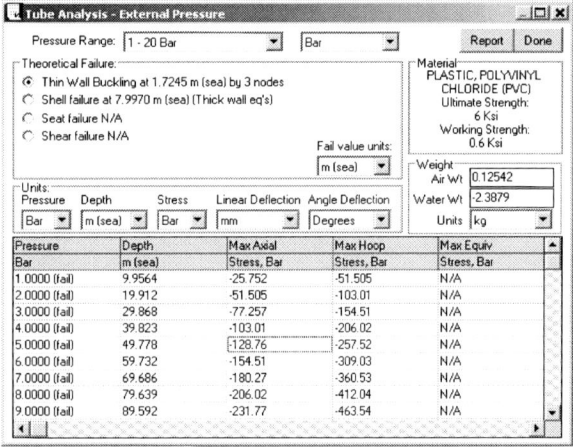

Fig. 6. The ratings of the enclosure made using "Under-pressure" software.

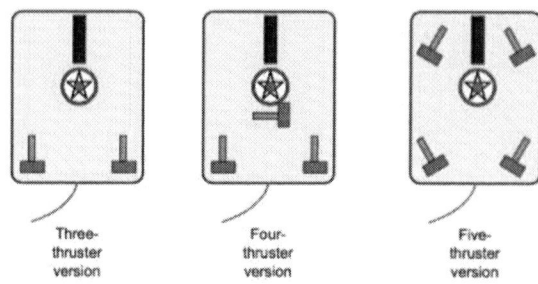

Fig. 7. Different placements for the motors.

E. Navigation and Positioning System; Sensors

Working remotely, the user needs to know the exact location and orientation of the ROV. This necessity is achieved by means of a IMU sensor and GPS module. The GPS gets the **NMEA** coordinates and using the reading from IMU, it can calculate the traveling course and the position of the unit, even if it is submerged.

F. Ground Control Station

Consists of an Arduino Mega board that receives data from the ROV and parses it, then sends it to a PC with LabView to be displayed in a custom made VI that show the orientation, depth and pressure helping the pilot maneuvering the unit.

The GCS also has to read the commands from a joystick and concatenate it in an array to send it down to the ROV.

IV. RESULTS

The first step in the development of this system was the development of a serial communication protocol (i.e. RS232, RS484, TCP/IP) which will enable the exchange of data streams between the ROV and the GCS. According to our previous research [1], we have chose to send data formatted in two arrays: one is the data sent from the ROV and the second is the data sent from the GCS. Each array consists of variables (i.e. values from the orientation on the xOyz axes, depth, joystick's axes positions, etc.) and then, the receiving microcontroller parses the array in integer values used for moving the robot or for the LabView interface.

The information sent by IMU sensor from the ROV are presented on the LabView interface (Fig. 8, Fig. 9, Fig. 10). The location of the robot is identified, the coordinates transmitted to the workstation being considered the 0 point where the movement into the water begins. The movement of the ROV is achieved by means of the brushless motors, the direction and the acceleration being transmitted by the joystick from the workstation.

The ROV communicates with the GCS from Arduino Mega board usind RS232 serial communication. The GCS's ARDUINO sends the data to be displayed on the PC using USB communication. The LabView interface show on the image and graphic the information about the ROV coordinated. The data are filtered and saved in a buffer on Arduino and then is transferred to the PC.

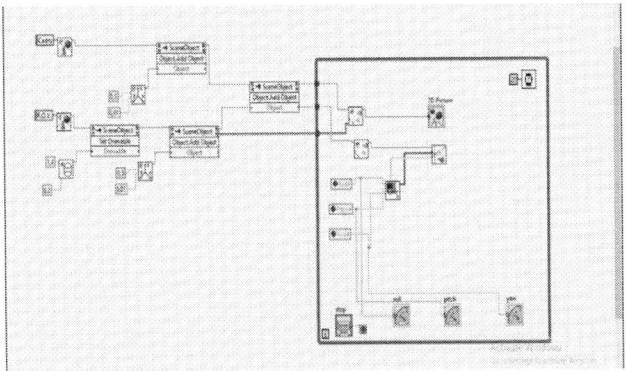

Fig. 8. Block diagram from LabView for ROV display

The LabView Block Diagram has the sub-vi for named get-data.vi in which data of IMU sensors are received and converted. The data protocol are used to sincronise the data in frame.

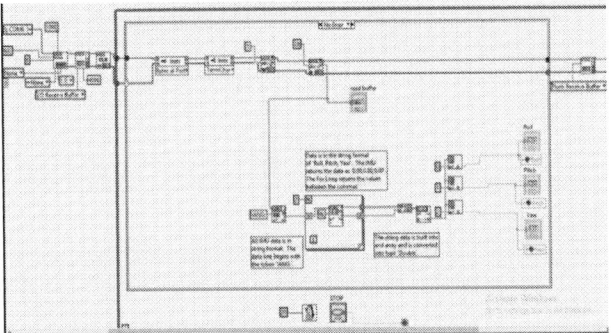

Fig.9. Block diagram from LabView to get data from Arduino

The data get from the IMU sensor are build in array and converted in double precision.

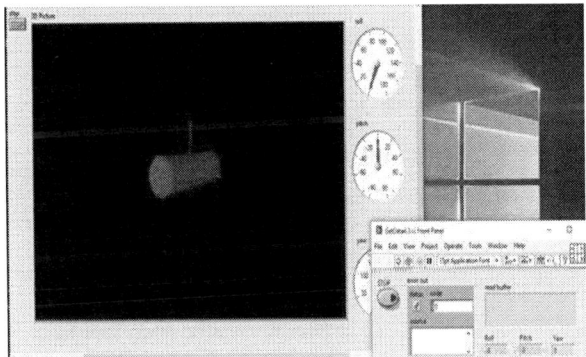

Fig. 10 Front panel of the user interface

The position in 3 D position and the angle of the ROV is displied in Fig 10.

V. CONCLUSIONS

During our research, we have managed to build a prototype for filming underwater within depths of 10 meters and knowing the orientation and the first position of the robot, the pilot does not have to watch over the ROV, going to search into a ship wreck or underwater cave.

The next steps into developing this unit are to change the material of which the enclosure is made and upgrade the motors used for movement underwater and to make the unit autonomous using FPGAs and Video Processing.

ACKNOWLEDGMENT

The authors would like to express their gratitude to AFCEA International for the financial support, and also to iREQUEST Concept Store for the support in the development of the ROV.

REFERENCES

[1] I. Lazar, A. Ghilezan, M. Hnatiuc, "Development of Tools and Techniques for Monitoring Underwater Artifacts", presented on ATOM-N 2016 Conferences, will be Published in SPIE volume.

[2] Ueno, M., Nimura, T., Ando, H., Maeda, K., Tamura, K., "On the descending motion of a deep-sea robot," Control Engineering Practice, Elsevier Ltd, 16(4), 446-456 (2008), ISSN: 0967-0661

[3] Donghwa, L., Gonyop, K., Donghoon, K., Hyun, M., Hyun-Taek, C., "Vision-based object detection and tracking for autonomous navigation of underwater robots", Ocean Engineering, PERGAMON-ELSEVIER SCIENCE LTD, 48, 59-68 (2012), ISSN: 0029-8018 I.S. Jacobs and C.P. Bean, "Fine particles, thin films and exchange anisotropy," in Magnetism, vol. III, G.T. Rado and H. Suhl, Eds. New York: Academic, 1963, pp. 271-350.

[4] Popa D., "Vehicule subacvatice telecomandate", ISBN: 973-7872-19-3.

[5] Robert D. Christ, Robert L. Werlni, "The ROV Manual, Second Edition", ISBN: 978-0-08-098288-5

[6] Nakajoh H., Hiroyuki O., et.al., "Development of work class ROV applied for submarine resource exploration in JAMSTEC", IEEE Explorer, ISBN 978-1-4577-2091-8/12, 2011

Real Time System for Extraction and Playback of an Instrumental Sound

Laurentiu Mihai Ionescu, Alin Gheorghita Mazare,
Ioan Lita, Nadia Belu

Dep. Electronics, Computers and Electrical Engineering
Faculty of Electronics, Communications and Computers,
University of Pitesti
Pitesti, Romania
ioan.lita@upit.ro

Adrian-Ioan Lita

Politehnica University of Bucharest
Bucharest, Romania

Abstract— **This paper presents an acquisition system for sound recording, typical spectral component identification and extraction of particular instruments from a soundtrack, based on each instrument's spectral components. This system is also capable of sound reproduction (playback) and its main feature consists of minimizing the delay between the recorded sound and the playback sound below 100ms. To achieve all these objectives, the proposed system consists of several components which will operate in parallel: the acquisition module which outputs the digital waveform of the sound and its FFT transform used for spectral representation; the comparison module, composed of associative modules which compares the closeness of the sound sample to certain templates which contain spectral representation of pre-recorded instruments, used to identify whether certain instrument is present; the spectral form extraction module which isolates a certain instrument from the original sound-mix, based on filtering the spectral components that don't belong to the respective instrument; the last module is the inverse FFT module, which converts the sound from spectral domain back to time domain. With the exception of the microphone, the analog-to-digital converter (ADC) and the digital-to-analog converter (DAC) capable of outputting the sound, all other operations are done by an SoC integrated system (Xilinx Zynq 7000).**

Keywords— *System on Chip (SoC), real time, sound recognition, sound playback*

I. INTRODUCTION

Analysis and processing of one-dimensional signals (such as sound) is done most often by software processing tools since the variety of algorithms allows flexibility and continuous improvement through development. This usually translates into an off-line sound analysis: the sound is recorded and at a later time is processed by the computer. There are several software solutions designed for sound recognition, but along with existing instruments new techniques for improving sound recognition algorithms are being researched, with applications in a large number of areas [1].

A class of applications require though real-time automated sound recognition. For applications requiring on-line sound analysis in order to identify sound patterns in real-time, hardware analysis tools are needed to be integrated in the area

where the sound is captured, resulting a „sound sensing" solution based on microsystems with classic microprocessor (e.g. ARM) [2] or digital signal processors [3]. These solutions have some limitations related to sequential programming and complexity of the calculations. Another solution can be using digital structures from reconfigurable circuits, or using certain solutions which combine many types of circuits, some for spectral processing, and others for identifying certain models [4].

Dedicated algorithms for sound sequences recognition already exist. In principle, they are divided into three main classes: Gaussian Mixture Models, Support Vector Machines and Deep Neural Networks [5]. Each of these imply using a certain amount of resources which must be taken into account when implementation using hardware structures is desired. The overall cost of the system must be taken into account as well.

In the field of analyzing the sound outputted by musical instruments there are already solutions which use classic microprocessor architectures [6]. Their purpose is to determine either the sound quality starting from component isolation and analysis [7] or the overall accuracy of the melody by comparison of certain points to different expected template patterns [8].

This article presents an alternative processing solution capable to identify and re-play certain sound sequences, using both DSP cores, as well as hardware processing structures – a hybrid structure, while keeping in mind a reduced hardware cost. This is possible due to the integration of many components – DSP modules, microprocessor and dedicated logic structures - into a single SoC (System on a Chip).

The overall presented structure uses digital signal processors (called here DSP modules) to calculate and play samples of the recorded sound's spectral representation, and hardware integrated modules which identify certain instruments that are desired to be extracted from the soundtrack. The experiments where done using soundtracks with more instruments and extracting the sound of only a certain instrument. The central component of the system is a System on a Chip – Zynq 7000 from Xilinx, which accommodates under the hood the following modules: a general purpose RISC processor (running Linux), which is used to select the chosen instrument to be extracted, DSP cores used for calculating FFT and inverse FFT, as well as an FPGA used for implementing the instrument template identifying modules.

Section II will present the overall system architecture, while Section III presents the experimental datasets and results.

II. SYSTEM ARCHITECTURE

The overall presented structure uses digital signal processors (called here DSP modules) to calculate and play samples of the recorded sound's spectral representation, and hardware integrated modules which identify certain instruments that are desired to be extracted from the soundtrack. The general architecture is presented in the figure below:

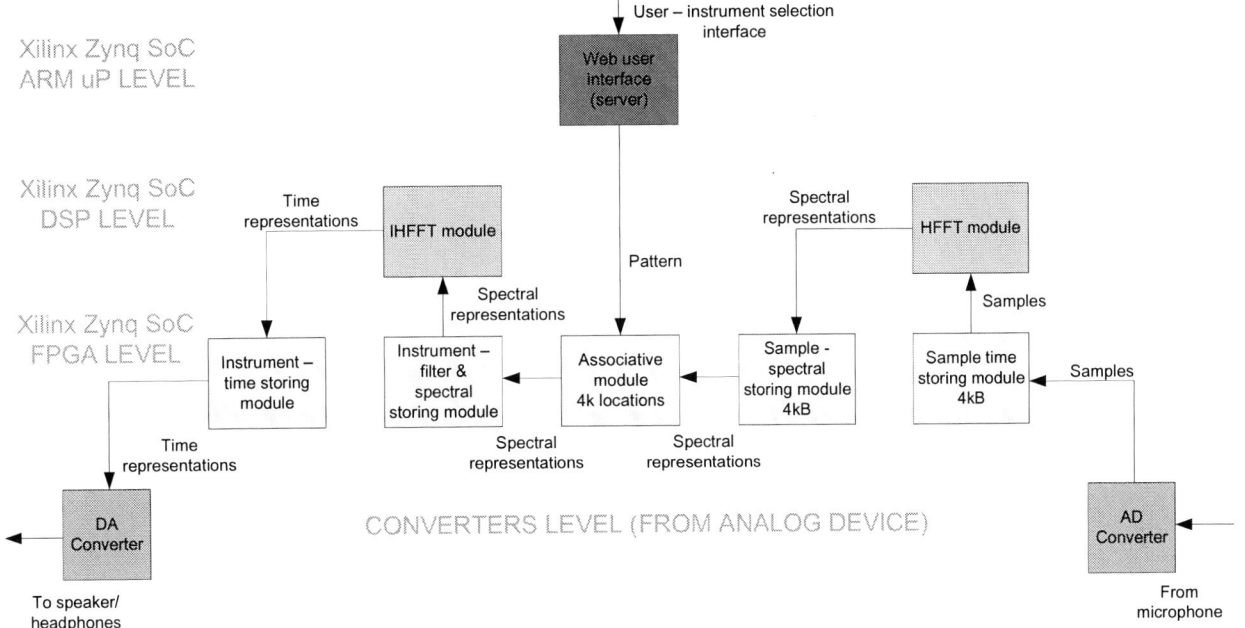

Fig. 1. System block diagram (multilevel hardware architecture)

As figure 1 depicts, the architecture was structured on 4 levels. The lowest level includes the components outside the SoC: the AD and DA converters. Analog Devices manufactures both converters used in this paper. The AD converter is SSM2603, which samples sound at a frequency of 96 KHz. The following levels are build inside the SoC Zynq 7000(XC7Z010):

The first level inside the SoC contains the custom hardware structures. These were implemented using the FPGA (based on Artrix 7 architecture). The custom hardware structures are the automated storing blocks for both the digital samples of sound over time and spectral representation, as well as the module responsible of recognizing patterns of sound.

In order to optimize the Fourier transformer module, memories are organized as 4K x 16. Synchronous high-speed SRAM memory was configured to work in bi-port mode, so that the blocks which need to store data and write the memory (such as the block which reads the AD converter) can work completely independent in terms of system frequency from the blocks which needs to read the memory (for example, the HFFT).

Another level implemented using the FPGA is the associative module. This uses an associative memory based on a Hopfield neural network. The main objectives are fast pattern detection on one side, and fast training. The associative memory is made with a network of comparators and sorting circuits.

The next level is represented by the DSP blocks. These are responsible with calculating the FFT and the IFFT. The DSP blocks already existing in the SoC XC7Z010 can process the transforms of the 4K x 16 samples stored in the internal memory.

The top level and the only level accessible to the user is the microprocessor level. The microprocessor is an ARM Cortex A9 which runs Linux operating system, on which a web based graphical interface application was developed to allow the user to select the musical instrument which is desired to be identified and isolated from the rest of the orchestra.

As figure 1 depicts, the architecture has a horizontal symmetry. The right side is responsible of acquiring and representing data, and the left side is responsible of filtering and outputting the signal. The central post is the instrument pattern recognition module, instrument selected through the web interface.

III. EXPERIMENTAL RESULTS

Experiments were conducted to identify, extract and playback desired instruments from a soundtrack, so that a sound specialist can monitor only a certain instrument at a time. The instrument can be extracted from any desired soundtrack. The experiments were conducted using three instruments: harpsichord, piano and trumpet, while the soundtrack also contains background noise (such as a theater at half break). The processing flow is depicted below:

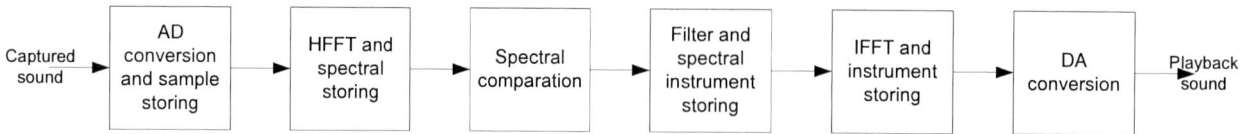

Fig. 2. Processing flow

A. Data set

The data set that was used in the experiments contains three studio recordings (with little to no noise) of the three instruments, from which spectral representation was extracted. The sample files were provided on Stanford University [9] website with instrument sound samples, and the spectral representation processing was done with MATLAB 2011 on a PC.

Using MATLAB, three other files were generated: each two instruments plus maximum 50 dB noise (coresponding to the noise level in a concert hall during break). The dataset consists of 6 files: 3 with original sound instruments and 3 with two instruments and noise mixed together.

B. Experimental results

When the experiments were conducted, three parameters were taken into consideration. The first is the response time to which a certain instrument is outputted as challenged by the input. Determination of the response time was conducted using the analysis and implementation tools of the Xilinx ISE Pack. The second parameter is related to the rate of success in detecting the instrument. The central component which identifies the spectrum of a certain instrument is, as shown above, the associative memory. This identifies the number of common points between the analyzed spectrum and the spectrum of a certain instrument. To experimentally visualize and determine the identification success rate, common spectral points between two instruments with different noise level were outputted. Common spectral points represents an undesired characteristic, which translates that on certain spectral zones two instruments sound similar and are hard to be separated one from another. The third parameter taken into consideration is the overall cost of the system.

The total processing time is maximum 100ms: this means that the playback sound will have a maximum delay of 0.1s from the original soundtrack. The processing time of each step of the flow is presented in the table below:

TABLE I.   PROCESSING TIMES FOR EACH MODULE

Step	Time	Step	Time
AD conv. & storing	42.7 ms	Filtering &storing	41 us
HFFT & storing	221 us	H – IFFT & store	250 us
Comparison	700 us	DA conv.	43 ms

Fig.3 Image with system while running

Along with the response time, another important parameter is the identification success rate. This highly depends on the associative memory modules, used for template matching. For soundtracks such as the one presented above, the success rate does not drop below 80%.

The total number of spectral points analyzed by the associative memory is 4096 (all spectral values stored). The analysis outputs the following common points between instruments in respect to the noise level power:

TABLE II.   COMMON POINTS BETWEEN INSTRUMENTS WITH DIFFERENT NOISE LEVELS

Noise level (W)	Trumpet-Piano	Trumpet-Harpsichord	Piano-Harpsichord
1.00E-07	76	2	0
3.00E-07	238	2	0
5.00E-07	420	2	0
1.00E-06	832	2	0
2.00E-06	1536	8	6

These points determine the detection probability of a certain instrument from a mix of two instruments, as follows:

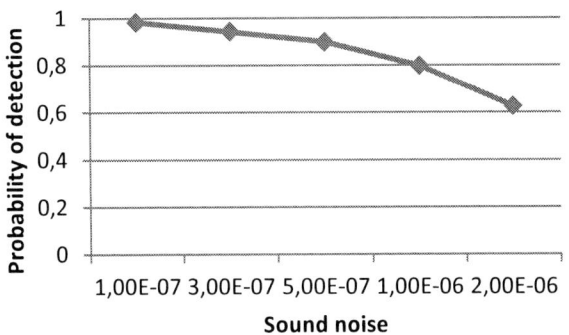

a) Trumpet – piano mix

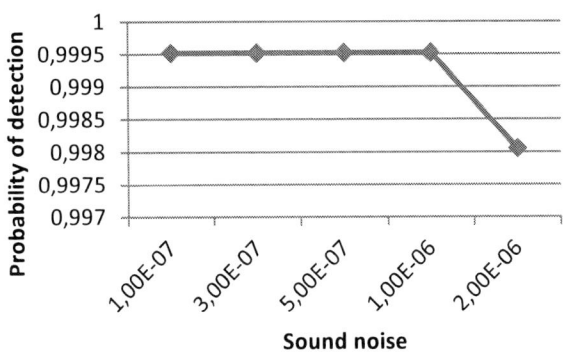

b) Trumpet- Harpsichord mix

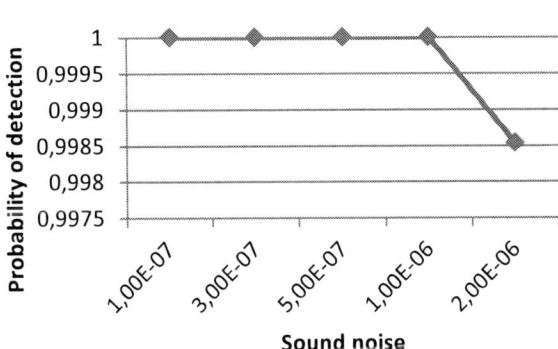

c) Piano- Harpsichord mix

Fig.4 Probability of detection for different instrument from a mixed sound

The third analyzed parameter refers to the overall cost of the solution. As figure 3 depicts, a Zynq 7000 system development kit provided by Digilent was used. The cost of such a system, which also include the AD and DA converters, as well as the SoC is 189$, which is a reduced cost for a high performance sound sensing system.

IV. CONCLUSIONS

The system was designed, implemented and experimented. It is a real time solution for determining the quality of music by separately reproducing certain musical instruments. The small response time allows audition in parallel with the orchestra. The performance of correctly detecting an instrument is very high, allowing high accuracy detection even with external noise (room noise), at a low price.

Further research directions include building a larger instrument library (both in quantity and quality) and conducting experiments using more complex sounds.

In a larger area, the system can be used for identifying certain sound spectrums in other areas, such as security and automotive.

ACKNOWLEDGMENT

The research that led to the results shown here has received funding from the project "Cost-Efficient Data Collection for Smart Grid and Revenue Assurance (CERA-SG)", ID: 77594, 2016-19, ERA-Net Smart Grids Plus.

REFERENCES

[1] D. Stowell, D. Giannoulis, E. Benetos, M. Lagrange, and M. D. Plumbley, "Detection and classification of acoustic scenes and events," IEEE Transactions on Multimedia, vol. 17, no. 10, pp. 1733–1746, 2015

[2] Tauhidur Rahman, Alexander T Adams, Mi Zhang, Erin Cherry, "BodyBeat: A Mobile System for Sensing Non-Speech Body Sounds", MobiSys'14, June 16–19, 2014, Bretton Woods, New Hampshire, USA, 2014

[3] Lianfu Han, Zhengguang Shen, Changfeng Fu, Chao Liu, "Design and Implementation of Sound Searching Robots in Wireless Sensor Networks", Sensors 2016, 16(9), 1550, 2016

[4] Siddharth Sigtia, Adam M. Stark , Sacha Krstulović , Mark D. Plumbley, "Automatic Environmental Sound Recognition: Performance Versus Computational Cost", IEEE/ACM Transactions on Audio, Speech, and Language Processing (Volume: 24, Issue: 11, 2016)

[5] Aris Tjahyanto, Diah Puspito Wulandari, Yoyon K. Suprapto, Mauridhi Hery Purnomo, "Gamelan instrument sound recognition using spectral and facial features of the first harmonic frequency", Acoustical Science and Technology, Vol. 36 (2015) No. 1 P 12-23

[6] T. Fujishima, „Realtime chord recognition of musical sound: A system using common lisp music", Proc. ICMC, pp. 464-467, 1999.

[7] O Romani Picas, H Parra Rodriguez, D Dabiri, „A Real-Time System for Measuring Sound Goodness in Instrumental Sounds", Society Convention 138, 2015

[8] A Arzt, A Widmer, „Real-time music tracking using multiple performances as a reference", Proc. of the International Society for Music, 2015

[9] https://ccrma.stanford.edu/~jos/pasp/Sound_Examples.html

Remote Communication Interface for Sound and Vibration Sensors

Daniel Alexandru Visan, Laurentiu Mihai Ionescu
Electronics, Communications, Computers Department
University of Pitesti
Pitesti, Romania
daniel.visan@upit.ro

Adrian Ioan Lita
Applied Electronics, Information Engineering Department
"Politehnica" University Bucharest
Bucharest, Romania
ioan.lita@upit.ro

Abstract — **In this paper are presented the results regarding the implementation of a communication interface dedicated for conditioning and remote transmission of measurement signals acquired from sound and vibration sensors. The interface contains two parts, an analog section, responsible with long distance signal transmission and a digital section that realize demodulation and analog-to-digital conversion of the received signal. The analog part of the interface uses the frequency modulation technique that is well known for its ability to operate in environments characterized by high levels of electrical perturbations, even when are realized high speed data transmissions. The acquisition of signals from sensors is realized with a resolution of 16 bits with a maximum speed of 200kS/s. In order to ensure the compatibility with various communication media and different types of sensors, the proposed circuit has adjustable carrier frequency in the range [0,01 − 0,5] MHz and offers also the possibility to select the carrier signal waveforms: square or sinusoidal wave. In addition, for increased versatility, the communication interface combines two alternative configurations, one based on phase-locked-loop (PLL) structure and the second based on quadrature oscillator with analogue multipliers. Both configurations were thoroughly tested for evaluating the performances of the proposed design. In comparison with other structures, the remote communication interface proposed in this work revealed good stability, accuracy and reliability. Also, the robustness and simplicity of the interface due to the use of frequency modulation technique represent other significant advantages.**

Keywords — communication interface, sensor, signal conditioning, frequency modulation.

I. INTRODUCTION

In almost every practical application the transfer of information from the output of the sensor to the measurement system represents a critical step in the implementation process because it implies very small signal levels and the interface has an increased sensitivity to interferences. These aspects are even more pronounced in the case of remote configurations of data acquisition and measurement systems that operate in harsh industrial environments [1].

In this context this paper approach the problem of implementing a reliable remote communication interface dedicated for sound and vibration sensors. The proposed design was divided in two parts, one responsible with

conditioning and long distance transmission of signals from sensors and the second part used for demodulation and analog-to-digital conversion of signals for facilitating the interconnection of the proposed system with a PC or other digital signal processing unit.

In the proposed configuration, because the communication is realized by using frequency modulation, no quantization errors are introduced, timing equipments are unnecessary and the communication requires less frequency bandwidth compared with digital transmission. Also the interface has a simplified structure, requires less conditioning of the input signals and this combined advantages lead to lower implementation costs of the system.

II. THE PHASE LOCKED-LOOP STRUCTURE AND FREQUENCY MODULATION TECHNIQUE

The remote communication interface uses a phase-locked-loop (PLL) structure combined with a quadrature oscillator based on analogue multipliers. The frequency modulated signal $s_{MF}(t)$ can be expressed in generally as:

$$s_{MF}(t) = A_0 \cdot \cos\left[\omega_c \cdot t + k_\omega \cdot \int_0^t u_m(t) \cdot dt + \varphi_c\right] \quad (1)$$

where $u_m(t)$ is the information signal, ω_c, φ_c, A_0 are the angular velocity, initial phase and amplitude of the carrier signal.

One of the main parameter that characterizes a transmission based on frequency modulation is the modulation index, β, defined as the ratio between frequency deviation, Δf, and the modulation frequency, f_m.

The expression of modulation index is:

$$\beta = \frac{\Delta f}{f_m} = \frac{k_f \cdot A_m}{f_m} \quad (2)$$

where k_f is the modulator's sensibility and A_m is the amplitude of the input signal applied to the modulator.

The phase locked-loop structure is a universal module frequently used for signal processing and synchronization purposes in data transmission equipments.

2016 IEEE 22nd International Symposium for Design and Technology in Electronic Packaging (SIITME)

In principle this type of circuit comprises a phase comparator that controls the frequency of a voltage controlled oscillator (VCO). The structure is characterized by an free oscillation frequency, f_{osc}, that is generated by VCO when no signal is applied to the input of the circuit, $u_{in}(t) = 0$.

The phase detector is sensitive to the phase differences between the input signal and the feedback oscillation generated by VCO. When the local generated signal and input oscillation have different frequencies a proportional voltage is applied to the input of the VCO.

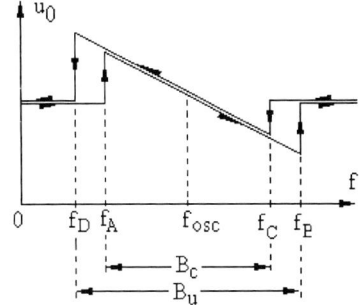

Fig. 1.   The basic structure of a phase locked-loop circuit.

Fig. 2.   The static transfer characteristic of a phase locked-loop circuit, considering the error as a function of frequency.

This voltage causes the variation of the VCO's frequency until it becomes equal with the frequency of the input signal. As can be seen from Fig. 2, the phase locked loop operates as a frequency control system. It is characterized by a lock range, B_u, and a capture range, B_c. Overcoming the lock range lead the system out of the synchronization and resynchronization is possible only if the signal enter in the capture range.

III.   THE PRESENTATION OF THE REMOTE COMMUNICATION INTERFACE

As was briefly mentioned before, the remote communication interface proposed in this paper is composed by two parts: the first part is based on analog signal processing and is intended for remotely connect the sound and vibration sensors to the data acquisition and measurement systems using frequency modulated signals transferred through metallic lines. The second part of the interface is responsible with demodulation and analog-to-digital conversion of the received signal from sensor. This section of the interface is necessary for proper acquisition and processing of the measured signals using a PC with a dedicated software application. In our design the modulated signal transmitted by the remote circuit

is demodulated and acquired with and NI USB 6212 data acquisition board characterized by a 16 bit resolution and a speed of conversion of 200kS/s. The signal transmission between vibration sensor and data acquisition is realized through frequency modulation. The carrier frequency is adjustable in the range of [0,01 – 0,5] MHz and the system offers also the possibility to select the carrier signal waveforms: square or sinusoidal wave. The main functional blocks of the proposed design are illustrated in Fig. 3. The transmission part the system uses a phase-locked-loop structure combined with a quadrature oscillator based on analogue multipliers. The selection of the transmission configuration is realized in the initial setup of the equipment and must be correlated with the demodulation side of the system. The quadrature oscillator structure is used for transferring the information from sensor to the data acquisition board using sinusoidal carrier. The voltage controlled oscillator, VCO_1, is connected to the transmission line when it is necessary the transfer of the signal from sensor to data acquisition board using a square wave carrier. For testing purposes, when the interface is operated on very short distances, it is also possible to transfer the signal without modulation [2], [3].

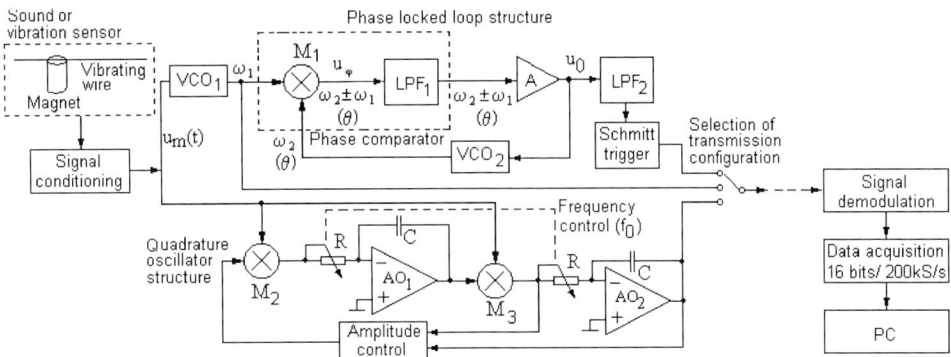

Fig. 3. The simplified block diagram of the remote communication interface for sound and vibration sensors.

IV. IMPLEMENTATION AND RESULTS

The performances of the analogue transmission part of the remote communication interface were evaluated through intensively tests regarding the accuracy of the transferred signals and the stability and reliability of the realized circuits.

In the Fig. 4 is presented the electric diagram of the frequency modulator based on quadrature oscillator structure.

The AD633 analog multipliers and TL082 operational amplifiers are the main active components used in the proposed design. The behavior of the modulator is deliberated guided, through the electronic switch represented by Zener diodes D_1 and D_2, towards a dumped or not dumped oscillating regime. Through this way is controlled and maintained a constant level of the amplitude of the modulated signal [4].

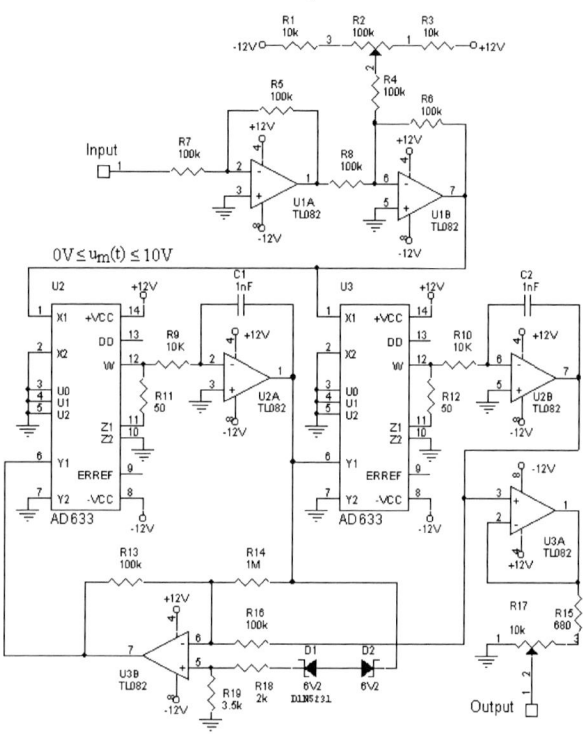

Fig. 4. The electronic diagram of the frequency modulator based on quadrature oscillator structure.

The frequency of the output signal for the modulator structure presented in Fig. 4 is controlled by the amplitude of the input signal. Some sample results obtained through simulation in Orcad Pspice are depicted in Fig. 5, 6 and 7. From these pictures we can remark the proper operation of the circuit.

For practical implementation, the multipliers used in the schematic must have a good precision and a relatively high slew rate and bandwidth. In the practical realization of the quadrature oscillator were used AD633 analog multipliers and TL082 operational amplifiers. As phase-locked-loop was chose LM565 general purpose circuit.

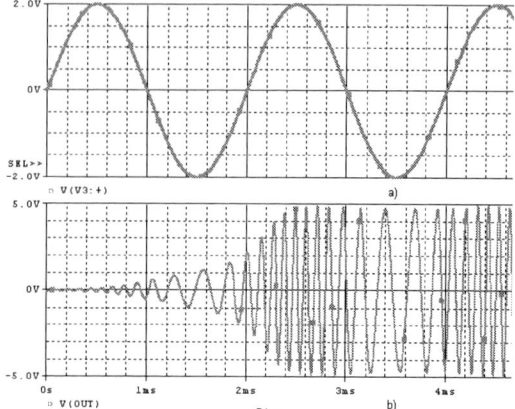

Fig. 5. a) The input test signal: $A = 2V_{pp}$, $f = 500Hz$; b) The output of the frequency modulator based on quadrature oscillator structure.

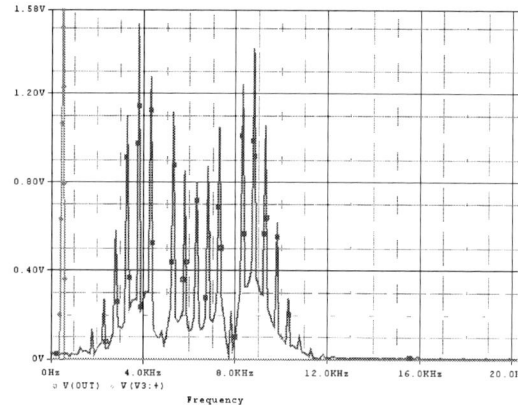

Fig. 6. The spectral analysis of the signal obtained at the output of the frequency modulator based on quadrature oscillator structure.

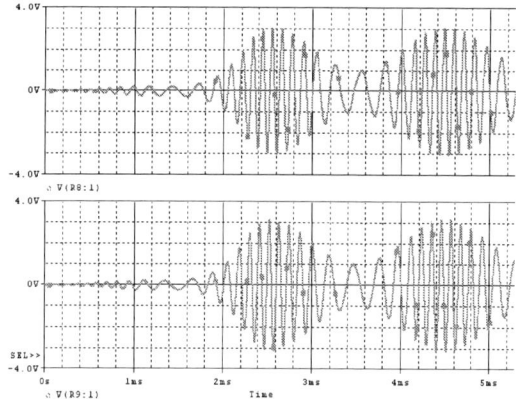

Fig. 7. The intermediate signals visualized at the output of the AD633 multipliers that compose the frequency modulator based on quadrature oscillator structure.

The four quadrants multiplier AD633 achieve a cumulated error of less than 2% considering signals in full scale range of operation. Also, the multiplier was choose because ensure relatively high operation frequency of 1 MHz and an acceptable slew rate of $20\,\text{V}/\mu\text{s}$.

The phase locked-loop circuit LM565 in combination with active filters based on TL 082 operational amplifiers was used as main elements in the implementation of the second configuration of frequency modulator for remote transmission of signals from sensors. The LM 565 has an frequency stability of the internal VCO of around $200\text{ppm}/^\text{O}\text{C}$ and a 0,2% linearity of the demodulated output. In the initial tests the PLL was the main limiting factor of the maximum frequency of the signal that can be accepted and processed from the vibration sensors but this aspect can be improved by choosing a more performing circuit [3].

In Fig. 8 can be observed the implementations of the electronic boards for the quadrature oscillator configuration and phase-locked-loop configuration, respectively.

Also, in Fig. 9 can be seen few sample results visualized with an oscilloscope connected to the input and output ports of the two implemented boards, one based on phase-locked-loop structure and the second based on quadrature oscillator with analogue multipliers. In both visualizations the oscilloscope was operated with the following deflection settings: vertical 2V/div. and horizontal $2\mu\text{s}/\text{div}$. As can be seen from this pictures waveform at the outputs of the implemented circuits are in accordance with the simulations results and prove the correct operation of the proposed interface. A further improvement in performances and a higher integration degree doubled by a increased reconfigurability of the structures can be obtained by using field programmable analog array circuits.

Fig. 8. The implementation of the communication interface using quadrature oscillator configuration (top) and phase-locked-loop configuration (bottom).

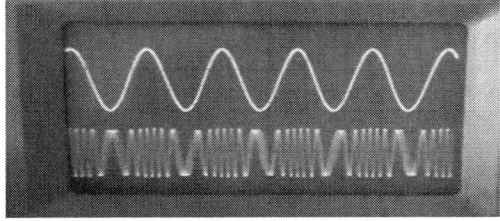

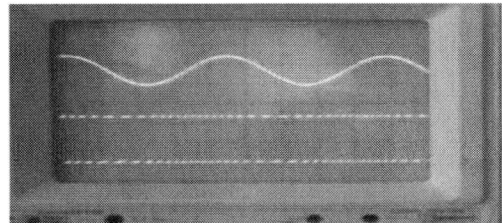

Fig. 9. Input and output waveforms for quadrature oscillator configuration (top) and phase-locked-loop configuration (bottom) The oscilloscope deflection settings: vertical 2V/div. and horizontal $2\mu\text{s}/\text{div}$.).

V.   CONCLUSIONS

Due to the very small signal levels and the increased sensitivity to interferences, the acquisition of information from the sensors represents an important problem in the implementation of accurate measurement systems.

In the proposed configuration, no quantization errors are introduced, timing is unnecessary and the communication requires less frequency bandwidth compared with digital transmission, because the communication is realized by using frequency modulation.

Also the proposed remote communication interface has a simplified structure, requires less conditioning of the input signals and this combined advantages lead to lower implementation costs of the system.

REFERENCES

[1]   L. Swathy, Lizy Abraham, "Analysis of vibration and acoustic sensors for machine health monitoring and its wireless implementation for low cost space applications", First International Conference on Computational Systems and Communications (ICCSC), pp. 80 – 85, 2014.

[2]   Andrei Drumea, Marian Blejan, and Ciprian Ionescu, "Differential inductive displacement sensor with integrated electronics and infrared communication capabilities," Advanced Topics in Optoelectronics, Microelectronics, and Nanotechnologies ATOM-N2012, pp. 841116-841116-7, November 2012.

[3]   Roberto Mugavero, Giovanni Saggio, Valentina Sabato, Mariano Bizzarri, "The multisensory integrated modules for training", 2014 International Carnahan Conference on Security Technology (ICCST), pp. 1 – 6, 2014.

[4]   Daniel Alexandru Visan, Ioan Lita, Marian Raducu, "FSK transmission circuit for remote sensors integration in distributed measurement systems", 2012 IEEE 18th International Symposium for Design and Technology in Electronic Packaging (SIITME), 25 – 28 Oct., pp. 237 – 240, 2012

FPGA-enabled Hardware Multitasking Applications in Energy Harvesting Laboratories

Octavian M. Machidon, Petru A. Cotfas, Daniel T. Cotfas

Department of Electronics and Computers
Transylvania University of Brasov
Brasov, Romania
octavian.machidon@unitbv.ro, pcotfas@unitbv.ro, dtcotfas@unitbv.ro

Abstract— **This work describes the technical details of a novel methodology which enables the parallel deployment and execution of hardware applications onto the same FPGA chip using dynamic partial reconfiguration. Our approach, which makes use of the PCAP (Processor Configuration Access Port) available on the latest generation of Xilinx Zynq Systems-on-Chip, allows the processing system to operate a fast partial reconfiguration of the programmable logic with little logic overhead. The advantages brought by this technique, like scalability, high performance and dynamic management of the programmable logic are being put to use in implementing laboratory works involving running the mathematical models of energy harvesting sources. This allows the same FPGA development board to be used by several student at the same time for running their applications. Thus, this is part of a larger research effort in which we aim to develop remote online hardware design laboratories with hardware multitasking support.**

Keywords—reconfigurable hardware; photovoltaic cell; simulation; hands-on laboratory.

I. INTRODUCTION

Hands-on laboratory work is a very important learning activity, especially in technical domains such as electronic engineering, where practical experiences play a key part in the learning outcome. There are many challenges that arise when trying to offer such hands-on activities to students, mostly related to achieving an efficient sharing and management of the hardware resources (which are usually limited) and also reaching a cost-effective approach when making available the latest technologies (which are usually expensive) for multi-student practice.

Some implementations [1] are based on a time-multiplexing approach, in which the same hardware device is being used by multiple users sequentially over a period of time, with a single user having total access to the platform at any given moment. However this approach lacks efficiency, since it usually limits either the time slots each student is being granted practice access, or the total number of students that can gain practical experience during a laboratory class.

A more recent approach [2] comes from the area of reconfigurable computing, which has become a very popular technology in the recent years with lots of application that leverage the adaptability and versatility that it brings on. The advances in this field also foster the use of the latest advances

in FPGA dynamic partial reconfiguration (DPR), a technology made available in the high-end FPGAs only recently that enhances the flexibility of these devices by allowing different modules running in the programmable logic to be swapped in and out during runtime, while the rest of the system operates undisturbed (as depicted in Fig. 1).

A DPR-enabled design is basically divided in two components: the static logic (which is loaded once at power-up and cannot be altered during runtime) and the reconfigurable logic – which is represented by one or multiple reconfigurable regions or partitions. A reconfigurable region can be reconfigured at any time during the circuit's operation, by transferring a partial bitstream file that describes a hardware module through a dedicated configuration port.

The solution described in this paper is based on this research direction and proposes an improvement with regard to the resource management in traditional hands-on laboratories, in the sense that, by making use of the dynamic partial reconfiguration feature of the FPGAs, the same hardware device can be used for the practice of multiple students in the same time.

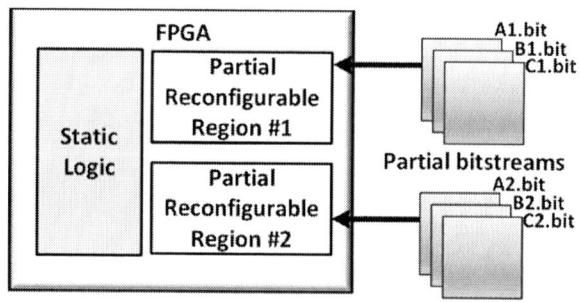

Fig. 1. Example of partitioning the FPGA in two partial reconfigurable regions

The FPGA accommodates multiple designs operating in parallel and independent from one another, thus achieving a form of hardware multitasking and consequently a more efficient use of the lab's hardware resources, contributing to exposing practical experiences to a wider range of students.

The students can deploy their digital designs on the FPGA board and communicate with them (write and read data) through a serial terminal running on a local computer that is connected also through Ethernet with the board.

The laboratory applications intended to benefit from our approach target Hardware-in-the-Loop (HiL) simulations of energy harvesting sources and also hardware-based design of mathematical models and algorithms used in the study of thermoelectric generator (TEG) or photovoltaic panels (PV).

II. SYSTEM ARCHITECTURE AND IMPLEMENTATION

For our work, we have chosen the Xilinx Zynq 7000 AP (All-Programmable) SoC (System-on-Chip) (Fig. 2) which integrates both a processing system (PS) (operating on the dual-core ARM A9 processor) and reconfigurable logic.

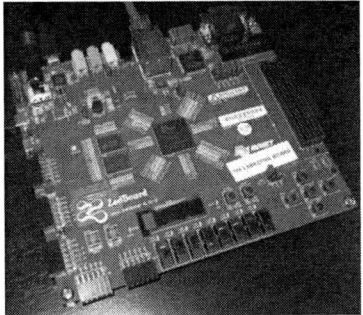

Fig. 2. Xilinx Zynq 7000 AP SoC

This synergy between hardware and software configurable resources provides high performance and versatility. An important asset available on the Zynq SoC is the possibility for the processing system to fully or partially reconfigure the programmable logic, this being one of the reasons for choosing this type of development platform.

We have developed an embedded C application running on the PS that is responsible with the operation of the entire system, including the communication sub-system via the Ethernet interface with a local computer for data transfer and complete or partial reconfiguration of the programmable logic.

A. Hardware architecture

The Zynq SoC's Processing System has been configured to operate at 667MHz on one of the ARM Cortex-A9 cores. Several peripherals have been integrated and instantiated:

- DDR controller (for interfacing the on-board DDR3 memory chip)

- Ethernet (for implementing the communication interface with the local computer allowing for configuration and data files to be transferred)

- UART (enabling a local terminal for debug and control).

These sub-systems are interconnected via the AXI (Advanced eXtensible Interface) protocol.

For our implementation we have used the PCAP (Processor Configuration Access Port), which enables the entire reconfiguration process to be handled by the PS, thus reducing the overhead and latency.

The Zynq's programmable logic is subject to a complete configuration only at startup (when the board is powered on).

The partial reconfiguration of any of the reconfigurable modules can be performed at any time during the device's operation, and it was implemented using a dedicated C method in the embedded application.

For our implementation we have set up two partial reconfigurable regions inside the Zynq's programmable logic (see Fig. 3), each being instantiated during the synthesis of the full bitstream with an instance of a custom created Verilog peripheral. The two PR regions and the peripheral have been designed as so to be able to accommodate the most common implementations that students will encounter during laboratory classes.

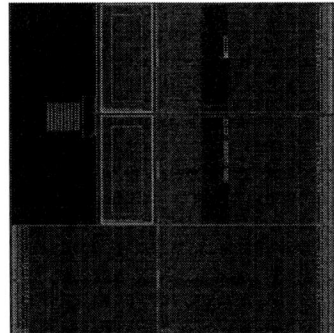

Fig. 3. Zynq layout with two PR regions

Also, a mechanism for individual data transfer to and from each reconfigurable module has been set up. This is based on a set of eight 64-bit software registers (4 input/4 output). These registers are available for read/write both to the internal logic of each reconfigurable module, and also to the embedded C application. Consequently, the C application further exposes these registers to the users by the means of the UART communication interface.

B. Software application

The embedded C application running on the Zynq's processing system is implementing four main functionalities:

- Configuring and handling a local file system

- Managing the device's Ethernet network communication

- Being in charge of the UART communication with the local computer

- Managing the entire partial reconfiguration process

A local 128MB memory-resident file system (MFS) has been implemented for storing the configuration files and also user I/O data. This file system, deployed using the Xilinx LibXil MFS component, is accessible from the C application by the means of specific function calls.

Another key part of the system is its network communication capabilities, based on the local Zynq on-board Gigabit Ethernet controller. Using the TCP/IP LwIP (Lightweight IP) Stack - dedicated for embedded networking [4] - we implemented on the Zynq a TFTP (Trivial File Transfer Protocol) Server.

2016 IEEE 22nd International Symposium for Design and Technology in Electronic Packaging (SIITME)

The server's main functionality is uploading partial bitstream configuration files from a computer to the local MFS on the Zynq device. This server has been deployed in an event-driven RAW API mode, using specific LwIP commands to set up callback functions that handle events (accepting connection, reading and writing data).

As mentioned above, the embedded application also handles the partial reconfiguration process. A dedicated method is responsible for initializing the XDcfg (Xilinx Device Configuration Interface) - enabling the PCAP and configuring the internal control register for partial reconfiguration mode -, clearing the specific DMA and PCAP interrupts, transferring the bitstream from the MFS (from a specific DDR3 memory address) to the programmable logic, and finally polling the previously mentioned interrupts to return when the transfer is complete.

C. Usage and applications

The students can thus deploy their designs on the FPGA and also communicate (transfer data, write and read values) through a serial terminal running on a local computer.

When designing and implementing their modules, students are being provided with design constraints regarding the number of available digital logic resources (like cells and bRAMs) and also with the top-level Verilog wrapper that is being instantiated in each reconfigurable peripheral of the PL. This last issue is important in order to ensure the correct overlay of the reconfigurable modules since they all must have the same I/O ports.

By respecting these constraints students are being guaranteed that their designs will easily be loaded onto the FPGA device.

The entire operation of the system is being controlled and monitored using a serial terminal (shown in Fig. 4) on the PC communicating via the UART connection with the Zynq board. The users are thus being provided a text-based interface that enables them to access the device for configuration and data transfer.

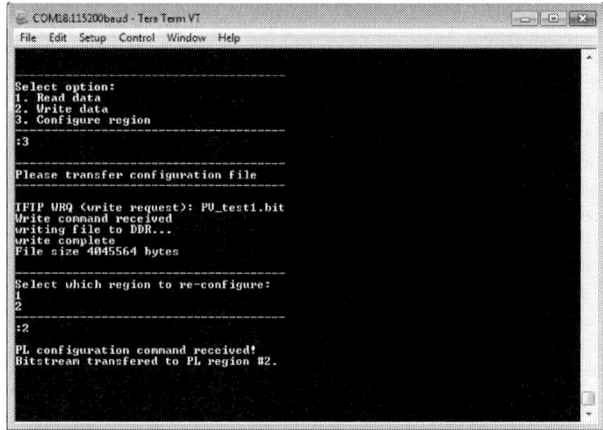

Fig. 4. Serial terminal running on the computer

This interface provides a menu with several options for the user to select. If the "Configure region" option is chosen, the system awaits for the transfer of a partial bitstream configuration file to the TFTP server. The user can upload the file to the Zynq board via the Ethernet connection from the local PC by the means of a TFTP client.

Following a partial bitstream upload, the user is further asked to specify which region is to be configured with that specific file. Following the re-configuration process, a status message is displayed on the console stating whether or not the process was successful.

Data transfer to and from each running module is achieved similarly, the user choosing the transfer type (read/write) and the specific region where his design is running.

This basic communication interface is extremely important in providing students with relevant feedback regarding both the successful configuration of the modules with their designs, but also, more importantly, enables them to perform experiments by sending sample data and receiving the processed results.

We have designed this hands-on laboratory setup with the intended goal for it to enhance applications like simulation and prototyping in the field of energy harvesting systems.

III. Results and Discussion

For the technical assessment of our solution we have configured the Zynq SoC with two partial reconfigurable regions. Further on, the models of thermoelectric generator (TEG) elements were developed and implemented in the Zynq SoC, being loaded as functional modules in both PR regions.

The TEG is an electric generator based on Seebeck effect. The TEG model is described by the following equations:

$$\alpha_{Sb} = 2 * V_m / \Delta T \qquad (1)$$

$$I_{SC} = \alpha_{Sb} \cdot \Delta T / R \qquad (2)$$

$$I = I_{SC} - U / R \qquad (3)$$

where α_{Sb} is the Seebeck coefficient of the TEG, V_m is the output voltage at the matched load (load resistance is equal with TEG internal resistance, $R_L=R$), $\Delta T = T_H - T_A$ is the temperature difference between the hot-side temperature and the cold-side temperature of the TEG, I_{SC} is the short-circuit current, I and U are the output current and voltage of the TEG. The calculus of the TEG parameters based on device datasheet are described in [5] and [6].

After this configuration, the system was validated from both functional and real-timing perspectives. The latter issue is particularly important for ensuring that the students' experience while working with this platform during laboratory class is based on a responsive - nearly real-time - feedback from the development boards with regard to module deployment and data transfer.

Hence, the timing analysis shows (see TABLE I) that the overhead induced by the DPR of the partial reconfigurable regions is minimal. This means that, on one side, the transfer of the partial configuration bitstream file from the local computer

978-1-5090-4446-7/16 $31.00 © 2016 IEEE   96   20-23 Oct 2016, Oradea, Romania

to the Zynq SoC via TFTP is fast enough (given the TFTP limitations, a transfer rate of aprox. 200Kb/s for a bitstream file size of 313kB).

On the other, the timing delays induced by performing DPR on the FPGA's reconfigurable partitions are very small – the process is perceived to happen virtually instantaneously by the users, total time required for the execution of the DPR method having an order of magnitude of a few milliseconds.

TABLE I

Operation	Duration
Partial bitstream transfer	1.62s
Loading partial bitstream	2.45ms

Following the timing validation of the system, we proceeded to the functional assessment. Using the loaded modules and performing various measurements using the serial console interface, the I-V characteristic of TEG has been studied in function of the temperature difference between cold and hot sides of the TEG (see **Error! Reference source not found.**).

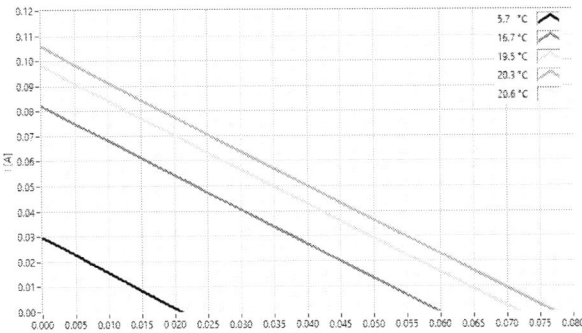

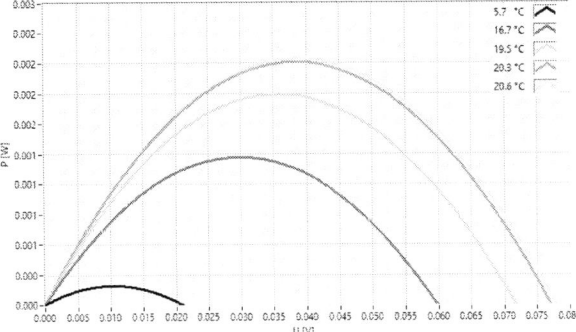

Fig. 5. The I-V characteristics of TEG obtained at different ΔT

In order to validate the implemented models, the comparison between the simulation and real data can be performed by the students. The real data for different TEG models and experiment conditions is available for download from the laboratory server.

IV. CONCLUSIONS AND FUTURE WORK

This paper presented an innovative approach meant to enhance the student's practical experience in electronics engineering hands-on laboratories. By making use of the dynamic partial reconfiguration feature of the latest generation of FPGAs, we managed to successfully achieve an efficient sharing of a single Zynq SoC development board by several students with their designs operating in parallel in the FPGA logic, thus allowing simultaneous multi-student practice with reconfigurable hardware platforms.

The students can deploy and communicate with their designs using Ethernet and a serial terminal running on a local computer connected to the FPGA board.

We have successfully validated this system by deploying and simulating models of thermoelectric generator elements and use these models for studying the TEGs I-V and P-V characteristics.

Given its functionalities and intended application range, our platform enables students to implement applications like real-time simulation of energy sources and also control circuit prototyping, thus contributing to a better understanding of the requirements and complexity of energy harvesting systems.

As future developments, we are currently working on improving the accessibility of the SoC by developing a friendlier interface for configuration and data transfer. Also, our ultimate goal is to integrate this system in a remote learning platform with advanced hardware multitasking support. This would enable multi-user access to reconfigurable hardware resources via the Internet for practicing with simulations and implementations of energy harvesting systems.

ACKNOWLEDGMENT

This work was supported by a grant of the Romanian National Authority for Scientific Research and Innovation, CNCS - UEFISCDI, project number PN-II-RU-TE-2014-4-1083.

REFERENCES

[1] V.J. Harward, J.A. Del Alamo, S.R. Lerman, P.H. Bailey, J. Carpenter, K. DeLong, C. Felknor et al. "The ilab shared architecture: A web services infrastructure to build communities of internet accessible laboratories." Proceedings of the IEEE 96, no. 6 (2008): 931-950.

[2] K Jozwik, H. Tomiyama, M. Edahiro, S. Honda and H. Takada, "Hardware multitasking in dynamically partially reconfigurable FPGA-based embedded systems," 2011 International SoC Design Conference (ISOCC), Jeju, 2011, pp. 183-186. doi: 10.1109/ISOCC.2011.6138678

[3] W. Lie, W. Feng-Yan. "Dynamic partial reconfiguration in FPGAs". In Third International Symposium on Intelligent Information Technology Application, vol. 2, pp. 445-448, 2009.

[4] A. Sarangi, S. MacMahon, U. Cherukupaly. "LightWeight IP application examples". Xilinx Corp., San Jose, CA. Application Note XAPP1026 v5.1; 2014.

[5] S. Lineykin, S. Ben-Yaakov, " PSPICE-Compatible Equivalent Circuit of Thermoelectric Coolers ", IEEE 36th Power Electronics Specialists Conference, DOI: 10.1109/PESC.2005.1581688, 2005;

[6] H. L. Tsai, J. M. Lin, "Model Building and Simulation of Thermoelectric Module Using Matlab/Simulink", Journal of Electronic Materials, 39, no. 9, 2105- 2111, 2010

Personalized Ring Oscillator-based True Random Number Generator Analysis using Non-Invasive Attacks

Andrei Marghescu[1], Daniel-Ciprian Vasile[1], Paul Svasta[1] and Emil Simion

"Politehnica" University of Bucharest, Romania,
[1] CETTI, Romania,
andrei.marghescu@cetti.ro

Abstract— Developing security products implies a great attention to every possible detail, from the design stage to the testing stage in order to obtain/create a good and secure product. It is well known that these products are most exposed to fault injection attacks. This paper describes a personalized True Random Number Generator, based on Ring Oscillators, implemented in FPGA logic, using a Zynq System on Chip development platform along, with a large amount of experimental statistical results. These results were obtained while acquiring data at the time where the generator was implied in different non-invasive attacks (temperature variations and electromagnetic influences over the integrated circuit). These tests emphasize the stability and security of the proposed True Random Number Generator, highlighting the fact that it is suitable for using in cryptographic applications.

Keywords—TRNG, cryptography, FPGA, security, side channel attacks

I. INTRODUCTION

True Random Number Generators are widely spread in different applications, from simulations, artificial intelligence to cryptography having the goal to provide nondeterministic sequences. Because of their critical usage (sensible areas – mainly in cryptography) they are commonly the target of various attacks. These attacks try to establish a pattern within the generated random sequence or to force the generator (by some means) to work deterministic.

Side channel attacks are widely spread among TRNGs [3], consisting of different invasive (altering physical components, having a permanent effect over them) or non-invasive techniques (creating a special working environment) that aim to make the circuit to act deterministic. The aim of this paper is to describe a personalized Generator and submit it to two non-invasive attacking techniques based on fluctuations in working temperature and electromagnetic field influence.

This paper is structured as follows: the second chapter presents the concept of True Random Number Generators, and how can we create one using Ring-Oscillators.

The third chapter describes the personalized approach on the Ring Oscillator-based True Random Number Generators. This description is focused on the functional scheme of the generator and explains how it works and how can someone use it to implement a TRNG.

The fourth chapter details how the implementation was done and how the tests were applied, presenting the statistical results. Finally, the paper ends with some conclusive remarks.

II. TRUE RANDOM NUMBER GENERATORS

True Random Number Generators (TRNG) are commonly used in cryptographic applications for their defining property of being unpredictable, making them the pillar of cryptographic key generators. They are based on a physically non-deterministic phenomenon like the de-synchronization of a large amount of interconnected oscillators or the jitter of them [1], which is further post-processed by applying a Randomness Extractor and a battery of statistical tests [4].

The most known Randomness Extractor is the one developed by Von Neumann (VN) which works with pairs of bits, outputting the first bit of the pair whose bits are different, dropping the pairs whose bits are the same.

The Randomness Testing Batteries are focused on finding different patterns among the generated sequence according to some well-known algorithms. Therefore the tests try to find a mechanism that can reproduce the output. The most known and used battery of statistical tests are developed by the National Institute of Standards and Technology [4].

One of the most known Ring Oscillator-based True Random Number Generator was developed by Sunar et al. [7] and is based on the interconnection of 114 free running Ring Oscillators. The interconnection is done by adding all of the signals from them modulo 2. This concept was further optimized (in terms of resource consumption) in [8] and this is the pillar of the scheme developed within this paper.

III. DESCRIPTION OF THE PROPOSED SOLLUTION

The personalized generator powers 4 Ring Oscillators (Fig. 1) consisting of 3, 5, 7 and 11 inverter gates each and works as follows:

- Each RO is connected to a Von Neumann Randomness Extractor – VN, which has the role to uniformly distribute the bits, and after that they are connected to an arbiter.

- The arbiter passes through each VN block and acquires valid bits and adds them modulo 2, forming the output bit.
- This output bit is XORed (XOR2) with the output of XOR1 (which has the ROs signals as input).

This scheme improves the classical ones based on desynchronization of multiple ROs (which just XORs the output of each RO), by increasing the complexity, while maintaining their security, optimizing the previous ones in terms of consumed hardware resources.

This technique prevents eventual influences within Ring Oscillators components.

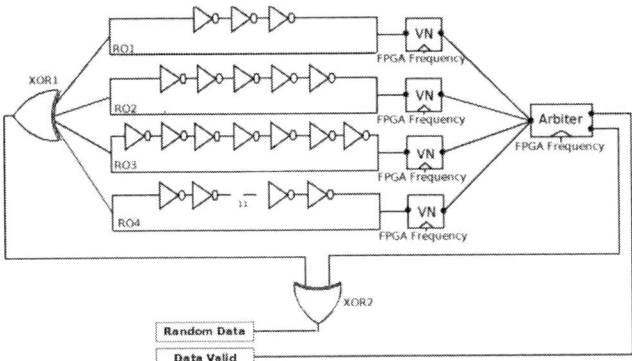

Fig. 1: Personalized TRNG scheme.

IV. IMPLEMENTATION AND RESULTS

For the implementation of this generator, it had been used the ZYBO Zynq System on Chip development board, powering both an ARM Cortex A9 processor and an FPGA. Within this scheme, the TRNG was implemented in the FPGA and the results were passed to the ARM processor which processes and sends them to the requester (in our case a laptop).

Implementing this generator using this particular Development Board the following oscillating frequencies were obtained: 96, 172, 203 and 243 MHz (Figures 2, 3, 4, 5).

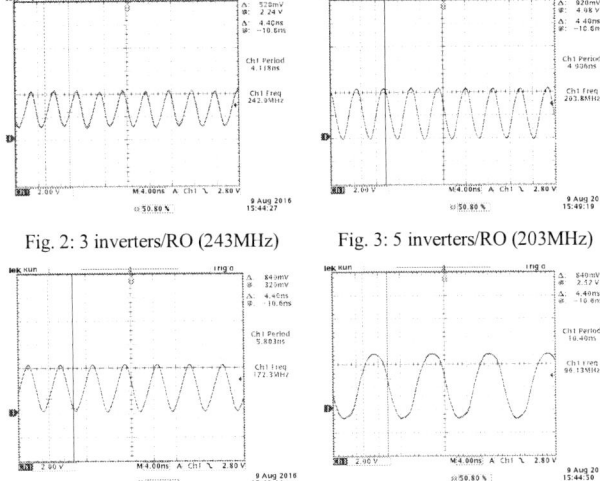

Fig. 2: 3 inverters/RO (243MHz)   Fig. 3: 5 inverters/RO (203MHz)

Fig. 4: 7 inverters/RO (172MHz)   Fig. 5: 11 inverters/RO (96 MHz)

The goal of this paper is to analyze the tolerance of the True Random Number Generator while supposed to the following non-invasive attacks:

A. Temperature change

Submitting the generator to different operating temperatures within 0 and 70 degrees Celsius, using a 10 degrees step, both at a stable operating temperature and a transit one (within the temperature change). In other words, the sequences were acquired once while the temperature is increasing (transits) and once while the temperature was stable at a fixed value. This test ensures us that generator is working properly inside this operating temperature range.

For the heating process it has been used the Espec SH-241 Temperature and Humidity Chamber.

B. Applying electromagnetic fields to the chip

A second test was applied to the FPGA circuit (Figure 6) in order to verify the stability of the Ring Oscillators when a variable magnetic field is induced in this circuit. In [5] there are presented a few cases of frequency injection attack against ring-oscillators-based true random number generators by applying a sinusoidal signal to the power supply (or ground sink connection). This attack reduced the entropy of ring-oscillators, thus it reduced the key space of a secure microcontroller.

We tested the RO-TRNG by applying a variable magnetic field using a small coil of 5 windings (11 mm in diameter) and an excitation signal of 20 dBmW. We made four assessments with frequencies close to natural oscillation frequencies of the four ring-oscillators (presented in Figure 2-5). The magnetic field will equally apply to every oscillator and to every cell of these oscillators. In frequency injection applied to the power supply, every cell of ring-oscillators will lock the phase, losing the jitter that is the main source of entropy.

We assume that small to medium magnetic fields will not affect the oscillators because of the low coupling between the near magnetic field and the inner structure of the oscillators. Even if a region of the FPGA chip is affected by the inducting field, the nearby regions would be affected only if the induced signal would travel through the power rail. This rail is connected to a power supply pin, externally decoupled, thus reducing the propagation of the induced signal.

The case of strong magnetic fields assessment is not the point of this article because in a regular system using RO-TRNG, the electronics would be installed in a metallic case that would substantially reduce this field. We are interested in testing the effect of small to medium fields accidentally created by faulty design of the circuit board in applications that use RO-TRNGs.

Each generated sequence was submitted to a battery of statistical tests, defined by the National Institute of Standards and Technologies [4] and the results are presented within Tables 1 and 2. Each table presents the statistical test that was applied to the data and the number of sequences that passed the test per total amount of them.

As we can see from the tables the generator provides good statistical results even if the hardware structure was supposed to "harsh" operating environments.

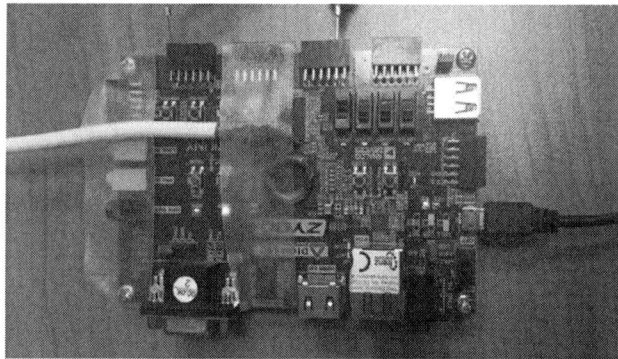

Fig. 6: Electromagnetic field applied to the chip

V. CONCLUSIONS

This paper presented a personalized True Random Number Generator Concept, based on the interconnection of 4 Ring Oscillators, obtaining low resource consumption and a high generation rate. Within chapter 2 it has been presented the concept of side-channel attacks and how can they be implemented.

Moreover, this TRNG was analyzed according to each consisting Ring Oscillator Frequency and was supposed to two sets of tests consisting of different operating temperature range and electromagnetic field demonstrating that it is a reliable generator from a statistical point of view.

ACKNOWLEDGMENT

This work was supported by the Romanian National Authority for Scientific Research (CNCSUEFISCDI) under the project PN-II-PT-PCCA-2013-4-1651.

REFERENCES

[1] Sunar, Berk, William J. Martin, and Douglas R. Stinson. "A provably secure true random number generator with built-in tolerance to active attacks." IEEE Transactions on computers 56.1 (2007): 109-119.

[2] Marghescu, A., Svasta, P., & Simion, E. (2016, May). Optimising ring oscillator-based true random number generators concept on FPGA. In Electronics Technology (ISSE), 2016 39th International Spring Seminar on (pp. 149-153). IEEE.

[3] Bayon, Pierre, et al. "Fault model of electromagnetic attacks targeting ring oscillator-based true random number generators." Journal of Cryptographic Engineering 6.1 (2016): 61-74.

[4] http://csrc.nist.gov/groups/ST/toolkit/rng/documents/SP800-22rev1a.pdf

[5] A. T. Markettos and S. W. Moore, 'The Frequency Injection Attack on Ring-Oscillator-Based True Random Number Generators'. In Proceedingsof Cryptographic Hardware and Embedded Systems (CHES) 2009,Lausanne, Switzerland, September 2009. Lecture Notes in ComputerScience 5747, Springer, pp. 317-331.

[6] Sunar, B.; Martin, W.J.; Stinson, D.R., "A Provably Secure True Random Number Generator with Built-In Tolerance to Active Attacks," in Computers, IEEE Transactions on , vol.56, no.1, pp.109-119, Jan. 2007;

[7] Marghescu, Andrei, Paul Svasta, and Emil Simion. "Optimising ring oscillator-based true random number generators concept on FPGA." Electronics Technology (ISSE), 2016 39th International Spring Seminar on. IEEE, 2016.

Table 2. Personalized TRNG scheme

Statistical Test	E.M. Field Frequency			
	100 MHz	170 MHz	200 MHz	240 MHz
Frequency	99/100	100/100	100/100	99/100
Block Frequency	99/100	100/100	100/100	99/100
Cumulative Sums	98/100	100/100	100/100	99/100
Runs	97/100	98/100	98/100	98/100
Longest Run	99/100	98/100	100/100	100/100
Rank	99/100	99/100	100/100	100/100
FFT	98/100	98/100	98/100	98/100
Non Overlapping	100/100	98/100	100/100	100/100
Overlapping	99/100	99/100	100/100	100/100
Universal	99/100	99/100	97/100	100/100
Approximate Entropy	100/100	98/100	100/100	98/100
Random Excursions	69/70	59/59	58/60	61/62
R.E. Variant	69/70	58/59	59/60	62/62
Serial	100/100	99/100	100/100	100/100
Linear Complexity	100/100	98/100	100/100	99/100

Table 1. Personalized TRNG scheme

Statistical Test	0 °C		10 °C		20 °C		30 °C	
	Transit	Stable	Transit	Stable	Transit	Stable	Transit	Stable
Frequency	98/100	100/100	100/100	98/100	99/100	100/100	98/100	99/100
Block Frequency	99/100	99/100	100/100	97/100	99/100	99/100	100/100	99/100
Cumulative Sums	99/100	100/100	100/100	97/100	99/100	100/100	100/100	99/100
Runs	97/100	100/100	98/100	97/100	100/100	100/100	100/100	97/100
Longest Run	98/100	100/100	99/100	99/100	100/100	100/100	99/100	98/100
Rank	100/100	99/100	100/100	98/100	99/100	100/100	97/100	98/100
FFT	99/100	99/100	99/100	97/100	98/100	98/100	98/100	100/100
Non Overlapping	99/100	100/100	100/100	99/100	100/100	99/100	98/100	98/100
Overlapping	99/100	100/100	98/100	100/100	97/100	98/100	100/100	98/100
Universal	100/100	100/100	99/100	99/100	98/100	99/100	99/100	100/100
Approximate Entropy	100/100	98/100	97/100	98/100	99/100	100/100	97/100	99/100
Random Excursions	62/63	63/63	63/64	59/59	62/63	68/69	60/62	70/70
R.E. Variant	63/63	63/63	64/64	59/59	63/63	67/69	62/62	70/70
Serial	100/100	99/100	100/100	99/100	99/100	100/100	100/100	100/100
Linear Complexity	99/100	99/100	99/100	100/100	98/100	98/100	100/100	99/100

Statistical Test	40 °C		50 °C		60 °C		70 °C	
	Transit	Stable	Transit	Stable	Transit	Stable	Transit	Stable
Frequency	100/100	100/100	96/100	100/100	100/100	97/100	96/100	99/100
Block Frequency	100/100	100/100	99/100	99/100	98/100	98/100	100/100	99/100
Cumulative Sums	100/100	99/100	99/100	100/100	100/100	99/100	96/100	99/100
Runs	100/100	97/100	98/100	99/100	99/100	97/100	99/100	100/100
Longest Run	99/100	99/100	100/100	100/100	97/100	100/100	99/100	99/100
Rank	98/100	98/100	100/100	100/100	100/100	98/100	98/100	100/100
FFT	99/100	99/100	100/100	99/100	100/100	100/100	99/100	99/100
Non Overlapping	99/100	100/100	96/100	97/100	100/100	97/100	99/100	100/100
Overlapping Template	99/100	100/100	100/100	98/100	98/100	100/100	100/100	99/100
Universal	99/100	100/100	100/100	100/100	96/100	97/100	98/100	97/100
Approximate Entropy	99/100	98/100	98/100	97/100	98/100	100/100	98/100	100/100
Random Excursions	58/60	57/58	58/58	66/66	58/58	66/66	62/62	58/58
R. E. Variant	60/60	58/58	57/58	65/66	58/58	66/66	62/62	57/58
Serial	98/100	100/100	98/100	97/100	98/100	100/100	96/100	100/100
Linear Complexity	99/100	100/100	100/100	99/100	97/100	100/100	99/100	100/100

Wireless Diagnosis and Monitoring System of Sensor Network from Civil Structures

S. Pop, V. Bande, I. H. Baciu

Applied Electronics Department, Technical University
Cluj Napoca, Romania
septimiu.pop@ael.utcluj.ro

Abstract—This paper is focused on the development of a wireless device used in diagnosis and monitoring of sensors networks from civil structures like hydro-energetic buildings. Those constructions are monitored with sensors and measurement systems spread over the entire construction's body. The measurement systems are connected through a RS485 network. The entire structure forms a data acquisition system that is usually controlled by a software application that runs on a computer. A daily management activity in hydro-energetic buildings consists in an automatic and a manual data collection. At the physical layer, in real life, the measurement devices and RS485 networks can be affected by a set of malfunctions that are dangerous signal integrity for the safety of a human operator. For that reason a manual measurement device based on wireless technology is useful. The issues of signals integrity of RS485 lines are detected with an electronic device that is connected to the lines. In additional, using communication protocol the data from each sensor can be collected. The panel interface of the human operator is a software application that runs on a smart mobile device that uses Bluetooth communication [2]. In the last years the mobile software applications has been greatly increased in industry [1]. By using a software application and a wireless mobile device a diagnosis and monitoring system is developed which is more safety and has a vast development space [3].

Keywords — wireless, Bluetooth, diagnosis, monitoring.

I. INTRODUCTION

The hydro-energetic building management is a health monitoring activity based on the information obtained from the sensors placed inside of buildings body. The hydro-energetic buildings are equipped with hundreds of sensors spread over the entire construction's body. In order to improve the buildings management activity, an automatic system is used to measure the sensors. The monitoring system has a hierarchical topology. At the lower level the sensors used to detect the physicals parameters are being placed and at the higher lever a PC and a software application will run. The task of these functional devices is to carry out the storage of the acquired data in such a way to get the building behavior history. Using a proper software tool, data can be displayed or processed. The information from the sensors, is used to evaluate whether the dam is performing as expected and whether it provides a warning of developing conditions that could endanger the safety of the dam. For example, at a concrete gravity dam,

increasing the uplift pressure, or decreasing drain flow may indicate that the foundation drains may need to be cleaned.

In order to measure the sensors, they are connected at a measurement device, as in the figure 1. Actually, because they are spread over the entire construction's body they are measured by devices spread also on a large area.

Fig .1 One measurement system

Furthermore, the measurement systems are connected together using RS485 network at a central unit which is a computer. In figure 2, the structure of the acquisition network inside a dam it is illustrated. In the picture below, each measurement system is named DCC.

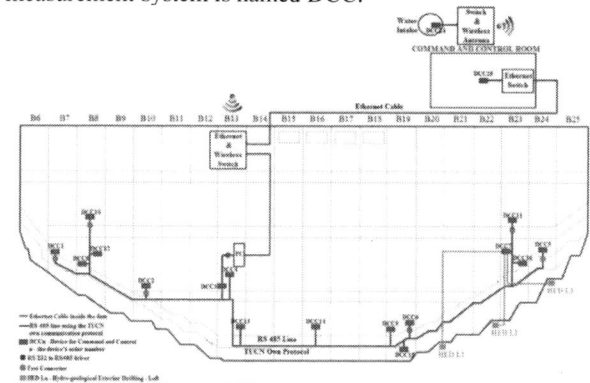

Fig 2. The structure of a monitoring network inside a dam [6]

In most situations, the monitoring system is place inside or around the dam spread over a large surface – even hundreds of square meters. The central unit is placed in the control room being far away from the monitored network. In the hydro-energetic buildings case, the monitoring system works in harsh conditions from the environment point of view. There are frequent situations when the humidity is excessive, the supply voltage varies beyond the standard values, breaks, open-circuits, short-circuits on the communication line, etc. All these situations can be serious problems in the acquisition system's functionality. The flaws can affect all the network or just a part of it. In the acquisition system, the flaws are being identified through the absence of the acquired data. In the figure below, the database which contains one sensor's measurements is being presented.

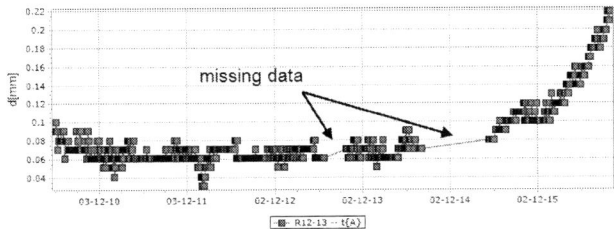

Fig.3 The database with the measurements for the R12-13 sensor – the relative displacement between two adjacent concrete blocks for a concrete dam

If the time periods during which data is missing are long, then there is more difficult to analyze one dam's behavior. For the network and also for the measurement system diagnosis, a portable and user friendly measurement device must be taken into consideration.

II. GENERAL DESIGN OF SYSTEM

The monitoring system uses the RS485 protocol for the communication between the measurement devices. This protocol is a specific one and it is based on the "master-slave" topology. The communication is being initiated by the "master" device and afterwards the line is cleared. Thus, on the communication line, a device that can initiate diagnosis test procedures, can be connected. This device must be reliable, user-friendly and capable to detect the following flaws: over-voltages, the absence of a voltage supply, open-circuit on the communication line, short-circuit, faulty measurement device.

- The over-voltages are being met in occasional situations, but are being persistent phenomena, which are causing functionality flaws dangerous for the integrated circuits and for the human operator. For the human operator protection, the diagnosis system must contain a galvanic separation between the communication line and the system's electronic part. This is the first step that the diagnosis system should made.

- The measurement systems connected through the RS485 line are galvanic separated. The integrated circuits that

assure the needed drivers for the communication are being supplied from the communication line. The absence of this voltage causes communication errors. This is the diagnosis system's second test.

- The short-circuit/open-circuit flaw is detected by testing the communication between different measurement devices from the network and also by verifying the communication line "sharing", as it is presented in the figure below.

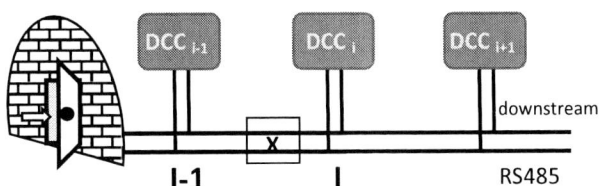

Fig. 4 Short-circuit on the communication line

Point **I-1**
- 1.a. test the current DCC$_{i-1}$ = FALSE, 2.a. open the line downstream, 3.a. test the current DCC$_{i-1}$ = FALSE => **DCC$_{i-1}$ = faulty.**
- 1.b. test the current DCC$_{i-1}$ =FALSE, 2.b. close the line 3.b. test the current DCC$_{i-1}$ =TRUE, 4.b. close the line, 5.b.test DCC$_i$ =FALSE => **DCC$_i$ =FALSE** ➔ go to point **I.**

Point **I**
- 1.a. test the current DCC$_i$ = FALSE, 2.a. open line downstream, 3.a. test current DCC$_i$ = FALSE => **DCC$_i$ = faulty or short-circuit line upstream.**

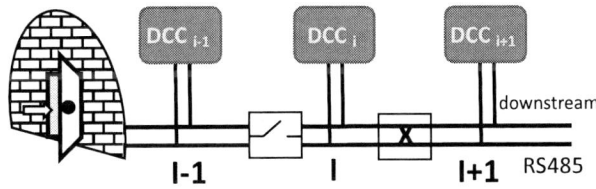

Fig. 5 Short-circuit and open-circuit of the communication line

Point **I-1**

- 1.a. test the current DCC$_{i-1}$ = FALSE, 2.a. open the line downstream, 3.a. test the current DCC$_{i-1}$ = FALSE => **DCC$_{i-1}$ = faulty.**
- 1.b.test the current DCC$_{i-1}$ = TRUE, 2.b. test the DCC$_i$ = FALSE, => **DCC$_i$ = FALSE** go point I

Point **I**
- 1.a. test the current DCC$_i$ = FALSE, 2.a. open line downstream, 3.a. test the current DCC$_i$ = FALSE => **DCC$_i$ = faulty.**
- 1.b. test the current DCC$_i$ = FALSE, 2.b. open the line downstream, 3.b. test the current DCC$_i$ = TRUE, 4.b close

2016 IEEE 22nd International Symposium for Design and Technology in Electronic Packaging (SIITME)

line 5.b, test DCC$_{i+1}$ =FALSE => **open-circuit line upstream and short-circuit line downstream.**
Even though this is a continuity test for the flaw detection, the "Ohm" method is not being used. With the portable diagnosis system we can move through the communication line. The test is being realized in different points, at the junction between the RS485 line and the measurement systems (DCC). The short-circuit/open-circuit flaws on the RS485 line are being determined through the analysis of the response obtained from the "slave" systems. From a junction point (point I) the upstream (I-1) and downstream (I+1). "slave" systems are being tested.

The testing algorithm is a complex one. It is built in such a way that through a single test crossing, the following flaws to be detected: short-circuit, open-circuit and communication between the measurement systems. In fig. 4 and fig. 5, short-circuit and open-circuit flaws are being graphically represented. The diagnosis system must be capable to analyze all situations occurred in different points. The obtained test report or the final and more complex report are obtained by analyzing the previous paragraph possible situations. Because the simplest mode of implementing the testing algorithm is the software procedure, the diagnosis system's central unit is a portable electronic device, such as: smartphone, tablet, pocket PC, etc. These mobile systems cannot be directly connected through the RS485 line, so an electronic module must be designed in order to overcome this problem. Beside the "bridge" function, the electronic module must supply 5V for other circuits and also must allow the detection of occasional over-voltages.

III. HARDWARE DESCRIPTION OF THE ELECTRONIC TESTING DEVICE

The block schematic of the electronic testing module which allows a portable device to be connected at the RS485 line is presented in the fig. 6. Physically, the RS485 line uses four wires: voltage supply (5V), ground, data + (A), data − (B).

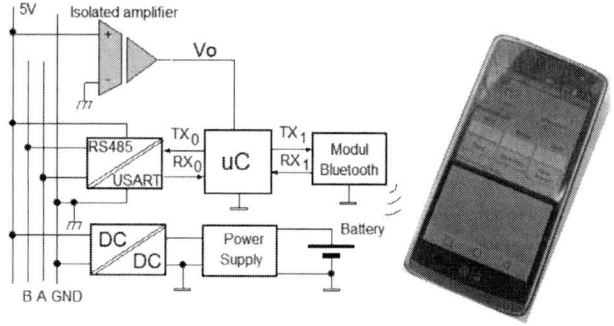

Fig.6 The block schematic of the electronic module

The central unit of the testing electronic module is the ATtiny1634 microcontroller. The microcontroller provides the "bridge" function and also the supplementary functions of the testing algorithm.

The smartphone connection (on which the diagnosis application runs) at the electronic module can be made using the Bluetooth interface. The Bluetooth interface is the most accessible wireless communication interface which allows mobile devices to connect at an electronic module [4], [5]. The microcontroller is wirelessly communicating with the testing application using a specialized Bluetooth module, ABT-BTM-222.

A. *The signal integrity test circuit, overvoltage detection*

The measurement systems connected at the communication line are being galvanic separated [6]. The RS485 drivers (SN75LBC184) are being supplied form the 5V line, voltage produced by an independent source. In reality, for the implemented acquisition systems used for building monitoring, the most common problems found are the absence of the 5V voltage or the over-voltages. The over-voltages are dangerous for both the transceiver circuits and the human operator. As in the test algorithm described in the previous paragraph, the first test step consists in the 5V line integrity monitoring. The voltage measurement circuit applies the procedure provided by Vishay [7] with the help of the IL300 dedicated circuit. This circuit is used for galvanic separation of communication line's signals. The electrical schematic is revealed in the figure below.

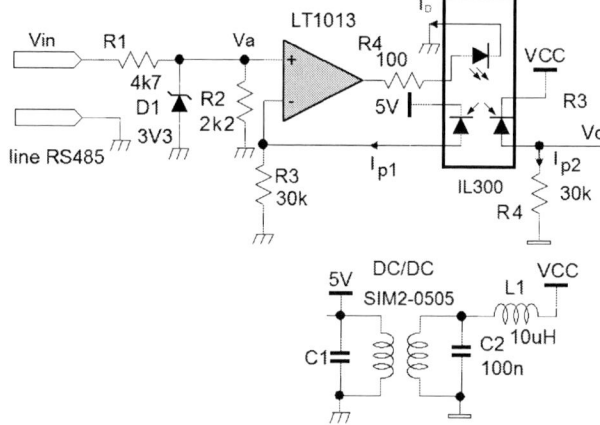

Fig.7 The electrical schematic of the analog testing circuit [6]

The electronic module is supplied through a SIM2-0505 DC-DC transformer. This voltage can be afterwards supplied to the rest of the circuit via a jumper.
The output voltage V$_o$ has the following expression:

$$V_o = K_3 \cdot \frac{R_4}{R_3} \ [V]; \qquad (1)$$

The IL300 transfer gain (K$_3$) is expressed as the ratio of the output gain (K$_2$) and the feedback gain (K$_1$). K$_1$ is the ratio of the input photodiode current (I$_{P1}$) and the LED's current (I$_D$), K$_2$ is the ratio of the output photodiode current (I$_{P2}$) and the LED's current (I$_D$). Best linearity can be obtained at drive currents I$_D$, between 5mA and 20mA [7]. For I$_D$=15mA, the transfer gain K$_3$ can be approximated at 1 (0.9958). For voltages lower than 5V, the circuit behaves as a resistive divider with the dividing factor $R_2/(R_1+R_2)$. For the input voltage equal with 5V, the output voltage is 2.24V. Basically, the over-voltage is limited. Any over-voltage beyond 5V is limited, at the output, at 3.3V. In fig. 6, the circuit response when, accidentally, we have as an input a 230VAC voltage, is being presented.

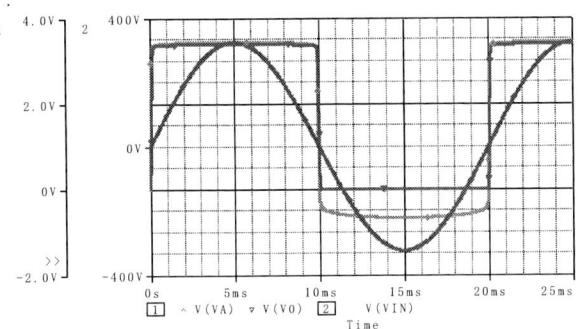

Fig. 8 The circuit response when over-voltages occur

With the help of circuit presented in fig. 7, the voltages from the $4 \div 6V$ domain are being measured with a 1% accuracy. In fact, the accuracy is not a critical parameter, because the circuits can function properly even with a 10% tolerance of the supply voltage.

B.   The RS485 Interface circuit

For the diagnosis device connection at the communication line a RS485 driver-circuit (PC816) is used – the same circuit used by the measurement system. This circuit is galvanic separated by the microcontroller using optocouplers.

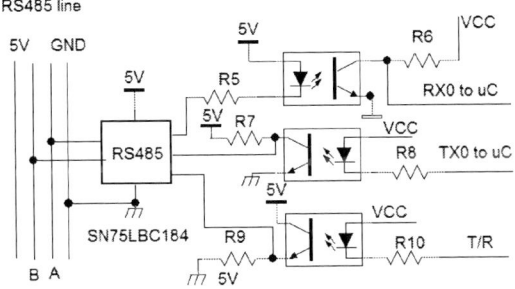

Fig.9 The galvanic separation of the transceiver circuit

The half-duplex communication is realized using a communication protocol.

IV.   HUMAN INTERFACE SOTWARE APPLICATION

The proposed diagnosis system consists of two components. Firstly, a hardware component is connected at the communication line under testing and secondly a software component implemented in the mobile device. The software component is represented by the user interface panel, which runs on a smartphone. Using this panel, the user launches the testing commands. The application is developed using the

MIT App Inventor environment. The App Inventor is a visual block language which works in cloud. The software application for Android can be built even in a web browser.

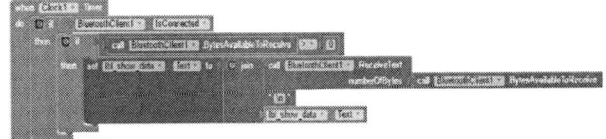

Fig. 10 The "Receive" function

The data receiving, Fig. 10, is controlled by a timer (*Clock1.Timer*), the received byte number is returned by the *BytesAvailableToReceive* block. The data is received in text format and displayed. The user panel (Fig. 6) contains data transmission functions, receiving data, analysis and data saving.

V.   CONCLUSIONS

The proposed diagnosis system is a very useful tool for the RS485 communication line implemented in the hydro-energetic buildings. Due to the harsh environment conditions, the flaws are very frequent and complex. The system is galvanic separated by the communication line, thus the human operator is protected. The user interface is developed using a virtual instrument and a software application runs on a smartphone. As future enhancements, the application can be further completed with software routines, such as: time domain (historical) analysis, sensor diagnosis, etc.

ACKNOWLEDGMENT

These results were obtained on research grants with SC. Hidroelectrica SA between 2015 and 2016.

REFERENCES

[1]   Yang Wang, " Wireless Sensor Networks in Smart Structural" Jia-Chin Lin, ISBN 978-953-307-274-6, Published: August 23, 2011 under CC BY-NC-SA 3.0 license

[2]   Zhen Huang " Wireless Monitoring and Control System Via Android Tablet PC" 2nd International Symposium on Computer, Communication, Control and Automation (3CA 2013) ISBN (on-line) 978-90786-77-91-8

[3]   J. P. LYNCH "Overview of Wireless Sensors for Real-Time Health Monitoring of Civil Structures " Source: Proceedings of the 4th International Workshop on Structural Control and Monitoring, New York City, NY, USA, June 10-11, 2004.

[4]   P. Vignesh "Bluetooth Based Patient Monitoring System" International Journal of Science and Research (IJSR) ISSN (Online): 2319-7064

[5]   Mrs. Pratibha Singh "A Modern Study of Bluetooth Wireless Technology" International Journal of Computer Science, Engineering and Information Technology (IJCSEIT), Vol.1, No.3, August 2011

[6]   V. Bande, S. Pop, D. Pitica "Smart Diagnose Procedure for Data Acquisition Systems Inside Dams" 23-26 Oct 2014, Bucharest, Romania, ISBN 978-1-4799-6962-3

[7]   www.vishay.com app. Note 50 Designing Linear Amplifiers Using the IL300Optocoupler

Continuous Respiratory Monitoring Device for Detection of Sleep Apnea Episodes

Cristian Rotariu, Ciprian Cristea, Dragos Arotaritei
Department of Biomedical Sciences
"Grigore T. Popa" University of Medicine and Pharmacy
Iasi, Romania
cristian.rotariu@umfiasi.ro

Radu G. Bozomitu, Alexandru Pasarica
Department of Telecommunications
"Gheorghe Asachi" Technical University
Iasi, Romania

Abstract—Because sleep disorders are common in a significant part of the entire population with diseases of the central nervous system, continuous monitoring of respiration during sleep has an important role in early diagnosis and treatment. The paper describes the research for the design and realization of a flexible, scalable and cost-effective medical monitoring device suitable to be used with patients suffering from sleep disorders, especially obstructive sleep apnea episodes. The monitoring device contains commercially available sensors for respiratory signal measurement, a data acquisition and processing module with microcontroller, and a Tablet PC. The respiration signal is acquired by using a commercially available piezoelectric thoracic belt and processing by the microcontroller in order to compute the respiratory frequency and to detect the sleep apnea episodes. On the Tablet PC a software application displays the patient's respiratory frequency and activates the alerts in the interface when an apnea episode is detected. A prototype of the described respiratory monitoring device has been developed, implemented and partially tested.

Keywords— monitoring device, sleep disorders, apnea detection.

I. INTRODUCTION

Sleep represents a naturally recurring state of rest for the mind, during the central nervous system is restored. Therefore, sleep plays an important role in healthy lifestyle by protecting the mental and physical health, increasing the quality of life. Sleep disorders are among the common conditions affecting contemporary patient. The prevalence of sleep disorders is estimated in Europe at 2% of women and 4% of men, having an important impact on their quality of life.

A common type of sleep disorder is represented by obstructive sleep apnea (OSA). The OSA is a common and serious disorder defined as a series of pauses in breathing for 10 seconds or more during sleep. Apneic episodes have many different possible causes, can last several seconds or even minutes. For a positive diagnosis of OSA the patient must present at least 5 apneic episodes per hour [1]. The absence of breathing causes an increase in the concentration of carbon dioxide in the blood, leading to awakening of the person. As soon as normal levels of carbon dioxide and oxygen in blood levels due to normal breathing is restored, the person falls asleep in place [2]. Unfortunately, given the specific condition

it is difficult to observe the patient, especially in the early stages, due to the side effects of the illness.

Sleep apnea can affect anyone, regardless of gender or age. However, it occurs predominantly in men, in people having obesity, age over 40, large diameter of the throat, tonsils, gastroesophageal reflux, allergies or sinus problems. Alcohol, sedatives or tranquilizers can cause apnea episodes due to the inhibitory effect on muscle tone in the upper airways [3]. Its effects generally include fatigue during the day, reduced reaction times, impaired vision or cognitive problems, and memory disorders [4]. Moreover, some studies show a possible increased risk of diabetes due to an increased number of apnea episodes [5]. OSA is also linked to one of the main factors of mortality today, namely cardiovascular diseases. Apnea episodes may lead to an increase in sympathetic tone, which in turn can cause cardiac ischemia [3].

Usually the complete diagnosis of OSA is performed by polysomnography, that is a multi-parametric test for recording a number of physiological parameters during sleep including heart rate, brain activity, arterial oxygen saturation or patient motion. The polysomnography is an expensive test, requires qualified medical personnel, however it is widely used in the study of sleep and as a diagnostic tool in sleep medicine.

The continuous respiratory long-time monitoring devices allow patients to be monitored from a distance, for example at their homes or specialized healthcare institutions, allowing sleep specialists to observe the patient. It also helps to avoid keeping the patient in hospital overnight with electrodes and sensors attached to his body.

II. MATERIALS AND METHODS

The proposed device architecture includes the following components, as they are represented in Fig.1: a) a piezoelectric thoracic belt attached on patient's chest and used to measure the respiratory signal; b) a custom developed module for signal conditioning containing low noise operational amplifiers; c) a data acquisition module - Arduino Leonardo board based on widely used low cost ATMega32 microcontroller with A/D convertors; and d) a Tablet PC running Windows 10 as operating system.

Fig. 1. Continuous respiratory monitoring device - overall architecture

The acquisition of the respiratory signal can be performed using various sensors and transducers including: thermistor based sensors [6], piezoelectric thoracic belts placed around the chest [7], microwave Doppler radars [8] or respiratory inductive sensors [9]. There are several software methods that process the electrocardiographic signals in order to detect the respiratory activity [10].

We have used the Pneumotrace II™ respiration transducer (Fig. 2), that generates a substantial, linear signal in response to changes in thoracic circumference associated with respiration. It contains a piezoelectric device that responds linearly to changes in length and it requires no excitation voltage. It has the following technical specifications: output signal range between 20 – 50mV for normal breathing, a capacitance of 2.2µF, and a resistance of 10MΩ.

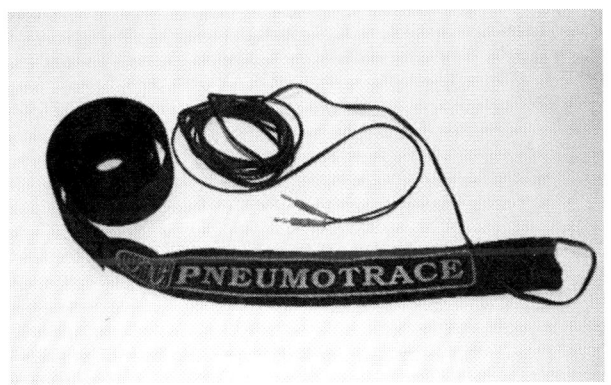

Fig. 2 Pneumotrace II™ respiration transducer

Signal amplification and filtration is performed by a custom developed hardware using low noise operational amplifiers. The schematic of the circuit has been developed using Orcad Capture (Fig. 3a) with PSpice simulator, and the PCB has been designed by using Altium Designer (Fig. 3b).

The circuit contains three colored leds and a buzzer, used to display information regarding the detection of apnea episodes (Fig. 4). The output of the circuit is connected directly to the A/D input of the microcontroller board, that also provide the power supply for the operational amplifiers. In order to set up the proper amplification for the amplifiers, several measurements of the transducer's response were performed. Then we have used the PSpice simulator in order to choose the values of the components. The obtained results for the

simulations performed on the respiratory and amplified signals are represented in Fig. 5.

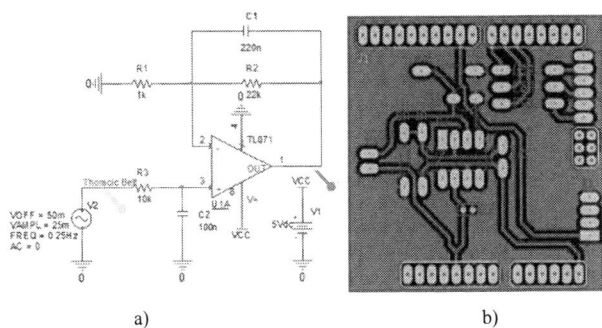

a)   b)

Fig. 3. a) Schematic and b) PCB of the custom developed hardware.

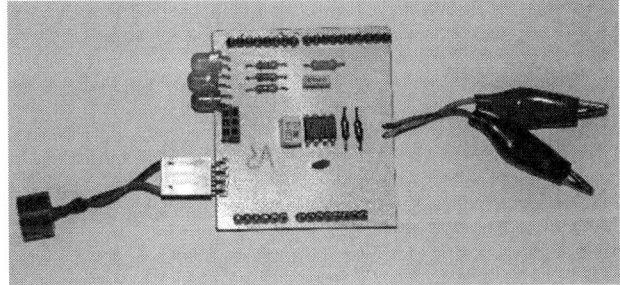

Fig. 4. Custom developed circuit for respiratory signal amplification.

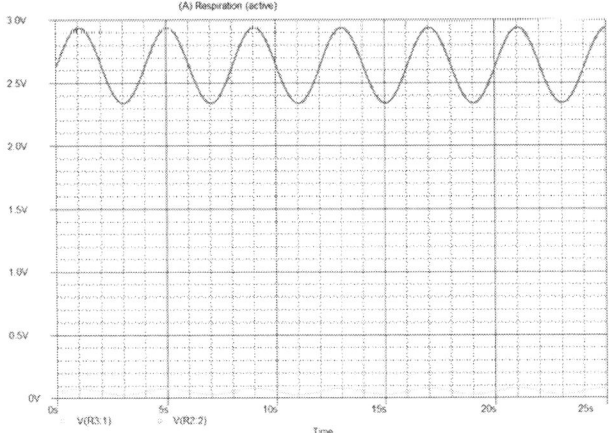

Fig. 5. PSpice simulation of the respiratory and amplified signals.

*2016 IEEE 22nd International Symposium for Design and Technology in Electronic Packaging (**SIITME**)*

The Arduino Leonardo is a complete development tool based on the ATmega32u4 microcontroller (Fig. 6).

Fig. 6. Arduino Leonardo microcontroller board.

The Arduino Leonardo has 20 programmable digital I/O pins that can be used as analog inputs (12-channels, 10-bit ADC), a 16 MHz crystal oscillator, a micro USB connection, a power supply jack, an ICSP header, and a reset button. The tool includes a USB-powered emulator to program and debug application in-system.

The flowchart of the firmware running on ATmega32u4 microcontroller is represented in Fig. 7. After a START command all the initializations are performed: ADC sample frequency, buffers for acquired and processed signals, serial port communications parameters, and the alert limits. Then the application reads the ADC values with a sampling frequency of 10Hz, perform the signal filtering with a digital band pass filter with cutoff frequencies of 0.01 – 1Hz, and computes the respiratory frequency by using a simple dual threshold peak detection algorithm. The algorithm also detects in the real time the missing of respiration.

The computed respiratory frequency and the detected apneic episodes are transmitted each time when they are detected to the Tablet PC by using the standard USB virtual serial port.

III. Results

The prototype of the continuous respiratory monitoring device for sleep monitoring (Fig. 8), as it was described above, has been implemented and partially tested.

A user-friendly Graphical User Interface (GUI), running on the Tablet PC (Allview Impera) has been developed for continuous respiration monitoring and detection of sleep apnea episodes. The interface has been developed using LabWindows/CVI (Fig. 9) to display the temporal waveform of respiratory frequency and to activate the alerts in the interface when a sleep apnea episode is detected for the selected patient.

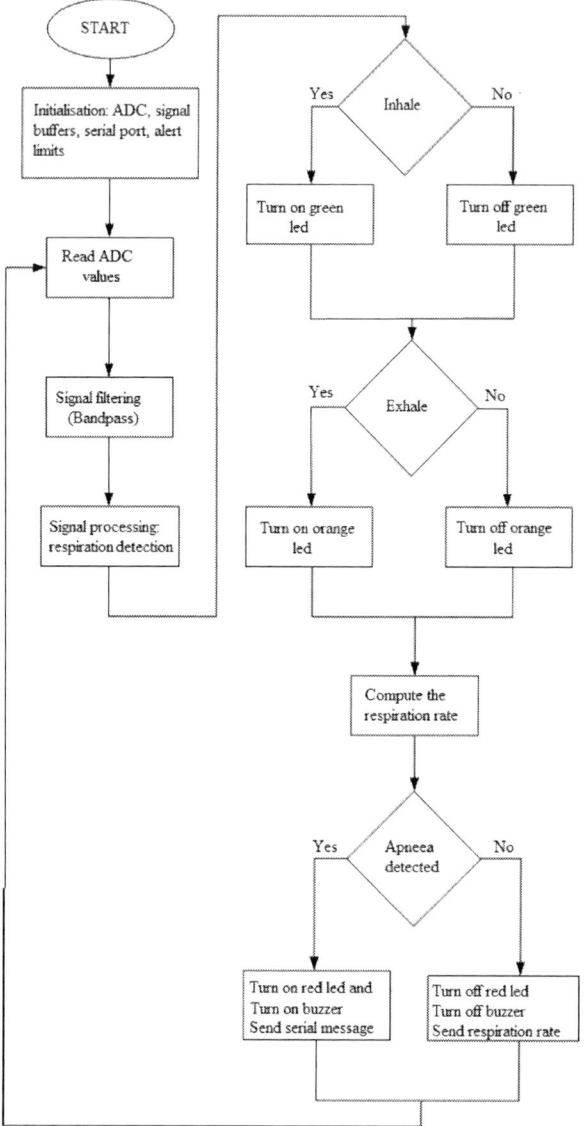

Fig. 7. Flowchart of the firmware running on microcontroller.

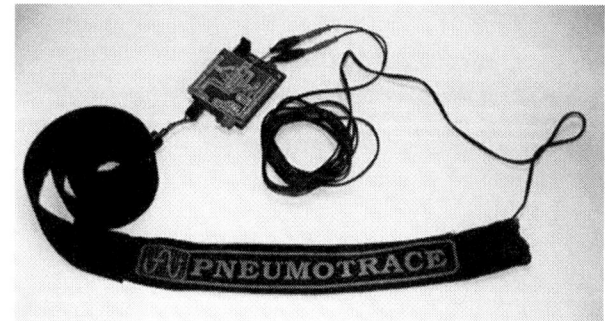

Fig. 8. The prototype of the respiratory monitoring device.

We have tested the device on 10 subjects that simulates 100 apnea episodes and the accuracy of the detection was 95%. We observed that the false detections were due to patient movement.

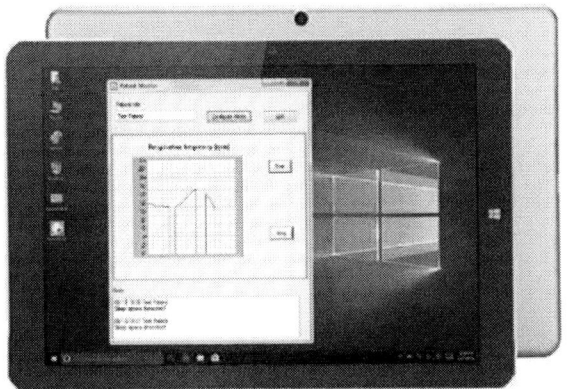

Fig. 9.   Graphical User Interface (GUI), running on the Tablet PC.

IV.   CONCLUSIONS

A prototype of a flexible, scalable and cost-effective device for continuous respiration monitoring and detection of sleep apnea episodes has been developed, implemented and partially tested.

The monitoring device is suitable for continuous long-time monitoring of respiration and detection of apnea episodes, as a part of a diagnostic tool used in sleep medicine.

The proposed device may be used with patients suffering from respiratory diseases, especially obstructive sleep apnea episodes, within their homes, as an alternative to medical supervision in healthcare institutions, with a detection degree of accuracy similar to the commercially available devices.

ACKNOWLEDGMENT

The work has been carried out within the program Joint Applied Research Projects, funded by the Romanian National Authority for Scientific Research (MEN – UEFISCDI), contract PN-II-PT-PCCA-2013-4-0761, no. 21/2014 (SIACT).

REFERENCES

[1]   "Sleep Apnea: What Is Sleep Apnea?", NHLBI: Health Information for the Public. U.S. Department of Health and Human Services. Mai 2009.

[2]   Green, Simon "Biological Rhythms, Sleep and Hyponosis". England: Palgrave Macmillan. pp. 85. ISBN 978-0-230-25265-3.

[3]   Longo DL, Fauci AS, Kasper DL, Hauser SL, Jameson J, Loscalzo J. eds., "Harrison's Principles of Internal Medicine", 18e. New York, NY: McGraw-Hill; 2012.

[4]   El-Ad B, Lavie P (2005), "Effect of sleep apnea on cognition and mood", International Review of Psychiatry (Abingdon, England) 17 (4): 277–82. doi:10.1080/09540260500104508. PMID 16194800.

[5]   Morgenstern M, Wang J, Beatty N, Batemarco T, Sica AL, Greenberg H (2014), "Obstructive sleep apnea: an unexpected cause of insulin resistance and diabetes", Endocrinology and Metabolism Clinics of North America 43 (1): 187–204. doi:10.1016/j.ecl.2013.09.002. PMID 24582098.

[6]   Jovanov E., Raskovic D., Hormigo R., Thermistor-based breathing sensor for circadian rhythm evaluation, Biomed Sci Instrument, 2001; 37: 493–497.

[7]   Ciobotariu R., Rotariu C., Adochiei F., Costin H., Wireless breathing system for long term telemonitoring of respiratory activity, Proceed 7th IntSympAdv Topics Elect Eng, 2011; 635 – 638.

[8]   Suzuki S., Matsui T., Kawahara H. et al, A non-contact vital sign monitoring system for ambulances using dual-frequency microwave radars, Med Bio Eng Comp, 2009; 47; 101 – 105.

[9]   Wu D., Wang L., Zhang Y.T. et al, A wearable respiration monitoring system based on digital respiratory inductive plethysmography, Eng Med Biol Soc 2009; 4844 – 4847.

[10]   Cerutti S., Bianchi A. M., Reiter H., Analysis of sleep and stress profiles from biomedical signal processing in wearable devices, http://embc2006.njit.edu/pdf/2010_Cerutti.pdf

2016 IEEE 22ndInternational Symposium for Design and Technology in Electronic Packaging (SIITME)

Design and Setup of Power Analysis Attacks

Mariana Safta, Paul Svasta, Mihai Dima and Andrei
Marghescu
Center for Technological Electronics and Interconnection
Techniques
"Politehnica" University of Bucharest
Bucharest, Romania
mariana.safta@cetti.ro

Mihai-Narcis Costiuc
Electronic, Information and Communication Systems for
Defense and Security Center
Military Technique Academy
Bucharest, Romania
narciscostiuc@gmail.com

Abstract—**This paper presents how to design and setup a** *Simple Power Analysis (SPA) attack* **and a** *Differential Power Analysis (DPA)* **attack on a smartcard and on a 32 bit microcontroller. SPA and DPA are power analysis attacks developed by Paul Kocher, Joshua Jaffe, and Benjamin Jun in 1995. These are non-invasive techniques which allow us to observe electrical patterns of different types of cryptographic algorithms, only by measuring the power consumption of the cryptographic devices. This paper presents the results obtained by measuring the power consumption of a secure smartcard and a 32 bit microcontroller programmed with the** *tiny AES128* **implementation and introduces the mathematical knowledge needed to extract the encryption keys based only on measurements. Using the** *DPA correlation coefficients* **algorithm, implemented in Matlab, we were able to extract the AES128 encryption key from captured power consumption traces. To obtain the measurements, an USB Scope was used. The interface to the smartcard was made using a Java platform to connect to the card reader.**

Keywords—SPA, DPA, smartcard, microcontroller, cryptography.

I. INTRODUCTION

Over the last years, the research community focused on finding and exploiting different mechanisms for extracting sensitive data from cryptographic devices. When talking about software solutions we refer to malware, mainly applicable on the hosting Operating System. Since hardware devices are more protected from these types of attacks, the researches focused on developing hardware exploits.

It is well known that almost any unprotected implementation of a cryptographic algorithm is insecure and may be easily broken even by just observing the power consumption traces [2], the processing time of the logical operations or the radiations that are emitted. No matter the strength of the cryptographic algorithm, if the implementation does not contain countermeasures for side channel attacks or it is not secure, the device may be considered vulnerable and a possible attacker can easily recover the encryption key.

In figure 1 is represented the general idea of a side channel attack. The main idea of this type of attack consists in exploiting any information leakage of the target cryptographic device. The possible leakages that can be exploit are electromagnetic radiation, power consumption, sound, visible

light, execution time and faulty outputs. All these can contain sensitive data that an attacker can capture and use further. There are many sources of leakages that can be used to implement side channel attacks, like timing attacks, electromagnetic attacks, fault attacks, power analysis attacks and other types.

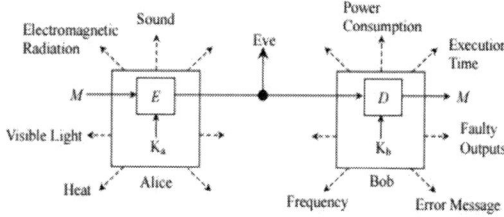

Fig. 1 Side Channel Attacks [8]

The main goal of this paper is to demonstrate how an attacker can obtain sensitive information by exploiting the data leakage from a cryptographic device, emulated within a microcontroller and a smartcard.

This paper is structured as follows: the second chapter describes the concept of side channel attacks and the state of the art of these. The third chapter will present in detail the fundamentals of the Differential Power Analysis (DPA) with the corresponding mathematical background.

The forth chapter presents the setup that was used for implementing a DPA attack and the results obtained by processing the information gathered from it. Finally, the paper will present some concluding remarks about the work described within this paper.

II. SIDE CHANNEL ATTACKS

Side channel attacks are based on exploiting the information leaked from cryptographic devices. Among side channel attacks, the most well known are power analysis attacks, like *Differential Power Analysis* (DPA) and *Simple Power Analysis* (SPA), timing attacks, electromagnetic attacks and fault attacks.

The first class – power analysis attacks – exploits the natural data leakage of devices, which can be measured by an oscilloscope and processed using specific algorithms. No

978-1-5090-4446-7/16 $31.00 © 2016 IEEE
110
20-23 Oct 2016, Oradea, Romania

matter how secure a cryptographic algorithm might be, its implementation on a chip may be insecure because of the data leakage that is unpredictable.

Another interesting type of side channel attack is timing attack. This attack is based on measuring the time needed by a logical operation to be executed. Countermeasures like executing all type of operations in x seconds have to be taken to avoid timing attacks so that a possible attacker would not obtain any valid information about the operations being executed. Timing attacks can be applied on any algorithm that has data depending on time variations. For example, RSA algorithm implementation is based on square and multiplies operations. If the value of a bit in the private key is 1, then we have square + multiply. Otherwise we have only square [3]. DES and AES traces are also easy to observe using a scope with high capabilities. The efficiency of these attacks is proven and is in our interest to explore them in as many ways as possible. In 2005, Daniel Bernstein have presented in his paper, *Cache-timing attacks on AES*, a timing attack on AES, recovering the encryption key from known plaintext timings of a network server [4].

Electromagnetic attacks are passive attacks performed by measuring electromagnetic radiations. Logical operations emit different electromagnetic traces, all having specific patterns that allow an attacker to extract sensitive information [1]. Because they are non-invasive, it is easy for an attacker to apply them without damaging the physical structure of the device under observation. The success of the attack depends on the signal to noise ratio and it's obvious that a low noise environment allows better measurements. This side channel attack also depends on the implementation of the cryptographic algorithm and not on the strength of the algorithm. There are two major classes of electromagnetic attacks and these are Simple Electromagnetic Analysis (SEMA) and Differential Electromagnetic Analysis (DEMA). SEMA is very efficient when it comes to asymmetric cryptography, while DEMA works well on symmetric cryptography. Zdenek Martinasek demonstrated in his paper [5] that it is quite easy to implement a SEMA attack on a device programmed with the AES, only by monitoring the amount of radiations. In 2016, Daniel Genkin developed an attacks based on electromagnetic radiations, that allowed the extraction of the ECDSA key from mobile devices [6].

Fault attacks are invasive, semi-invasive and non-invasive. These attacks rely on inducing faults in the normal mode of operation of a cryptographic device. Faults can be hardware or software. Most common fault attacks are clock glitches and voltage spikes. Another well studied fault attack is based on optical induction. Using this technique Sergei Skorobogatov and Ross Anderson managed to set and unset individual bits of a SRAM memory in a secure microcontroller.

This paper focuses on the *differential power analysis attacks* and describes the setup needed to perform a DPA attack on a secure 32 bit microcontroller and on a VISA smartcard.

III. POWER ANALYSIS

There are 3 important things to do to succeed in implementing a power analysis attack and these are:

- the measurements of the power consumption must be done with high accuracy;

- the encryption algorithm must be known;

- a set of plaintexts or cipher texts must also be known.

It is necessary to do a lot of measurements so that the amount of information gathered to be enough for finding the encryption key.

As described in [7], the physical mechanism of power analysis is based on the way a CMOS digital circuit works. Any change of state of a CMOS gate can be measured on the V_{DD} or V_{SS} pins. When the gates are clocked at the same time they all change their state, dissipating more power. The dissipated power of a circuit can be measured using a small resistor, typically 50 Ω. Any input transition in a digital circuit induces output transitions that cause a current flow on the V_{DD} or V_{SS} pins.

The mathematical model of the power consumption relies on the fact that the power consumption at time x is equal to the sum of the power dissipated of all gates at same time [7].

In Equation (1) is represented a simplified mathematical model of power consumption:

$$P(t) = \sum_g f(g,t) + N(t) \qquad (1)$$

The function $f(g,t)$ represents the power consumption of the gate g at the time t and the function $N(t)$ represents the noise components.

To correctly estimate the values of the power consumption, estimators must be use. There are some well known methods to construct estimator. There are:

- Maximum-Likelihood method;

- Using the empirical moments.

Both methods lead to the same results. The probability distributions can be distinguished if their moments can be distinguished.

A practical method to calculate the power consumption of a microcontroller having implemented a cryptographic algorithm, like AES, is based on the following procedure:

1. Considering the first byte of the plaintext, the effect of its MSB on power consumption is analyzed. If the value of it is "1", it goes to the class of ones. Else, it goes to the class of zeroes.

2. The means of these two classes are processed.

3. After processing the means, the difference of them is calculated. The resulting difference contains a few peaks that reveals the moments when the power consumption depends on the value of the MSB.

This procedure is also called the *correlation coefficients* algorithm. It is described in details in [2], where the authors present their studies on 8-bit microcontroller having implemented the AES128 algorithm. To analyze the power consumption, they studied the effect of the MSB for 1000 plaintexts with the corresponding traces. With these, they formed 2 groups of ones and zeros. After calculating the means of these groups, they processed the difference of them.

In figure 2 are represented the two means and the corresponding difference of them:

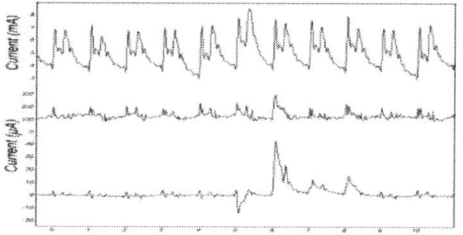

Fig. 2 Power consumption measurements [1]

The next step in finding the encryption key from the power analysis refers to the structure of the AES algorithm. AES128 has 128 key bits. The plaintext is exclusive-ored with the key. The result is processed with the SubBytes function. The output of this function depends only on the input plaintext and on the encryption key. SubBytes is a byte orientated function so applying the same procedure as described before, we can find each encryption byte. Consider the MSB or the LSB of the first byte resulted from the SBOX. This value depends on the key byte, so there are only 255 possible values of the byte and this can be guessed. For each possible value of the key, the means and the difference are recalculated and the result is plotted. The plot corresponding to the key match has a spike at the corresponding byte value.

IV. IMPLEMENTATION AND RESULTS

The purpose of this paper is to introduce a way to setup and implement a *Differential Power Analysis* attack on a 32 bit microcontroller and on a secure smartcard.

Usually these attacks are applied on 8 bit microcontrollers, because AES works on bytes, using an assembler implementation and triggering the exact moment to start capturing data. In this case, the novelty comes with the AES implementation that was used - *tiny-AES128*, developed in *C* programming language, the 32 bit microcontroller used - STM32F407VG, without any cryptographic accelerator and the banking card.

In Fig. 3 is represented the setup measurement that was used for measuring the power consumption of the microcontroller. A development board, STM32F4 Discovery, which has the 32 bit microcontroller mentioned above, was used.

Fig. 3 Microcontroller measurements setup

On the GND pins of the microcontroller, a 50 Ω resistor was wired. Using a 6000 series PicoScope, with a bandwidth of 250 MHz and a resolution of 8 bits at 5 GS/s, the power consumption of the microcontroller was measured by attaching the scope's probes at the resistor's pins.

The microcontroller was programmed using IAR Embedded Workbench for ARM 7.30 compiler. The *tiny-AES128* implementation is available for download on any dedicated site and is a version of the AES implementation optimized for low resource integrated circuits. The development board was connected to a PC and the scope was connected both to the PC and to the development board. The encryption key that was set is *E1 8E C0 E5 D7 21 30 27 07 02 98 9B EB BB 4D 7E*. The key was sent to the microcontroller using a RS232 cable. 1000 plaintexts were encrypted using the encryption key and 1000 traces resulted.

After obtaining these traces, them and the plaintexts were loaded in Matlab, where they were processed using the *correlation coefficients* algorithm, which is described in section III. For each key byte, a key guess was done using the algorithm mentioned. Two groups of ones and zeroes were created based on the value of the LSB of the first byte resulted after the SubBytes operation. A graphic was plotted when a peek representing the key guess was detected. For example, for the first byte of the encryption key, the following graph was plotted:

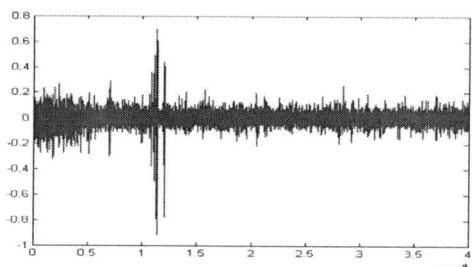

Fig. 4 Resulting plot of the first key byte

The first key byte is 225 (0xE1 in hex representation) and this value is shown in the command window.

The graph for the second key byte contains 2 significant peeks, but only one of them (the highest) reveals the exact match:

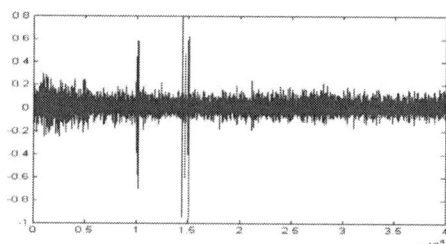

Fig. 5 Resulting plot of the second key byte

The corresponding value is shown in the command window and it has the value 142 (0x8E in hex representation).

This procedure is repeated successfully for every key byte, until the entire encryption key is revealed. In this situation, the DPA attack worked with no difficulty.

The next setup measurement was done using a secure smartcard, more precisely a credit card. It is difficult to distinguish the encryption algorithm implemented in it since there's no available information regarding the structure of the integrated circuit contained or of the cryptographic algorithm.

Unfortunately, a DPA attack couldn't be applied because by monitoring the traces of the power consumption, only noise was seen. Noise is a countermeasure used to prevent side channel attacks. By adding noise to the algorithm performing the encryption, it becomes more difficult to extract the information about the encryption algorithm using a DPA/SPA attack.

SPA revealed some periodic signals but no information could be extracted. The tools and the setup that was prepared are represented in Fig. 6:

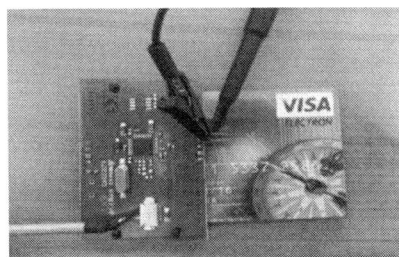

Fig. 6 Setup measurements for smartcard

In Fig. 6, a card reader Omnikey 3121 was used. To measure the power consumption of the smartcard, the GND lines of the PCB representing the card reader, were connected to a 50 Ω resistor.

Using the scope, measurements were taken by connecting its probes to the resistor's pins. A Java application was used to send APDU commands to the chip inside of the smartcard, but with no success. The traces obtained by monitoring the power consumption are represented in the following picture:

Fig. 7 Power consumption of the smartcard

Besides of noise, there are 3 places in the trace, where some processing takes place, but since there's no information about what is inside of the chip, there's no possibility to exploit the power consumption information.

If the card would have been programmable, then it would have been much easier to get some information and even find the encryption key of the algorithm implemented in the chip.

DPA attacks are usually applied to smartcards that are programmable and they have a high rate of success. But in this case, it was not possible to apply a side channel attack.

V. CONCLUSIONS

The field of side channel attacks is in continuous development because of the proved efficiency of this type of attacks. No matter the strength of the cryptographic algorithm implemented in a device, if the implementation is not properly protected, then there are leakages that can be exploited.

By applying power analysis attacks we can explore the vulnerabilities of the cryptographic algorithms implementations inside of different chips and determine countermeasures to prevent these attacks. In this paper we will give details about the DPA correlation algorithm and details about the setup needed to apply power analysis attacks.

In this paper, a study of the state-of-the-art of side channel attacks was made and also was described the setup needed to be done for a successful DPA attack to take place. It's very important to have knowledge in statistics, mathematic, programming, and electronics to be able to implement this attack.

The attack implemented on the 32 bit microcontroller was done successfully, obtaining the encryption key. It was proven that this type of attack can be applied on any commercial microcontroller and on a C/C++ implementation of the AES algorithm. The next step consist in studying the secure smartcards and finding methods to extract the noise from the power consumption traces, so that algorithm processing to be observed and analyzed correctly.

Further work will focus on developing new techniques to apply DPA attacks on all kind of embedded circuits no matter the countermeasures implemented.

REFERENCES

[1] Paul Kocher, Joshua Jaffe and Benjamin Jun,"Differential Power Analysis", 1995;

[2] Mangard Stefan, Oswald Elisabeth, Popp Thomas, "Power Analysis Attacks Revealing the Secrets of Smart Cards", Springer, 2007;

[3] Pierre-alain Fouque, Sebastien Kunz-Jacques, Gwenaelle Martinet, "Power Attack on Small RSA Public Exponent", Proceedings of 2006 Cryptographic Hardware and Embedded Systems, Yokohama, Japan, October 10-13, pp 339-353.

[4] Daniel J. Bernstein, "Cache-timing attacks on AES", 2005

[5] Zdenek Martinasek "Simple Electromagnetic Analysis in Cryptography", International Journal of Advances in Telecommunications, Electrotechnics, Signals and Systems, 2012

[6] Daniel Genkin, Yuval Yarom, Eran Tromer "ECDSA Key Extraction from Mobile Devices via Nonintrusive Physical Side Channels", 2016

[7] Manfred Aigner and Elisabeth Oswald, "Power Analysis Tutorial"

[8] YongBin Zhou, DengGuo Feng "Side-Channel Attacks: Ten Years After Its Publication and the Impacts on Cryptographic Module Security Testing"

2016 IEEE 22nd International Symposium for Design and Technology in Electronic Packaging (SIITME)

Machine-to-Machine Communications for Cloud-Based Energy Management Systems within SMEs

George Suciu, Octavian Fratu
Telecommunication Department
University POLITEHNICA of Bucharest
Bucharest, Romania
george@beia.ro

Lucian Necula, Adrian Pasat, Victor Suciu
R&D Department
BEIA Consult International
Bucharest, Romania
lucian.necula@beia.ro; adrian.pasat@beia.ro;
victor.suciu@beia.ro

Abstract— Nowadays, in order to reduce costs and ensure a proper working environment for employees, Small and Medium Enterprises (SMEs) concern themselves with the adoption of technologies and methodologies that could potentially help them efficiently monitor and manage resources. This paper starts with an overview on the benefits that Machine-to-Machine (M2M) communications can provide to the business environment, since the technology can be used in a wide range of applications in order to ensure monitoring and optimal control functionality. The main purpose of the paper is to present authors' conceptual model of a Cloud energy management system based on M2M communications from sensors, which aims to help companies monitor and reduce energy costs while improving comfort at the workplace.

Keywords— *Machine-to-machine, M2M, Energy Management, Cloud, SME, Data Mining*

I. INTRODUCTION

In the past decades, monitoring the energy consumption within companies of all sizes and activity domains became a decision-making tool of great importance. Companies use data regarding their energy consumption in order to support the provision of more efficient services to the consumers and to implement energy efficiency measures, without losing sight of the environment [1].

Automation provides users complete control over their home or work environment while bringing energy cost to a minimum and enhancing comfort. Automation relies on the integration of centrally/remotely controlled appliances.

In the context of environmental policies and high energy costs, many smart home solution providers tried to adapt their products and services for the business sector, promising to help companies achieve a greater energy efficiency and enhance comfort within the working environment [2]. Based on this, we present our conceptual model of a Cloud energy management system relying on M2M communications between different network components.

Traditional energy monitoring and management systems heavily rely on gateways as a mean to transfer data from different types of sensors and actuators to the user. Thus, the data communication between the user and the remote terminals is indirectly achieved through an intermediary device – the gateway [3]. To overcome problems caused by bottlenecks at

the gateway's level, this paper introduces M2M technology as a viable solution for direct communication between devices and users within Cloud-based energy management systems.

The rest of the paper is organized as follows: Section II presents related work in the field of M2M communications and energy management systems, Section III describes the conceptual Cloud energy management platform relying on M2M communication, while Section IV concludes the paper.

II. RELATED WORK

A. M2M communications

In the last decade, M2M communications draw research community's attention mainly due to the emergence of wireless communications which can provide the infrastructure for M2M communications, due to progress achieved in developing software that allows devices to autonomously operate and due to the evolution of technology behind sensors and actuators used to collect information for M2M systems [4].

Communications between smart devices is usually characterized by low mobility, low data rate and low power consumption.

To interconnect smart devices (M2M components) while ensuring a covering transmission range, short-range communication technologies, such as Bluetooth, IrDa (Infrared Data Association), UWB (ultra-wideband) are being replaced by medium-range ones, such as ZigBee, Z-Wave, Thread and Wi-Fi.

Due to the fact that the cellular network is the most widely spread wireless network around the globe, it provides the infrastructure for developing M2M communications between remote devices (sensors, actuators, etc.) distributed over a wide geographic area.

B. Energy Management Systems using Cloud Business Models

At international level there are companies that provide building automatization and modernization to enhance energy efficiency, security and comfort within homes and working environments.

WINS [5] is monitoring and controlling technology with applicability in environmental monitoring, transportation, health care, manufacturing and security. Compared to wireline

978-1-5090-4446-7/16 $31.00 © 2016 IEEE   114   20-23 Oct 2016, Oradea, Romania

sensors and actuators systems, WINS can be embedded and distributed at a lower cost.

DEXcell Energy Manager [6] is an energy management tool that provides advanced utility bill tracking, widget-based dashboards, energy patterns, benchmarks and cost analysis. The platform supports the integration of different protocols and devices from several manufacturers.

Engage [7] is an energy management platform developed by Efergy which provides real-time statistics related to energy usage and associated costs, thus making the users understand their energy consumption habits. The user gets access to the platform services through a web based dashboard.

Cisco Energy Management Suite [8] is a platform that allows its users to manage energy usage of devices connected to the network by providing data regarding energy use, costs, savings, pollutant emissions and inefficient systems and operational practices.

ZigBee Smart Energy [9] is a standard that allows users to develop and deploy technologies for monitoring, managing, informing and automating resource consumption. Among other functionalities, it provides metering support, demand response and load control, pricing information, event scheduling, alarming and tunneling of manufacturer specific protocols.

BEMOSS ™ [10] is an energy management platform designed to enhance sensing and control of devices in commercial buildings. It specifically aims to optimize energy consumption and provide support for demand response (DR) programs implementation. It also aims to accelerate the development of market-ready products such as embedded energy management systems.

These solutions available on the market aim to solve energy management problems. More than ever, consumers are looking for new ways to reduce waste and pollution while increasing the overall efficiency. Smart energy management technologies play a major role in making this possible by providing relevant resource consumption information to the user.

III. CLOUD ENERGY MANAGEMENT PLATFORM RELYING ON M2M COMMUNICATION

Using the benefits that M2M communications and Cloud environments have to offer, we are presenting a conceptual model of a Cloud platform that can help SMEs monitor and manage energy consumption, thus lowering costs and saving resources.

The system can potentially be integrated within the existing electric network infrastructure. Using the information provided by the sensors and on the instructions sent to the actuators, the control system combines human with software based control to ensure automation, thus complying with users' needs. The automation network ensures that all M2M components can exchange information regarding the status and control of the system. The architecture heavily depends on the computational capabilities of the devices that build the solution, for example home networks having lower performance [11]. To address the requirements of resource-dependent devices and IoT (Internet of Things) scenarios, M2M communications require the

integration and convergence of different communication systems and protocols, as depicted in Fig 1.

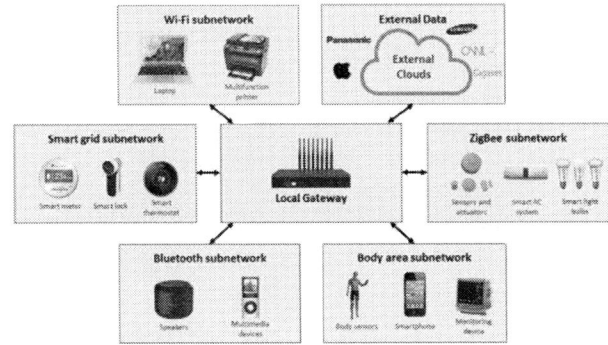

Fig. 1. Proposed M2M network architecture

The M2M network is a heterogeneous network which relies on a Gateway which manages the whole network and connects the network with the outside world through Internet. The Gateway is also responsible of access control, security management, multimedia conversion and QoS/QoE management.

The conceptual platform aims to monitor data in real time by using a wide variety of sensors capable of measuring environmental parameters (temperature, water leak, ventilation, lighting, etc.) and the amount of energy consumed in building. Monitored parameters can be successfully archived for a further analysis or presented in different graphical forms that would comply with users' needs. As presented in Fig. 2, for the gathering of environmental data, the platform will use smart sensors deployed inside the SME and a Cloud Middleware to import data from external systems.

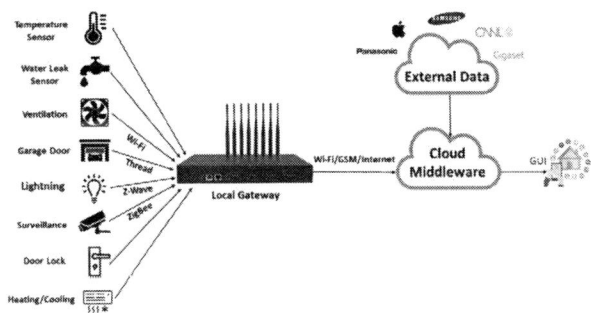

Fig. 2. M2M communication for energy-aware resource monitoring and management

Beside the energy consumption, the platform measures comfort parameters such as the weather overview, intensity of solar radiation, temperature, relative humidity, air pressure, precipitation level, wind intensity, air quality etc. thus increasing comfort and safety in working environments.

In addition to that, the platform provides to the user the means to remotely control and survey the building by the means of an online web application. Through machine learning, the Cloud platform will be able to detect movement patterns and typical interactions and use this information to automate decisions. In addition to the data collected directly by sensors, the platform also aims to import data from third party commercial energy management Clouds (Samsung, Gigaset, Apple, Panasonic, OWL etc.), thus ensuring support for wide variety of sensors and other appliances.

In order to provide cost efficiency and flexibility for the deployment of the conceptual platform, we considered the integration of wireless M2M communications to connect devices located in remote areas even with limited accessibility. A fast and cost effective interconnection between M2M devices can be achieved by using medium-range communication technologies as ZigBee (IEEE 802.15.4), Wi-Fi (IEEE 802.11), Z-Wave and Thread.

When compared to Wi-Fi and Bluetooth, ZigBee, Z-Wave and Thread devices only require a few milliseconds to wake up from sleep states. Table I presents different characteristics of Wi-Fi, Bluetooth, ZigBee, Z-Wave and Thread.

TABLE I. COMPARISON BETWEEN COMMUNICATION PROTOCOLS

Characteristic	Wi-Fi	Bluetooth	ZigBee	Z-Wave	Thread
Frequency bands	2.4GHz; 5GHz	2.4GHz	2.4GHz; 868MHz; 915MHz	2.4GHz; 868,4MHz; 908,4MHz	2.4GHz
Operating range	< 100 m	< 10 m	< 35 m	< 30 m	< 30 m
Maximum number of devices (QoS)	32	7	65.000	230	300
Network topology	Tree; Peer-to-Peer	Tree; Peer-to-Peer	Tree; Cluster Tree; Mesh; Star	Mesh	Mesh
Bandwidth (kb/s)	11.000	720	250	100	250
Encryption	CCMP128	AES64; AES128	AES128	AES128	AES128

For the proposed platform, ZigBee is a viable solution due to its low power requirement (both for transmission and reception of data) and simplicity of network configuration and management. ZigBee provides a communication range of 10-100 meters while maintaining a low power consumption of 1-100 mW. ZigBee allows the creation of a cluster tree, mesh or star topology network, thus ensuring a great flexibility of M2M devices configuration.

Machine learning will allow the platform to make predictions and decisions based on inhabitants' activity patterns, thus optimizing comfort, security and productivity within the work environment. The proposed platform will rely on context awareness as a key element for maximizing comfort, safety and energy efficiency while minimizing user's explicit interaction with the environment. By using prediction algorithms to automate interactions between the user and the environment, the system removes or at least diminishes the need for manual control of devices. The automatic actions can sometimes be detrimental to the user if the action needs to be undone or if it causes any damage. To eliminate the occurrence of such events, the prediction logic should be accompanied by episode discovery prediction which identifies significant episodes within users' event history. A significant episode is an

event that occurs at a regular interval or it is a response to an initial event called "trigger". System automatization can be achieved based on the discovered pattern significance and the predictive accuracy of upcoming events. Devices will be able to automatically figure out how they are supposed to work together and only trusted devices will be able to join the network and exchange data with other network devices [12].

IV. CONCLUSIONS

The paper presents a conceptual model of a Cloud energy management platform relying on M2M communication which aims to help companies monitor and reduce energy costs and adopt cleaner energy sources of energy while improving comfort at the workplace.

The benefits of adopting energy management solutions may also consist of major improvements on how the company is perceived by its stakeholders and its sales.

As future work, we intend to provide support for a wide range of sensing devices from various producers which use protocols such as Wi-Fi, Bluetooth, ZigBee, Z-Wave and Thread for M2M communication.

ACKNOWLEDGMENT

The work has been supported in part by UEFISCDI Romania through the project Power2SME and under grants no. 20/2012 "Scalable Radio Transceiver for Instrumental Wireless Sensor Networks - SaRaT-IWSN", grant no. 262EU/2013 „eWALL" support project, grant no. 337E/2014 "Accelerate" project and by European Commission by FP7 IP project no. 610658/2013 "eWALL for Active Long Living - eWALL".

REFERENCES

[1] M. Schulze, et.al. "Energy management in industry–a systematic review of previous findings and an integrative conceptual framework", Journal of Cleaner Production, vol. 112, pp. 3692-3708, 2016.

[2] P. Kumar, C. Martani, L. Morawska, L. Norford, R. Choudhary, M. Bell, M. Leach, "Indoor air quality and energy management through real-time sensing in commercial buildings", Energy and Buildings, vol. 111, pp. 145-153, 2016.

[3] Yang, Xiang, and Hui-hong Wang. "The design and implement of embedded M2M smart home system." Communication Software and Networks (ICCSN), 2011 IEEE 3rd International Conference on. IEEE, 2011.

[4] Niyato, Dusit, Lu Xiao, and Ping Wang. "Machine-to-machine communications for home energy management system in smart grid." IEEE Communications Magazine 49.4 (2011): 53-59.

[5] S. Park, S.W. Hong, E. Lee, S.H. Kim, and N. Crespi, "Large-scale mobile phenomena monitoring with energy-efficiency in wireless sensor networks", Computer Networks, vol. 81, pp. 116-135, 2015.

[6] A. A. Serrano, "Systems for energy efficiency management of users in office buildings", Universitat Politecnica de Catalunya,pp.1-20, 2015.

[7] A. Santos, C. Resende, R. Marques, and A.C. Lima, "EnAware: A Comprehensive and Scalable Energy Management and Awareness Solution", Fraunhofer Portugal AICOS report, pp. 1-6, 2014.

[8] B.L. Capehart, and T. Middelkoop, "Handbook of web based energy information and control systems" The Fairmont Press, Inc.2011.

[9] C. Gezer and C. Buratti, "A ZigBee Smart Energy Implementation for Energy Efficient Buildings", Vehicular Technology Conference (VTC Spring), IEEE 73rd, pp. 1-5, 2011.

[10] W. Khamphanchai et al., "Conceptual architecture of building energy management open source software (BEMOSS)," IEEE PES Innovative Smart Grid Technologies, pp. 1-6, 2014.

[11] Zhang, Yan, et al. "Home M2M networks: architectures, standards, and QoS improvement." IEEE Communications Magazine 49.4 (2011): 44-52.

[12] S. K. Das, D. J. Cook, A. Bhattacharya, E. O. H. Iii, and T.-Y. Lin, "The role of prediction algorithms in the mavhome smart home architecture," IEEE Wireless Communications, December 2002.

Mouse and Display Driver on a Single Microchip Tested on FPGA and Built for an ASIC

Roland Szabo
Applied Electronics Department
Fac. of Electronics and Telecom., Politehnica Univ.
Timisoara, Romania
roland.szabo@upt.ro

Aurel Gontean
Applied Electronics Department
Fac. of Electronics and Telecom., Politehnica Univ.
Timisoara, Romania
aurel.gontean@upt.ro

Abstract—**This paper presents the creation of mouse and display drivers on a single microchip. First it was created the mouse driver on PS/2 interface and after it was created the display driver on VGA port. The drivers were implemented in hardware using VHDL on an FPGA. After this they were converted from VHDL code to Verilog code using tools from Mentor Graphics, it was created the layout of the ASIC. After this it was created the mouse and display driver on a single microchip.**

Keywords—ASIC; display; driver; FPGA; microchip; mouse; PS/2; Verilog; VGA; VHDL.

I. INTRODUCTION

This paper presents the whole creation of a single microchip with VGA and PS/2 drivers for display and mouse [1].

The goal was to create a desktop screen and move the mouse on it [2]. The whole system started to get shape after the drivers were created for PS/2 and VGA [3], [4]. After that the background needed to be colored and the mouse pointer needed to be drawn. For simplicity a square was drawn for the mouse pointer and it was colored with a different color than the background [5]-[7].

The whole system was made using the NEXYS 2 development board form Digilent in VHDL code on Spartan-3E FPGA.

After this the code was converted with Mentor Graphics tools to Verilog code, the layout of the microchip was made to finalize the whole project in a standalone ASIC, which can do the desired task: to create a desktop on a computer screen, where it is displayed the movement of a mouse with a mouse cursor [11]-[13].

The NEXYS 2 board uses 10 lines of data from the FPGA to create an 8-bit color VGA port and two standard synchronization lines (HS – Horizontal Sync, VS – Vertical Sync). The color signals use a resistive divider circuit of 75 Ω. The VGA display creates eight signal levels of the red and green lines, and four signal levels of the blue line.

A VGA controller circuit must generate Vertical Sync signals – VS, Horizontal Sync – HS signals and the coordinate

delivery of video data based on a pixel clock. Pixel clock defines the time available for displaying information to a single pixel. VS signal defines the refresh frequency of the screen, after this, all the information on the screen is redrawn. Minimum refresh frequency is a function of the phosphor of the screen and the intensity of electronic spot. Basically refresh frequency is in the range of 50-120Hz. For a display of 480 lines by 640 pixels per line, using a pixel clock of 25 MHz and a refresh of 60 +/- 1 Hz, the timings of the signals are shown in Fig. 1. Sync pulse width times for intervals front and back porch (these intervals are times pre- and post-synchronization, during which information cannot be displayed). This information is based on observations taken from real VGA displays.

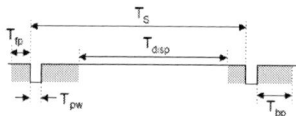

Symbol	Parameter	Vertical Sync			Horiz. Sync	
		Time	Clocks	Lines	Time	Clks
T_S	Sync pulse	16.7ms	416,800	521	32 us	800
T_{disp}	Display time	15.36ms	384,000	480	25.6 us	640
T_{pw}	Pulse width	64 us	1,600	2	3.84 us	96
T_{fp}	Front porch	320 us	8,000	10	640 ns	16
T_{bp}	Back porch	928 us	23,200	29	1.92 us	48

Fig. 1. Temporization for a 640x480 resolution.

A VGA controller circuit (Fig. 2) decodes the counter output Horizontal Sync, which is controlled by the pixel clock, to generate horizontal synchronization times. This counter can be used to locate any pixel on a given line. Similarly, the output of the vertical synchronization counter that is incremented with each pulse of HS can be used to generate the time of vertical synchronizing VS and the counter may be used to locate any given line. These two counters (which are continuously operating) can be used to address the RAM.

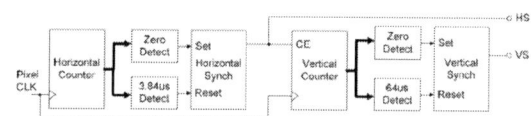

Fig. 2. VGA controller circuit.

The authors would like to thank the AFCEA Chapter from POLITEHNICA University of Bucharest for its support.

The 6 pin mini-DIN connector can be used to connect a mouse or a keyboard. Most of these can be powered from 3.3 V, but the old ones from 5 V. The 3.3 V or 5V supplies should be set with a jumper on the board.

Both the mouse and keyboard is using a 2-wire communication bus (clock and data) to communicate with the host device. Both uses words of 11 bits, which includes start, stop bits and an odd type parity bit, but data packets are organized differently. The keyboard allows bidirectional transfer (the host device can illuminate the keyboard's LEDs). Bus Timing is shown on Fig. 3. Clock signals and data are generated only when there is data transfer, otherwise they are kept on idle on "1" logic. On the FPGA can be implemented a PS/2 interface.

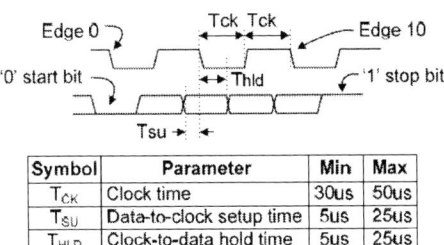

Symbol	Parameter	Min	Max
T_{CK}	Clock time	30us	50us
T_{SU}	Data-to-clock setup time	5us	25us
T_{HLD}	Clock-to-data hold time	5us	25us

Fig. 3.   Timing of the PS/2 signal.

The mouse generates a clock signal each time is moved; otherwise these signals remain on "1" logic. Whenever the mouse is moved, it sends three 11-bit words to the host device. Each 11-bit word containing a "0" bit of start, 8 data bits (LSB first), and then an odd type of parity bit, and it is finished with a stop bit on "1" logic. Each transmitted data includes 33 bits, where bits 0, 11, 22 are "0", the start bits and bits 11, 21, 33 are stop bits on "1" logic. The three 8-bit data fields contain data for movement as shown on Fig. 4. Data is valid on the falling edge and the clock frequency is between 20 and 30 KHz.

Mouse movement involves relative coordinate so right movement generates a positive number and left movement a negative number in the X field. An upwards movement generates a positive number and a downwards movement a negative number in the field below Y. XS and YS are signs of bits, where "1" is the sign of a negative number. XY and YY is the overflow of motion, when it is set to "1". Fields R and L indicates left and right button on the mouse when it is set to "1" logic.

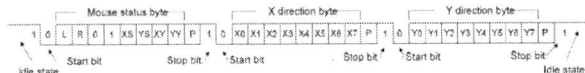

Fig. 4.   Data format for mouse.

II.   PROBLEM FORMULATION

There was the NEXYS 2 development board, a PS/2 mouse and VGA display available.

The challenge was to create the ASIC which can control

these two devices and to create a simplified desktop system where the mouse movement can be seen on a screen. The whole system could be similar to a computer desktop, where it can be seen the mouse as cursor and it can be made some operations executed when buttons of the mouse are clicked.

III.   PROBLEM FORMULATION

First the NEXYS 2 board was studied, after the I/O ports and the timing diagram of the used protocols.

The understanding of the timing diagram was a crucial part of making everything to work.

On Fig. 5 there is the block diagram of the experimental setup. There is a display connected to the NEXYS 2 board on VGA interface and a mouse connected to the same NEXYS 2 board on PS/2 interface. There is no PC; the NEXYS 2 board plays the role of a PC, by creating an environment where a mouse movement can be shown on a display.

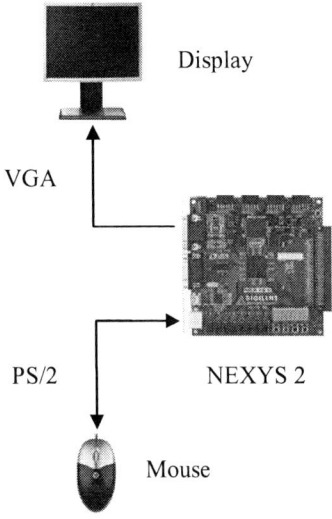

Fig. 5.   Block diagram of the experimental setup.

Finally it was created a microchip with the whole system. The microchip implemented the VGA and PS/2 drivers as sown on Fig. 6.

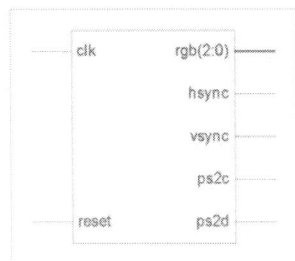

Fig. 6.   VGA and PS/2 drivers on a single microchip.

If the microchip is unfolded it can be seen its structure. On Fig. 7 there is the structure of the microchip. Inside the created microchip there are 3 other microchips, one of the PS/2 protocol, a D-latch for the 3 RGB colors and another microchip for the horizontal and vertical synchronization.

If the circuits are unfolded more, it can be seen that the circuit is not simple at all.

These integrated circuits have inside quite complicated circuits and some of them are combined, there are not only circuits separately for mouse and circuits separately for display, these things needs to go along each other so they need to be combined. Upon this these is also a microchip which draws the mouse cursor, in our case a green square, this is the graph circuit.

These circuits were actually generated with Xilinx ISE tools after compiling the VHDL codes.

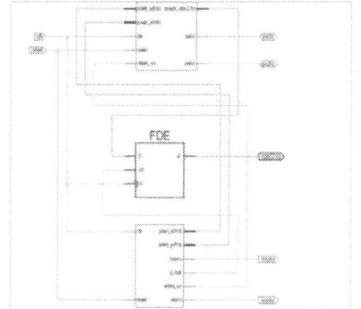

Fig. 7.   The structure inside of the VGA and PS/2 driver microchip.

On Fig. 8 there is the mouse controlling circuit. This circuit enables the movement on the screen and the synchronization between the drawn square and the mouse output. This circuit basically combines together the movement of the mouse and the graphical part for the user to see the mouse cursor.

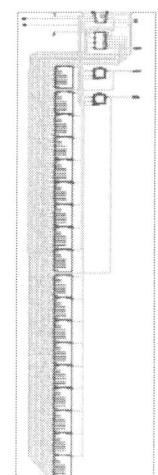

Fig. 8.   The mouse controlling circuit.

On Fig. 9 there is the PS/2 RX controller circuit which enables the sending the receiving of the data on the PS/2 interface.

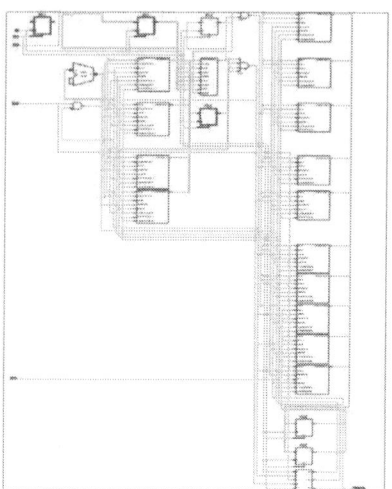

Fig. 9.   The PS/2 RX controller circuit.

On Fig. 10 there is the PS/2 TX controller circuit which enables the sending of the data on the PS/2 interface.

The PS/2 circuit communication circuits are basically the driver of the PS/2 interface. This protocol is also used for the keyboard too. Even in the newer USB keyboard and mice the same PS/2 protocol is used, but with minor adaptations for the USB interface.

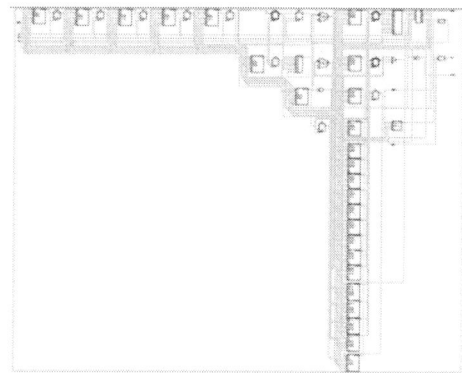

Fig. 10. The PS/2 TX controller circuit.

On Fig. 11 it can be seen the structure of the graph circuit. This circuit actually draws the mouse cursor, the green square and paints the background of the desktop (white in this case).

This is circuit which needs to be changed if the look of the mouse pointer or the color of it is needed to be changed, or the color of the background of the desktop is needed to be replaced.

From this circuit it can be changed the background by putting an image in the background.

It can be also displayed some other icons which can be clicked with the mouse.

It can be also made some bigger objects which can move in a certain way when they are clicked with the mouse.

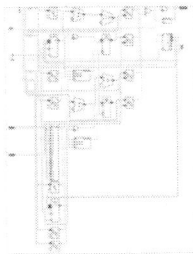

Fig. 11. The structure of the graph circuit, the drawing of the mouse cursor and the coloring of the desktop.

On Fig. 12 is presented the result of the output of the microchip. It can be seen the picture of a CRT display which shows the desktop with white background and the mouse cursor which is displayed as a green square. All this is running on an FPGA board.

Until now it was implemented the function to change colors of the background and of the mouse cursor. There were implemented 8 colors easily (red, green, blue, cyan, magenta, yellow, black white) and even more colors with combinations. There was also implemented the function to change the shape of the cursor from a square to a spot or to a different shape, for example an arrow-like shape. The shape was need to be loaded binary in the ROM memory as a binary shape. The graph circuit than can take the shape from the ROM memory and display it as the pointer of the mouse.

It can be seen how complex can it be only a mouse driver to build. It can be imagined how hard it can be to create multiple interfaces and PC microcontrollers. Only for the mouse there are separate circuits for the RX part and for the TX part of PS/2 port, this makes only the PS/2 driver. After this there is needed a mouse controller circuit which uses the PS/2 driver. Finally there is needed a graphical circuit which actually draws the cursor of the mouse and the background of the display. This circuit combines the VGA display driver and the mouse driver with the PS/2 driver.

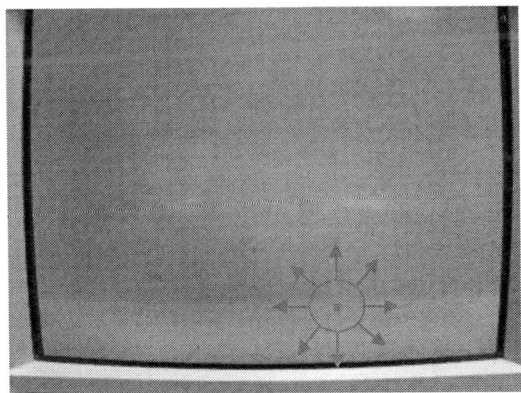

Fig. 12. The output of the created microchip, the square as the mouse cursor can be moved with the mouse and displayed on the CRT display.

IV. CONCLUSION

As it could be seen it was created a microchip which can act as the base of a graphical desktop system. It was created a desktop which displayed the mouse movement on a display with a cursor.

The system can have many improvements, but is the base of the microchip which can control a VGA display and a PS/2 mouse.

It can be added a lot o functions like clickable objects on the desktop, but the best improvement would be to extend the drivers to other interfaces too. Future plan is to add to the system also a PS/2 keyboard too. Further enhancement would be to port the microchip on another development board from Digilent, like the ATLYS board and create the mouse and keyboard drivers on USB interface and the display drivers on DVI and on HDMI interfaces.

Finally after the work will be finished, the plan is to actually create an ASIC and to package the silicon die to actually create a physical microchip.

ACKNOWLEDGMENT

The authors would like to thank the AFCEA Chapter from POLITEHNICA University of Bucharest for its support.

REFERENCES

[1] Ye Xien, Ye Zhiquan, "Implementation of PS/2 Mouse and Keyboard Based on Windows CE," International Forum on Computer Science-Technology and Applications, Chongqing, vol. 3, 2009, pp. 318-321.

[2] Punj Pokharel, Binod Bhatta, Anand D. Darji, "Optimized drivers for PS/2 and VGA using HDL," IEEE International Conference on Computer Science and Automation Engineering, Shanghai, vol. 3, 2011, pp. 262-266.

[3] Guoping Zhang, Man-de Xie, "Design of visual based-FPGA Ping-Pang game with multi-models," Second Pacific-Asia Conference on Circuits, Communications and System, Beijing, vol. 2, 2010, pp. 31-34.

[4] Deep Vardhan Bhatt, D. Du Toit, Gerhard P. Hancke, "Design of a controller for a universal input/output port," IEEE International Instrumentation and Measurement Technology Conference, Graz, 2012, pp. 647-652.

[5] Kyu-Sam Sam Lim, Kon-Woo Kwon, Heeju Ju Park, Jeong-Hun Hun Kim, Suki S. Kim, Jun-Jea Sung, Kwang-Hyun Hyun Baek, "Design an infrared wireless optical mouse system and a dual-band infrared receiver," 15th IEEE International Conference on Electronics, Circuits and Systems, St. Julien's, 2008, pp. 810-813.

[6] Noriyuki Aibe, Moritoshi Yasunaga, "Reconfigurable I/O interface for mobile equipments," IEEE International Conference on Field-Programmable Technology, 2004, pp. 359-362.

[7] Liu Kai, Yang Yuliang, Zhu Yanlin, "Tetris game design based on the FPGA," 2nd International Conference on Consumer Electronics, Communications and Networks, Yichang, 2012, pp. 2925-2928.

[8] R. Szabó, A. Gontean, "Image Acquistion with Linux on FPGA," 22nd Telecommunications Forum, Belgrade, 2014, pp. 1007–1010.

[9] R. Szabó, "Creation of the Chips Placement Game with Backtracking Method in Borland Pascal," International Symposium on Electronics and Telecommunications, Eleventh Edition, Timişoara, 2014, pp. 85-88.

[10] R. Szabó, A. Gontean, "Pong Game on FPGA with CRT or LCD Display and Push Button Controls," Federated Conference on Computer Science and Information Systems, Warsaw, 2014, pp. 735-740.

[11] R. Szabó, A. Gontean, "Creating a Serial Driver Chip for Commanding Robotic Arms," Federated Conference on Computer Science and Information Systems, Kraków, 2013, pp. 671-674.

2016 IEEE 22ⁿᵈ International Symposium for Design and Technology in Electronic Packaging (SIITME)

Creation of a Fight Game in Borland Pascal with the Possibility to be Ported on an FPGA

Roland Szabo

Applied Electronics Department
Fac. of Electronics and Telecom., Politehnica Univ.
Timisoara, Romania
roland.szabo@upt.ro

Aurel Gontean

Applied Electronics Department
Fac. of Electronics and Telecom., Politehnica Univ.
Timisoara, Romania
aurel.gontean@upt.ro

Abstract—**This paper presents the creation of a fight game. The whole game was created using Borland Pascal programming language and after it was finished it was ready to be ported on an FPGA. It was used the graph unit to draw all the parts of the fighters. The body parts of the fighters are quite complex, starting from circles to trapezoids. The fighters have also eyes and mouth. It was also made the functionality for them to move and to hit. After a fighter gets a hit from a punch, it can be observed a decrease of his health; this can decrease until it will go to 0, which means the fighter lost the fight.**

Keywords—*computer's intelligence; fight game; graph unit; platform creation; shape drawing; structure movement.*

I. INTRODUCTION

In this paper it was planned to create a fight game in the Borland Pascal programming language for porting it to FPGA.

The Borland Pascal programming language was chosen, because it is quite simple to work with the graph unit and draw different shapes.

There are predefined shapes for drawing usual geometric objects and there is possibility to draw arbitrary shapes too [1].

The fight game is very interesting to create because of its complexity of movements and drawing [2].

In a normal arcade game the character can only run and jump, but in a fight game the character can do different body movements. In arcade game the character can do different tasks, but only optionally changing its body parts' positions, but in fight games changing body parts' positions is compulsory [3].

II. PROBLEM FORMULATION

The task was to create an interesting game but yet very simple to use and fun to play.

It was chosen the creation of a fight game, because the focus is on the characters and not on the background like on arcade games. This added a little simplicity to the project, because it was not need to create a fancy background which changes time to time.

The fighters need to be drawn and it is needed to create the

The authors would like to thank the AFCEA Chapter from POLITEHNICA University of Bucharest for its support.

model of their movements.

Basically the goal was to create two fighters in white karate kimono which can fight with each other, hit each other and lose health if they are hit. The fighter which loses the most health will lose the fight too. In order to make a difference between the two fighters in case they change position, so like in karate competitions, one of them has red and the other has blue belt.

The other goal was to add computer's intelligence too, this way a player can play against a computer.

This fight game is the base of the next project, which is intended to implement this fight game on a single ASIC, the goal is to create a game on a chip. First it is needed to port the game from software to FPGA and then to create the layout of the chip and create the silicon die of the chip with the game.

III. PROBLEM SOLUTION

First it was created a plan, what is the goal and what are the movement combinations.

It was decided to create the whole background in black for simplicity; it can be said that the fighters fight in dark, as this is usual for fights to be at night.

This can also make simplicity in the algorithm, because for the player to change positions we can just overlay on it a big black square and draw it a little bit shifted, according to the position where is intended to be moved. It can be used the overlaying black square on the fighter's hands or on other moving parts of the fighter to simulate movements.

The fighter's skin is colored in yellow, their kimonos are white or darker shades of white and to be able to make difference between them, one has red and the other has blue belt. It was also made the kimono of one player light gray, this way to make easier difference between them.

The palms of the hand are circles and the head is oval. The necks, legs, hands and some parts of the belts are parallelograms. The bodies and some parts of the belts are squares. The feet are trapezoids.

Basically there were created 3 functions one to create trapezoids, one to create squares, one to create parallelograms, one to create circles and one to create ovals. After this these functions were called and their parameters were configured which was the color which filled them and the size of them, like the size of their borders and in case of circle the centre and the radius.

On the top of the screen it can be seen the health of the players, which are red dots and after one fighter it is hit, the number of dots decreases until no more dots are left and this it is known the fighter which lost the match.

On Fig. 1 it can be seen a print screen of the fight game.

On Fig. 2 it can be seen a picture of the LCD screen made with a photo camera when the fight game was played in full screen.

As it can be seen, it was created a simple, yet fun to play game, which is focused on the complexity of the characters and not on the background.

Next it will be presented the character movement procedure in Borland Pascal programming language. This program part has many procedures which are called; these procedures are mostly drawing procedures which draw the parts of the characters.

```pascal
procedure move;
  begin
  d:=readkey;
  if d=chr(115) then
    begin
      if (x2<=470) then
        begin
          clearviewport;
          b1:=450;
          b2:=451;
          a1:=a1-9;
          a2:=a2-25;
          figure2(a1,a2);
          y1:=450;
          y2:=451;
          x1:=x1+50;
          x2:=x2+66;
          figure1(x1,x2);
        end;
    end;
  if d=chr(97) then
    begin
      if (x1>=100) then
        begin
          clearviewport;
          b1:=450;
          b2:=451;
          a2:=a2-25;
          a1:=a1-9;
          figure2(a1,a2);
          y1:=450;
          y2:=451;
          x1:=x1-70;
          x2:=x2-54;
          figure1(x1,x2);
```

```pascal
        end;
    end;
  if d=chr(54) then
    begin
      if (a2<=470) then
        begin
          clearviewport;
          y1:=450;
          y2:=451;
          x1:=x1-45;
          x2:=x2-29;
          figure1(x1,x2);
          b1:=450;
          b2:=451;
          a2:=a2+61;
          a1:=a1+77;
          figure2(a1,a2);
        end;
    end;
  if d=chr(52) then
    begin
      if (a1>=100) then
        begin
          clearviewport;
          y1:=450;
          y2:=451;
          x1:=x1-45;
          x2:=x2-29;
          figure1(x1,x2);
          b1:=450;
          b2:=451;
          a2:=a2-59;
          a1:=a1-43;
          figure2(a1,a2);
        end;
    end;
  if d=chr(32) then
    begin
      x1:=x1+35;
      x2:=x2+2;
      y1:=y1+17;
      y2:=y2+18;
      c:=white;
      hand1(x1,y1,x2,y2);
      x:=4;
      y:=4;
      c:=yellow;
      x1:=x1+4;
      y1:=y1+6;
      circle(x1,y1,x2,y2);
      x1:=x1+6;
      x2:=x2+20;
      y1:=y1+4;
      y2:=y2+10;
      delay(200);
      c:=black;
      hand1(x1,y1,x2,y2);
      y1:=y1+6;
```

```pascal
    x1:=x1-20;
    c:=yellow;
    circle(x1,y1,x2,y2);
    x1:=x1-25;
    x2:=x2-22;
    y1:=y1-13;
    y2:=y2-8;
  end;
if d=chr(48) then
  begin
    a1:=a1-35;
    a2:=a2-2;
    b1:=b1+17;
    b2:=b2+18;
    c:=lightgray;
    hand1j(a1,b1,a2,b2);
    a:=4;
    b:=4;
    c:=yellow;
    a1:=a1-4;
```

```pascal
    b1:=b1+6;
    circlej(a1,b1,a2,b2);
    a1:=a1-6;
    a2:=a2-20;
    b1:=b1+4;
    b2:=b2+10;
    delay(200);
    c:=black;
    hand1j(a1,b1,a2,b2);
    b1:=b1+6;
    a1:=a1+20;
    c:=yellow;
    circlej(a1,b1,a2,b2);
    a1:=a1+25;
    a2:=a2+22;
    b1:=b1-13;
    b2:=b2-8;
  end;
end;
```

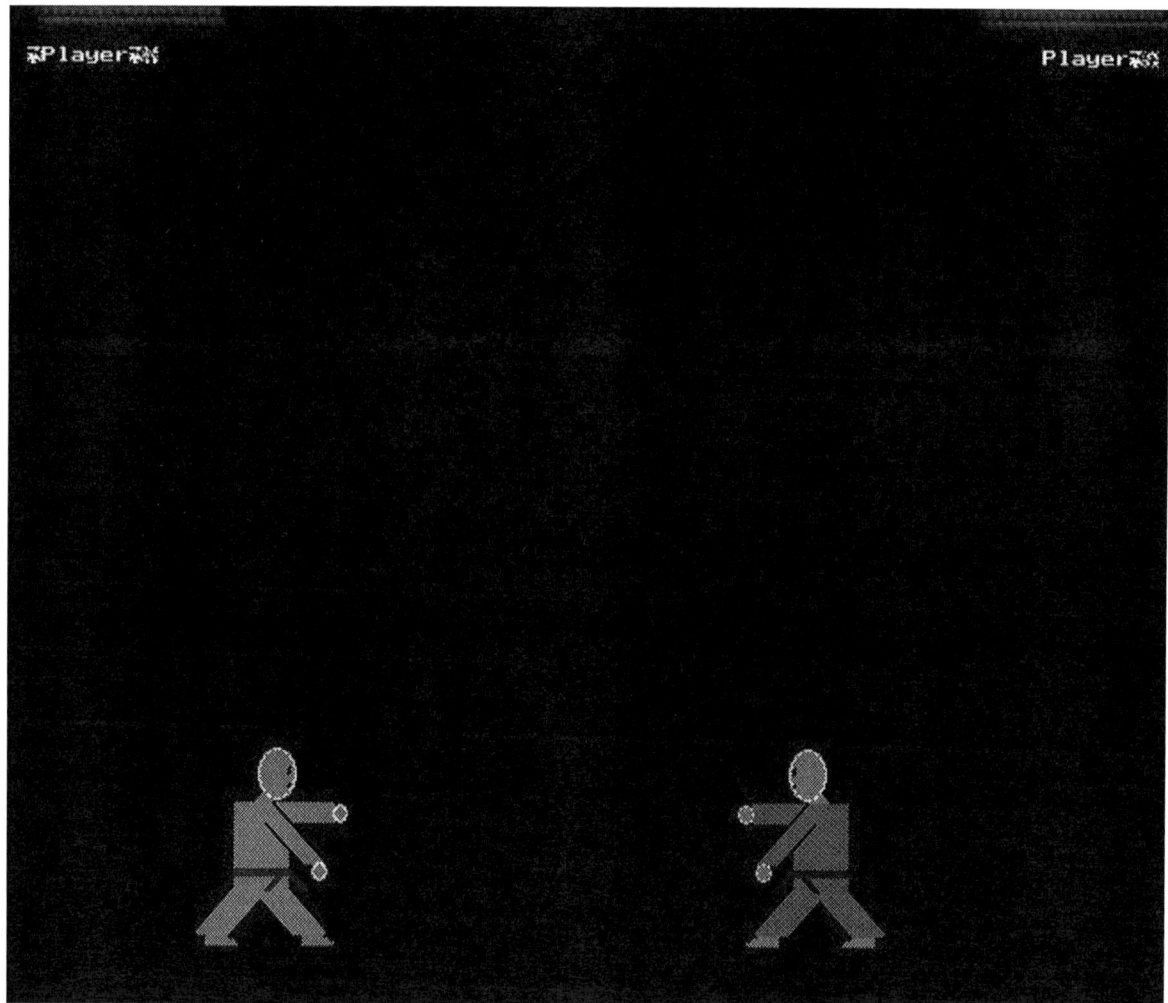

Fig. 1. Print screen of the fight game made in Borland Pascal.

2016 IEEE 22nd International Symposium for Design and Technology in Electronic Packaging (SIITME)

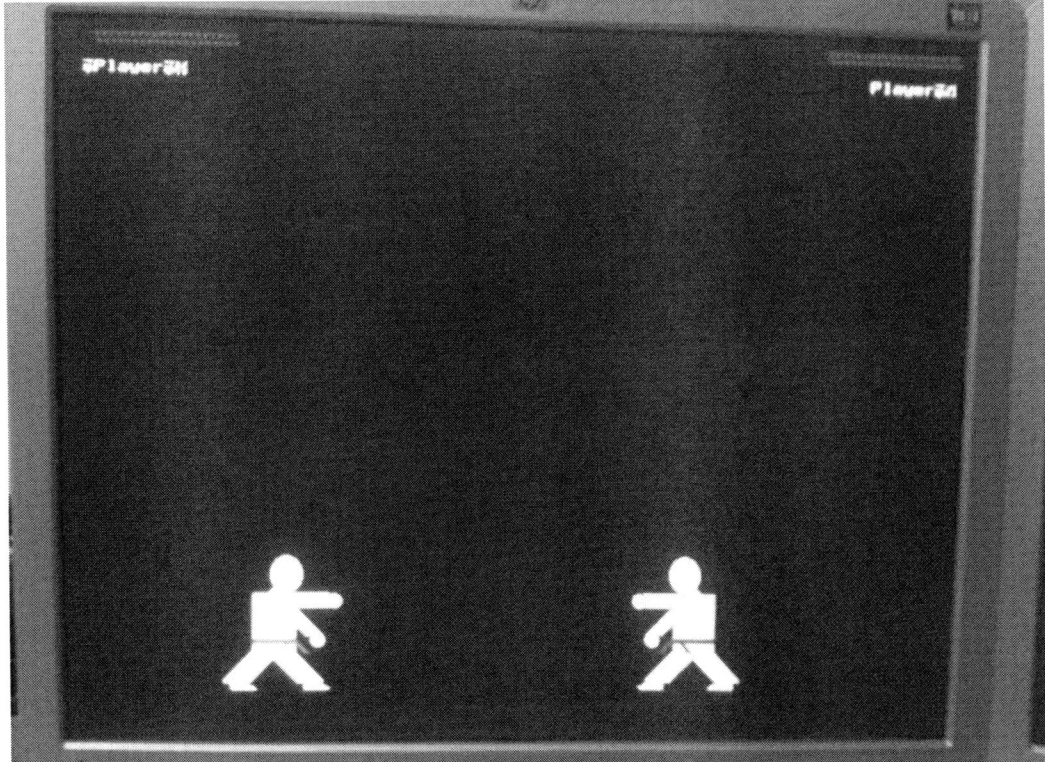

Fig. 2. Picture made with photo camera of the fight game created in Borland Pascal when displayed in full screen on an LCD monitor.

IV. CONCLUSION

As it could be seen, it was created a full 2D fight game. The characters were drawn from geometrical objects and they were moved with functions. It was created functions to draw trapezoids, parallelograms, circles and ovals by configuring their parameters the characters could be built up. Their parameters were size and color configuration.

It was used the black square overlaying technique to move different parts of the fighters.

Finally there were created 2 fighters in white kimono with blue and red belts and yellow skin. The background in black and the health level is drawn with red dots, the number of these decreases when a fighter gets a hit until he loscs the fight.

The next goal is to create the game in VHDL language, to port it to FPGA.

After this with the tools from Mentor Graphics the code can be converted in Verilog language, after this it can be created the layout of the silicon die and it can be packaged, this way it can be created a microchip with the fight game, this way it will be obtained a game on a chip, like a TV game cassette.

ACKNOWLEDGMENT

The authors would like to thank the AFCEA Chapter from POLITEHNICA University of Bucharest for its support.

REFERENCES

[1] Jun-Ichi Kushida, Iori Nakaoka, Kazuhisa Oba, Katsuari Kamei, "Learning System Using Hierarchical Fuzzy ART for Two-Player Games," Fourth International Conference on Innovative Computing, Information and Control, 2009, pp. 1019-1022.

[2] Huai-Che Lee, Chia-Ming Chang, Jui-Shiang Chao, Wei-Te Lin, "Realistic Character Motion Response in Computer Fighting Game," Ninth IEEE International Symposium on Multimedia, 2007, pp. 169-175.

[3] Nicolas Pronost, Franck Multon, Qilei Li, Wei-Dong Geng, Richard Kulpa, Georges Dumont, Interactive Animation of Virtual Characters: Application to Virtual Kung-Fu Fighting," International Conference on Cyberworlds, 2008, pp. 276-283.

[4] Simardeep S. Saini, Christian W. Dawson, Paul W. H. Chung, "Mimicking player strategies in fighting games," International Games Innovation Conference, pp. 44-47.

[5] S. Saini, P. W. H. Chung, C. W. Dawson, "Mimicking human strategies in fighting games using a Data Driven Finite State Machine," 6th IEEE Joint International Information Technology and Artificial Intelligence Conference, vol. 2, 2011, pp. 389-393.

[6] R. Szabó, A. Gontean, "Pong Game on FPGA with CRT or LCD Display and Push Button Controls", Federated Conference on Computer Science and Information Systems, 2014, pp. 735-740.

[7] R. Szabó, "Creation of the Chips Placement Game with Backtracking Method in Borland Pascal", International Symposium on Electronics and Telecommunications. Eleventh Edition, 2014, pp. 85-88.

[8] Takashi Taneichi, Masashi Toda, "Fighting game skill evaluation method using surface EMG signal," 1st Global Conference on Consumer Electronics, 2012, pp. 106-107.

2016 IEEE 22nd International Symposium for Design and Technology in Electronic Packaging (SIITME)

DSP Based Interconnection Circuit of the Renewable Energy Sources to a Smart Grid

General presentation

Nistor Daniel Trip
Department of Electronics and Telecommunications
University of Oradea
Oradea, Romania
dtrip@uoradea.ro

Marius Ovidiu Neamțu
Department of Electronics and Telecommunications
University of Oradea
Oradea, Romania
oneamtu@uoradea.ro

Abstract—**This paper presents general considerations on a DSP based control circuit that can be used to connect different types of renewable energy sources that supply electrical energy into a smart grid, where the droop control method is involved for an efficient energy management. The main purpose of the circuit is to control the amount of energy transferred from the renewable energy sources to the smart gird, following in the same time and taking into account the smart grid frequency. When the management system of the smart grid uses droop control technique to allow the activation of different renewable energy sources connected to the grid, the grid frequency shift is small. In order to determine with high accuracy the smart grid frequency and also to assure a proper PWM control of the power inverter, the authors take into considerations a digital signal processor that offer to this problem a convenient practical solution.**

Keywords—DSP; digital filter; droop speed control; renewable energy sources;

I. Introduction

There is a great interest in the use at a high scale the renewable energy sources [1] especially those that produce electrical energy: solar energy converted in electrical energy with photovoltaic panel, geothermal energy converted in electrical energy with thermo-electro generators, to mention only few of them. As a remark here, the thermo-electro generators can be, in general, of two types: with thermal engines that produces mechanic torque and then this mechanical torque drive a synchronous motor [2], [3] or with solid state thermo-electric generators, or on brief TEG [4].

In many situations, renewable energy sources are used as building blocks for smart grids. In the case of the isolated smart grid [5], the management system of the available energy can use different kind of control algorithms, since in this case the grid is no more a conditioning factor. In the case when the management system of the smart grid uses droop speed control technique [6], to allow the activation of different renewable energy sources connected to the grid, the utility frequency has a key role. The frequency shift can determine the activation or deactivation of certain renewable energy sources. In this context, this paper presents a possible interconnection circuit that control the energy transfer from a renewable energy sources to the smart grid – see Fig.1.

As one can observe, the digital signal processing DSP control circuit accomplishes a digital filtering operation of the mains voltage, determines the utility frequency, and imposes then the results of a droop speed control algorithm to a PWM command block.

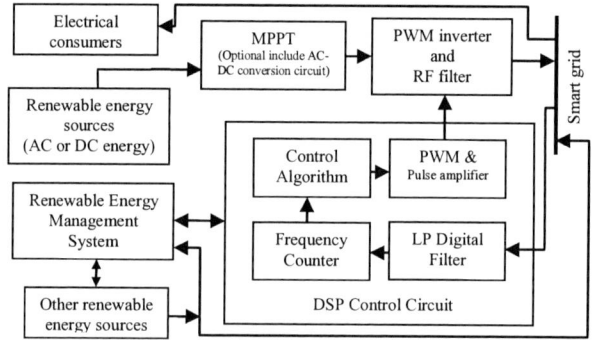

Fig. 1. Block diagram of the proposed circuit

The actual technology offer valuable processors that support the above mentioned tasks. For example, the processor TMS320F28087 [7] has important features that can be exploited in this specific application, such as: high efficiency 32 Bit CPU, special instructions for signal processing, analog-to-digital converter, PWM modules, capture module, a high resolution set of timers, and so on.

Each component of the DSP control circuit is implemented as subroutine of a main program that follows instructions from the main energy management unit. Taking into account this fact, in the next part of the paper the authors will discuss some aspects regarding the implementation of the specific tasks.

II. Principle of Operation

A smart grid comprises in addition to the renewable energy sources an energy management system, see Fig.1. The management system is responsible with the control and good operation of all renewable resources integrated into the grid. In general, the smart grid may comprise a main renewable energy source that has the highest probability to operate even in

978-1-5090-4446-7/16 $31.00 © 2016 IEEE   126   20-23 Oct 2016, Oradea, Romania

difficult operation conditions. As an example we can mention thermo-electro generators that exploit the geothermal energy. Such sources contain a synchronous motor that operates as an electric energy generator. In such situation, when the electrical energy consumption increases, the grid frequency of the grid suffers a small decreasing since the synchronous motor speed decreases. The energy management system detects this change and tries to activate other renewable sources to solve the energy demand. For this purpose, the energy management unit has a communication bus used to transfer commands and information with control unit of the renewable energy sources. When the communication cannot be established between the main unit of the smart grid and the renewable energy source, a possible control method of the system could be based on the droop speed control technique. It means that, at a certain value of the grid frequency, imposed before by the operator, an additional renewable energy source starts its operation, providing energy into the smart grid. If the additional energy is not enough and the grid frequency continues to decrease its value, other renewable energy sources will start their operation. When the electrical energy consumption decreases, the grid frequency increase its value and successively, in reverse order of the starting, the additional renewable energy sources will pass in standby mode. One can observe that in this mode of operation it is very important to know exactly the value of the grid frequency, problem discussed later in the paper.

III. IMPLEMENTATION MATTERS

A fraction of the grid voltage is brought to the analog-to-digital converter of the processor by means of a scaling circuit and an optocoupler to assure the isolations against the grid. The scaling circuit has to adjust the range of the grid voltage to the full range of the analog-to-digital converter input, range comprised between 0 and 3.3 V. For this reason, a voltage divider and a voltage shifter are used as one can see in the Fig. 2. This match can be designed with the help of Matlab / Simulink program [8].

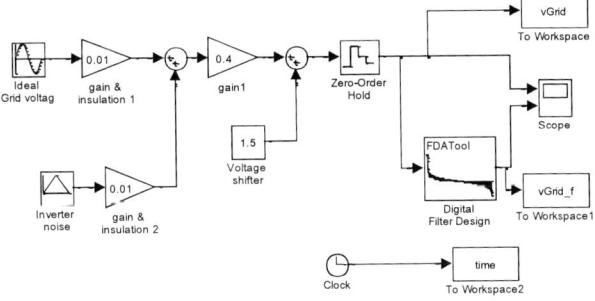

Fig. 2. Test circuit to design the filter.

Since the analog-to-digital converter of the processor has a resolution of 12 bits, the equivalent digital value of the input voltage [7], digitized at any one time, can be obtained with (1).

$$Digital\ value = 4096 \cdot \frac{Input\ analog\ voltage}{3.3} \qquad (1)$$

However, if the input voltage overpass 3.3 V, the maximum digital value is limited to 4095, due to the next binary coding

condition: $N = 2^n - 1$, where N is the digital value or an integer number that can be represented in binary code with the help of n bits.

After the digitization, the input signal is processed further with a FIR algorithm. This step is compulsory since on the grid are many perturbing signals, especially if the grid includes different types of inverters. The authors prefer FIR filtering due to the fact that these digital filters offer a good stability and have a predictable phase delay. The design of such filter may be accomplished with Simulink / FDA Toolbox [8], see Fig. 2. This design tool offers the possibility to choose different types of digital filters, with different characteristics. The software will calculate automatically the coefficients of the filter and the user can export them for example in Q15 format as a header file. In this case, the FIR filter has a band pass characteristics, see Fig. 3, where the cut frequencies are $F_{C1} = 40$ Hz and $F_{C2} = 60$ Hz. The attenuation of the filter has to be more than 40 dB outside the band pass of the filter.

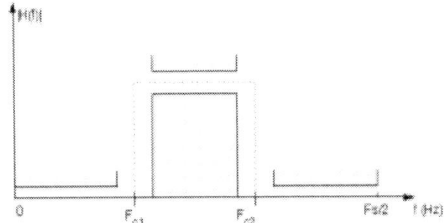

Fig. 3. Band pass filter characteristics imposed with Simulink / FDA Toolbox.

It is well known that the order of the filter influence its response. However, for a high FIR order, an important amount of resources such as data memory is used. This filtering also introduces a significant phase delay. Due to the fact that the processor kept in sight has a high computation power, the processing time and the memory resources are negligible. Fig.4, Fig. 5 and Fig. 6 present the main characteristics of a FIR with the next design parameters: order of the filter is 128, Hamming windowing, sampling frequency 24 kHz and the pass band comprised between 40 and 60 Hz, obtained with FDA Tool.

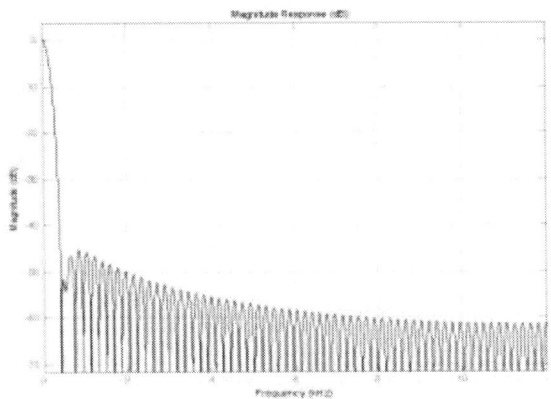

Fig. 4. Magnitude response in dB for a FIR, the order of the filter is 128, sampling frequency of 24 kHz and Hamming windowing.

2016 IEEE 22nd International Symposium for Design and Technology in Electronic Packaging (SIITME)

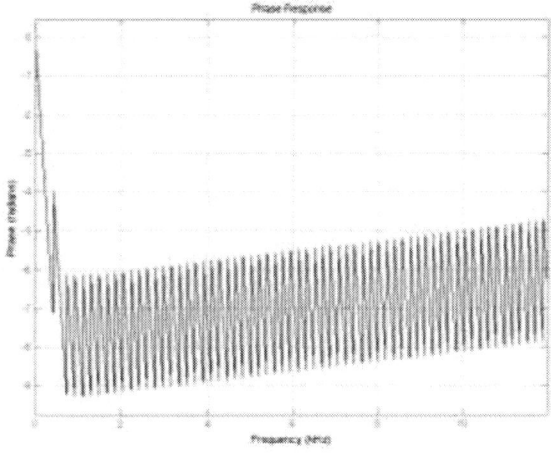

Fig. 5. Phase response for a FIR, the order of the filter is 128, sampling frequency of 24 kHz and Hamming windowing.

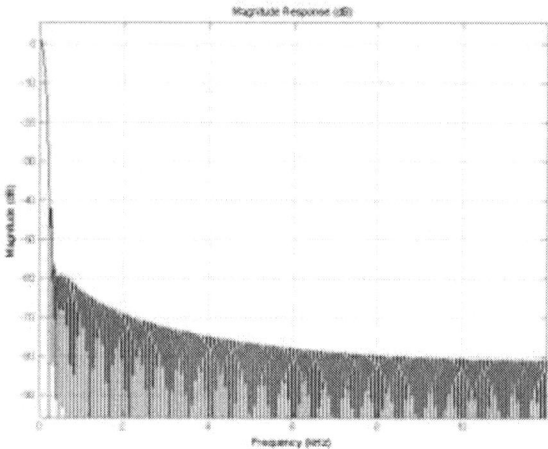

Fig. 7. Magnitude response in dB for a FIR, the order of the filter is 256, sampling frequency of 24 kHz and Hamming windowing.

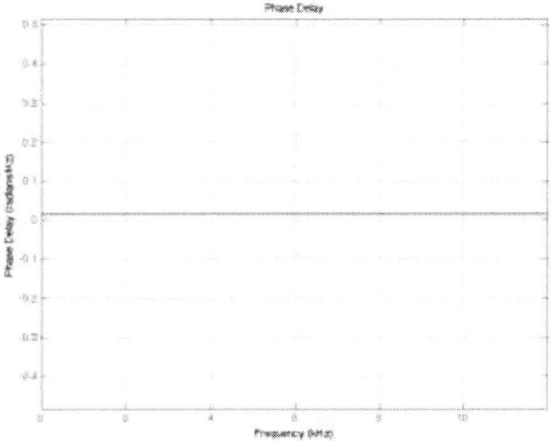

Fig. 6. Phase delay for a FIR, the order of the filter is 128, sampling frequency of 24 kHz and Hamming windowing.

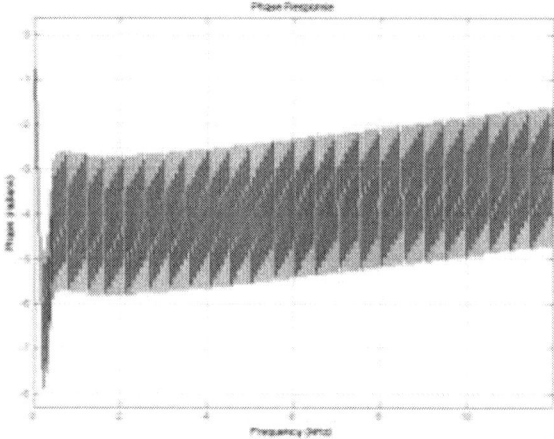

Fig. 8. Phase response for a FIR, the order of the filter is 256, sampling frequency of 24 kHz and Hamming windowing.

The digital filter response can be improved if the filters order becomes 256. The main characteristics of the digital filter are now better, as one can see in the Fig.7, Fig.8 and Fig.9.

The filtering is more efficient then in the previous case even the phase delay of the last filter is lower. This thing is not so important since we need to determine in fact the grid frequency. This thing becomes very important when we wish to use this circuit to synchronize the power inverter, see Fig. 1, with the grid. The efficiency of the digital filter and in the same time the efficiency of the filtering algorithm that will be implemented on the DSP can be evaluated by simulation with the help of the circuit shown in Fig.2. The simulation results are depicted in Fig.10 and Fig.11. First signal diagram mentioned in Fig.10 and Fig.11 represents a proportional value of the voltage grid, at the DSP input, when it is affected by a possible 1000 Hz triangular perturbation. This perturbation could be introduced in the smart grid by an inverter that has not a good output filtering circuit.

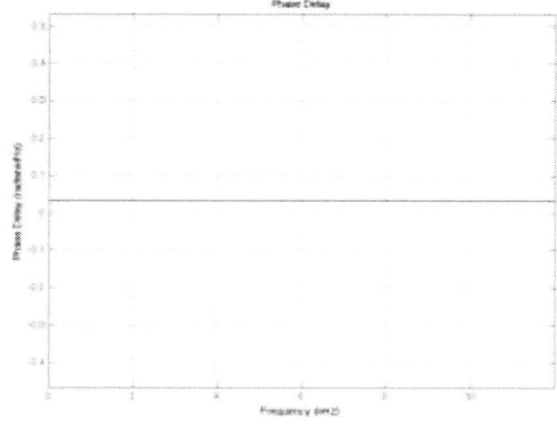

Fig. 9. Phase delay for a FIR, the order of the filter is 256, sampling frequency of 24 kHz and Hamming windowing.

2016 IEEE 22nd International Symposium for Design and Technology in Electronic Packaging (SIITME)

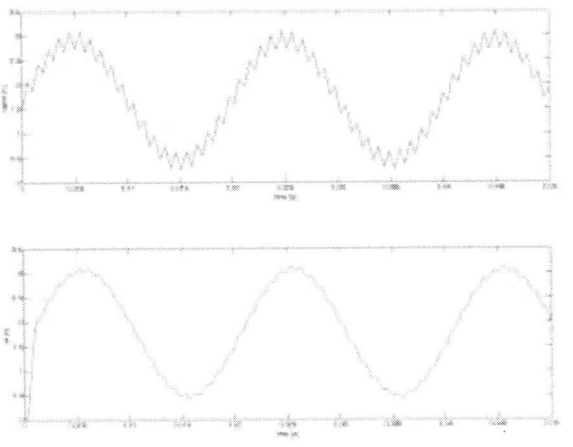

Fig. 10. Simulation results for a FIR, filter order 128, sampling frequency of 24 kHz, Hamming windowing and band pass between 40 and 60 Hz.

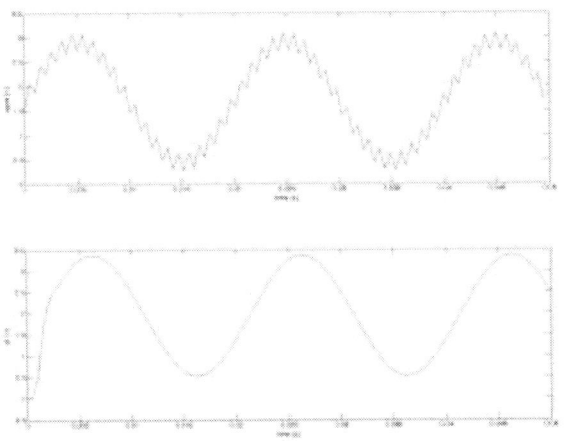

Fig. 11. Simulation results for a FIR, filter order 256, sampling frequency of 24 kHz, Hamming windowing and band pass between 40 and 60 Hz.

Once it has been obtained the clean proportional grid voltage signal, the next task is to determine the exact value of the grid frequency. An efficient method to determine this parameter is to count the number of the sampling periods during the two consecutive peaks of the filtered signal. Since the sampling frequency is 24 kHz, the measurement precision in this case is around 0.002, value that is good enough for this application. The algorithm will increment the content of an internal register when a result of the sampling process is generated. Based on this indexed value it is determined the grid frequency with a good precision. Further, at a frequency f1, smaller then the grid frequency without any load, the processor will start the inverter that interconnects the renewable energy source to the smart grid, see Fig.1. Then, the inverter continues to work for smaller or higher grid frequencies, until its value reach an imposed frequency f2, higher then f1, when the power

inverter is turned off again. At the start of the renewable energy source, the algorithm must assure a soft start option. That means that the duty cycle of the PWM command pulses for the power stage will be increased slowly from zero up to a nominal value required by application.

IV. CONCLUSIONS

The authors present some issues of principle on the use of digital signal processing control circuits to interconnect renewable energy sources to a smart grid. The droop speed control technique is also mentioned and put in value. The authors show then some design and implementation considerations, and finally describe simulations results. The design process pays a special attention to the digital filter since its characteristics may influence the value of the utility frequency. The higher the noise level, the lower the utility frequency precision. On the other hand, a complex filtering algorithm cannot increase without the limit the utility frequency precision. Taking into account these remarks, the authors use a digital filter that offers a good stability for the operation conditions.

ACKNOWLEDGMENT

This paper is supported through the programme "Parteneriate in domenii prioritare – PN II", by MEN – UEFISCDI, project no. 53/01.07.2014.

REFERENCES

[1] Aldo V. Da Rosa, "Fundamentals of renewable energy processes", Elesevier Academic Press, 2005.

[2] A. Setel, C. Antal, Dana Bococi, M. Gordan, „Considerations on the design of a low-power electric plant and analysis of influence factors", 13th International Conference on Engineering of Modern Electric Systems (EMES), pp. 1 - 4, DOI: 10.1109/EMES.2015.7158399, 2015.

[3] Aurel Setel, Mircea Gordan, Cornel Antal, Dana Bococi, „Use of geothermal energy to produce electricity at average temperatures", 13th International Conference on Engineering of Modern Electric Systems (EMES), pp.1-4, DOI: 10.1109/EMES.2015.7158398, 2015.

[4] P. A. J. Stecanella; M. A. A. Faria; E. G. Domingues; P. H. G. Gomes; W. P. Calixto; A. J. Alves, „Eletricity generation using thermoelectric generator - TEG", IEEE 15th International Conference on Environment and Electrical Engineering (EEEIC), pp. 2104-2108, DOI: 10.1109/EEEIC.2015.7165502, 2015.

[5] A. Grama, T. Patarau, E. Lazar, D. Petreus, "Estimating the Size of the Renewable Energy Generators in an Isolated Solar-Biodiesel Microgrid with Lead-Acid Battery Storage", Journal of Electrical and Electronics Engineering, Vol. 8, No. 2, pp. 15-18, October, 2015.

[6] T. Pătărău, D. Petreuș, R. Etz, E. Lázár, "Small Signal Induction Generator Model Connected in a Frequency-Droop Controlled Renewable Energy Microgrid", 39th International Spring Seminar on Electronics Technology – ISSE2016, Pilsen, Czech Republic, May 18-22, pp. 157-158, 2016.

[7] ***, TMS320F2802x Piccolo™ Microcontrollers, Texas Instruments Inc., data sheet SPRS523K, June 2016.

[8] http://www.mathworks.com/products/simulink/

2016 IEEE 22nd International Symposium for Design and Technology in Electronic Packaging (SIITME)

Improved Tamper Detection Circuit Based on Linear-Feedback Shift Register

D. C. Vasile, A. Marghescu, P. Svasta

CETTI

University POLITEHNICA of Bucharest

Romania

ciprian.vasile@cetti.ro

Abstract — **The paper presents an improved method to detect tamper intrusions based on an active circuit. It is composed of a logical part, a microcontroller, capable of generating pulses that follow the rule of a linear-feedback shift register (LFSR), and an analogical part made of a mesh network, used to cover secure modules, and a pulse forming circuit. Pulses resulted from this forming circuit are analyzed by the microcontroller to determine the durations between pulses and the durations of pulses. The novelty of this method is that the pulses are generated synchronously at both ends of the mesh network in order to prevent any attempts of an attacker to break the wire of the mesh network and to simulate the generation of pulses.**

Keywords—tamper, active, LFSR, mesh, security.

I. INTRODUCTION

Secure modules are logic systems that process and store secure data and they are used in many cryptographic devices such as: Point of Sale (POS), Hardware Security Module (HSM), cipher machines and so on. Electronic circuits inside them, that process secure data, must be protected against intrusions in order not to allow an unauthorized person to gain access to the secure data or to influence the correct functioning of processes in order to gain benefits. The most usual method to protect these circuits is to cover them entirely with a special foil or printed circuit board (PCB) containing a copper mesh, fine meanders of conductive material. This cover is connected to a tamper detection circuit that continuously verifies the integrity of the mesh and in case of intrusion attempt it takes the appropriate actions not to reveal the secrets (erase memory devices). The detection is made by two methods: passive and active. The passive method uses the current injection in the mesh and detects variation of it: either the dropping or the raising of the current. This method is less effective because it doesn't detect small variations of the current. It only detects the continuity of the electric circuit. The active method is more efficient and consists of probing the mesh in order to detect the continuity of the electric circuit and its length by sending pulses through the electric circuit and detecting the response at the other end of the mesh. If the delay differs from the correct value the detection circuit triggers the tamper event. This method is very efficient only if the copper mesh is properly designed. In this kind of active tamper detection circuits, as presented in [1],[2], an attacker will be able to reproduce and to delay pulses if he gains access to a well delimited area of copper traces that can be bypassed.

The active detection method proposed in this paper consists of probing the mesh by simultaneously sending pulses from both ends of it and analyzing the superposed pulses at each end of it. The analyzed pulses depend on the continuity and the length of the mesh and also depend on the duration of the pulses, essentially the way they overlap.

II. DESCRIPTION OF THE DETECTION CIRCUIT

The detection circuit is composed of a microcontroller, a mesh network and a pulse forming circuit. The microcontroller used in the experimental work is STM32F407, which is an ARM Cortex®-M4 processor running at 168MHz. It is part of a development board, STM32F4-DISCOVERY, that provides all the peripherals of the microcontroller it needs to run (power source, crystal oscillator, program loader/debugger) and connectors to access all usable pins of the microcontroller. This development board is the core of the experimental circuit, the mesh network and the pulse forming circuit are directly connected to this board. The schematic of the tamper detection circuit is presented in Fig. 1.

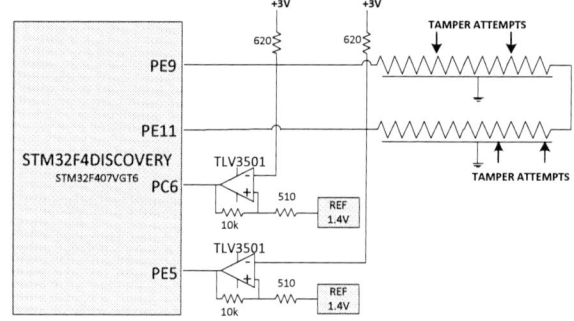

Fig. 1. The reference schematic of the tamper detection circuit.

The microcontroller generates pulses at both ends of the mesh network in order to confuse an attacker that wants to cut the mesh and simulate the pulses in both directions. The output pins PE9 and PE11 are connected to the mesh network and are configured as PWM output channels 1 and 2 of the Timer 1. The driving configuration of the output buffers is open drain.

The mesh network is a 13 x 25 cm PCB with copper meanders of 0.3 mm in width and spaced at 0.35 mm. The board is printed on one side and the other side is entirely copper foil connected to the ground of the development board.

978-1-5090-4446-7/16 $31.00 © 2016 IEEE   130   20-23 Oct 2016, Oradea, Romania

The pulse forming circuit is designed using the high speed comparator TLV3501 with added hysteresis in order to increase the noise immunity. The enlarged threshold region is equally distributed around the reference voltage of 1.4V.

Formatted pulses are collected with input pins PC6 and PE5 and analyzed with Timer 8 and 9, in PWM input mode.

III. MODE OF OPERATION

A. Generating probing pulses

Pulses are generated in accordance with the output of a pseudo-random number generator based on a linear-feedback shift registers (LFSR). LFSR generators can produce pseudorandom bit sequences with high period based on a shift register and a feedback function. The feedback function can be determined from a number of output bits that is only two times the length of the shift register [3]. In practice more complex configurations are used.

To increase the security of the LFSR based generators, cascaded configurations were developed using complex interconnections. The most known scheme is the Gollmann Cascade. As presented in [4], a modified Gollmann Cascade increases the complexity and security of the generator. The scheme is presented in Fig. 2 and works as follows:

- There are a number of n LFSRs;

- Each LFSR outputs a "state" consisting of an address (ADDR – the address of the output of the next LFSR), a CLK bit(specifying that the next LFSR will run once or will run twice before outputting the result), and the output bit OUT;

- The LFSRs (excluding the first one, which has a fixed output ADDR and CLK) depend on the previous ones;

- The output of the generator is a XOR operation over all of the LFSR's output bits.

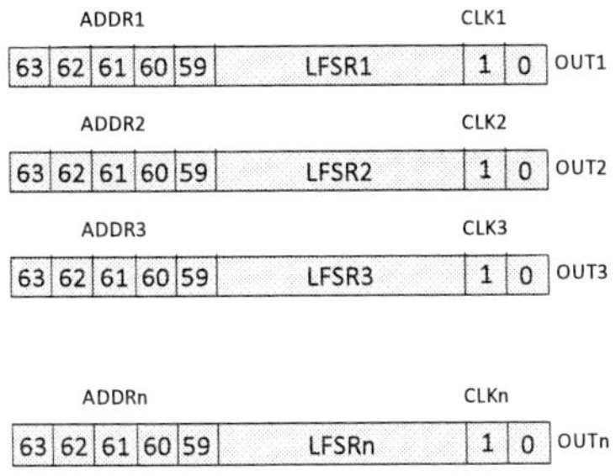

Fig. 2. LFSR-based Pseudo-Random Number Generator.

In [4] is experimentally demonstrated that using 7 LFSRs characterized by 64th degree polynomials, the statistical properties of the output are very good. This generator is used to decide which pulses are active on outputs PE9 and PE11, configured as alternate function 1. Timer 1, which controls these outputs, is configured in output PWM mode with frequency of 250 kHz. It is synchronized by the internal core clock of 168 MHz. At each time period of 4 μs, Timer 1 enters the interrupt routine where the program logic decides to generate or not a pulse depending on the output of the LFSR.

If the LFSR provides a zero then the outputs of the Timer 1 are disabled, otherwise the LFSR is clocked two times and the resulted two bits will determine the following cases:

- If the first bit is logic 1, then pulses generated by PE9 (TIM1_CH1) will have duration D1 and pulses generated by PE11 (TIM1_CH2) will have duration D2 $\pm \Delta t$, the sign depends on the second bit of the LFSR;

- If the first bit is logic 0, then pulses generated by PE11 will have the duration D1 and pulses generated by PE9 will have duration D2 $\pm \Delta t$, the sign depends on the second bit of the LFSR.

The relation between D1 and D2 is that D2 = D1 + d, where d is the delay introduced by the mesh network, in this particular case $d = 167$ ns. D1 equals to about three times d. Δt is experimentally chosen in order to accurately be detected by the microcontroller and has a value of 59 ns. Considering the case when PE9 generates one pulse of duration D1 and PE11 generates one pulse of duration D2, signals captured at PE11 and transformed in pulses by the forming circuit feed the PE5 pin. There are two possible pulses resulted from the composition of the local pulse and the remote pulse: one pulse of duration D1 + d + Δt and the other of duration D1 + d. The latter pulse results from the composition of the local pulse with duration of D1 + d − Δt and the remote pulse with duration of D1, delayed with d. At this point, an attacker will not be able to determine the value D1 because at PE9 it will be masked by the pulse generated from PE11. At PE11 the rising edge of the pulse is varying with $\pm \Delta t$ around the rising edge of the delayed pulse D1, masking it from being determined. In this way, the attacker that wants to cut the mesh network and to reproduce the pulses, in order to send them to the forming circuits, will not be able to detect the rising edge of the original pulses. This situation comes from the mesh connection mode with the open drain output buffers of PE9 and PE11 to obtain the AND logic function. The principle described above is presented in Fig. 3.

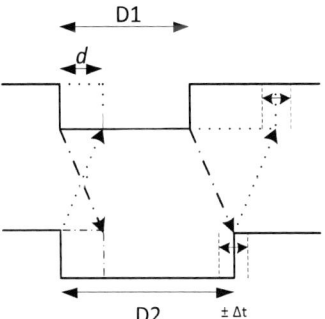

Fig. 3. Synchronization of probing pulses.

An attacker would face two steps in his challenge to cut the mesh network and simulate the pulses:

- Firstly, he must determine the parameters that characterize the pulses and the whole delay between the ends of the mesh network. The rising edges of the pulses are affected by the integral effect of the mesh network.

- Secondly, he must be able to know which channel of Timer 1, PE9 or PE11, would generate D1 and D2 pulses.

B. Mesh network as a transmission media

The conductive mesh network is characterized by the following parameters: the length of the conductive circuit and the distributed resistance and capacitance. These parameters affect the delay and the edges of the pulses. The mesh network may be characterized by a distributed RC model as presented in Fig. 4.

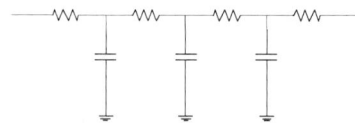

Fig. 4. The distributed RC model of the mesh network.

For this model the propagation delay can be computed as:

$$t_p = 0.38RC \tag{1}$$

and is defined at 50% of the final value, with R and C the total resistance and capacitance of the mesh network [5].

The equivalent lumped RC model of the mesh network (in a T configuration) connected to outputs PE9 and PE11 (configured as open drain) is presented in Fig. 5.

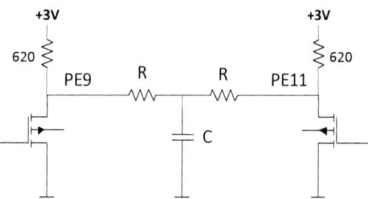

Fig. 5. The equivalent schematics of the mesh connected to PE9 and PE11.

For the experimental test, the pull-up resistors were chosen to have a resistance of 620 Ω, in order to be at least three times higher than the series resistance of the mesh network and to be as small as possible in order to reduce the charging time of the equivalent capacitor of the mesh network. The particular mesh network used in the experiment can be approximated as the T circuit presented in Fig. 5 with each resistor R of 91.4 Ω and the capacitor C of 2.4 nF. Considering these values and the expression (1), the computed propagation delay for the mesh network in this experiment is 167 ns. This value is verified by the measurements with the oscilloscope.

The voltage variation of the rising edge of the pulses follows the expression presented in [5]:

$$v_C = V_{cc}\left(1 - e^{-\frac{t}{R_e * C}}\right) \tag{2}$$

where V_{cc} is the voltage of the power supply and R_e is the equivalent resistance between capacitor C and the power supply. For the experiment, the expression is defined as:

$$v_C = 3\left(1 - e^{-\frac{t}{853.68 * 10^{-9}}}\right) \tag{3}$$

C. The pulse forming circuit

The pulse forming circuit is made with the high speed comparator TLV3501. Taking into account that the rising edges of the pulses are slow-moving and noisy, the comparator output can cause an undesirable ripple as input signals move through the switching threshold. In order to increase the noise immunity, the comparator was added an external hysteresis circuit that increases the hysteresis range to about 150 mV, based on resistors values presented in Fig. 1. This range is centered on the reference voltage. The value of the reference voltage (1.4V) is chosen neither to be too close to the lower rail, because of the pulse switching ripple, nor to be too close to the upper rail in order to miss the end of the pulse (the next pulse could switch to logic 0 before the full raise of the equivalent capacitor charge voltage). The two comparators used for each signal on pins PE9 and PE11 will form the pulses and provide them for analysis to pins PC6 and PE5. The formatted pulses have longer durations than D1 and D2 $\pm \Delta t$ but they are proportional to these values.

D. Analyzing detected pulses

The formatted pulses are analyzed by Timer 8 channel 1 (TIM8_CH1) and Timer 9 channel 1 (TIM9_CH1) connected to pins PC6 and PE5, configured as alternate function 3. These timers are configured in PWM input mode and they detect the period and the duration of pulses. Both timers are configured to generate interrupts at the falling edge of the pulses that travels through the mesh network, corresponding to the rising edges of pulses formatted by the TLV3501 comparators. Inside the interrupt routines, the period and the duration are compared to the expected values and if they are not equal, the microcontroller will generate the tamper event. Timer 8 and Timer 9 are clocked at the frequency of 168MHz, just like the Timer 1. This means that the resolution for detecting delays is minimum 5.9 ns.

IV. EXPERIMENTAL RESULTS

The tamper detection circuit used in the experimental testing is presented in Fig. 6.

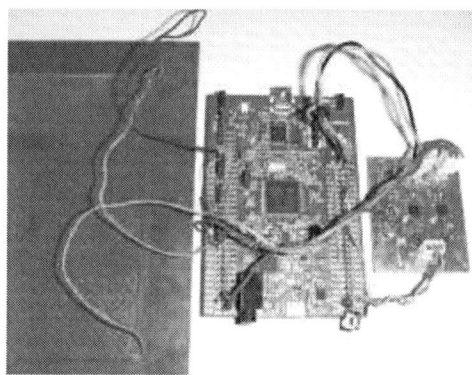

Fig. 6. The experimental tamper detection circuit.

This tamper detection circuit doesn't allow an attacker to determine the pulses parameters in order to cut the mesh network and generate fake pulses as to mislead the tamper detection circuit. Figure 7 presents an example of the pulses generated and injected in the mesh network (channel 1) and the same pulses inside the mesh at about 50% of its length (channel 2). The variation $\pm \Delta t$ of pulses is observable in this figure. Channels 3 and 4 show the formatted pulses.

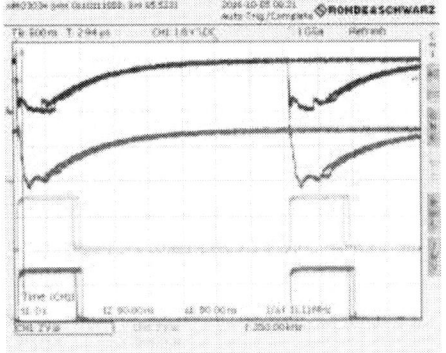

Fig. 7. Pulses at generator (channel 1), inside the mesh (channel 2) and formatted (channels 3 and 4).

Any intervention to the mesh network would produce one of the following possible cases:

1) Duration between pulses would be affected. In this case if the variation is longer than 5.9 ns, the values detected by the timers 8 and 9 will trigger the tamper event. Pulses on pins PE9, PE11, PC6 and PE5 are presented in Fig. 8 (from top to bottom). Delays between pulses depend on the output of the LFSR generator.

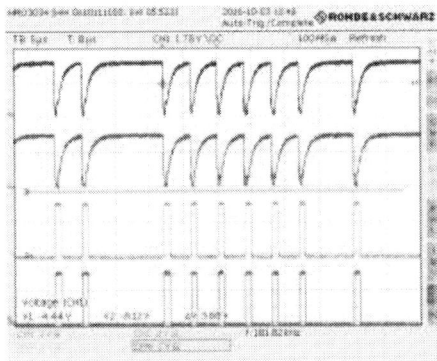

Fig. 8. Image of the probing (channels 1 and 2) and the formatted pulses (channels 3 and 4).

2) Duration of pulses would be affected by the modification of the effective capacitance of the mesh network. Using expression (3) for a variation of time of 5.9 ns (resolution of timers 8 and 9), the resulted capacitance is about 26 pF. Experimenting with a capacitor of 22 pF in parallel with the mesh network resulted in generating the tamper event, as presented in Fig. 9 where channel 4 of the oscilloscope shows the tamper event signal triggered inside the interrupt routine. This signal follows the formatted pulses as the interrupt routine starts at the end of these pulses. Presence of noise triggers the alarm for slightly lower capacitance than calculated (26 pF).

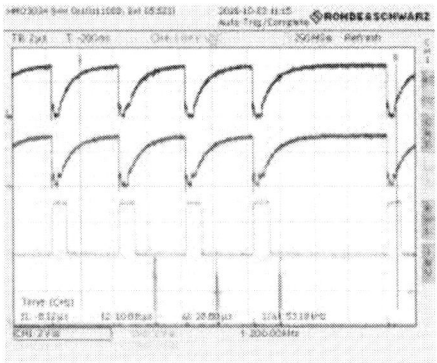

Fig. 9. Simulation of capacitance modification (22 pF) in the mesh network and triggering the tamper event.

V. CONCLUSIONS

The active tamper detection circuit designed and presented in this paper improved the known active mechanisms to detect probing and intrusion attempts against the protective mesh cover of secure modules. The method to generate and compose the pulses for the mesh network does not allow an attacker to determine the parameters of pulses and to simulate the normal functioning of the detection circuit.

The tamper detection circuit used in experimental tests reached the following performance:

- It detects variations of durations between pulses greater than 5.9 ns, limited by the internal clock of the microcontroller;

- It detects variations of duration of pulses greater than 5.9 ns;

- It detects a capacitive overload of at least 26 pF;

- It detects the cutting of mesh network and any intervention that might modify in duration and amplitude the signals through the mesh network.

Future works on this subject could improve the timings and the ability to detect even smaller capacitive loads (e.g. oscilloscope probe) by using a faster processing system and optimize it for minimal consumption in battery powered devices. Also, it is important to isolate and translate the ground level in order to have a better detection effect when the mesh network is probed by an attacker.

REFERENCES

[1] M. Arora, R. Pandey, P. Sareen, P. Bhargava, "Tamper Detector for Secure Module", U.S. Patent No. 8689357 B2, April, 2014.

[2] Atmel Corporation, "AT10458: SAM L22 Tamper Detection using RTC Module ", Application Note AT10458, San Jose, USA, August, 2015.

[3] B. Schneier, "Applied Cryptography, Second Edition", John Wiley & Sons, 1996.

[4] A. Marghescu, P. Svasta, E. Simion, "High Speed and Secure Variable Probability Pseudo/True Random Number Generator Using FPGA", 2015 IEEE 21st International Symposium for Design and Technology in Electronic Packaging (SIITME), October, 2015, (pp.323-328), IEEE.

[5] J. Rabaey, A. Chandrakasan, B. Nikolic, "Digital Integrated Circuits, A Design Perspective, Second Edition", Ch. 4, Pearson, 2003.

High Reliability Wireless Sensor Node for Bee Hive Monitoring

M. G.Vidrascu, P. M. Svasta
Center for Technological Electronics and Interconnection
Techniques (UPB-CETTI)
Politehnica University of Bucharest
Bucharest, Romania

M.Vladescu
Optoelectronics Research Center (UPB-CCO)
"Politehnica" University of Bucharest
Bucharest, Romania

Abstract—Since the ancient times apiculture relies one hundred percent on the bee-keeper's skills and experience. He constantly checks the hives to perform the required activities at the right moment. The rising demand for apiculture products leads to larger and larger apiaries, and checking every hive by hand becomes costly and time consuming. This project suggests a durable solution for bee hive monitoring. The prototype can be used by unqualified personnel, with minimum maintenance.

Keywords— WSN, EDLC, sensor node, prototype

I. INTRODUCTION

This prototype was developed solely for practical reasons, so bee-keepers can benefit from the advantages of modern technology. To make sure the device will truly ease the bee-keepers' work, a meeting was established with a local bee-keeper from Buzau region, Palici village. The owner manages 90 hives, and checking each and every one of them every day is time-consuming hard work. Required activities are conducted by the bee-keeper based on his experience, but some of the detective work can be done automatically. Disease, swarming and abnormal behavior of the bee family can be indicated by changes in the environmental parameters. The result of a lengthy (but useful) discussion is a short list of parameters which, when out of normal values, indicate that an intervention is needed. Commercial solutions are currently available [1][2][3]. However, most of them use either primary cells (batteries [1]) or rechargeable cells (with limited charge-discharge cycle number and lower performance in cold weather), so maintenance activities are required. Also, this prototype is modular (more sensors can be added) and monitors more parameters than other systems.

II. DESIGN REQUIREMENTS

The main feature of the device is ruggedness and reliability. It must operate in harsh environmental temperatures (-20 °C to 60°C) for as long as the energy storage device still holds charge (20 years according to the specifications [4]). Another feature is the remote accessibility. This paper presents only the wireless node, but the system also includes a data collector and software infrastructure (still under development). Measured data will be available anywhere via internet. Another feature is modularity: the ability to add or remove sensors, based on the bee-keeper's needs. It is easier to attain the required modularity by digital means, so the sensors, although analog at low level, will be interfaced using a digital bus. A very common interface is the I2C (Inter-Integrated Communication) bus. It requires only 2 lines (one for serial data and one for clock signal) and allows up to 122 different devices on the same bus (with the 7 bit addressing mode). So, I2C sensors for temperature, relative humidity, weight, sound and motion are required.

A. Temperature and relative humidity measurement

These parameters are especially useful in the winter, to determine the dew point temperature. In the cold season, the bees are crowded together in a ball to preserve heat; they do not ventilate the hive like they do in the warm season. If the relative humidity (RH) inside becomes too high, dew will form on the ball's outside layer. During frosts, the dew will freeze and the bees from the outer layers will die. Also, mold will develop on the hive's walls, favoring the apparition of disease. If the bee-keeper knows that dew is about to form inside the hive, he can ventilate it, reducing the RH.

In the summer, under normal conditions, bees ventilate and cool the hive, so RH fluctuations are small. A higher value can indicate rainy weather (a rain sensor can confirm the readings) or disease. RH will rise if the bee family is affected by the American foulbrood, a bacterial disease locally known as "loca", affecting the larvae.

SHT21 from Sensirion was selected. Its greatest advantage is the extremely low energy consumption: 300uA in measurement mode and 0.15uA in sleep mode. Temperature is measured with a resolution of 14 bits, and RH with 12 bits,. The precision is ±2% for RH and ±0.3°C for temperature. The sensor comes in a 3x3mm Dual Flat no Lead (DFN) package; a carrier board with a connector is required [5]. Multiple such sensors are required for a hive (at least 2), but the address cannot be modified, so an I2C address translator will be used. TCA9543A from Texas Instruments is a dual I2C address translator, and it creates 2 more I2C buses. This allows the installation of 3 identical sensors.

B. Weight measurement

Hive weight is important for production. Normally, the bee-keeper weighs hives to determine if bees are actively gathering. But it is mostly useful for pastoral bee-keeping. The bees forage during the day, but during the night they de-humidify and ventilate the gathered nectar, which is very fluid in its initial state. If the trailer leaves after a large quantity of nectar was gathered, the nectar will drip from the cells to the hive's bottom and it will be wasted. In this process, part of the family might die, drawn in the fluid. According to the bee-keeper discussions, transportation of the hive becomes risky if the bees gathered more than 300-500g of nectar. A common hive weighs from 20-25kg in the spring up to 70-80kg in the autumn. A scale capable of withstanding this weight, having 100g resolution must be designed. Hives in an apiary are placed on a flat, somewhat level surface, but not perfectly level. There are two ways to remove the influence of tilt. Four load cells (similar to the ones in figure 1) can be attached in each corner of the hive or a single distributed load cell can be constructed using stand-alone half-bridge tensometric stamps (figure 2), also one in each corner. The first method is more precise than the other, but the second one is cheaper (but not by much).

 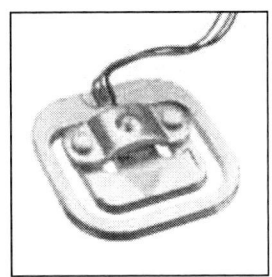

Fig. 2: Complete load cell, Wheatstone bridge assembly [6].

Fig. 2: Stand-alone half-bridge tensometric stamp [7].

No matter what solution will be used, an analog front-end is required for signal conditioning. The voltage from these sensors is in the millivolt range, so an amplifier is mandatory. Also, the bridge requires excitation voltage, as accurate as possible. There are many solutions to meet these requirements, but a simple human scale mechanism is perfect for this application. A completely integrated solution is HX711 from Avia Semiconductor. It includes an excitation voltage power supply, a programmable low noise amplifier (with selectable gain of 32, 64 or 128) and a 24-bit analog-to-digital converter. It has a digital interface, operating from 2.6V to 5.5V, in the industrial temperature range of -80° C to +85° C, with low power consumption: only 1.5mA in active mode and less than 1 μA in sleep mode. [8]

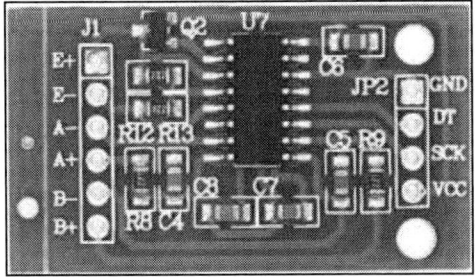

Fig. 3: HX711 24-bit amplifier and ADC for load cells. [9]

C. Tilt sensor

Tilt information is mostly useful to indicate that the hive sits level. The bee-keeper will be immediately informed if the hive starts to tilt to a side, or if it was knocked over by some strong wind or wild animals (in remote places). It might also indicate theft. There are two possible ways to determine dangerous angles of tilt: a simple tilt switch, connected to an interrupt GPIO of the micro-controller, or a low cost dual axis accelerometer. The first one has the obvious advantage of cost and the second one fits better in the I2C bus, maintaining the modularity feature. Also, the price of low precision MEMS sensors is constantly dropping, so cost difference might not be as high as one might think, especially at higher volumes.

 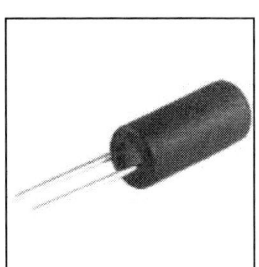

Fig. 4: Cheap MEMS inertial sensor module [10].

Fig. 5: Ball tilt switch [11].

D. Microphone

Among disease, another problem for the bee-keepers is swarming. If the bee family grows large enough not to fit in the hive and there are lots of resources, the family prepares to leave, searching for a better place. Obviously, if no counter-measures are taken, after several days, the bee-keeper will find an empty box instead of a productive hive. Fortunately, bees do not swarm instantly, they take 3-4 days to prepare, and noise is the most obvious sign of swarming (followed by reduced gathering activity). The calm buzz of a normal situation is replaced by a loud hum (drones are humming); the intensity and the frequency changes can be detected using a microphone. Hive temperature will also rise during swarming preparations. Noise is even more intense in the evening before departure (most often after 9 pm).

During winter, the hive is relatively quiet. Loud buzzing during this period indicates that the hive is running out of food. In extreme cases, the bees will exit the hive, searching for food and will die due to the cold weather or they will slowly starve to death inside (leading to complete silence). This situation can be avoided if the bee-keeper is warned at the first buzz.

Fig. 6: SMD MEMS microphone with PDM output [12].

A noise sensor (i.e. a microphone) is very useful in this situation. A classic microphone outputs a small signal, easily influenced by noise. It is also large and it requires an amplifier, increasing energy consumption. A digital output MEMS microphone is more appropriate and any micro-controller can decode the PDM output (Pulse Density Modulation) of a MEMS microphone.

E. Micro-controller and radio module

The micro-controller will directly interface with the sensors. Since the entire process is relatively slow (the data rate is low), a high computational capability is not necessary. It only has to read data from the I2C bus (which is also slow, running at a maximum clock frequency of 400 KHz), from the scale amplifier, measure the energy storage device voltage and to interface with the radio module. Also, it must provide some general input/output pins for the tilt switch and other future devices.

The main requirements are low power consumption, low power modes (in which most of its peripherals are disabled), a hardware I2C interface, and integrated analog-to-digital converter.

The radio module will also be selected based on its energy consumption in active and in stand-by mode, as this component will have the highest current draw (several hundred milliamperes in transmit mode). The operating frequency is also important; ISM band and other free bands are preferred.

However, a new component appeared on the market and it integrates both a micro-controller and a radio transceiver. It is the ESP8266, available as a chip or as an assembled module (figure 7).

Fig. 7: ESP8266 assembled module: processor, flash memory and WiFi on a single board. [13]

Its core is a TENSILICA L106 Diamond series 32-bit processor running at 80MHz, with multiple GPIOs and interfaces: I2C, I2S, SPI, SDIO, and UART. Among the ultra-low energy consumption (10uA in Deep-Sleep mode and 0.5uA in Power-Off mode), its greatest advantage is that it runs the 802.11 b/g/n/e/I protocols (WiFi). The entire communication and addressing protocol is already available, easing the design process. The module can be configured as a client (to connect to a specific wireless network), server or access point (meaning that it broadcasts its own network),

allowing direct connections from other devices. Using an internet browser (on any WiFi enabled device), data can be read from any specific node, in the form of a webpage, containing all the sampled data. [14]

It also runs at 3.3V, allowing direct sensor interfacing.

F. Power supply

If correctly designed, the lifetime of a wireless sensor node is limited by the energy storage device. Rechargeable batteries have a limited number of charge-discharge cycles (500-2000, depending on the chemistry and operating conditions) and the capacity drops with every cycle. Deep discharge will also reduce the capacity. Even more, they have poor performance during cold weather and the good ones are far more expensive than the standard types. However, super-capacitors perform very well at low temperatures (leakage current drops dramatically with temperature), can recharge very fast and have hundreds of thousands of charge-discharge cycles. Over-discharge problems are completely eliminated, too. Since good quality super-capacitors have a life time of 20 years, they are perfect candidates for this sensor node. [4][15]

The node will be operated outdoors, so solar power is the obvious choice to charge the capacitor. The capacitor has a far lower energy density than a rechargeable battery, so every minute of sunshine matters. A single cell solar panel is desired, so an efficient boost charger is necessary. Super-capacitors are currently rated at 2.7V, but the sensors and micro-controller need 3.3V. A low voltage start-up boost converter will power the entire node [15]. A prototype (figure 8 and figure 9) has already been developed and tested. More details and results can be found in reference [15].

Fig. 8: The first, handmade, power supply prototype: charger and 3.3V, step-up converter [15].

Fig. 9: The power supply board for the sensor node [15].

III. RESULTS

Using the information collected so far, a block schematic was developed (figure 10).

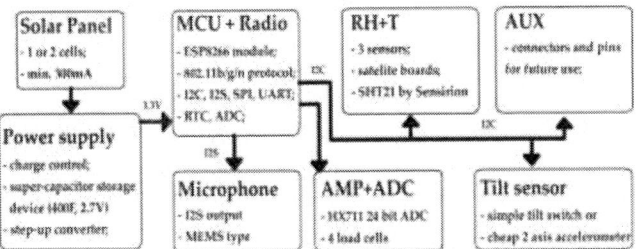

Fig. 10: Block schematic of the hive monitoring node.

Most of the sensors need to be placed somewhere inside the hive, so satellite boards will be developed, just like the one for the temperature and relative humidity (figure 11). The microphone will also have its own board. These boards will have their own enclosures to protect them from dirt and from the bees' activities (mostly propolis deposits).

Fig. 11: Satellite board for the relative humidity and temperature sensor.

The tilt sensor fits easily on the board. The load cell amplifier is already assembled on a PCB, so it will be placed somewhere inside the enclosure, as far away as possible from power circuits and antenna. For future developments, all ESP8266 module pins are brought out to standard pin headers, to easily attach other sensors and devices.

To further reduce the power consumption, a switch (a P-channel MOSFET transistor) powers all the sensors. The switch will be toggled by the micro-controller. Even more, as the data update rate is low (a complete sensor reading at 5-15 minutes), an ultra-low power timer footprint was included on the board. TPL5110 from Texas Instruments consumes only 35nA and it can wake up a processor or a power supply (using its enable pin) at regular intervals, ranging from 100ms to 7200s. This timer is wired to enable the 3.3V boost converter at regular intervals [16].

A stacked architecture is desired. On the bottom level sits the storage capacitor. Next is the power supply board, containing the MPPC charger, the 3.3V step-up converter, the I2C address translator and several connectors for the sensors and for the logic board. The next level is reserved for the logic board. It contains the transceiver, the break-out pin headers and the power switch for the sensors. The stack ends at the top with a solar panel, consisting of one or two solar cells. The entire stack will slide on horizontal guides placed on the walls of the enclosure and the wires will exit through a lateral wall.

A small batch of PCBs was manufactured for testing. To reduce manufacturing cost, the power and the logic board are joined together in the design, but they will be separated after assembly. Figure 12 shows the complete board (80mm wide and 80mm long), and the stacked configuration (after separation) can be seen in figure 13.

Fig. 12: Partially assembled sensor node PCB.

Fig. 13: Stacked PCBs: power board on the bottom and the logic board on top [15].

Fig. 14: PCB stack and a 400F capacitor.

Reference [15] showed that the entire node can be powered using a single solar cell (with much better performance with two cells), and it can also last several days of cloudy weather, depending on the data rate and the output power of the radio transceiver. Using the low power timer, the energy drain will be even lower, because the 3.3V step-up converter will be turned on only when needed.

CONCLUSIONS

This sensor node design is driven by precise information from the bee-keepers and the need to improve bee-keeping activity. The result is a dedicated, easy to use, compact, modular, maintenance-free hardware, embeddable in any hive configuration. Information sampled by the sensors can be sent, stored, and reviewed anywhere in the world using the internet infrastructure and it is also available on site, using any WiFi enabled device.

The next step is to write the software for sensor reading and sending data to the data collector (which is currently under development). Further testing of the power supply is required, especially during winter and low lighting conditions. The first tests will not be made on real hives, but in other outdoor conditions (as a weather station, as greenhouse monitor etc.) The reason for this is that if something fails it can be easily repaired, without disturbing a bee family (especially during winter). For statistical reasons, several nodes (10 available at the moment) will be put to the test, to determine what (if any) and how often (if ever) something fails.

REFERENCES

[1] "Arnia: remote hive monitoring", http://www.arnia.co.uk/, 04.10.2016.

[2] "Solutionbee", http://solutionbee.com/, 04.10.2016.

[3] "Bee Hive Monitor: An open-source bee-hive monitoring system", https://openenergymonitor.org/ emon/beemonitor , 04.10.2016.

[4] ***, "Supercapacitors: XV Series", Cooper Bussmann, component datasheet, 2014.

[5] ***, "SHT21 - Humidity and Temperature Sensor IC", component datasheet, Sensirion, may 2014.

[6] "YZC-133 Kitchen scale load cell", http://www.banggood.com/YZC-133-5kg-Kitchen-Scale-Electronic-Load-Cell-Weighing-Sensor-p-982666.html, 10.10.2016.

[7] "50kg Half Bridge Weight Scales Sensor", http://www.banggood.com /4Pcs-50kg-Half-Bridge-Weight-Scales-Sensor-Resistance-Strain-Human-Scale-Load-Sensor-p-1053172.html, 10.10.2016.

[8] ***, "HX711 - 24-Bit Analog-to-Digital Converter (ADC) for Weigh Scales", Avia semiconductor, component datasheet.

[9] https://software.intel.com/en-us/iot/hardware/sensors/hx711-analog-to-digital-converter, 01.09.2016.

[10] "6DOF MPU-6050 3 Axis Gyro With Accelerometer", http://www.banggood.com/6DOF-MPU-6050-3-Axis-Gyro-With-Accelerometer-Sensor-Module-For-Arduino-p-80862.html , 11.10.2016.

[11] "Ball Tilt Switch, 45/10degree", http://www.globalsources.com/gsol/I/ Roll-ball/p/sm/1089650733.html, 11.10.2016.

[12] ***, "MP34DT01-M: MEMS audio sensor omnidirectional digital microphone", component datasheet, ST Microelectronics, 03.09.2014.

[13] http://iot-playground.com/blog/69-esp8266, 12.10.2016.

[14] ***, "ESP8266EX", component datasheet, Espressif Inc., 2016.

[15] M. G. Vidrascu, P. Svasta, M. Vladescu, "Maintenance-free, super-capacitor based WSN power supply", Proceedings of the 2016 International Conference "Advanced Topics in Optoelectronics, Microelectronics and Nanotechnologies", Constanta, Romania, August 25 − 28.

[16] ***, "TPL5110 Nano-power System Timer for Power Gating", component datasheet, Texas Instruments Inc., 2015.

Coupled surface plasmon resonance on gold nanocubes - investigation by simulation

Attila Bonyár

Department of Electronics Technology
Budapest University of Technology and Economics
Budapest, Hungary
bonyar@ett.bme.hu

Géza Szántó, István Csarnovics

Department of Experimental Physics
University of Debrecen
Debrecen, Hungary

Abstract—The refractive index sensitivity of coupled plasmonic nanostructures, namely gold nanocubes in various arrangements, were simulated with the MNPBEM Matlab toolbox. The size of the cubes, the distance between the particles were the running parameters. It was found that the enhancement factor (which characterize the increase in the peak shift for multi-particle arrangements compared to single-particle models) is an exponential function of (D/a) where D is the gap between the particles and a is the edge length of the cube. It was also found that significant plasmonic coupling effects starts below 0.5 D/a for cubical nanoparticles.

Keywords— plasmonics, nanocubes, refractive index

I. INTRODUCTION

The refractive index sensitivity of localized surface plasmon resonance (LSPR) based sensors is depending on several parameters, including the material type, size and shape of the particle and also the spatial arrangement of multi-particle systems [1]. It is already proven that with the proper nanostructures in the proper arrangement the sensitivity of LSPR (considering molecular or biosensing applications) can reach the sensitivity of classic Kretschmann-configuration based SPR devices on the market [2]. The general aim of our research group is the development of technologies, which would enable the cost-effective fabrication of highly ordered nanoparticle systems on large surface areas (several cm^2). Hence, one motivation of the presented simulations is to study the connection between the particle size/shape/arrangement and the bulk refractive index sensitivity of the nanoparticle based sensor, by using gold nanocubes in this particular case.

The MNPBEM Matlab toolbox, which utilizes the boundary element method (BEM) approach, provides a convenient way for the simulation of coupled plasmonic nanostructures [3]. Besides the advantageous short running times compared to other finite element methods, it also enables the relatively simple inclusion of substrates in the model [4].

II. MODELING AND SIMULATION PARAMETERS

Although in this work only gold nanospheres are investigated, the MNPBEM Toolbox supports the convenient setting up of particles with elementary shape (sphere, rod, torus, and cube). The investigated nanocube arrangements were created with the 'tricube' function and are illustrated in Fig. 1.

The changing parameters were the edge of the nanocubes (a) and the distance between the particles (D). A plane wave excitation was used with light propagation in the Z and light polarization in the X directions (see Fig. 1). For the evaluation of the plasmonic behavior of the particles the resulting extinction cross sections were used. To calculate the bulk refractive index sensitivity, the media surrounding the nanoparticles was changed between air ($n = 1$) and water ($n = 1.33$). The sensitivity of a single particle (S, [nm/RIU]) is defined as the shift of the extinction peak divided by the refractive index change of the media. For multi-particle arrangements the enhancement factor is defined as the absolute peak shift of the multi-particle arrangement (due to refractive index change) divided by the absolute peak shift of the single particle model. In this way the enhancement factor quantify the increased peak shift which originates from the interparticle coupling effects, compared to the single particle model. In this current work, only simple, two-particle arrangements are investigated.

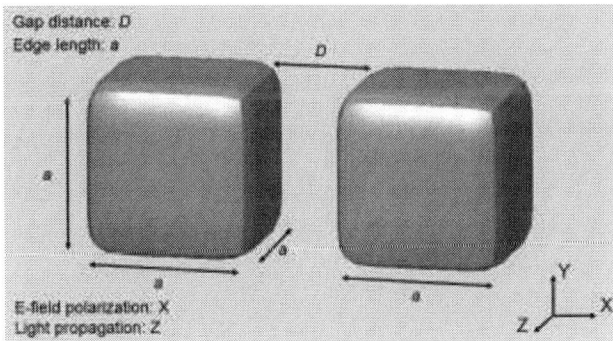

Fig. 1.   Illustration of the investigated gold nanocube arrangement. D: gap distance between the two nanocubes, a: edge length.

Two dielectric functions with tabulated values are available for the plasmonic simulation of gold particles, based on the optical constants of Johnson and Palik, respectively. Although the peak shift (between air and water) obtained with the constants are nearly the same, the Palik dielectric function yields systematically shorter peak wavelengths, and also secondary peaks could be observed, which increase significantly when the distance between particles is decreased in multi-particle arrangements (data not shown). Taking these observations into consideration, the Johnson dielectric function was used for further calculations.

III. RESULTS AND DISCUSSION

A. Solver optimization

The solver of the MNPBEM Matlab Toolbox offers two simulation approaches for the investigation of our nanocube arrangements. Using 'Retarded simulation', the full Maxwell equations are solved on the arrangement. Although this approach can be considered as more precise it is also more time consuming than the 'Quasistatic simulation' which is developed to provide a faster approach. Although the developers claim that the quasistatic approach usually give accurate and reliable results for nanoparticles below 100 nm, they also encourage users to compare the two solvers on their particular problem.

Comparing the extinction cross section spectra obtained by the retarded and quasistatic solvers, the most notable difference is the appearance of smaller, secondary peaks in the spectra obtained by retarded simulations. As can be seen in the results presented in Fig. 2 and 3 the appearance and extent of these secondary peaks is also depending on the size and distance between the nanocubes, on the grid density (the number of mesh divisions per cube edge) and also, on the refractive index of the media.

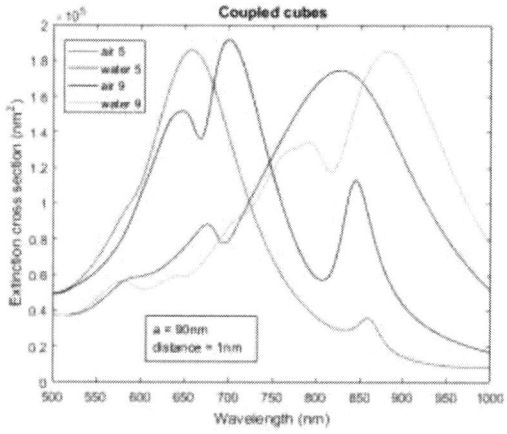

Fig. 2. Extinction cross section spectra of nanocubes with 90 nm edges (*a*) and 1 nm gap (*D*) simulated in air and water environment with 5 and 9 grid density (divisions/edge).

Based on Fig. 2 and 3 it can be said, that decreasing the gap between the particles, increasing the size of the cube, the grid density or the refractive index of the media all increase the amount of secondary peaks in the spectra, which can possibly be related to the appearance of higher plasmonic resonance modes. A problem with the secondary peaks presented in Fig. 2 is that by scaling the gap distance between the particles it is possible that the dominant peak changes (dominance transition between multiple peaks) in function of the distance. By defining the sensitivity as the absolute difference between the dominant peaks (simulated in air and water), this abrupt change in dominant peak position can cause jumps in an otherwise continuous sensitivity/distance function, as can later be seen in Figs. 4, 5, 6.

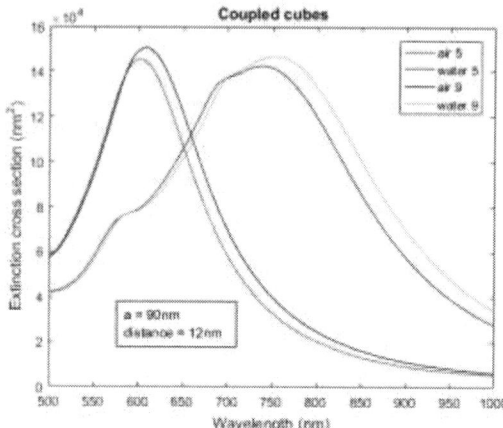

Fig. 3. Extinction cross section spectra of nanocubes with 90 nm edges (*a*) and 12 nm gap (*D*) simulated in air and water environment with 5 and 9 grid density (divisions/edge).

We have to note, that although it is possible to observe small secondary peaks in the case of the quasistatic solver as well (especially at small gap distances), they are usually too small compared to the dominant peak to be considered as problems. It can also be seen in Figs. 2 and 3 that increasing the grid density from 5 to 9 increased the obtained refractive index sensitivities (the absolute difference between the peaks in water and air) to a small extent. Since denser grids significantly increase the running time with the more time-consuming retarded simulations, 9 divisions/cube edge was used in the following investigations.

B. Dependence of the enhancement factor on D/a

Fig. 4 presents the obtained enhancement factor values in function of the gap distance between the nanocubes, for cube edge lengths from 10 to 90 nm. The results of the retarded and quasistatic solvers are also compared in this figure. The jumps which can be observed below 10 nm for the results obtained with the retarded solver can be attributed to the discussed transition of the dominant peak between multiple peaks in an otherwise exponential distance-dependence.

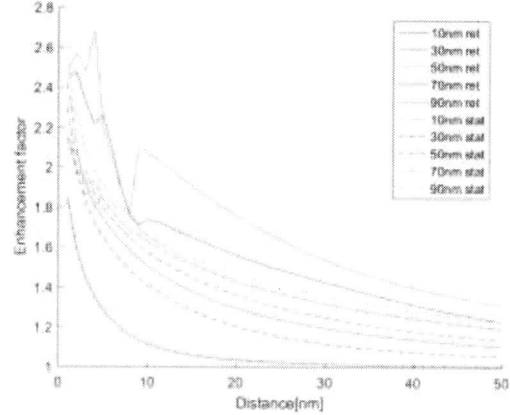

Fig. 4. Dependence of the enhancement factor on the distance between the nanocubes, for the retarded and quasistatic solvers.

The difference between the retarded and quasistatic solvers is more pronounced for the larger nanocubes, where the retarded solver yields significantly higher enhancement factors. Fig. 5 illustrates the same results in a dimensionless way, in the function of D/a. It is interesting to note that in this representation, by using the quasistatic solver, the enhancement factor curves decrease by increasing the cube size, while with the retarded solver it increases for larger cubes. Based on the results it can be stated, that in order to achieve a significant field enhancement effect and enhanced sensitivity compared to the single-particle model, the nanocubes should be very close to each other, specifically, on a relative scale of distance/edge length (D/a) the significant enhancement starts below 0.5. If the distance between the particles is substantially large (e.g. $D/a > 0.5$), the results converge into the theoretical shift for single-particles (enhancement factor = 1).

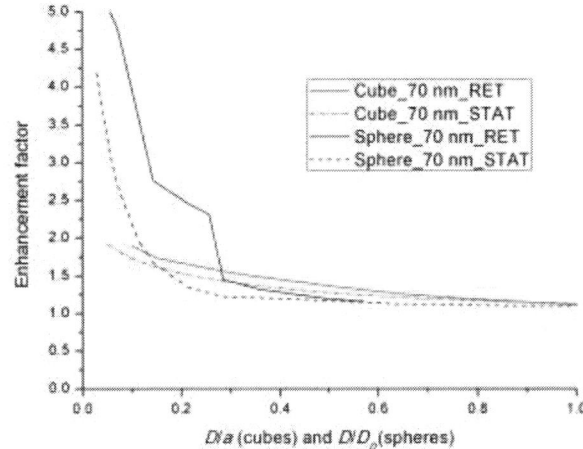

Fig. 6. Dependence of the enhancement factor on D/a (for nanocubes) and D/D_0 (for nanospheres) for the for the retarded and quasistatic solvers.

IV. CONCLUSIONS

The effect of size and gap distance on the plasmonic behavior (especially plasmonic coupling) of paired nanocubes was investigated through simulation with the MNPBEM Matlab toolbox. Calculations with both the retarded and quasistatic solvers of the toolbox showed, that the enhancement factor (which characterize the increase in the peak shift for multi-particle arrangements compared to single-particle models) is an exponential function of (D/a) where D is the gap between the cubes and a is the length of the edges. By comparing the results with simulations made on spherical nanoparticles we found that spheres have significantly (2-3 times) higher enhancement factor below 0.2-03. D/D_0 (where D_0 is the particle diameter), compared to the cubes.

ACKNOWLEDGMENT

This research was supported by the European Union and the State of Hungary, co-financed by the European Social Fund in the framework of TÁMOP 4.2.4. A/2-11-1-2012-0001 'National Excellence Program'. Attila Bonyár is grateful for the support of the János Bolyai Research Scholarship of the Hungarian Academy of Sciences.

REFERENCES

[1] M.H. Tu, T. Sun, K.T.V. Grattan, "LSPR optical fibre sensors based on hollow gold nanostructures" Sensors and Actuators B 191, 37–44. 2014

[2] A. Dimitrev, Nanoplasmonic Sensors, Springer, ISBN 978-1-4614-3932-5, 2012.

[3] U. Hohenester, A. Trügler, "MNPBEM – A Matlab toolbox for the simulation of plasmonic nanoparticles" Computer Physics Communications 183, 370-381, 2012.

[4] J. Waxenegger, A. Trügler, U. Hohenester, "Plasmonics simulations with the MNPBEM toolbox: Consideration of substrates and layer structures" Computer Physics Communications 193, 138-150, 2015.

[5] A. Bonyár, "Simulation of the refractive index sensitivity of coupled plasmonic nanostructures", Procedia Engineering, in press, 2016.

[6] K.-H. Su, Q.-H. Wei, and X. Zhang, "Interparticle Coupling Effects on Plasmon Resonances of Nanogold Particles" Nano Letters 3(8), 1087-1090, 2003.

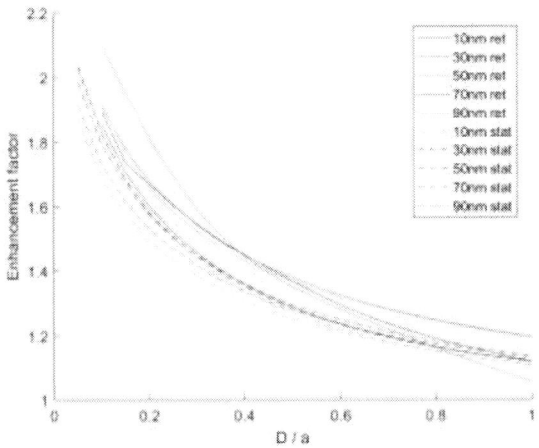

Fig. 5. Dependence of the enhancement factor on D/a for the retarded and quasistatic solvers.

C. Comparison with nanospheres

Fig. 6 compares the results simulated on the nanocubes with previous simulation results obtained by using spherical nanoparticles. The paired nanospheres were simulated with the exact same parameters, by using the gap distance (D) and the particle diameter (D_0) as equivalent running parameters. It can be seen, that there is also a significant difference between the retarded and quasistatic simulations for nanospheres as well, along with the marks of dominant peak transition. The results indicate, that although the enhancement starts earlier (by decreasing D/a) for the nanocubes, below 0.3-0.2 D/D_0 the spherical particles have significantly higher enhancement. The obtained enhancement factor curves can be approximated as exponential functions of the gap between the particles. This exponential decay is in good agreement with the theory of quantum tunneling of photons through the gap during evanescent coupling, and also with previous experimental results [5]. The decay length is depending on the shape of the particles, both on the opposing surface areas and on the curvature of the particles. The investigation and generalization of the effect of nanoparticle shape/area on the enhancement factor will the next step in our work.

Enhancing Thermal Capabilities of Component Packaging

A. Fodor, G. Chindris, and D. Pitica

Applied Electronics Department
Technical University of Cluj-Napoca
Cluj-Napoca, Romania
alexandra.fodor@ael.utcluj.ro

Abstract—**Heat transfer from a component to the ambient can be optimized, primarily by reducing the thermal resistance from the main source of heat to the ambient. At package level, reducing thermal resistance can be achieved through adding embedded thermal pads into the package, increasing the number of vias in the package's substrate, and using materials with high thermal capabilities. The research presented in this paper aims to bring further optimizations to component packaging, by showing that making the proper electrical connection inside the IC package can bring improvements to the thermal blueprint of the electronic system.**

Keywords— component packaging, thermal optimization, simulation

I. Introduction

The easiest way to redirect heat away from an electronic component is to ensure that the thermal resistance from the heat source to the ambient is minimal. Thus, IC manufacturers have developed various designs to reduce the thermal resistance from junction to the ambient. An example of optimizing thermal capabilities of component packaging by reducing thermal resistance is Texas Instruments' PowerPAD Thermally Enhanced Package. It provides greater design flexibility and increased thermal efficiency in a standard size device package [1]. A similar concept as the PowerPAD has been employed also by various manufacturers of integrated circuits. Linear Technologies developed thermally enhanced leaded plastic packages which include exposed pads for various component packaging (LQFP, TSSOP, SOIC). The purpose is to provide lower thermal resistance path between the device junction and exterior of the case [2]. ON Semiconductors released thermally enhanced SO8-FL [3] as an improvement to the standard SO8-FL package, proving that the SO8-FL TE package has a lower junction temperature at a given power. This advantage became more prevalent as air flow and heat sink are added.

II. Related Work

Thermally enhanced packages have been widely discussed and employed for various types of ICs. Through [4], Vishay introduces a new family of power MOSFET packages, which are leadless and have the drain connected to a large exposed copper pad on bottom of the package for the purpose of dissipating heat, called PowerPAK. In terms of thermal capabilities, the PowerPAK family brings small size and improved thermal performance to surface mount power MOSFETs.

As mentioned previously, Texas Instruments brought their own research on enhancing thermal capabilities of IC packages by developing the PowerPAD package. It was employed for components in standard TSSOP and TQFP packages and the results show significant decrease in the overall temperature of the IC during operation, by reducing drastically the junction-to-ambient resistance [6].

Beside the aforementioned packages, Texas Instruments performed studies on QFN devices as well (SO-8 footprint), in [5], where the DUAL COOL concept is developed for enhancing thermal capabilities of the package. The package design considerations enable heat conduction to the top surface of the microelectronic package through the use of a high thermal conductivity path which reduces by more than a factor of ten the junction-to-top thermal resistance compared to standard solutions [5].

III. Additional Thermal Enhancements at Package Level

In addition to having a thermal pad embedded in the casing of an IC, it is important also the electrical role the thermal pad has. Hence, thermal capabilities of a component can be improved just by connecting the thermal pad to the proper function. With reference to this, several configurations were studied and simulated using SolidWorks, for DPAK and SOIC devices that have a thermal pad embedded in the casing, configurations that differ only through the electrical designation the thermal pad has.

A. DPAK Simulation Considerations and Results

Usually, the DPAK component is a variant of casing for voltage regulators that have pins for input and output voltages, and a ground pin normally connected via the thermal pad to the rest of the system. For voltage regulators, approximately all the current flows from the input pin to the output one. In this respect, two scenarios were simulated, with a DPAK component having the thermal pad firstly connected to the ground (as found normally) and secondly, to the output pin.

The two scenarios can be found in Figure 1. A current of 20 A was set from input to output to accelerate the simulation effects, and the only cooling scheme is conduction, without any airflow or radiator, with an initial ambient temperature of 25 °C. The proper materials were set for all components used: FR4 for the PCB, copper for the pins and thermal pad, and typical TSOP for the plastic casing. Other simulation parameters are presented in Table I.

TABLE I.   SIMULATION DOMAIN PARAMETERS

Parameter	Value
Ambient temperature	25 °C
Pressure	101325 Pa
Air flow speed	0 m/s
Initial solid temperature	25 °C
Humidity	0%

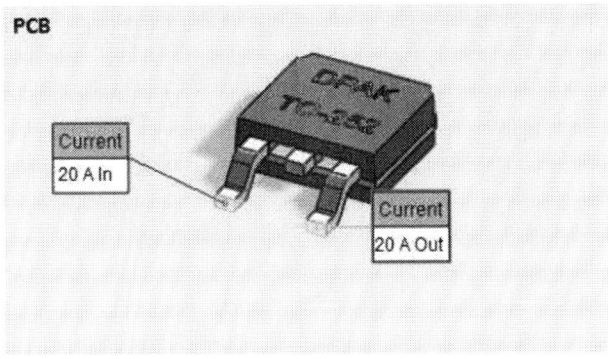

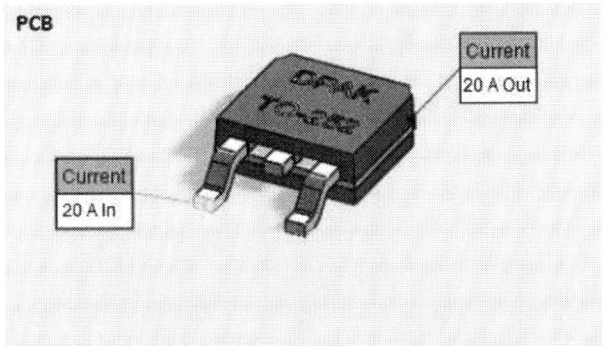

Fig. 1. Simulation scenarios for enhancing thermal capabilities of a DPAK component, with: GND pin connected to the thermal pad (top) and OUT pin connected to the thermal pad (bottom).

The simulation was set to run 200s in each of the two scenarios described above, and the obtained curves can be observed in Figure 2.

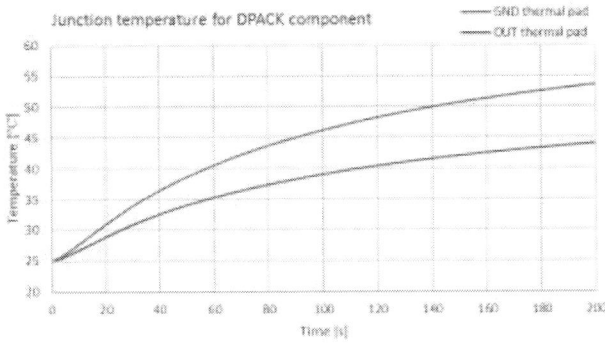

Fig. 2. Junction temperature comparison for DPAK component: thermal pad connected to GND pin vs. thermal pad connected to OUT pin.

The graph in Figure 2 shows that the temperature the junction reaches at the end of the 200s simulation time is approximately 10°C lower when the thermal pad is connected to the pin that has a greater current capability.

B. SO8 Simulation Considerations and Results

In addition to the DPAK study, the behavior of connecting the thermal pad to the pin that delivers the highest current was evaluated. Generally, all thermally enhanced SO8 packages are built to follow the same footprint and the same pinouts as the standard SO8 components [7]. Many IC manufacturers mention that the thermal pad is exposed and it is mandatory to connect it to ground. The assumption made in subchapter A is applied for SO8 components as well – connecting the thermal pad to the pun that delivers the highest current improves the thermal dissipation of the component.

To test the validity of the assumption, two simulation scenarios were developed, as seen in Figure 3. In the first scenario, the current flows from input pin to the output pin, with the ground pin connected to the thermal pad, as in the standard SO8 component, and in the second scenario, the current is divided between in-out and in-thermal pad.

The materials used in the simulation are FR4 for the PCB (with two signal layers and two plane layers), typical SOIC material for the casing, copper for the pins and thermal pad, and silicon for the die. The initial ambient temperature was set to 25 °C.

Fig. 3. Simulation scenarios for enhancing thermal capabilities of an SO8 component, with: GND pin connected to the thermal pad (top) and OUT pin connected to the thermal pad (bottom).

For consistency purpose, a simulation time of 200s was also set for the SO8 experiment. The resulted graph is presented in Figure 4.

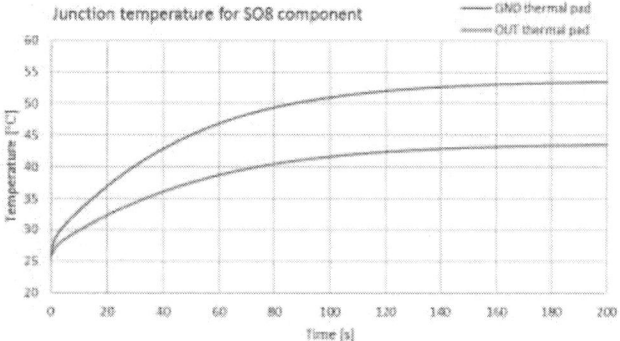

Fig. 4. Junction temperature comparison for SO8 component: thermal pad connected to GND pin vs. thermal pad connected to OUT pin.

The 10 °C decrease in temperature is kept also in case of the SO8 simulation, at the end of the 200s simulation time. When the output pin is connected to the thermal pad, there is a significant reduction of the temperature the junction of the component reaches.

IV. EXPERIMENTAL MEASUREMENTS

The measurement setup was developed using two voltage regulators in DPAK packages: MC33269 from ON Semiconductor (output connected to thermal pad) and LD39300 form ST (ground connected to thermal pad). The latter has a maximum supply voltage of 6.5V, this voltage being chosen as the input voltage for both measurement setups. All other relevant parameters, specified in datasheets, are described in Table II.

TABLE II.   THERMAL PARAMETERS FOR MEASURED COMPONENTS

Parameter	LD39300	MC33269
Output voltage	3.3V	3.3V
Thermal resistance junction-ambient	100 °C/W	92 °C/W
Thermal resistance junction-case	8 °C/W	6 °C/W
Maximum output current	3A	0.8A
Operating junction temperature range	-40 to +125 °C	−40 to +150 °C
Operating ambient temperature range	-40 to +125 °C	−40 to +150 °C
Thermal shudown OFF	+170 °C	-
TAB connection	GND	OUT

A 7.5Ohm resistance was chosen for the load, resulting a load current of 380mA for each configuration.

Thermal measurements were performed using CEM Thermal Imager DT-980. Both voltage regulators were supplied at 6.5V, the measurements being performed when thermal equilibrium was reached.

The captured images can be seen in Figures 5 and 6. To be noted that the measurements were done at an ambient temperature of 25 °C, with no heatsink attached to the components.

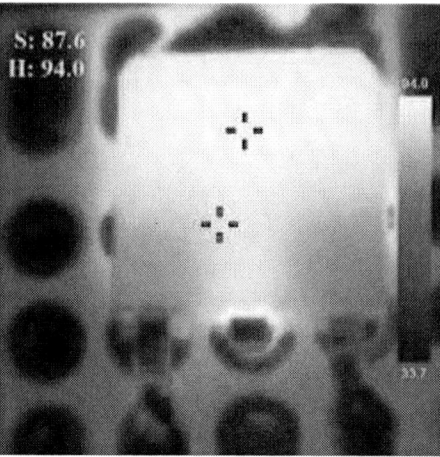

Fig. 5. Chip temperature for LD39300 voltage regulator (GND connected to thermal pad)

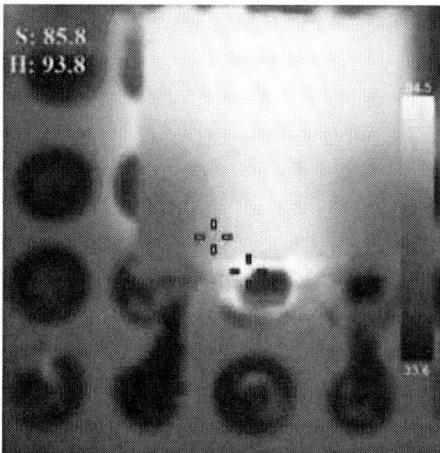

Fig. 6. Chip temperature for MC33269 voltage regulator (OUT connected to thermal pad)

The differences in temperature are minor, in case there is no additional cooling for the component. Although they do not show promising results, the measurements are relevant to our assumption: the junction-to-ambient and junction-to-case thermal resistances differ from one component to the other, therefore even if the temperatures the components reach are approximately the same, the component that has the output connected to the thermal pad has a better thermal performance.

The measurements were repeated with heatsink soldered to both components. A copper sheet of 2.5cm x 2.5cm was used to better dissipate heat, and the thermal images can be seen in Figures 7 and 8.

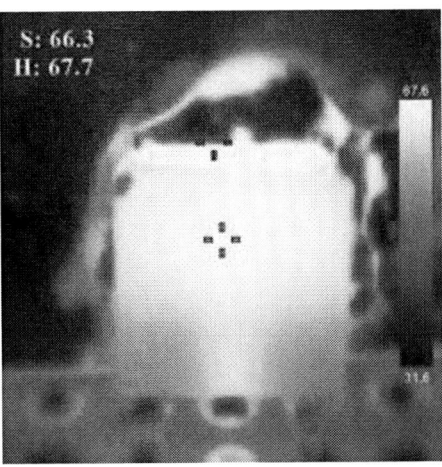

Fig. 7. Chip temperature for LD39300 voltage regulator (GND connected to thermal pad) with heatsink connected to the back of the component

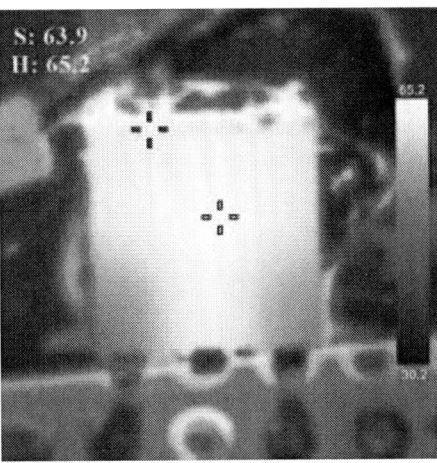

Fig. 8. Chip temperature for MC33269 voltage regulator (OUT connected to thermal pad) with heatsink connected to the back of the component

Adding a heatsink to the measurement setup increases the advantage the MC33269 component has, in terms of thermal performance, the temperature being approximately 2.5 °C lower for MC33269 voltage regulator.

Beside the thermal images captured during the operation of each regulator, an evaluation of the capability of dissipating heat to the ambient was performed, by measuring the decrease in temperature each regulator has when it stops functioning.

Thus, both regulators were brought to a temperature of approximately 60 °C, then the supply was removed, and the temperature decrease the regulators had during 50s was measured. To be noted that both regulators were connected to heatsinks during this measurement.

The obtained graph can be seen in Figure 9, where the advantage the MC33269 has in thermal dissipation capabilities is visible, reaching a temperature with 5 °C less than LD39300.

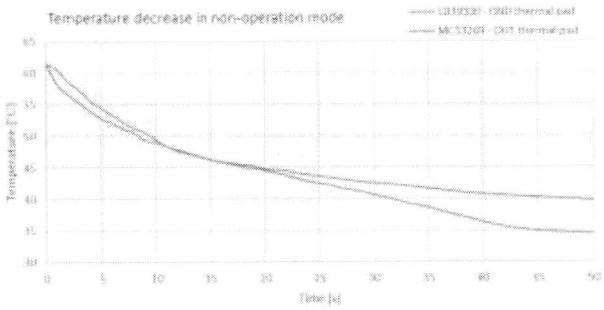

Fig. 9. Comparison between the decrease in temperature for both voltage regulators, in non-operation mode.

V. CONCLUSIONS

This research is based on the assumption that enhancing thermal capabilities of a component package can be improved not only by adding a thermal pad that increases thermal capabilities of a certain package, but connecting the thermal pad to the pin delivers the highest current. In that respect, simulations were developed for two types of packages, DPAK and SO8, and the results are consistent with our assumption: connecting the thermal pad to the pin that delivers the highest current decreases the temperature the components reach with approximately 10 °C. Additionally, measurements were performed using Thermal Imager DT-980, for two voltage regulators in DPAK packages, and two setup configurations, without and with heatsink. In no-heatsink case, the measured temperatures, when the components reached thermal equilibrium, were approximately the same, but with a slight advantage to the one which had the output pin connected to the thermal pad, due to its lower junction-to-case and junction-to-ambient resistances. Adding heatsinks to both components increases the aforementioned advantage, the temperature being approximately 2.5 °C lower for the component with the thermal pad connected to the pin that delivers the highest current. The cooling curves were also measured, during 50s measurement period, and they prove that the thermal dissipation is better for the component having the output pin connected to the thermal pad.

REFERENCES

[1] Texas Instruments, PowerPAD™ Thermally Enhanced Package, Application Report SLMA002G – November 1997 – January 2011

[2] Linear Technologies, Application Notes for Thermally Enhanced Leaded Plastic Packages

[3] ON Semiconductor, AND9014/D - Guide to Thermally Enhanced SO8- FL

[4] Wharton McDaniel, "Thermal performance of power MOSFET packages", Conference Proceedings, International IC Đ Taipei

[5] J. A. Herbsommer, J. Noquil, C. Bull and O. Lopez, "Novel thermally enhanced power package," *Applied Power Electronics Conference and Exposition (APEC), 2010 Twenty-Fifth Annual IEEE*, Palm Springs, CA, 2010, pp. 398-400.

[6] Milton L. Buschbom, Mark Peterson, Shih-Fang Chuang, David Kee, and Buford Carter, "PowerPAD™ - A Method To Create Thermally Enhanced Plastic Package Solutions for Semiconductors", SMI Conference, San Jose, California, 25 August, 1998

[7] Vishay Siliconix, Application Node AN821, PowerPAK SO-8 Mounting and Thermal Considerations

Analysis of Crosstalk Effects on Single Ended Signal Lines Crossing Split Reference Planes

Marcel Manofu, Cătălin Negrea

Interior Instrumentation Human-Machine Driver Interface
Continental Automotive
Timisoara, Romania
marcel.manofu@continental-corporation.com

Roxana Vlăduţă

Telecommunication Systems and Technologies
Maritime University of Constanta
Constanta, Romania

Abstract—The electromagnetic consequences of splits in the reference plane of signal lines include signal reflections, crosstalk and electromagnetic interference (EMI). This paper deals with the effects of a split reference plane on the crosstalk level between single ended (SE) signal lines crossing the split. Coupling mechanisms are analyzed by investigating microstrip and stripline configurations for different plane split geometries through electromagnetic simulation and time-domain reflectometry (TDR) measurements. The influence of the split on crosstalk is evaluated by examining the coupled electromagnetic energy at the near and far ends of the victim signal line. Design rules for minimizing crosstalk levels in crossing splits scenarios are discussed by correlating split length and width to observed electrical effects.

Keywords—*printed circuit board; split plane; crosstalk; time-domain reflectometry.*

I. INTRODUCTION

In high-speed multilayer printed circuit board design, power and ground plane partitioning represents a critical factor. It enables the delivery of multiple supply levels required by various interfaces and maintains noise isolation between different regions of the PCB [1], [2]. Isolated power and ground planes, so called 'power islands', are used to isolate noisy or sensitive circuits [3].

Plane partitioning can be an effective way of improving EMI, especially by means of minimizing noise injection into the cabling system, therefore in the design process there is a tradeoff between signal integrity (SI) and electromagnetic compatibility (EMC) performance.

Power and ground planes also act as reference planes for signal lines serving as a path for the return currents to flow [4]. However, the separation of planes creates splits which constitute discontinuities in the current return paths of different signal lines crossing them and generate undesired electromagnetic effects. The resulted consequences comprise of signal reflections, crosstalk and radiation.

Due to the continuous increase in operational frequencies and routing density demands of PCBs, traces crossing plane splits became a scenario hard to avoid that should be taken into account for high-speed digital design. Crosstalk level evaluation is an important aspect in the signal integrity analysis with significant impact on channel performance.

The authors would like to express their gratitude to AFCEA International for the financial support and TENSOR SRL for technical support in using ANSYS Software.

Fig. 1 presents an example of a common PCB layout with splits crossed by signal lines.

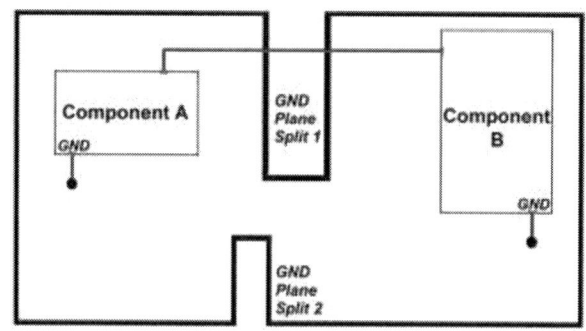

Fig. 1. PCB with GND splits resulted from plane partitioning

Several papers present a study of the splits effects in the reference planes including crosstalk explanation based on the numerical slot waveguide model in correlation with TDR measurements of various microstrip and stripline configurations [1], SPICE and 3D full-wave simulations of microstrip lines crossing over two splits [3], and coupling coefficient for different trace geometries and split widths [5]. Far-end crosstalk measurements and 3D Transmission Line Matrix/Method (TLM) simulations of microstrip lines of a single split scenario are presented in [2]. Conclusions in these papers are valid only for certain applications, being focused on printed circuit board substrate and traces geometry, rather than providing insights on crosstalk modeling for various split scenarios. In the before mentioned articles there is a lack of direct correlation between plane split geometrical parameters and the magnitude of the near-end (NEXT) and far-end (FEXT) crosstalk coupled between traces crossing the split.

In this paper we have investigated the crosstalk effects caused by the presence of a split in the ground plane of coupled single ended signal lines. NEXT and FEXT waveforms are evaluated through time domain electromagnetic simulations using a 3D TLM solver as well as TDR measurements taken on test boards built for this purpose. Fig. 2 presents such a PCB test coupon highlighting split length (L), width (W) and the separation between the 2 traces (S).The split is shorted at one end as to consider a real design situation and is positioned at equal distance from signal lines ends.

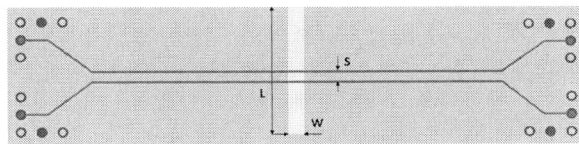

Fig. 2. Top view of a test PCB coupon

Our focus is to define a set of guidelines usable for single ended interface routing when a tradeoff between acceptable crosstalk and plane separation has to be made

II. CROSSTALK

This chapter presents numerical solutions to calculate near-end and far-end crosstalk between coupled transmission lines with a continuous reference plane. The coupling mechanism determined by the plane split is described and a model for the analysis of coupled transmission lines crossing a split is presented.

A. Near-end and far-end crosstalk in transmission lines

The NEXT between two signal lines, also defined as the near-end crosstalk coefficient (k_b) as well as the reverse crosstalk, has a pulse width twice as long as the time delay (TD) for the coupled region line. NEXT peak value is reached only if the TD is greater than half the rise time of the aggressor signal and is given by [6]:

$$NEXT = \frac{v_b}{v_a} = k_b = \frac{1}{4}\left(\frac{C_{mL}}{C_L} + \frac{L_{mL}}{L_L}\right) \quad (1)$$

FEXT pulse width and peak value scale with the rise time (RT) of the signal injected into the aggressor line and with the coupled length between the two traces (L_{en}), respectively. FEXT magnitude is related to the far-end coupling coefficient (k_f) as defined by [7]:

$$FEXT = \frac{v_f}{v_a} = \frac{L_{en}}{RT} \times k_f = \frac{L_{en}}{RT} \times \frac{1}{2v} \times \left(\frac{C_{mL}}{C_L} - \frac{L_{mL}}{L_L}\right) \quad (2)$$

, where v_b and v_f are the voltage noises at the near-end and far-end, respectively, of the victim trace, and v_a is the voltage level on the aggressor line. Signal propagation speed is denoted by v, C_L and L_L represent the capacitance and inductance per length, while C_{mL} and L_{mL} represent the mutual capacitance and inductance per length, measured in pF/mm and nH/mm, respectively.

The distributed circuit model for two coupled transmission lines is presented in fig. 3.

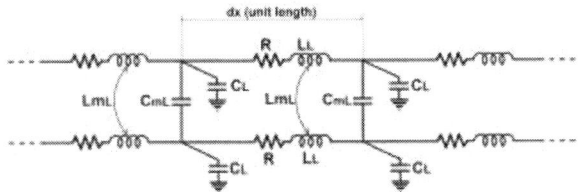

Fig. 3. Equivalent circuit model for two coupled transmission lines

B. Crosstalk coupling through slotline mode conversion

Slot waveguide properties were closely examined in the last decades, and it is understood that when the input signal edge reaches a split in the reference plane, part of the electromagnetic energy undergoes a mode conversion resulting into a slot wave which propagates in a non-transelectromagnetic (TEM) mode along the slot [1]. The generated slotline mode couples through reciprocal mode conversion to the other traces crossing the split.

Fig. 4 presents the equivalent circuit model of two signal lines crossing over a split. The energy conversion between microstrip/stripline and slotline mode is modeled through current controlled current sources I_{X1} and I_{X2}, while the reciprocal conversion is described by voltage controlled voltage sources V_{X1} and V_{X2}. The characteristic impedance of the line and of the slotline are Z_0 and Z_g, respectively. C_x represents the capacitance due to fringing fields between signal line and split edges, and L_x models the trace section located above the split [3].

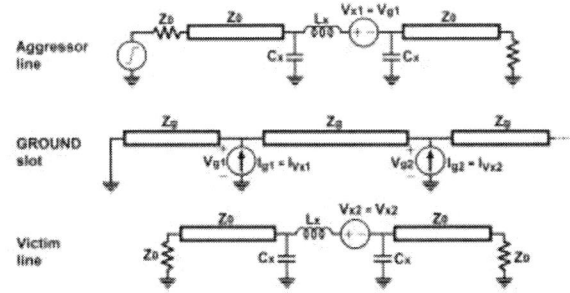

Fig. 4. Equivalent circuit model of two signal lines crossing over a split

III. IMPEDANCE DISCONTINUITIES CAUSED BY SPLITS

In order to understand better the impact of the split in the discussed scenario, the impedance profile of the microstrip and stripline traces for different split widths was analyzed.

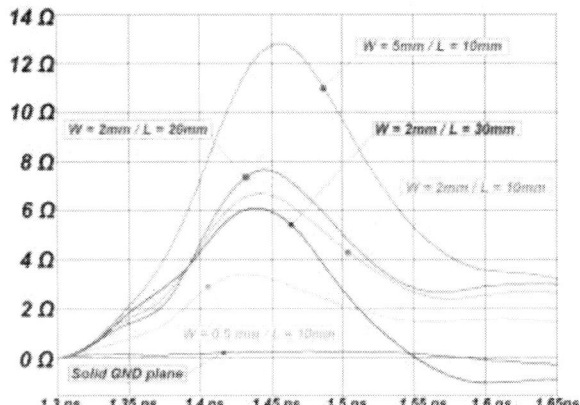

Fig. 5. Relative impedance variation for microstrip lines

Fig. 5 presents the variation of the microstrip impedance in the split region for different split sizes. It is observed that the absence of the plane has an inductive behavior, thus increasing the local impedance of the trace. The effect is less pronounced on the stripline configuration, as shown in fig. 6.

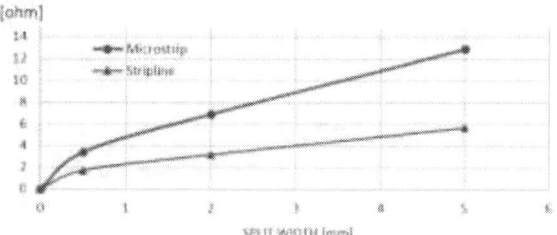

Fig. 6. Impedance discontinuity magnitude vs split width

IV. TDR MEASUREMENTS

A. Measurement setup description

For the experimental study of the discussed crosstalk effects, PCB test coupons were realized. The geometry was calculated as to result 50 ohm target impedance for the microstrip and stripline, in order to match the impedance of the TDR measurement port. The TDR step signal has 250mV amplitude and a rise time of 12 ps (10% to 90%).

The coupled length between the traces is maintained at 130mm on every test coupon. The variations in the test PCBs include: trace spacing of 100um and 350um, split width of 0.5mm, 2mm, 5mm and the split length of 10mm, 20mm and 30mm.

Fig. 7 shows a picture of a test PCB, with the attaching cables to the TDR.

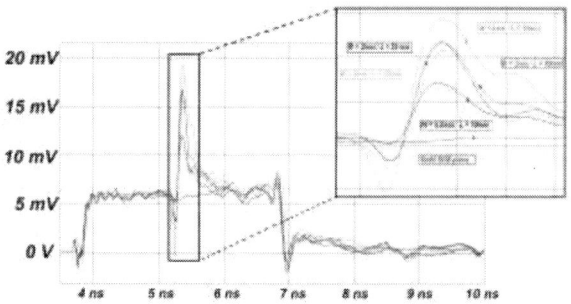

Fig. 7. Test coupon with attached TDR measurement cables

B. Near-end Crossatlk

Figure 8 shows the measured noise signal at the near end of the victim for different split width and length, both for microstrip and stripline.

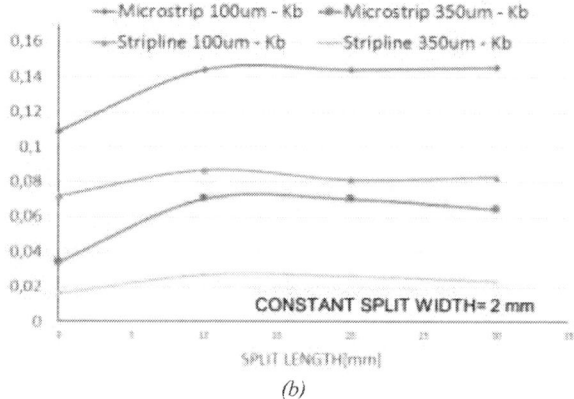

The variation of the near-end coupling coefficient for each trace spacing scenario is presented as a function of split width and length in fig. 9. The coupling coefficients for the scenarios with the continuous ground are plotted at the "0" value of the "Split Width" axis.

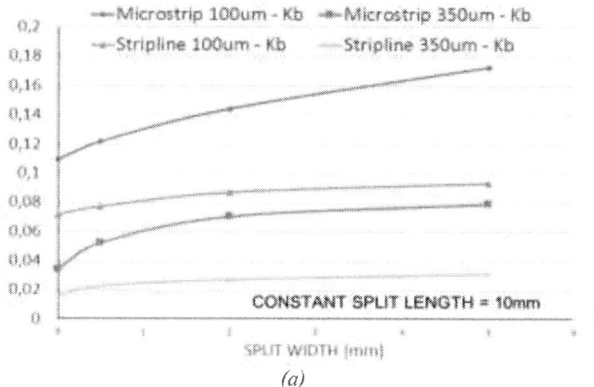

(a)

(b)

Fig. 9. Near-end coupling coefficient variation with (a) split width and (b) split length

It can be denoted that the larger the width of the split, the greater the coupling coefficient becomes.

C. Far-end Crossatlk

Far-end crosstalk waveform for 350um spacing microstrip is presented for different split geometries in fig. 10.

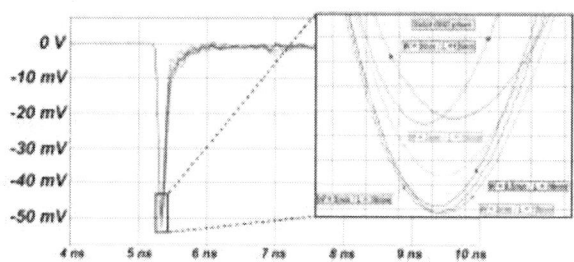

Fig. 10. FEXT measured voltages for microstrip traces at 350um spacing between the traces

Fig. 8. NEXT measured voltages for microstrip traces at 350um

Fig. 11 presents the far-end coupling coefficient, obtained from TDR measurements, for both microstrip and stripline configurations as function of split width and length.

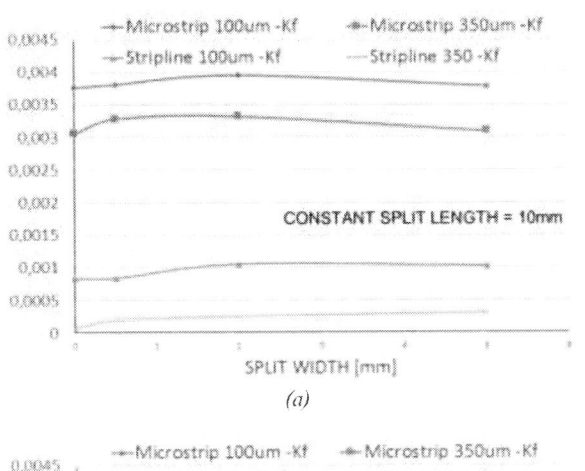

(a)

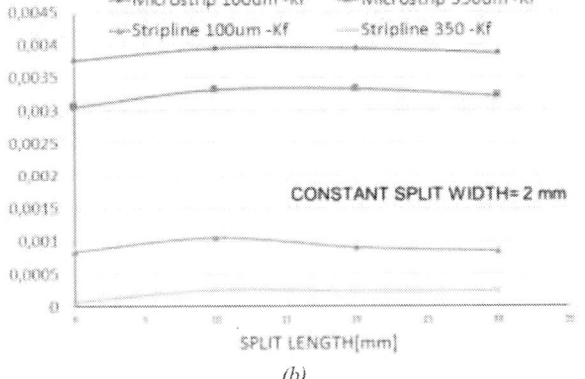

(b)

Fig. 11. *Far-end coupling coefficient variation with (a) split width and (b) split length*

V. MODEL EXTRACTION AND 3D EM SIMULATION

For the simulation the PCB test coupons were extracted using a 3D TLM solver and Touchstone files were generated, that were used in a time domain simulation.

The basic modeling used in the transient simulation is presented in fig. 12.

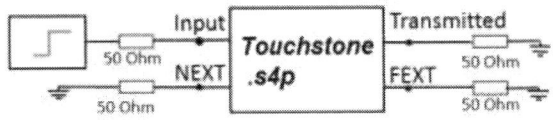

Fig. 12. *Simulation model diagram*

Based on the simulated results, the near-end and far-end coupling coefficients were calculated. Figure 13 presents the near-end coupling coefficients, while the far-end coupling coefficients are presented in fig. 14.

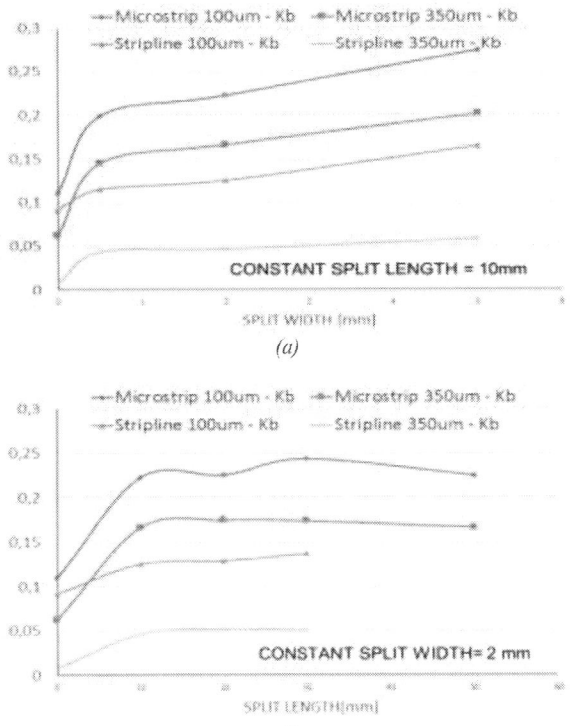

(a)

(b)

Fig. 13. *Near-end coupling coefficient variation with (a) split width and (b) split length obtained by simulation*

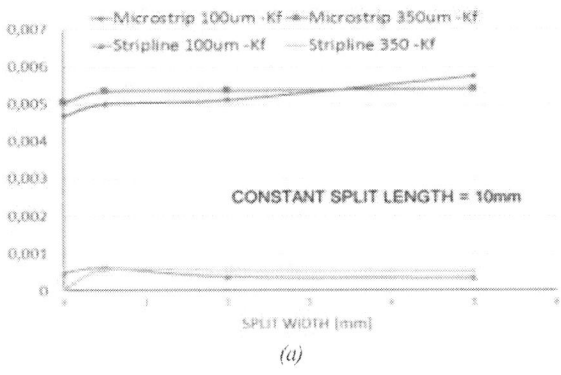

(a)

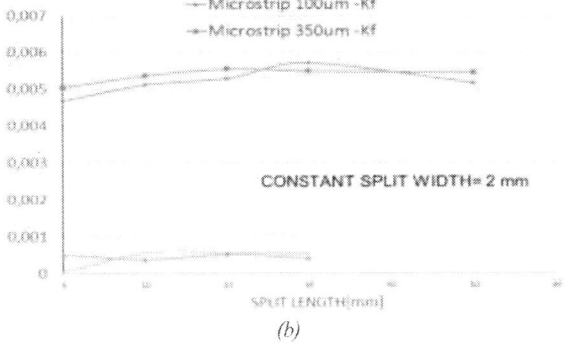

(b)

Fig. 14. *Far-end coupling coefficient variation with (a) split width and (b) split length obtained by simulation*

The coupling coefficient values both for microstrip and stripline show a similar trend compared with the TDR measurements, with slightly higher values. The higher values can be attributed to the cables and system losses associated with TDR measurements.

VI. CONCLUSIONS

The near end coupling coefficient was found not to have significant variation for split lengths greater that 10mm. It has been observed that for split widths smaller than 2mm the near-end coupling variation is increased, compared to splits greater that 2mm, where the variation is smaller and the slope decreases. The separation between the traces has also an impact on the slope of the near-end coupling coefficient. The far-end coupling coefficient presents small variation due to split geometry, below 1%.

The presented results can be used in design scenarios where a tradeoff between SI and EMC performance is required. It has been shown that there is a strong correlation between split geometry and NEXT magnitude, but for the FEXT there has not been observed too much difference.

REFERENCES

[1] Jingook Kim, Heeseok Lee, Joungho Kim, "Effects on Signal Integrity and Radiated Rmission by Split Reference Plane on High-Speed Multilayer Printed Circuit Boards", IEEE Transactions in Advanced Packaging, Vol. 28, No. 4, pp. 724-735, November, 2005.

[2] Juan Chen, Weimin Shi, Adam J. Norman, Ponniah Ilavarasan, "Electrical Impact of High-Speed Bus Crossing Plane Split", IEEE International Symposium on Electromagnetic Compatibility, Vol. 2, pp. 861-865, August, 2002;.

[3] Haw-Jyh Liaw, Henri Merkelo, "Signal Integrity Issues at Split Ground and Power Planes", IEEE Electronic Components and Technology Conference, pp. 752-755, May, 1996

[4] Dror Haviv, "Analysis of Single-Ended and Differential Striplines Crossing Split Reference Planes in PCBs", IEEE International Symposium on Electromagnetic Compatibility, pp. 156-162, August 2013.

[5] Jing Ni, Li Shufang, Qiu Xiaofeng, "Coupling Issues Between Two Signal Lines at Split Ground Planes", Asia-Pacific Conference on Environmental Electromagnetics, pp. 465-468, November, 2003.

[6] Howard W. Johnson, Martin Graham, "High-speed Signal Propagation: Advanced Black Magic", Prentice Hall Professional, 2003.

[7] Eric Bogatin, "Signal and Power Integrity - Simplified (2nd Edition)", Prentice Hall Signal Integrity Library, 2010.

Analysing of Half-Bridge Inverter Using the Simulink Platform

I. H. Baciu, S. Pop and V. Bande

Applied Electronics Department, Technical University of Cluj Napoca, Romania
ionel.baciu@ael.utcluj.ro

Abstract— **Evolution technologies and increasing consumption constrain us to think of reducing resource consumption. This cannot be achieved except through improvements of existing systems and the invention of new systems. One of these systems is also the induction heating. This system has revolutionized heating systems. If until now was using a simple electrical resistance heating, the new systems use induced current effect intro electric conductive surface. Using these systems has shown that the efficiency of the new systems is approximately 30% higher. This paper presents a theoretical study of an induction heating system whose order is made through an half bridge inverter circuit.**

Keywords—Half-Bridge Inverter, ZVS, ZCS.

I. INTRODUCTION

Inverters are a class of circuitry which controls the power in an AC load supplied from a current source or voltage. The amplitude and the frequency of the output signal obtained can be constant or variable depending on the application [1]. Analysis of half bridge inverter with the load as an oscillating circuit is important because is often used in induction heating systems. Inverter switching elements is a signal having frequency close to the resonant frequency of load. Given that, the load is inductive, current will lag behind voltage. This allows the emergence of a recovery arrangement in which some of the power accumulated in load is transmitted via one of the two diode conduction (D1 and D2) to power supply (figure 1).

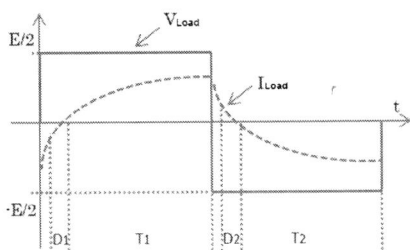

Fig. 1. Waveforms of current and voltage on an inductive load [1].

II. FUNCTIONALITY DESCRIPTION

Functionality of this circuit can be described based on figure above. We can observe four important intervals of time,

which was described in an older paper [5], and two very short interval of time whose equations I will describe now.

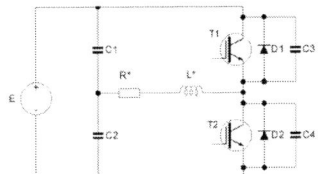

Fig. 2. Half Bridge Inverter circuit.

First interval of time is represented by the moment in which T_1 transistor is in conduction mode.

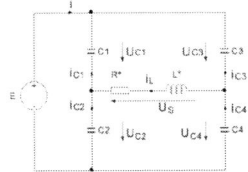

Fig. 3. Second interval of time.

Second interval of time is the moment in which the both transistor branches are blocked. C_3 and C_4 will give the load current, continuing the previous time. C_3 will charge at +E and C_4 will discharged at 0V. The differential system equation are presented below (1).

$$\begin{cases} \dfrac{di_{L^*}}{dt} = \dfrac{E - u_{C2} - u_{C3} - R^* \cdot i_{L^*}}{L^*} \\[2mm] \dfrac{du_{C1}}{dt} = \dfrac{i_{L^*}}{C_1 + C_2} \\[2mm] \dfrac{du_{C2}}{dt} = -\dfrac{du_{C1}}{dt} \\[2mm] \dfrac{du_{C3}}{dt} = \dfrac{i_{L^*}}{C_3 + C_4} \\[2mm] \dfrac{du_{C4}}{dt} = -\dfrac{du_{C3}}{dt} \end{cases} \tag{1}$$

Third interval of time start at the moment in which voltage on C_4 decrease under 0.6V. This is the moment in which D_2 enter in conduction mode. In this interval of time the voltage on C_4 became 0.

Fourth interval of time is the moment in which T_2 enter in conduction mode.

Fifth interval of time is equivalent with second interval but the difference lies in changing roles between C_3 and C_4. The differential system equation are presented below (2).

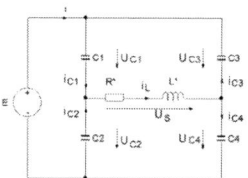

Fig. 4. Fifth interval of time.

$$\begin{cases} \frac{di_{L^*}}{dt} = \frac{E - u_{C1} - u_{C4} - R^* \cdot i_L}{L^*} \\ \frac{du_{C1}}{dt} = \frac{i_{L^*}}{C_1 + C_2} \\ \frac{du_{C2}}{dt} = -\frac{du_{C1}}{dt} \\ \frac{du_{C3}}{dt} = \frac{i_{L^*}}{C_3 + C_4} \\ \frac{du_{C4}}{dt} = -\frac{du_{C3}}{dt} \end{cases} \quad (2)$$

The sixth interval of time start when the voltage on C_3 decrease under 0.6V. This causes the conduction of D_1.

In that older paper, based on mathematical description, I realize too an MATLAB interface in which I simulate the behavior of this circuit. That interface was obtained based on differential system equations. I will use this simulation to compare the results obtained with Simulink and PSpice.

Simulink is an extension of MATLAB that allows us to create a quickly and accurately models for dynamic systems. A Simulink model can include continuous or discrete components [2].

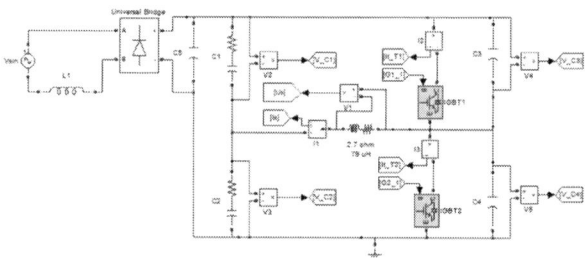

Fig. 5. Half Bridge Inverter Simulink circuit.

For simulation of Half Bridge Inverter circuit the command were analyzed for several types: command with a periodic signal of constant frequency and variable duty cycle; command with a variable frequency signal (ZVS and ZCS); and order by trains of pulses, which can be applied to both the control systems mentioned above.

III. ORDER WITH A PERIODIC SIGNAL

The power circuit was performed with an alternating voltage having a maximum value of 310V and frequency of 50Hz. To improve the power factor was connected in series

with the power supply, an inductance of 100uH, calculated according to the equation (3).

$$L_1 = \frac{U_{L_1}}{\frac{\Delta i_{L_1}}{\Delta t}} \quad (3)$$

I obtain that value If I consider the voltage across approximately 1V and requires current passing through it of 0.1A in a period of 10us.

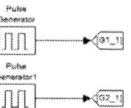

Fig. 6. Command signal generator.

For this type of control I considered three situations. Command with a signal period of 45us, 50us and 55us considering the resonance frequency 15354 Hz (R = 2.7 Ω and L = 79 uH). I have done so to highlight the change in value of the load current and voltage.

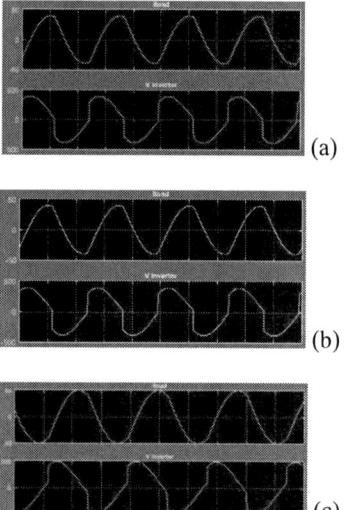

Fig. 7. I load and U load for 45us, 50us and 55us.

Analyzing these waveforms conclude that the amplitude of current and voltage induction coil is directly proportional to the growth period of the control signal.

IV. ORDER WITH DETECTION OF ZERO VOLTAGE CROSSING

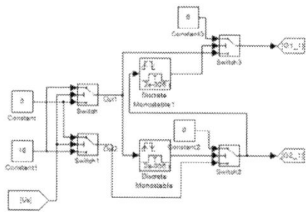

Fig. 8. Order with ZVS.

The circuit on which to analyze this type of operation is the same as before. What changes is the control circuit (figure 8). It is noted that the control signals are generated based on the zero crossing of the voltage.

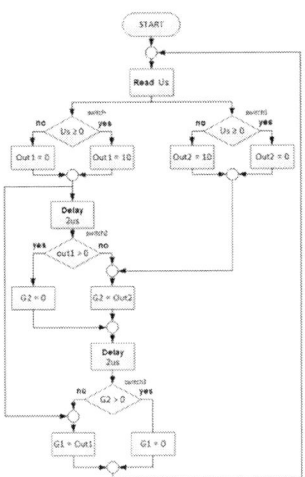

Fig. 9.   Functional diagram.

For a better understanding of this type of control we have achieved a functional diagram of it.

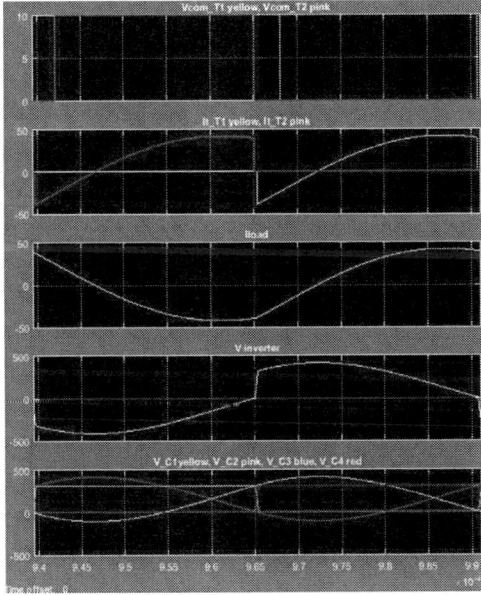

Fig. 10. The simulation results obtained for ZVS comutation.

The conclusions which can be drawn by analyzing these waveforms are:

- Unlike the previous situation in which switching was done randomly, depending on the time control signal, in this case the order is depending by the currently passing through zero of voltage on resonant circuit;

- Variation in the load can be controlled only by interrupting supply control signal. This can be accomplished with the command with pulse trains.

V.   ORDER WITH DETECTION OF ZERO CURRENT CROSSING

Functional diagram is the same with the previous case. In that diagram we have a simple modification. U_S is changed with I_S.

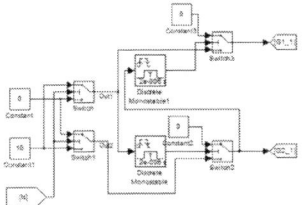

Fig. 11. Order with ZCS.

From this simulation we can observe some instability given by the glitch from the current throw the commutation elements.

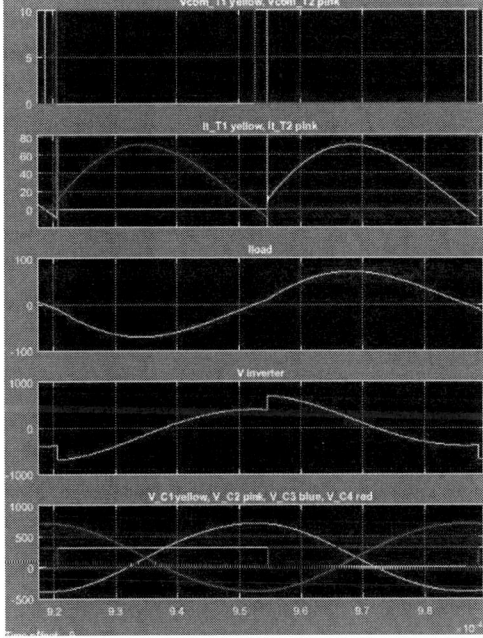

Fig. 12. The simulation results obtained for ZCS comutation.

VI.   ORDER WITH TRAIN OF PULSES

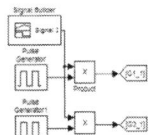

Fig. 13. Order with Train of Pulses.

This type of command is commun because can be applied to all orders discussed above.

According to the power that needs to be transmitted during the entire heating load we generate a signal ENABLE / DISABLE pulse command on a certain time.

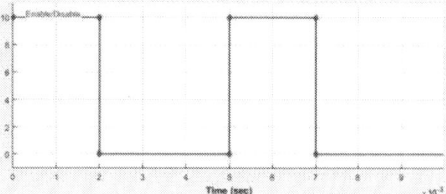

Fig. 14. Enable/Disable signal.

The simulation obtained for a situation like this is presented below.

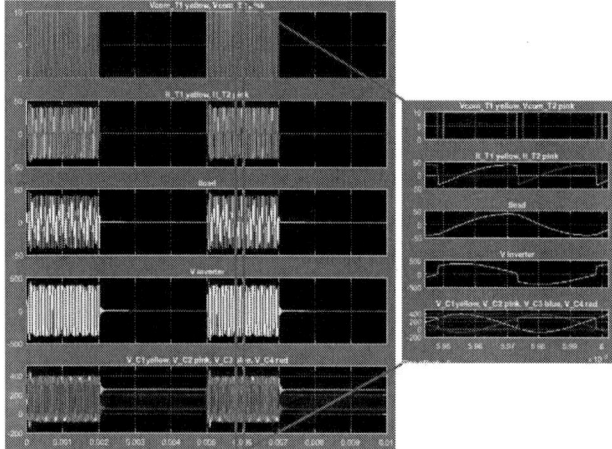

Fig. 15. Simulation results obtained for order with train of pulses.

VII. CONCLUSIONS

This paper is dedicated to the study of the induction heating system provided with half bridge inverter. The switching of the inverter elements is done through a signal having the frequency equal to the resonance frequency of the circuit.

The control for the transistors is done by applying signals to their gate. Between the conduction times of the transistors, a dead time exists (Td). This method prevents the simultaneous conduction of the two transistors.

The circuit was analyzed also using Simulink platform. Four command methods were studied:

- PWM command mode;
- Zero-crossing voltage command mode (ZVS);
- Zero-crossing current command mode (ZCS);

- Pulse train command mode.

A power analysis was performed. According to it, we concluded that in case of PWM command mode, for a high efficiency, the switching frequency must be greater than the resonance frequency.

TABLE I. VALUES OF PERIOD CONTROL SIGNALS USED FOR SWITCHING PWM SIGNAL, POWER LOAD, POWER SWITCHING ELEMENTS AND CALCULATION OF EFFICIENCY.

Frequency [Hz]	Period [us]	Dead time [us]	Conduction time [us]	Power on load [W]	Power on T1 = T2 [W]	Efficiency
40000,00	25,00	2,00	10,50	775,00	12,00	98,48
30303,03	33,00	2,00	14,50	1252,00	22,10	98,27
25641,03	39,00	2,00	17,50	1851,00	34,30	98,18
22222,22	45,00	2,00	20,50	2752,00	50,30	98,21
20000,00	50,00	2,00	23,00	3967,00	64,80	98,39
19230,77	52,00	2,00	24,00	4545,00	69,59	98,49
18181,82	55,00	2,00	25,50	5760,00	70,20	98,80
16000,00	62,50	2,00	29,25	9210,00	55,50	99,40
15750,00	63,49	2,00	29,75	9240,00	53,32	99,43
15600,00	64,10	2,00	30,05	9272,00	53,02	99,43
15500,00	64,52	2,00	30,26	9384,00	71,27	99,25
15354,00	65,12	2,00	30,56	9598,00	207,27	97,89
15000,00	66,60	2,00	31,30	9454,00	883,20	91,46
14000,00	71,43	2,00	33,71	7724,00	993,84	88,60
12000,00	83,33	2,00	39,67	3359,00	966,00	77,66

To confirm the mathematical relations and the simulation results obtained in Simulink I realize a comparison between this and simulation obtained with Pspice and Matlab. Comparing the results obtained by the three methods we can say that the results obtained in Simulink platform with the other simulators are in a margin of tolerance of up to 2%.

TABLE II. CURRENT AND VOLTAGE COMPARED VALUES.

Matlab	Simulink	Pspice
i_{Lmax} = 39,94 A	i_{Lmax} = 39,48 A	i_{Lmax} = 39,53 A
u_{Smax} = 391,12 V	u_{Smax} = 389,994 V	u_{Smax} = 389,47 V

REFERENCES

[1] N. Palaghita, D. Petreus, C. Farcas – Electronica de putere, partea a II-a, Editura Mediamira Cluj-Napoca, 2004, ISBN 973-713-039-1; (references)

[2] James B. Dabney, Thomas L. Harman, Mastering Simulink, Pearson Prentice Hall, 2004, ISBN. 0-13-142477-7.

[3] „Induction Heating System Topology Review", AN9012 Fairchild Semiconductor.

[4] O. Pop and A. Taut - "Analysis and simulation of power inverter with load variation for induction heating applications", 33rd International Spring Seminar on Electronics Technology, ISSE 2010, pp.378-382.

[5] S. Lungu, I. H. Baciu – "Comparison between different method to obtain the solution for differential equations of half bridge inverter", 31st International Spring Seminar on Electronics Technology, ISSE 2008, pp.562-565.

Analysis of LEDs Thermal Properties

N. Bădălan (Drăghici), P. Svasta

University "Politehnica" of Bucharest, Romania, Centre of Technological Electronics and Interconnection Techniques, UPB-CETTI

E-mail: niculina.badalan@cetti.ro

Abstract— One of the challenges of designing LED lighting systems is to ensure proper thermal management. It is known that during nonradiative combination phenomena[1-3] heat is generated in the active region of the LED but due to the high thermal resistance between this region and air much of the heat is conducted to the substrate. The problem is not that heat is generated but that it is not removed as fast as it is produced. In order to find solutions to this issue it is important to know as many parameters of the materials used to produce an encapsulated LED. This paper proposes an equivalent diagram corresponding to the thermal transient state, taking into account as many parts as possible of a power LED assembly including resistance and thermal capacitance of the embedded protection diode and calculates some time constants which occur on the chip-to-air via lens path. There are many equivalent schemes for the transient regime but they do not take into account the path followed by heat from chip to lens and very few calculate thermal resistances and thermal time constants. Knowing the thermal parameters intervening in transient regime, and implicitly knowing the thermal time constant, contributes to knowledge of the dynamic behavior of the LED but also provides information about the structure of the thermal device and this in turn will lead to a solution to provide adequate thermal management of the LED-based lighting systems.

Keywords—power LED, thermal resistance, thermal capacitance, thermal time constant.

I. GENERAL LED NOTIONS

LED operating conditions are influenced by manufacturing technology as well as by the technology used in making the chip. Semiconductor manufacturing technology has a direct impact on maximum acceptable junction temperature, while chip manufacturing technology directly impacts the minimum and maximum direct or pulsed current or pulses that can sweep across the chip. Capsule and lens material influence maximum operating temperature and maximum handling temperature (during assembly and installation) [1]. Materials and construction come together to determine the thermal behavior of the LED during operation [1]. On the other hand, operating conditions are strongly influenced by the way the LED is used in the application: thermal interface material, if a PCB is used or an active or passive heat sink.

To improve LED efficiency a heterogeneous structure and a transparent substrate are used. This means that the active region is manufactured of another material than the rest of the substrate, and this material has a high reflective index and a small band gap.

LED manufacturers do not provide a great deal of information regarding LED structure, material, physical, mechanical or thermal properties.

LEDs used for general lighting (ambient light) should produce white light. At the moment there are no LEDs that emit white light directly [8]. The way white light is obtained influences the structure of the LED module and consequently the thermal behavior during operation.

According to Ref.19 there are four methods used for obtaining white light (Fig. 1).

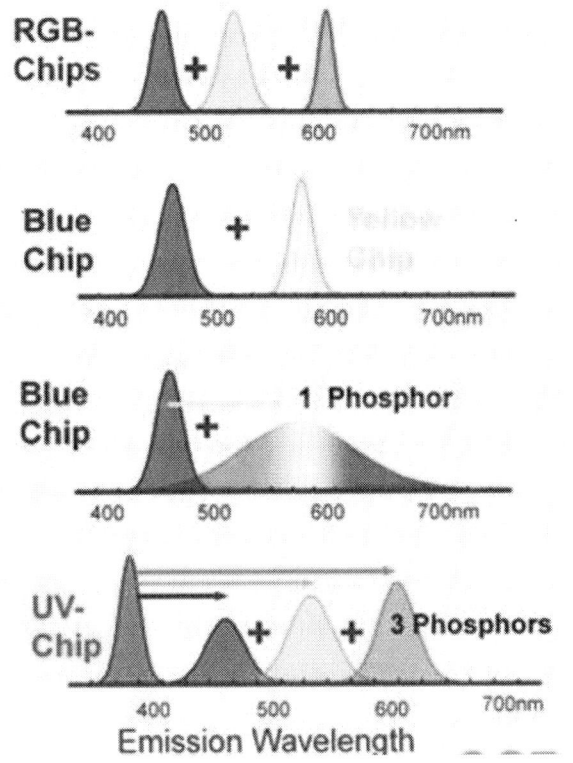

Fig. 1: White LED approaches[19]

One method is to generate white light by mixing the light from three LED chips: one red, one blue and one green. [19]

Another method uses blue and yellow LED chips mixed together in a certain ratio to produce white light [19].

The third method uses a blue colored LED and yellow phosphorus. This is the most common method used for generating white light. In this case the conversion can take place at the chip-level by depositing a layer of yellow phosphorus on top of the LED (Fig. 2) or may occur by scattering yellow phosphorus particles in a material that is poured on top of the blue LED (Fig. 3) [19].

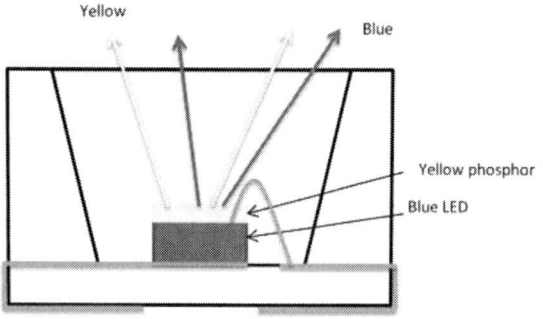

Fig. 2. Chip-level conversion

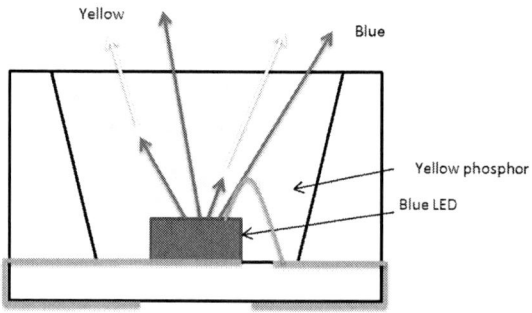

Fig.3. Volume conversion

A fourth method uses an ultraviolet LED to excite three types of phosphor: red, green and blue.

To characterize the steady state of an encapsulated LED it is enough to know its thermal resistance.

To characterize the transient thermal regime both the thermal resistance and thermal capacity of the encapsulated LED should be taken into account [17].

To determine the junction temperature different estimative methods are used that study the LED from outside [13-15,20] or from the inside [3-5, 9-11].

II. EXPERIMETAL SET-UP AND RESULTS

In Fig.4 (layers not to scale) the schematic layout of a LED system with heat sink is shown.

The LED studied in this work produces white light and was obtained from a blue LED on top of which a yellow phosphorus layer was deposited [6-8,18]. A more homogenous white color is obtained through this method.

Fig.5 shows the thermal equivalent circuit for the transient state for the assembled LED from Fig.4, assuming that each material has a thermal capacity and thermal resistance.

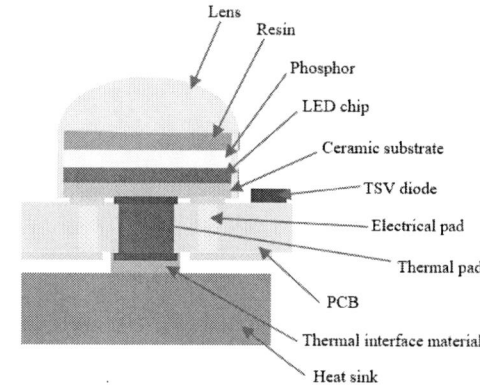

Fig. 4: Schematic layout of an LED system with heat sink

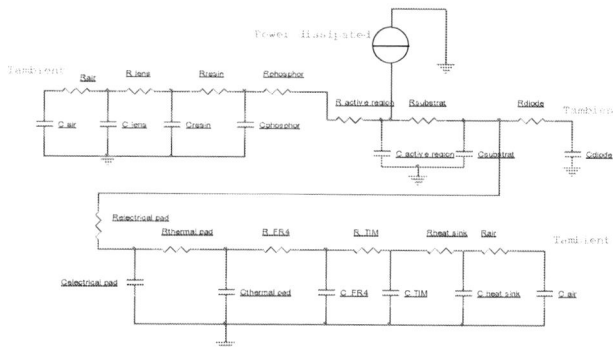

Fig. 5: Equivalent thermal circuit for transient state

We presume a hypothetical round model of a LED for ease of calculations.

We assume that the active region has the same area as the substrate so that the substrate will conduct heat but will not function as a heat spreader.

It is known that thermal resistance for 1D material is:

$$R_{th} = \frac{d_{substrate}}{K_{th} \cdot A_{substrate}} \qquad (1)$$

Where:

- $d_{substrate}$ is the thickness of the substrate along the heat transfer direction,

- $A_{substrate}$ is the cross-section area of the substrate,

- k_{th} is the thermal conductivity of the substrate material. And thermal capacity is:

$$C_{th} = c_{th} \cdot \rho \cdot d_{substrate} \cdot A_{substrate} \qquad (2)$$

Where c_{th} is the specific heat and ρ is the mass density.

Thermal time constant is equal to the product of thermal resistance and thermal capacity:

$$\tau_{th} = R_{th} \cdot C_{th} \qquad (3)$$

It was reported in Ref.2 that the calculated thermal time constant is 3.5ms for a GaInN LED chip with a 200µm thick sapphire substrate, k_{th}=0.350 Wcm^{-1}K^{-}1, c_{th}=0.760Jg^{-1}K^{-1}, ρ=3.98gcm^{-3} and emitting at 395nm. It was reported in Ref.2 that the thermal time constant for the 1D material does not depend on $A_{substrate}$.

If the active region has a thickness of 1µm and: k_{th}=1.30Wcm^{-1}K^{-1}, c_{th}=0.490Jg^{-1}K^{-1}, ρ=6.15gcm^{-3} we can compute the time constant for the active region: $\tau_{active\ region}$=23ns. Since this value is small, it is often neglected in calculations.

If the silicon lens has a thickness of 0.6mm and: k_{th}=0.31Wcm^{-1}K^{-1}, c_{th}=1.37Jg^{-1}K^{-1}, ρ=1.02gcm^{-3} we can compute the thermal constant for silicon lens: $\tau_{silicon\ lens}$ =16ms.

Measurements were performed with the thermal camera on two light sources that use power LEDs in order to visualize their thermal maps. The first was outfitted with an 18W LED which was equipped with a passive heat sink. The second had a 39W LED and was fitted with a Nuventix SynJet active type heat sink. Measurement results are presented in Fig. 6 and Fig. 7.

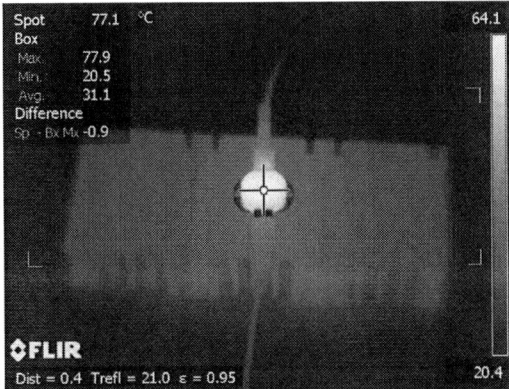

Fig. 6. Thermal map

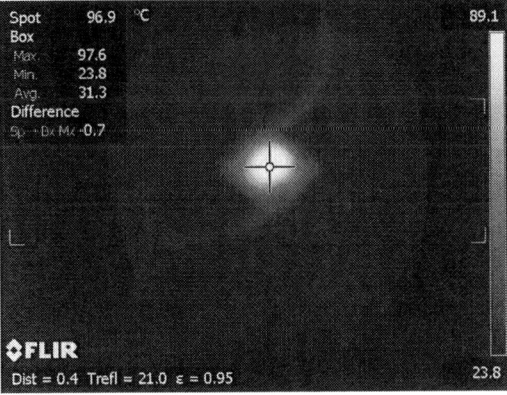

Fig.7. Thermal map

We can see that the highest temperature is on the lens, 77.9°C for the first LED and 97.6°C respectively for the second. This is because the heat propagates in all directions not only to the bottom side of the LED, but the top side fails to transfer heat to the ambient and heat transfer cannot be facilitated through additional methods.

Subsequently although some of the heat reaches the lens in a short time, the lens fails to transfer this heat to the ambient due to its low thermal conductivity.

Another material found in the manufacturing process of white LEDs is the phosphor layer. Thickness, uniformity and density of the phosphor layer affects the wavelength (and by default the color) emitted by the LED but since it is a powder it does not significantly influence the thermal time constant of the LED assembly.

The TSV protection diode is electrically connected in parallel with the LED and is positioned on the same PCB with it. In normal operating conditions, it can be considered as an LED cooling element or can be used to determine the operating temperature of the LED [12,15].

Depending on dissipated power but also on operating conditions, LEDs can be mounted on various types of PCBs (Printed Circuit Boards), such as FR4 with plated holes, ceramic substrate, flexible PCBs on metal base, copper metal core PCB [16].

For the analyzed model we assumed that FR4 with plated holes was used. Usually holes with a 0.5 diameter and a 25 µm thickness, an overall plating of 42 µm and a PCB thickness of 0.8 mm are used for the via design[16].

Often a radiator is outfitted under the LED module. It can be made of aluminum (usually) or copper. Sometimes besides the cooling function it may also have the role of mechanical protection in case of soldering or desoldering the LED from the cooling system (for very high power LEDs) or from the PCB (for high power LEDs).

In general most of the materials used for the bottom side have dimensions and thermal properties (thermal resistance, thermal capacity and thermal conductivity) chosen so that no heat flow bottle neck phenomenon occur.

Although in normal conditions there are no exceptional requirements in terms of processing, regardless of intended use it is recommended for the heat to be dissipated by suitable means to ensure proper thermal management.

III. Conclusion

A model for an equivalent circuit for the transient state was proposed and hypothetical thermal constants were computed for three materials used in encapsulated LEDs. The implemented model is for emitter type power LEDs. For the equivalent diagram for chip on board or RGBW assemblies it will undergo significant changes.

Measurements were performed with the thermal imaging camera on two luminaires fitted with passive respectively active heat sink and it was observed that the greatest temperature is recorded on the lens; as opposed to the bottom side, the lens does not have additional means of heat transfer is not made of material that can ensure a good heat transfer to the environment. When choosing materials optical properties are paramount.

It was found that the thermal properties of an LED are influenced by the thermal capacity and thermal resistance of the materials that form the structure and not by dissipated power.

Knowing the thermal parameters contribute to the characterization in terms of dynamic and transient thermal behavior of an LED, and implicitly to finding solutions to ensure proper heat management.

REFERENCES

[1] http://www.osram-os.com/Graphics/XPic1/00165234_0.pdf/ Thermal %20Consideration %20of% 20Flash%20LEDs.pdf;

[2] Q. Shan, Q. Dai, "Analysis of thermal properties of GaInN light-emitting diodes and laser diodes," Journal of applied physics 108, 2010;

[3] N. C. Chen, Y. K. Yang, Y. N. Wang, and Y. C. Huang, "Heat flow in AlGaInP/GaAsAlGaInP/GaAs light-emitting diodes, " Appl. Phys. Lett. 90, 181104, 2007;

[4] C. Lasance, A. Poppe, "Thermal Management for LED Applications," Springer Publishers, Chapter 1, pp. 3-72, 2014;

[5] E. F. Schubert and J. K. Kim, "Solid-state light sources getting smart," Science 308(5726), pp. 1274–1278, 2005;

[6] N. Wei, T. C. Lu, F. Li, W. Zhang, B. Y. Ma, Z. W. Lu, and J. Qi, "Transparent Ce:YAG ceramic phosphors for white light-emitting diodes," Appl. Phys. Lett. 101(6), 061902, 2012;

[7] S. Nishiura, S. Tanabe, K. Fujioka, and Y. Fujimoto, "Properties of transparent Ce:YAG ceramic phosphors for white LED," Opt. Mater. 33(5), pp. 688–691, 2011;

[8] K. H. LEE1 and S.W. Ricky LEE, "Screen-printing of yellow phosphor powder on blue light emitting diode(LED) arrays for white light illumination," Proceedings of IPACK2007 ASME InterPACK '07, Vancouver, British Columbia, CANADA, July 8-12, 2007;

[9] M. Vidrascu and M. Vladescu, "Programmable pulsed current driver for high power LEDs applications", Design and Technology in Electronic Packaging (SIITME), 2014 IEEE 20th International Symposium for, Bucharest, 2014, pp.123-127;

[10] M. Branzei and M. Vladescu, "Aspects on thermophisycal properties of interconnection structures for power LEDs applications", Design and

Technology in Electronic Packaging (SIITME), 2015 IEEE 21st International Symposium for, Brasov, 2015, pp. 83-86;

[11] I. Plotog; M. Vladescu, "Power LED efficiency in relation to operating temperature," Proc. SPIE 9258, Advanced Topics in Optoelectronics, Microelectronics, and Nanotechnologies VII, 92582O, 2015;

[12] M. Vladescu and B. Mihailescu, "Power LEDs operating temperature measurement using the protection reverse diode", Design and Technology in Electronic Packaging (SIITME), 2014 IEEE 20th International Symposium for, Bucharest, pp. 129-130, 2014;

[13] Y. Xi, E.F.Schubert, "Junction–temperature measurement in GaN ultraviolet light-emitting diodes using diode forward voltage method", Applied Physics Letters, Volume 85, Number 12, September 2004;

[14] A. Keppens, W.R.Ryckaert, G.Deconinck, P.Hanselaer, "High power light emitting diode junction temperature determination from current-voltage characteristics", Journal of Applied Physics 104, 093104, 2008;

[15] M. Weilguni, J. Nicolics, R. Medek, M. Franz, G. Langer and F. Lutschounig "Characterization of the Thermal Impedance of High-Power LED Assembly based on Innovative Printed Circuit Board Technology", IEEE, Proc. of 33th International Spring Seminar on Electronics Technology - ISSE 2010, Warshaw, Poland, May 12th – 16th 2010, D14, pp. 1 – 6, 2010;

[16] http://www.osram-os.com/Graphics/XPic7/00165069_0.pdf/Details %20to%20the%20Assembly%20and%20Solder%20Pad%20Design%20 of%20the%20OSLON,%20OSLON%20SSL%20and%20OSLON%20S quare%20Family.pdf;

[17] P. Mashkov, B. Gyoch, S. Penchev and H. Beloev ,"Method for in-situ power LEDs' junction temperature measurements", Proc. of 33th International Spring Seminar on Electronics Technology - ISSE 2012, Bad Aussee, Austria, 9-13 May 2012, pp. 95-100;

[18] V. Bachmann, C. Ronda, and A. Meijerink, "Temperature quenching of yellow Ce3+ luminescence in YAG: Ce," Chem. Mater. 21(10), pp. 2077–2084, 2009.

[19] https://ledlight.osram-os.com/wp-content/uploads/2013/01/OSRAM-OS_LED-FUNDAMENTALS_Basics-of-LEDs_v1_09-01-10_SCRIPT.pdf

[20] C. Ionescu, A. Drumea, N.D. Codreanu, and Al. Vasile, "Thermal Investigations on High Power LED's," Proc. of SPIE, vol. 8411, pp. 84112Y-84112Y-6, November 2012.

Stability Evaluation Method Using Phase Response Measurements

Radu BELEA, Silviu EPURE

Department of Electronics and Telecommunications,
"Dunărea de Jos" University of Galaţi,
Galaţi, Romania
radu.belea@ugal.ro

Abstract—In almost all linear applications the internally compensated operational amplifiers have no stability problems. But, some particular circuits as "the capacitive load amplifier" or "the composite amplifier" may have ringing step response. The paper proposes a new technique by which the integrated circuit user can estimate the open-loop Bode diagrams of the amplifier configured with a negative feedback loop. The method has three advantages. First, the phase measurements are executed directly on the op-amp application circuit. Second, the application designer can calculate the open-loop transfer function dynamic model, and the amplifier phase margin. Third, if there are stability issues, the circuit designer has information necessary to calculate the op-amp frequency compensation network.

Keywords—negative feedback, Bode diagrams, phase measurements, phase response, phase margin, stability evaluation.

I. INTRODUCTION

In high frequency analyze of an integrated operational amplifier configured with a negative feedback loop there are two points of view: the IC designer point of view and the IC user point of view.

The IC (Integrated Circuit) designer knows the true values of all component parameters (transistor parameters, resistances, capacitances, parasitic capacitances, etc.) used in the analogue integrated operational amplifier schematic. So the IC designer can calculate the dynamic model of each gain stage, and multiplying that transfer functions he get the whole amplifier transfer function. The IC designer point of view is well treated in the analogue integrated circuits design books, as: [1], (Gray et al., forth edition, 2001) or [2], (Razavi 2001). Some relevant examples are:

- In [1], the chapter 7.2 "Single-Stage amplifiers", uses the Miller effect approximation to calculate the frequency response of common emitter and the common source amplifiers. Also this method is used with the half-circuit method in analyse of differential amplifiers.

- In [1], in the section "Frequency Response of Voltage Buffers" is a detailed analyse of frequency response of op-amp output stage;

- The cascode amplifier and the active load differential amplifier frequency response is the subject of [1], chapter 7.3 "Multistage Amplifier Frequency Response";

- An extended example is [1], chapter 7.4 "Analysis of the Frequency Response of 741 Op-Amp";

- The same subjects are repeated [2], chapter 6 "Frequency Response of the Amplifiers", but the assay is oriented to the analogue CMOS integrated circuits technology.

The IC user knows only the DC open-loop gain, and UGBW (Unity Gain Bandwidth) as these parameter values are specified in IC datasheet. This information is not enough to calculate the frequency compensation network required if the operational amplifier application circuit has stability problems. In this situation, the op-amp user must choose an appropriate dynamic model for the designed application and must identify the op-amp dynamic model of on him own measurements.

The paper is a study from the user's point of view for small-signal and high frequency behavior of an internal compensated operational amplifier. The text is organized as follows: Chapter II and III are a recall of dynamic system models definition used for parametric identification, Chapter IV presents the electronic circuits that are subject of the experiments, Chapter V describes the phase measuring procedure, Chapter VI is an example of identification program results, Chapter VII presents the identification results for the open loop transfer function model and Chapter VIII resumes the conclusions. The inverting amplifier dynamics is discussed in the appendix, as control engineering subject.

II. BACKGROUND

Due to high DC voltage gain, it is practically impossible to make direct measurements of voltage or phase in the open-loop operational amplifier. So further the test circuit is an evaluation kit with an internally compensated op-amp, configured with a negative feedback loop, with or without capacitive load. In control engineering terms, the tested circuit is a LTI (Linear Time Invariant), SISO (Single Input Single Output) dynamic system.

We denote $G_{CL}(\omega)$ and $\varphi_{CL}(\omega)$ the measured gain and phase response. Also we denote $H_{CL}(s)$ the estimated parametric model (the open-loop TF). Results that $\hat{G}_{OL}(\omega)$ and $\hat{\varphi}_{OL}(\omega)$ is the estimated open-loop gain model and estimated open-loop phase model.

The paper idea. From negative feedback theory results that measurement errors are propagated in the following way:

- The voltage negative feedback, decrease the influence of the $G_{OL}(\omega)$ value, to the $G_{CL}(\omega)$ value. Because of this desensitization, results that the $G_{CL}(\omega)$ measuring errors produce larger errors to the estimate $\hat{G}_{OL}(\omega)$;

- The voltage negative feedback changes the critical frequencies of phase response, but there are no major changes in function shape and not desensitize phenomena occurs.

- So, we believe that the measured $\varphi_{CL}(\omega)$ and the estimate $\hat{\varphi}_{OL}(\omega)$ have similar errors.

The proposed method consists in two steps: first the experimenter uses a DDS (Direct Digital Synthesis) function generator to measure the amplifier phase response, and then process the resulted data with a Matlab program that use a "last square" minimization method to reduce the distance between measurements and a parametric model of open-loop TF. In this way the experimenter have an indirect measure of phase margin and can calculate the compensation network components.

The stability problem. The dynamic behavior of the 2nd order linear system TF is:

$$H_2(s) = \frac{G_{CL}}{T^2 s^2 + 2\xi T s + 1} = \frac{G_{CL}\,\omega_c^2}{s^2 + 2\xi \omega_c s + \omega_c^2} \tag{1}$$

In (1) G_{CL} is the DC gain, T is TF time constant, ξ is the TF damping coefficient and $\omega_c = 1/T$ is the corner frequency. If we know the damping coefficient value we get a complete behavior of the 2nd order linear system dynamics:

- $\xi > 0.7$, the step response is slow;

- $\xi < 0.7$, the step response is ringing and the frequency response has a maximum at the resonance frequency;

- $\xi = 0.7$, the step response is critically damped.

The cut-off frequency (denoted f_c or ω_c) is the close loop upper bandwidth limit frequency.

The phase margin is a stability indicator resulting from open-loop Nyquist diagram or from open-loop Bode diagrams. The phase margin is the difference:

$$G(\omega_c) = 1 \quad \Rightarrow \quad \gamma = 180° - |\varphi(\omega_c)| \tag{2}$$

The phase margin stability indicator is suitable for any order transfer function stability characterization:

- $\gamma > 63°$, the step response is slow;

- $\gamma < 63°$, the step response is ringing;

- $\gamma = 0°$, the dynamic system is a sine wave oscillator;

- $\gamma < 0°$, the dynamic system is unstable (only if system order is greater then 2).

For 2nd order system (see [4] pp 185), the phase margin expressed in radians, γ_2, can be calculated from the damping coefficient is:

$$\gamma_2 = \arccos \frac{1}{2\xi^2 + \sqrt{4\xi^4 + 1}} \tag{3}$$

With (3) we calculate $\gamma = 63°$ the corresponding value for critical damping $\xi_{critic} = 0.7$.

III.   PARAMETRIC MODELS USED IN IDENTIFICATION

All experiments analyzed in this paper refers to the Fig. 2 inverting amplifier, where $v_g(t)$, the signal generator voltage and $v_{out}(t)$ the op-amp output signal.

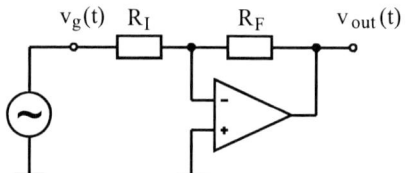

Fig. 1.   The inverting amplifier schematic

A detailed study that conducts to the inverting amplifier open-loop TF is presented in the paper appendix. The expression of this transfer function is:

$$H_{OL}(s) \cong \frac{1}{|G_{CL}|} \cdot \frac{G_{OL}}{(s\,T_1 + 1)\,(s\,T_2 + 1)\,(s\,T_3 + 1)\ldots} \tag{4}$$

where G_{OL} is the op-amp dc gain G_{CL} is the inverting amplifier dc gain, and T_1 is the dominant pole tome constant, and T_2, T_3 … are the secondary poles time constants. The TF (4) poles frequencies, measured in Hz, are calculated with formula:

$$f_n = \frac{1}{2\pi T_n}, \quad n = 1, 2, \ldots \tag{5}$$

Let be $f_{min} \ldots f_{max}$ the frequency interval on which we make measurements. For example, the section 4 experiments are made for 5...10 frequencies in the interval $(0.1\ldots2)\,f_c$, where f_c is the corner frequency (see section 2). This means that the identified model must be a good approximation of $H_{OL}(s)$

transfer function only in within the inner band $f_{min} \ldots f_{max}$. So, if $f_1 \ll f_{min}$ from (4) results that we can use the following approximation:

$$H_{OL}(s) \cong \frac{1}{s\,T_{int}\,(s\,T_2+1)\,(s\,T_3+1)\ldots} \quad (6)$$

where T_{int} is the equivalent integrator time constant:

$$|G_{CL}| = \frac{R_F}{R_I} \quad \Rightarrow \quad T_{int} = \frac{|G_{CL}|}{G_{OL}}\,T_1 \quad (7)$$

If $f_2 \gg f_{max}$ from (6) we get the 2nd order model:

$$H_{OL}(s) \cong \frac{1}{s\,T_{int}\,(s\,T_\Sigma+1)},\text{ where } T_\Sigma = T_2+T_3+\ldots \quad (8)$$

Also $f_3 \gg f_{max}$ from (6) we get the 3rd order model:

$$H_{OL}(s) \cong \frac{1}{s\,T_{int}\,(s\,T_2+1)\,(s\,T_\Sigma+1)},\text{ where } T_\Sigma = T_3+T_4+\ldots \quad (9)$$

The approximations used in (8) and (9) are well known in control engineering (see [6], the compensator design using Kessler variant of module criteria).

The purpose of the study is to evaluate the inverting amplifier phase margin starting from the close loop phase measurements. We use the following parametric models:

- In section 5 is identified the close-loop, 2nd order, parametric model (1) and the phase margin is calculated with equation (3);

- In Chapter VI is identified the open-loop, 2nd order model (8), or the 3rd, order model (9). The phase margin is calculated with the same Matlab program that does the parametric identification.

IV. EXPERIMENTAL CONDITIONS

In the following sections all experiments are performed on Fig. 1 schematic. It is an inverting amplifier built with the internally compensated op-amp TL72, configured with a negative feedback loop. At the amplifier output can be connected a capacitive load. Table I is a list of the experiments described in the paper.

TABLE I.   THE EXPERIMENTAL CONDITIONS

Experiment	R_I (kΩ)	R_F (kΩ)	G_{CL}	C_L (pF)
#1	1	10	-10	0
#2	1	1	-1	0
#3	10	1	-0.1	0
#4	1	1	-1	500

The simplest low stability example. Consider the Fig. 1 inverting amplifier where input $v_i(t)$ is a 200 kHz, square wave voltage signal. The $v_i(t)$, and $v_o(t)$, signals are measured with a two channel digital oscilloscope. In Fig. 2 are presented the measured time diagrams for #2 and #3 conditions.

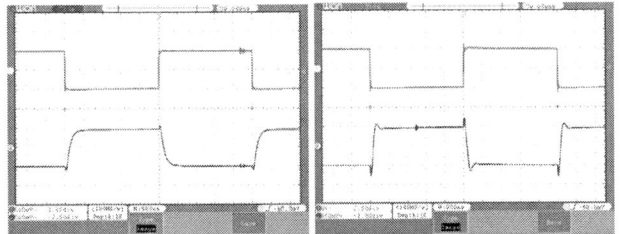

Fig. 2.   Inverting amplifier step response for $G_{CL} = -1$ and $G_{CL} = -0.1$

Comparing the Fig. 2 time diagrams we observe:

- If $G_{CL} = -1$, the output amplifier signal is a square wave with damped fronts (see the lower left time diagram);

- If $G_{CL} = -0.1$, the output amplifier signal is a square wave with ringing front response (see the lower left time diagram). In this case amplifier a phase margin less then 63°, but we don't know the exact phase margin value.

V. PHASE MEASURING METHOD

The Fig. 3 phase measurement setup consists of: OWON AG1012F, (two channel function generator), an oscilloscope working XY in mode as phase null indicator, two coaxial cables of the same length, terminated on the 50 Ω characteristic impedance and a 10x oscilloscope probe connected at Y channel oscilloscope input. The 10x probe input capacitance is about ten times lower than the oscilloscope input capacitance.

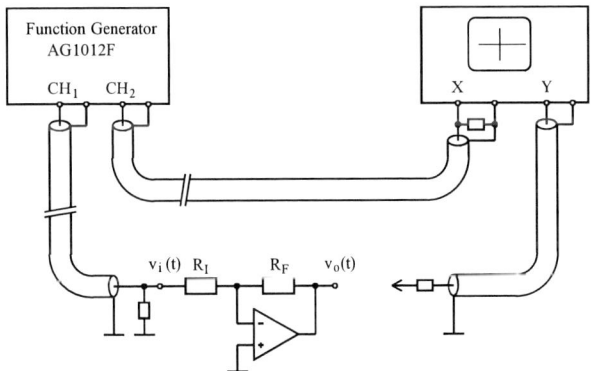

Fig. 3.   The phase measurement setup

The measuring principle. For a specified frequency, all the Fig.3 signals are coherent. The function generator can generate two sine wave signals with same frequency, same amplitude and adjustable phase shift in range 0...360° with a resolution of 0.1°. If the two coaxial cables lengths are equal, then the signals $v_i(t)$ and the oscilloscope X channel input signal are "in phase" for any frequency. At a phase measurement, the experimenter manually adjust, the CH_2 phase shift signal until the oscilloscope Lissajous figure indicates 180° phase shift. All the voltage signal phase measurement must be performed with the same 10x probe.

The phase response measurements are taken in 5...10 points in the interval (0.1 f_c...2 f_c), where f_c is the corner frequency (see section II). For one measuring point of the the amplifier phase response, the following operations are required: set the measuring frequency (i.e. f_k, $k = 1...N_k$), measure the phases of the $v_o(t)$ and $v_i(t)$ signals and calculate the amplifier phase shift with formula:

$$\varphi_{CL}(f_n) = \varphi(v_o(t))\big|_{f=f_k} - \varphi(v_i(t))\big|_{f=f_k} - 180° \qquad (10)$$

VI. FIRST EXPERIMENTAL RESULTS

The phase measured data and identification results are summarized in Fig.4. The measurements are performed for experiments #1, #2 and #3 (see Table I).

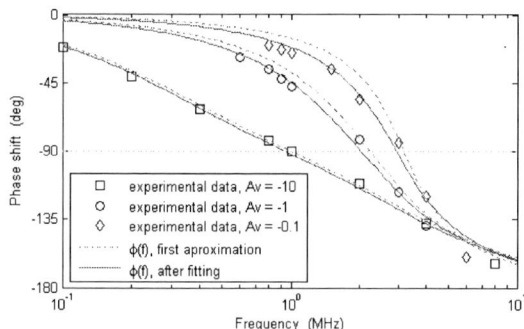

Fig. 4. Experimental data used in phase identification

Parametric identification. Let use in identification the simplest 2nd order, linear system, transfer function (1). The identified parametric model is implemented in "CL_System2" call-back function:

```
function vPhi = CL_System2(x, vFreq)
wc = 2*pi*x(1);
csi = x(2);
Bs = [wc^2];
As = [1, 2*csi*wc, wc^2];
HH = freqs(Bs, As, 2*pi*vFreq);
vPhi = unwrap(angle(HH))*180/pi;
```

The function formal parameters are "x", the parameter vector and "vFreq" the vector of frequencies. The function "vPhi" return the estimated phase vector. The function identified parameters are: "wc" the corner frequency and "csi" the damping coefficient.

The experimental data are organized in two vectors "fVect" and "PhiVect". The identification is done by "lsqcurvefit" (last square curve fitting) Matlab, optimisation Matlab function that reduces the distance between "PhiVect", the experimental measured phase vector and approximation phase vector calculated with "CL_System2" model:

```
Xin = [fc, Csi];
[Xout, resnorm] = lsqcurvefit(@CL_System2,
Xin, fVect, PhiVect);
```

Where "@CL_System2" is the address of the call-back function, "Xin" is the input parameter vector, "fVect" and "PhiVect" are the experimental data vectors, and "Xout" is the optimization resulting parameter vector.

The Matlab "lsqcurvefit" function computes the next $f(x)$: $R^n \rightarrow R$, objective function:

$$f(x) = \sum_{k=1}^{N_k} \left(\varphi_k - \hat{\varphi}_k(x) \right)^2 \qquad (11)$$

where $x \in R^n$ is the parameter vector, N_k is the number of measuring points, φ_k is the measured phase, and $\hat{\varphi}_k(x)$ is the estimated phase computed by "CL_System2" call-back function. The optimum value and the $f(x)$ residual value are:

$$rez = \min_x f(x) \qquad \Rightarrow \qquad x_{out} \qquad (12)$$

The optimization algorithm parameter "Xin" is the starting point in the parameter space R^n, and returns "Xout", the optimum point position. The chosen "Xin" is composed from:

- "fc" as the frequency that polygonal line joining measured points intersects the constant phase angle line of -90° (see Fig. 4);

- "Csi" = 0.7, the critical damping of the 2nd order dynamic system.

In Fig. 4, $\varphi(f)$ function computed with "Xin" parameter vector is represented with dotted red line and in Fig. 4 legend is named as "$\varphi(f)$, first approximation". Also, $\varphi(f)$ function computed with "Xout", is plotted with blue line.

Table II contains the identification result parameters list: G_{CL} is the close loop gain (column 1), ξ is the identified damping coefficient (column 2), f_c is the identified the 2nd order transfer function corner frequency (column 3), and γ_2 is the calculated phase margin with equation (3).

TABLE II. EXPERIMENTS #1, #2 AND #3, PARAMETER IDENTIFICATION RESULTS

G_{CL}	ξ	f_c	γ_2
10	1.806	0.941 MHz	84.9°
1	0.769	2.037 MHz	66.3°
0.1	0.593	2.932 MHz	51.2°

Comments:

- For parametric identification to be successful it is better that thee frequency measurements samples the $f_{min}...f_{max}$ interval in N equidistant points on the logarithmic frequency axis.

- From Fig. 4 and Tab. 2 results that the graphical information used in identification are: phase response inflexion point frequency (depending on f_c parameter), and the inflexion point tangent slope (depending on ξ parameter). So it is good to eliminate measured points located in the asymptotic phase response region (phase is less then 20°).

VII. IDENTIFICATION OF OPEN-LOOP MODELS

In this chapter we identified the open-loop models (8) and (9) TF and the (see Chapter III). The experiment #2 (inverting amplifier without capacitive load, $G_{CL} = -1$) identification results are plotted in Fig. 5.

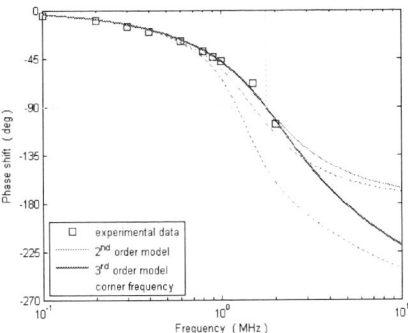

Fig. 5. Experiment #2, $G_{CL} = -1$, $C_L = 0$

The experiment #4 ($C_L = 500$ pF, $G_{CL} = -1$) identification results are plotter in Fig. 6.

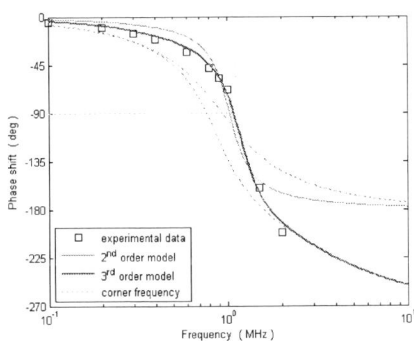

Fig. 6. Experiment #4, $G_{CL} = -1$, $C_L = 500$ pF

In Fig. 5 and 6, there are represented the next information:

- the (f_i, φ_i) measured points have square marks;

- the TF (8), 2nd order identified model, is plotted with thin red thin line;

- the TF (9), 3rd order identified model, is plotted with thin blue thick line;

- with dotted red line are represented the "Xin" model used in each optimization.

A synthesis of optimization results is in Table III. The first three columns indicate the experimental conditions (Table I experiment number, the identified TF and the graphic results). The next columns are:

- f_{int} is the integrator mode cut frequency (see equations (4) and (6) in Chapter III). The f_{int} has the same information as "the dominant pole frequency";

- is the pole frequency corresponding to T_Σ time constant or, f_2 is the lowest frequency of secondary poles.

- f_2 is the 3dr pole frequency.

TABLE III.   THE EXPERIMENTAL CONDITIONS

Exp.	TF	Fig.	f_{int} (MHZ)	f_2 (MHZ)	f_3 (MHZ)	rez
#2	(8)	5	1.31	2.37	∞	113
#2	(9)	5	1.27	4.85	4.85	78
#4	(8)	6	2.67	0.39	∞	2148
#4	(9)	6	1.38	1.71	1.72	176

VIII. CONCLUSIONS

In figures 4, 5 and 6 we observe the small distances between measured points phase and the identified phase response. This means that the principle stated in the background section entitled "The paper idea" is correct. Consequently the proposed method may be used as follows:

- If there are stability problems (for example: ringing step response), the application engineer can find the information necessary to calculate the op-amp frequency compensation network;

- The phase measurements are done direct on the application circuit with stability problems;

- With measurements taken at the end of the manufacturing cycle, the IC design engineer can verify the design parameters stipulated.

Acknowledgments

The authors thank Professor Emil Ceanga who read the text carefully and gave us some useful suggestions. Also we thank Professor Laurenţiu Frangu that created the possibility to do a detailed study on the possible applications of the vectorial voltmeter. The research was done under the "Research in Electronics Centre" from "Dunarea de Jos" University of Galati.

References

[1] Paul R. Gray, P. J. Hurst, S. H. Lewis and R. G. Meyer. "Analysis and Design of Analog Integrated Circuits". Fourth edition, John Wiley & Soons, Inc., New York, 2001.

[2] Behzard Razavi, "Design of Analog Integrated Circuits", chapter 10, pp. 345-369, McGraw-Hill, New York, International edition 2001.

[3] Grayson King, "Op Amp Driving Capacitive Loads", Analog Dialogue, Volume 31, November 2, 1997, Analog Devices.

[4] Sergiu Călin. "Regulatoare automate". Editura didactică şi pedagogică Bucureşti 1976.

[5] Viorel Mînzu, E. Ceangă. "Bazele sistemelor automate", Editura didactică şi pedagogică Bucureşti 2002. Chapter 10, pp 179-187.

[6] E. J. Mastascusa. "Operational Amplifier Stability", www.facstaff.bucknell.edu/mastascu/econtrolhtml/CourseIndex.html

Appendix

Let be Fig a.1 operational amplifier. If is used in linear operating region. This op-amp has only three time varying voltage signals: $v_{out}(t)$, $v_{+in}(t)$ and $v_{-in}(t)$. Let denote $V_{out}(s)$, $V_{+in}(s)$ and $V_{-in}(s)$ the Laplace transform of these signals.

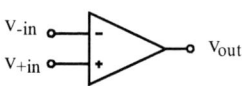

Fig. a.1. The ideal op-amp symbol

If G_{OL} is the dc open loop gain, the Fig. a.1 op-amp transfer function is:

$$A(s) = \frac{V_{out}(s)}{V_{+in}(s) - V_{-in}(s)} = \frac{G_{OL}}{(sT_1 + 1)(sT_2 + 1)(sT_3 + 1)\ldots} \quad (\alpha.1)$$

where $\quad T_1 \gg T_2 > T_3 > \ldots \quad$ and $\quad f_n = \frac{1}{2\pi T_n} \quad (\alpha.2)$

In equation ((α.1), T_1 is the dominant pole time constant; respectively f_1 is the dominant pole frequency (measured in Hz). If the poles time constants are ordered as in ((α.2), all the secondary poles frequencies are greater than f_1.

In the one pole per gain stage hypothesis op-amp TF function is 3rd order. The TF parameters are: G_{OL} is the open-loop gain, T_1 is the time constant of common emitter transistor, T_2 is the time constant of differential amplifier T_3, is the time constant of output emitter follower. If we analyze the dynamics of the active load of differential amplifier, or the dynamics of other active bias circuits we find other secondary poles.

Let be the Fig. a.2 inverting amplifier Laplace equivalent schematic.

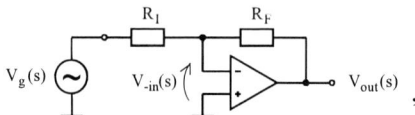

Fig. a.2 The inverting amplifier

For the node where inverting input is connected we write the write equation:

$$\frac{V_{out}(s) - V_{-in}(s)}{R_F} + \frac{V_g(s) - V_{-in}(s)}{R_I} = 0 \quad (\alpha.3)$$

In fig. 1 $V_{+in}(s) = 0$, so from equation (α.1) $V_{-in}(s)$ can be expressed as:

$$V_{-in}(s) = -\frac{V_{out}(s)}{A(s)} \quad (\alpha.4)$$

From equation (α.3) ant (α.4) results:

$$\frac{V_{out}(s) + \dfrac{V_{out}(s)}{A(s)}}{R_F} + \frac{V_g(s) + \dfrac{V_{out}(s)}{A(s)}}{R_I} = 0 \quad (\alpha.5)$$

$$V_{out}(s)\left(\frac{1}{R_F} + \frac{1}{R_F A(s)} + \frac{1}{R_I A(s)}\right) = -\frac{V_1(s)}{R_I} \quad (\alpha.6)$$

$$V_{out}(s) = -\frac{A(s)}{1 + \dfrac{R_I}{R_F}\big(A(s) + 1\big)} V_g(s) \quad (\alpha.7)$$

Consequently the close-loop TF is:

$$H_{CL}(s) = \frac{V_{out}(s)}{V_g(s)} = \frac{-A(s)}{1 + \dfrac{R_I}{R_F}\big(A(s) + 1\big)} \quad (\alpha.8)$$

The idea of pervious demonstration is taken from E. J. Mastascusa in [6]. Only the (α.7) last fraction terms are arranged in a different manner. Because of $|A(s)| \gg 1$, The following approximation is obvious:

$$H_{CL}(s) \cong \frac{-A(s)}{1 + \dfrac{R_I}{R_F} A(s)} \quad (\alpha.9)$$

If we denote in (α.9): $H_D(s) = -A(s)$, the direct path transfer function, and $\beta = -R_I/R_F$, the feedback factor, we recognize all blocks of the negative feedback dynamic system topology. So from (α.9) and (α.1) result the open-loop transfer function expression $H_{OL}(s) = \beta\, H_D(s)$:

$$H_{OL}(s) = \frac{1}{|G_{CL}|} \cdot \frac{G_{OL}}{(sT_1 + 1)(sT_2 + 1)(sT_3 + 1)\ldots} \quad (\alpha.10)$$

In Fig. a.2 schematic, we observe that R_F resistance peak voltage from output circuit, and return a reaction current input node. There is another method of analysis that transforms the circuit in Fig. a.2 in an equivalent scheme corresponding to the current-mixing voltage-sampling feedback amplifier topology. The result is the same expression for (α.9) and (α.10) equations.

Simulation & Modelling of a Tungsten Filament with COMSOL for Electrothermal Process

Sergiu Cadar

Department of Analytical Instrumentation Research
INCDO-INOE2000, Research Institute for Analytical
Instrumentation
Cluj-Napoca, Romania
Sergiu.cadar@icia.ro

Etz Radu, Toma Patarau, Dorin Petreus, Fonou Serge
Maxime

Department of Applied Electronics, Technical University of
Cluj-Napoca
Cluj-Napoca, Romania

Abstract— **This paper presents the modeling and simulation of a tungsten filament using COMSOLE. The simulation results are compared with experimental data. The study consist in analyzing temperature variations for tungsten filaments (commercial bulb, 12V and 21W), correlated to low and high temperatures of metallic filaments used in electrothermal evaporation processes. The temperature measurement of metallic filaments is a complex process. A lot of factors have to be taken into account, like the resistance temperature dependency, the temperature behavior and the properties of the metal used to produce it, the emissivity and so on. The thermal effects can cause different problems or can introduce unwanted outcomes. One of the important issues is the modeling of this phenomena and interactions to build a reliable mathematical model for the behavior of a filament based on these factors. The main source of all heat transfers is the filament and there are different methods to determine its temperature such as optical methods and resistivity method.**

Keywords— *simulation, modelling, COMSOL, tungsten, filament.*

I. INTRODUCTION

The temperature measurement of metallic filaments is a complex process. A lot of factors have to be taken into account such as the resistance of the filament, the temperature behavior and the properties of the metal used to produce it (the emissivity, the current intensity etc.). The thermal effects can cause different problems or can introduce unwanted outcomes. One of the important issues is the modeling of these phenomena and the interactions to build a reliable mathematical model for the behavior of a filament based on these factors. There are different methods to determine metallic filaments temperature such as optical methods and resistivity methods. Tungsten filaments have to be calibrated from electric measurements in order to determine the temperature as a function of current intensity and the filament's resistance as a function of temperature. The voltage on the filament also varies with temperature. The calibration procedure is developing and has to be more accurate to be able to realize a proper model for the interaction of emission with a closed chamber for example.

An accurate measurement of temperature is difficult for filaments having a very high temperature above 2500K due to a spectral and thermal dependence of emissivity coefficient for different wavelengths. The middle turns of the helix shaped filament becomes hotter than the turns at the ends, so the filaments highest temperature could be measured somewhere in the middle. It would be easier to calibrate filaments if a reliable model would be available where the current could be specified in order to know exactly the highest temperatures that could occur. There are also some analytical approaches involving heat transfer balance between components of the bulb and radiosity and Monte-Carlo Method involving estimation of spatial and spectral integrations and density probability functions (modeling radiative effects) [1][2].

Also having such a model there is no need for additional expensive temperature monitoring sensors, being able to determine it from other parameters (such as resistivity or resistance of the wire if we know exactly how it changes with temperature), the filament being the sensor too.

The filament is basically a resistance which could be used for quasi-simultaneously heating and temperature sensing. Since most properties (biological, chemical and physical) depend on temperature, a temperature control is crucial for many applications such as microelectronic devices pressure sensors, implantable systems, chemical sensors, bio-MEMS, memories, material science and more [3].

Tungsten-coils are used and studied as compact electro-thermal vaporization devices for plasma-optical emission spectrometry in chemistry. Electro-thermal vaporization is a method used in atomic spectroscopy for some time [4-6].

II. THERMAL PROPERTIES OF USED METALS

Pure tungsten has some amazing properties like having the highest melting point (3695 K), lowest vapor pressure, great tensile strength, lowest coefficient of thermal expansion of all metals and a high level of electrical conductivity. Because such properties it is the most commonly used material for light bulb filaments. Tungsten is also ductile and has a high resistivity.

The resistance R for a temperature T is defined as in (1).

$$R(T) = \frac{\rho(T) \cdot l(T)}{S(T)} \qquad (1)$$

where S in case of a circular section is $(\pi d^2)/4$.So at room temperature the resistance would be (2).

$$R_0 = \frac{\rho_0 \cdot l_0}{\frac{\pi \cdot d_0^2}{4}} \qquad (2)$$

Another characteristic is the thermal expansion coefficient denoted by β which is given as a percentage of the filament's length l_0 at room temperature [17]. As a consequence of thermal expansion the length and diameter of the filament depends on temperature as described by (3).

$$R(T) = \frac{\rho(T) \cdot \left[l_0 + \frac{\beta \cdot l_0}{100} \right]}{\frac{\pi}{4} \cdot \left[d_0 + \frac{\beta \cdot d_0}{100} \right]^2} \qquad (3)$$

Dividing both sides of (3) by equation (2) (R0) the geometry dependent part, l_0 and d_0 will be eliminated and we obtain the following resistance ratio:

$$\frac{R(T)}{R_0} = \frac{100 \cdot \rho(T)}{[100 + \beta] \cdot \rho_0} \qquad (4)$$

If the temperature T does not vary too much, a linear approximation of resistivity can be used described by (5).

$$\rho(T) = \rho_0 \cdot [1 + \alpha \cdot (T - T_0)] \qquad (5)$$

where α is the temperature coefficient of resistivity from measurement data and can be found in literature for most metals (for Tungsten α = 0.0045 and ρ_0= 5.6 μΩcm , T_0=293K – room temperature). In general, electrical resistivity of metals increases with temperature. If a solid metal is non-magnetic the electrical resistivity's temperature dependency is mainly from the electron-phonon interaction. There are metals with linear resistivity variation with temperature above 100-200K and there are some materials with non-linear variations of resistivity with a very close to linear variation, but if accuracy is important it cannot be described by such equations in order to have a real mathematical model. Tungsten is such a material. The relationship between resistivity and temperature is best described by a power relationship:

$$\rho(T) = \rho(0) \cdot \left(\frac{T}{T_0} \right)^{\mu} \qquad (6)$$

where:

- $\rho(0)$ is the residual resistivity at almost 0K degrees, for tungsten 0.06052 nΩ·m;

- T_0 is a near zero temperature, where the residual resistivity was measured, 1K;

- μ is a constant specific to such materials with non-linear variation of resistivity, for tungsten its value is 1.203.

Based on these equations a model was implemented in COMSOL software and experiments were made to validate this model. The model is very important because the behavior of the filament is of high interest in fields like the atomic spectrometry analysis. Here, different materials are used to build the filaments some of them being very expensive like Rh. The behavior of the filament is important at low and high temperatures because there are at least two steps involved in evaporating a sample using metallic filaments. The first takes place at low temperatures, under 100°C for a time period of minutes needed to dry the sample under study and the second one at temperatures higher than 1500°C for just a couple of seconds for the sample to evaporate. Because the temperatures and the temperature distribution in the filament are important for the analysis of the sample a mathematical model of the filament is useful. This model helps to improve the design of the filaments and also decreases development costs by avoiding trial and error experimental design.

III. THE FILAMENT MODEL

The filament used to create the simulation model is a solid coiled filament with 38 turns, major radius 0.4 mm, minor radius 0.075 mm and axial pitch 0.25 mm. These physical dimensions can be changed anytime, if another filament needs to be evaluated.

COMSOL Multiphysics is chosen because is a finite element analysis, solver and simulation software, having many modules in each physical domain, thus one can add the physical effects needed for his model study.

For this model the traditional computational modeling workflow has been used, which involves creating geometry and defining all the necessary materials, constants and physics for that geometry. Each model is composed of meshes and different solvers are available, like time dependent solver or steady state solver.

The next steps need to be followed in order to build the filament model:

- Import or draw the device or system's geometrical model;

- Set materials for geometry, material data or relations from the same files using constant or temperature-dependent properties;

- Decide and implement the best description of the heat transfer for the system from a range of interfaces (sub-modules) that may or may not depend on other physics coupled to the system, in this case heat transfer is coupled with electric currents (Joule Effect);

- Include any other physical effects that are coupled with the effects of heat transfer;

- Define conditions and constraints on boundaries;

- Mesh the system and then use the meshes between different simulations;

- Run the solving process, with an appropriate solver and settings for the analysis being performed;

- Process and visualize results, and present them on graphs and figures.

For the presented model a helix of solid type was created setting the right dimensions and then it was put inside a sphere. The material used is tungsten in this case so two tungsten materials are needed to be defined, one for the boundary (or the surface) of the helix and one for the solid part because both the radiation and conduction is under study. Also two materials are needed for the sphere, Glass for its surface and Argon for its interior. Every object has more sub-sections, each labeled, so one can choose what phenomena to simulate on each part of an object (volumes and surfaces). The presence of fluid/gas in the system automatically requires introduction of convection into the heat transfer application and energy contribution through pressure work and viscous effect.

For modeling radiative heat transfer as real as possible, emissivity has to be taken into account and the ideal black-body model cannot be used. The grey-body model was chosen with surface emissivity from material library. The surface of the filament will take the temperature values from heat transfer in solids, calculated there, from the source and processes further as radiation, so as reference temperature variable T has to be set. In order to have a working model an inward heat flux and a convective cooling is necessary. Doing so the system will stabilize around a certain temperature, else tungsten will melt.

For convective cooling the surfaces are defined as glass. Room temperature is set as the external temperature and as heat transfer coefficient the value of air is given, thus we have a convective cooling on the surface of the light bulb in room temperature by air.

In order to introduce a current into the filament the electric current module has to be properly set. The two ends of the helix have to be selected and set manually as boundary sections. The current conservation has to be defined because it makes the interconnection with the heat transfer model. The calculated heat is set as the operating temperature.

In COMSOL a model can be studied and simulated only if a mesh is correctly defined and is smooth enough. The filament has small dimensions, therefore its mesh has to be much finer then the surface of the sphere for example. When defining the mesh, an element's size can be calibrated for Fluid dynamics, Plasma and General Physics. The available resolutions for the mesh are minimal element size dependent, thus knowing the minimal and maximal element sizes of the geometry helps defining correctly the mesh Fig. 1.

Time dependent and steady state configurations are used to see the temperature over time variation in steady state. Also dependent variables like the voltage on the filament were introduced to study the resistance variation as function of the filament temperature.

The model, Fig. 2, was built based on the configurations presented in this section and the obtained results are presented in the next section.

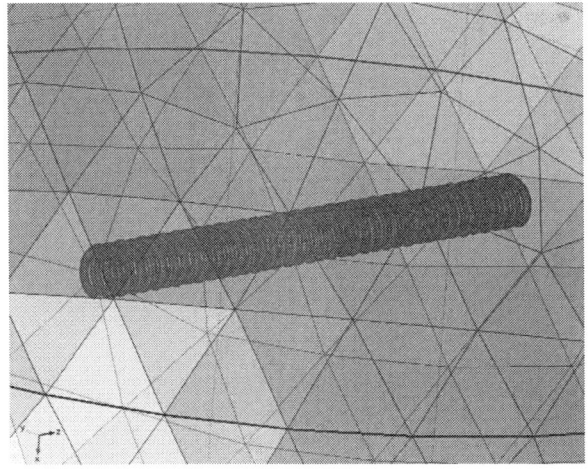

Fig. 1. The mesh defined in COMSOL.

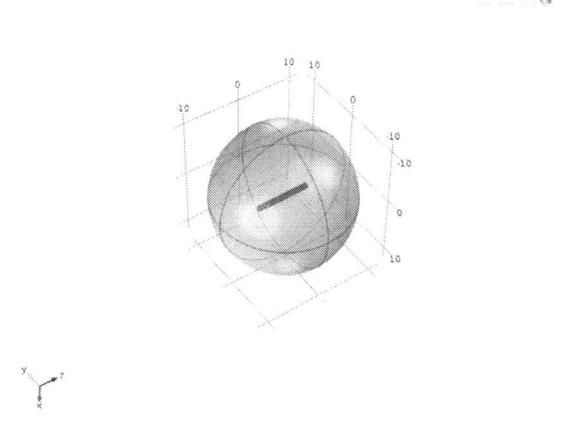

Fig. 2. The sphere around the filament in COMSOL Multiphysics.

IV. RESULT OBTAINED WITH THE MODEL

The resistance variation with temperature, Fig. 3, and the temperature variation with time Fig. 4 are plotted selecting using a Point Graph, so the values are given at a point in the filament. Another important aspect is the temperature variation of the filament on all its points as presented in Fig. 5. As one can see from the graphs the temperature domain corresponds for the first step used in plasma spectrometry analysis, the sample drying process, mentioned in section II.

The same types of plots are presented for the high temperature domain corresponding to the second step, the evaporation of the sample. The temperature domain covered by the second filament model is for high values, reaching approximately 2500K (~2200°C).

In literature the mathematical model for $(R(T))/R_0$ (Kelvin), is similar to the proposed model except a non-linear equation was used in this case for the resistivity Fig. 6 to get more realistic behavior for Tungsten. In literature the curves are determined only based on the mathematical models and are not compared with the experimentally obtained ones.

2016 IEEE 22nd International Symposium for Design and Technology in Electronic Packaging (SIITME)

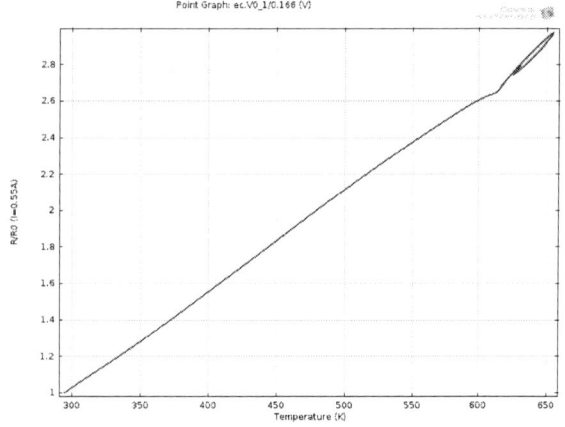

Fig. 3. R/R0 in COMSOL with 0.55A.

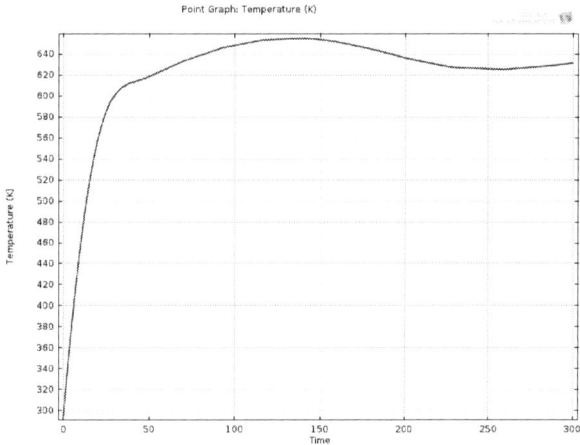

Fig. 4. Temperature variation with Time in one point

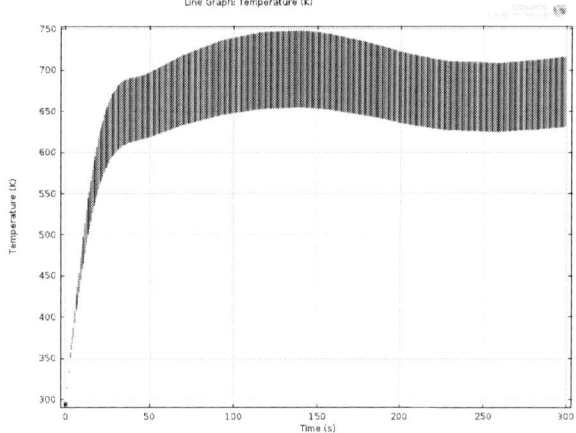

Fig. 5. Temperature variation on the whole filament surface.

The data collected experimentally is used in section 5 for the $(R(T))/R_0$, power as temperature function, voltage as temperature function and current as temperature function.

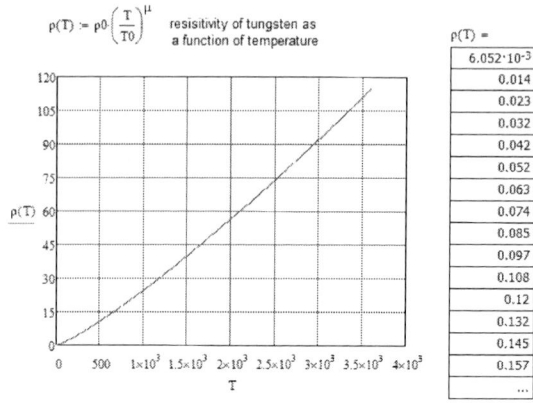

Fig. 6. Resistivity as a function of temperature (6).

In COMSOL was also possible to visualize the thermal radiation, to see how argon takes temperature over from the filament. Argon around the filament has almost the temperature of tungsten gradually decreasing towards the surface of the sphere, reaching around 500K from 2500K Fig. 7. The coldest part of the system is the outer surface of glass, being around 425K. This is a surface 3D plot with two plan slices in the middle crossing the filament through z-x and a-y plans. Using 1D plot we can view temperature as a function of time on a point graph and on a line graph too. Point graph shows the temperature at the ends of the filament and line graph on the entire outer surface of the filament.

It can be observed that the temperature difference between the middle of the filament and its ends is around 500°K. Also in COMSOL it was possible to test different filament dimensions and settings and it has been noticed that the more turns the filament has the higher is its temperature and the smaller the dimensions the lower is the current for the same temperature Fig. 8.

In order to validate the model implemented in COMSOL experiments with a tungsten filament were made.

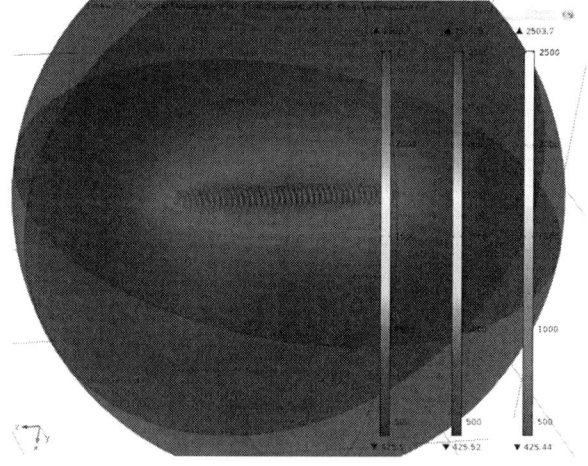

Fig. 7. Temperature radiation from the filament through Argon.

978-1-5090-4446-7/16 $31.00 © 2016 IEEE   168   20-23 Oct 2016, Oradea, Romania

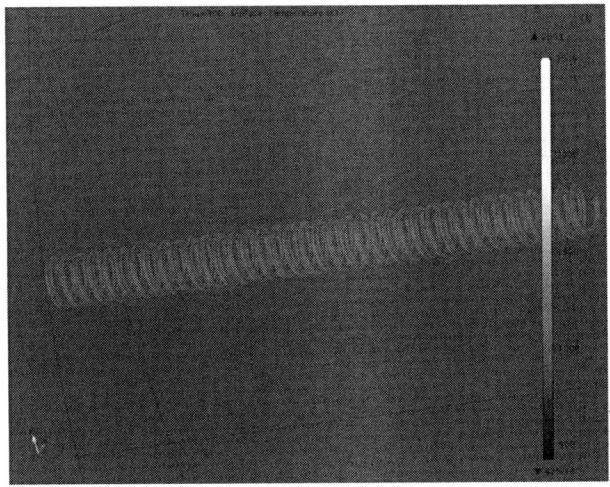

Fig. 8.   Thermal distribution on the filament.

The results of the experiments are presented in the next section and are compared with the ones obtained with the model.

V. EXPERIMENTAL SETUP

Two noncontact infrared temperature sensors from Optris, a stand with x, y, z axis position settings, a digital microscope, a support stand for the filament, a power source and a laptop were used for the temperature measurements Fig. 9. One of the pyrometers can measures temperatures between 50°C and 400°C, while the other one from 800°C to 2200°C. The model of the first pyrometer is 3ML-CF1 and the second one is an: Optris CTlaser 1MH1 Pyrometer. A great advantage of these pyrometers is that they can measure the filament temperature through the glass bulb. The heat sensor for the first one is an infrared thermometer with 2.3 µm spectral response for measurements of metals, metal oxides and ceramic metals. The optical resolution of the pyrometer is 33:1, while the temperature resolution is 0.1 K.

The first step in the measurements was to find the minimum current from which the experiment can start. As long as the pyrometer can measure temperatures from 50°C, the current where the filament was starting to change its temperature from 50°C had to be found. This current is 200 mA for a 53.4°C temperature. The current was gradually increased with 10 mA step every 20 s in order to collect the data, voltage and temperature, needed for the COMSL model comparison.

Fig. 9.   The experimental setup.

The last temperature measurement was at 390.2°C corresponding to a 0.54A current before the scale of the first pyrometer was exceeded.

The next step was finding the value of current to be applied to get a temperature of 800 degrees and its value to get a temperature of 2200 degrees (the two ends of the scale) the second pyrometer. The lower value was 0.64 amps for 800.7 degrees Celsius and 1.78 amps for 2199°C. The power source displayed the voltage across the filament for each current setting so it was possible to limit the voltage as a safety measure to protect the filament while the current is varied on a scale.

During experiments also was noticed using an electronic microscope the same behavior as in COMSOL. The middle of the filament due to the nearby turns was hotter than its ends.

Knowing the current, voltage and temperature at each measurement grade a chart was completed and the resistance of the filament was calculated for each raw of the chart. The available plots from measurement data are: resistance as a function of temperature Fig. 10, voltage as a function of temperature and temperature as a function of current. The plots in COMSOL are resented in Section IV and can be compared with the experimentally ones.

The values from the mathematical model made in Mathcad correspond very well with the COMSOL with the experimental results too. However in reality the shape of the curve is a bit different.

For the 800-2200°C temperature domain the variation of temperature with current is convex, but between 50 and 400 degrees it is concave. Thus between the two domains of temperature we had to estimate (400-800° Celsius) the variation, but as a final result we got the same shape from experimental measurements as in reference [1].

An exponential increase of the power versus temperature can be noticed Fig. 11 (the filament is rated at 21 Watts), this is the reason why COMSOL cannot show correctly the stabilization temperature for higher currents. Its calculation system works with powers calculated from current densities but the resistance of the filament can be simulated on the entire temperature domain, regardless the currents, obtaining correct values.

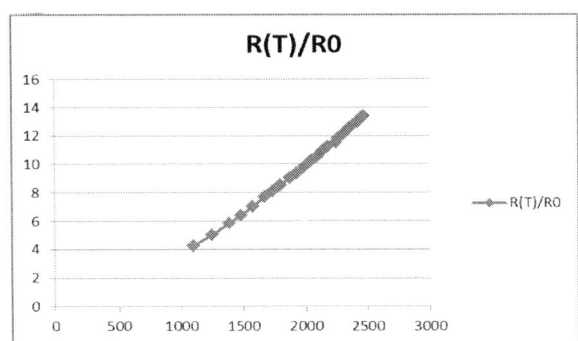

Fig. 10. Resistance as a function of temperature from experimental results for the second pyrometer (800-2200 degrees Celsius) in Kelvin as it is in COMSOL.

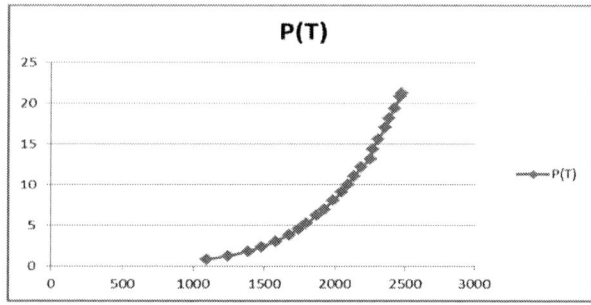

Fig. 11. Power as a function of temperature.

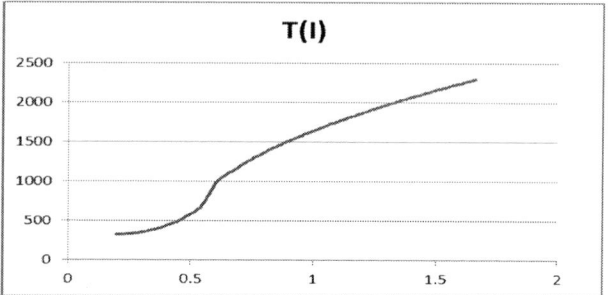

Fig. 13. Temperature as a function of current for both domains.

VI. RESULTS AND DISCUSSIONS

The results obtained with the two pyrometers were combined after the experiments were finished. Between 500 and 800°C the incandescence process takes place; unfortunately this domain couldn't be measured with the two pyrometers (50-400°C, 800-2200°C), thus from 400°C to 800°C the temperature values had to be approximated based on the voltage and current values read from the power supply. In Fig. 12 is presented the resistance of the filament as a function of temperature on the full scale from 300° to 2300°K. Its values correspond well with simulated and calculated values for temperatures above and below incandescence; also in those domains the curve is almost linear. Between 500°C and 800°C where the incandescence takes place, the temperature rises faster, but after 800°C the slope of the curve will become similar to the slope before incandescence. This result is the most important one, because the resistance of the filament can be model in this way to see how it will increase at high operating temperatures.

In Fig. 13 is presented the temperature as a function of current. Before and after incandescence the shape of the curve is different, for lower domain it is convex and for higher domain is concave, as a consequence during incandescence the temperature rises faster as a function of current.

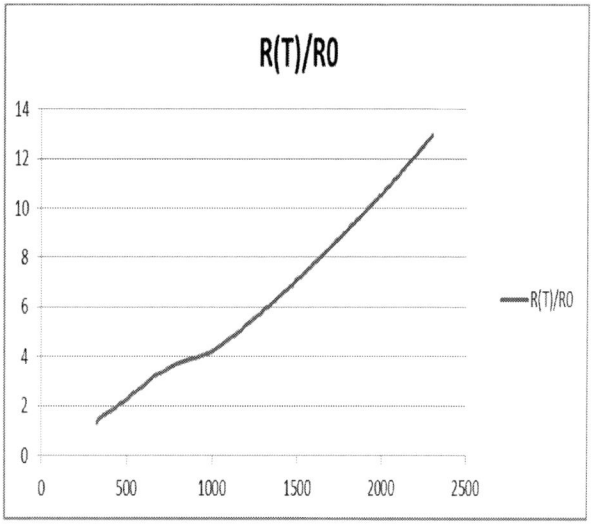

Fig. 12. Resistance as a function of temperature over both temperature domains, from 300 to 2300 Kelvin.

VII. CONCLUSIONS

It can be observed after a comparison between the simulated 3D models in COMSOL Multiphysics, the derived mathematical model constructed in Mathcad and the results of the experimental measurements that the incandescence domain is not simulated by the software and it is linearized obtaining a continuous almost linear curve. In COMSOL the voltage and current doesn't vary the same way as in reality, but calculating the resistance the same shape and values for $(R(T))/R_0$ are found, except the domain where incandescence starts.

The mathematical model for the incandescence part of the phenomena couldn't be correctly modeled, being a complex process, but overall the behavior of the resistance as function of temperature can be correctly modeled over the domains of interest, which helps a lot for designing a digitally controlled switch mode power supply that can calculate resistance values and adjust voltage and current respectively based on a temperature-resistance control method.

ACKNOWLEDGMENT

This work was supported by a Grant of the Romanian National Authority for Scientific Research, CNDI–UEFISCDI, Project number PN-II-PT-PCCA-2011-3.2-0219 (Contract no.176/2012).

REFERENCES

[1] Charles de Izarra and Jean-Michael Gitton "Calibration and temperature profile of a tungsten filament lamp", in European Journal of Physics 31 (2010) 933-942.

[2] M. Dauphin, S. Albin, M. El Hafi, Y. Le Maoult, F. M. Schmidt "Towards thermal model of automotive lamps", 11th Intern. Conf. on Quantitive InfraRed Thermography 2012.

[3] Christian Falconi "Systematic design of micro-resistors for temperature control by quasi-simultaneous heating and temperature sensing" Sensors and Actuators B 179 (2013) 336-346.

[4] Xiandeng Hou et al. "Tungsten Coil Devices in Atomic Spectrometry: Absorption, Fluorescence and Emission" Analytical sciences January 2001, vol. 17.

[5] Summer N. Hanna "Tungsten Coil Electrothermal Vaporization for Atomic Spectroscopy" Doctor of philosophy Chemistry, Wake Forest University Graduate School of Arts and Sciences, Winston-Salem, North Carolina 2011.

[6] Summer N. Hanna et al. "Design of a compact, aluminum, tungsten-coil electrothermal vaporization device for inductively coupled plasma-optical emission spectrometry" Microchemical Journal 99 (2011) 165–169.

[7] Lide R D 1991-1992 CRC Handbook of Chemistry and Physics 72nd edition, p. 10-286.

An Improved Method for the Electrical Parameters Identification of a Simplified PSpice Supercapacitor Model

Ionuţ Ciocan, Cristian Fărcaş, Alin Grama
Department of Applied Electronics,
Technical University of Cluj-Napoca,
Cluj-Napoca, Romania
ionut.ciocan@ael.utcluj.ro, cristian.farcas@ael.utcluj.ro

Adrian Tulbure
Department of Engineering
"1 Decembrie 1918" University of Alba Iulia
Alba Iulia, Romania
aditulbure@uab.ro

Abstract—**The increasing complexity of the electronic devices and the challenges involved in satisfying the needs of higher power and energy determine us to pay an important attention on supercapacitors behaviour study. Even if their energy density is about ten times lower than the energy density of the batteries, supercapacitors offer new alternatives for applications where energy storage is needed. This paper discusses one simplified PSpice model for supercapacitors. The proposed method for computing the parameters of the equivalent electrical circuit is based on experimental data achieved in different test conditions. The improved parameter identification method used provides a satisfying accuracy if the profiles of the charge/discharge/self-discharge of the supercapacitors are known.**

Keywords—supercapacitor; electrical parameters identification; PSpice model

I. BATTERIES VERSUS SUPERCAPACITORS

Supercapacitors can be a good replacement of usual batteries in applications requesting power burst, quickly charging, temperature stability, and safety properties such as: immunity to shock and vibration [1], [2].

When comparing supercapacitors with common batteries [3], or with usual capacitors [4], the main difference consists in energy and power density. A supercapacitor has a higher power density than a battery, but it has a significantly lower energy density, as shown in Fig. 1 [5].

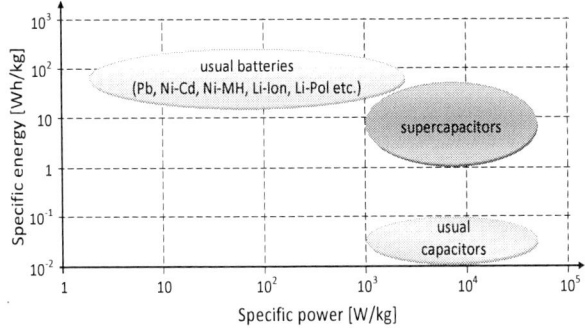

Fig. 1. Specific energy and power for different energy storage devices.

Supercapacitors or Electrical Double Layer Capacitors (EDLC) have specific benefits when using them in renewable energy resources systems. Some of these advantages are listed below, according to [6]:

- *lack of maintenance*: In contrast to battery, supercapacitors theoretically require no maintenance. This greatly reduces system cost over time and allows the storage system to be located in places impractical for chemical battery systems.

- *longevity*: Because supercapacitors store charge physically rather than chemically, cycling has virtually no effect on their capacity or longevity. Twenty-year life is easily achieved by proper selection of materials and control of operating parameters.

- *environmentally benign*: EDLCs do not employ toxic materials, and thus present no environmental threat in manufacture, transport, or disposal. They do not outgass in use and present no threat of explosion.

- *high discharge rate capability*: Superapacitors can be discharged at very high rates without damage. High rates, however, reduce the delivered energy of the unit.

Another comparison between batteries and supercapacitors parameters, adapted from [4] and [5], is given in Table I. One battery can store a large amount of energy in a relatively small volume and weight, but, as the power requirements have grown, the battery's capability seems to have been exceeded. In this context, supercapacitors main duty is to provide higher specific power and longer cycle use until replacement.

TABLE I.   A COMPARISON BETWEEN BATTERIES AND SUPERCAPACITORS

Property	Batteries	Supercapacitors
Storage method	Faradic reactions	Electrostatic interactions
Cycle life	Mechanical stability, chemical reversibility	Side reactions (>>1000000)
Energy level	High (Alkaline - 60Wh/kg, Li-Ion - 140Wh/kg)	Limited (1-10 Wh/kg)
Power level	Limited by mass transport (0.3-1.5kW/kg for Li-ion)	Limited by electrolyte conductivity (1-10kW/kg)
Charge rate	Kinetically limited	High (same as discharge)

II. Equivalent Electrical Circuit Modelling

Common models that describe the behavior of the usual capacitors are inadequate when modelling supercapacitors. The double layer capacity may be considered in parallel with the capacitor impedance, due to the charge transfer reaction. According to [5], this approach is not quite similar with the experimentally obtained characteristics. Therefore, in [5] and [6], different models consisting of three R-C branches (one of them with a voltage-dependent capacitance) are used for achieving a better fit to the real EDLC behavior.

Some simplified equivalent circuit models for EDLCs are presented in [7] and [8]. In this paper an improved method for computing the parameters of a simplified model is proposed. In [7] the parameters identification is realized by using only one charge/discharge characteristic. This paper proposes the use of two different charge/discharge sequences. For developing the model, the PScap350 produced by the Econd Ltd. was charged and discharged using the experiemtal setup shown in Fig. 2. The electronic load was used for constant current discharge of the supercapacitor and the professional data-logger was used for monitoring the voltage across it. Table II presents the main parameters of the PScap350 supercapacitor.

TABLE II.   Econd PScap350 Specifications

Parameter	Value
Nominal Voltage	14 [V]
Nominal Capacitance	350 [F]
Mass	24 [kg]

The equivalent PSpice circuit used for modelling the supercapacitors is shown in Fig. 3. The R_{esr} resistor represents the equivalent series resistance of the supercapacitor. The C capacitance is responsible for modelling the main energy storage/providing processes, the R_{sd} resistor models the self-discharge processes and R_{fit}-C_{fit} branch is used for improving the accuracy of the dynamic processes during charge/discharge of the supercapacitor.

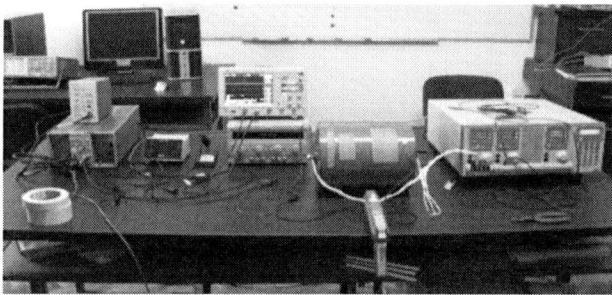

Fig. 2.   The experimental bench used for determining the model parameters.

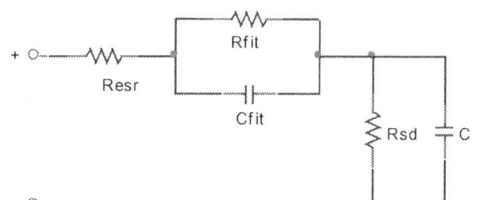

Fig. 3.   Equivalent circuit used for modelling the PScap350 supercapacitor.

III. Parameters Identification Method Proposed

For computing the parameters of the supercapacitor PSpice equivalent circuit, two experimental charge/discharge scenarios were used. In the first case, the initial PScap350 supercapacitor voltage was about 12.32V, and it was charged for 5 minutes using a constant current of 3A. Then, the supercapacitor was allowed to self-discharge for about 28 hours, monitoring the voltage across it once per second. The data achieved in this charge/self-discharge scenario are presented in Fig. 4.

Second, the PScap350 was charged at 10A constant current for 5 minutes. The initial voltage of the supercapacitor was about 0.7V, and the self-discharge time was 7.5 minutes. Then, the PScap350 was forced to discharge at constant current (20A) until the voltage across it reaches again the value of 0.7V. The experimental data obtained this time are shown in Fig. 5.

The equivalent capacitance of the supercapacitor is given by:

$$C_{sc} = \Delta Q / \Delta V = \int_{t_0}^{t_1} i(t)dt \Big/ \Delta V = I \cdot \Delta t / \Delta V \qquad (1)$$

where I is the charge current, $\Delta t = t_1 - t_0$ is the monitoring time interval, and ΔV is the voltage variation on the supercapacitor.

If one considers the set of measurements from Fig. 4, in the first 120 seconds, the value of the charge current is 3A and the voltage variation on the PScap350 supercapacitor is about:

$$\Delta V \approx 13.34V - 12.32V \approx 1.02V \qquad (2)$$

Results an equivalent C_{sc} capacitance of about:

$$C_{sc} \approx 3A \cdot 120s / 1.02V \approx 350F \qquad (3)$$

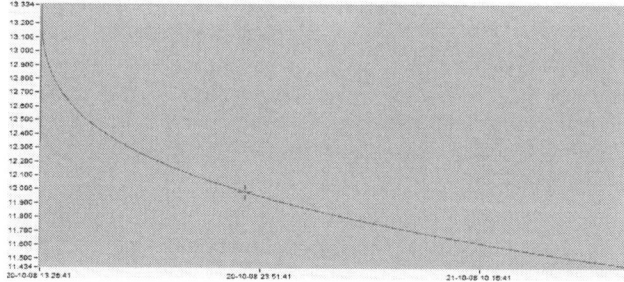

Fig. 4.   Voltage across the supercapacitor for the first set of measurements.

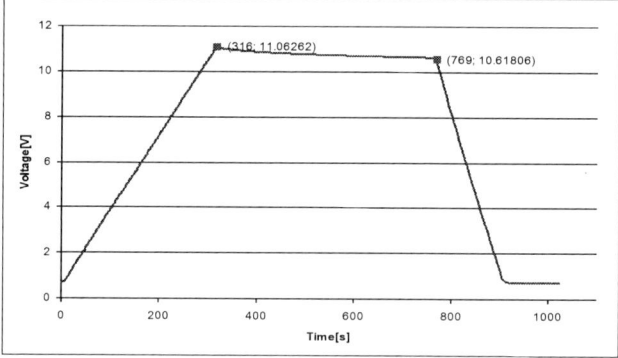

Fig. 5.   Voltage across the supercapacitor for the second set of measurements.

The C_{fit} capacitance should be 10 times higher than the main C capacitance of the model, for not influencing too much the value of the equivalent C_{sc} of the supercapacitor [10]. Results that the following relations can be written:

$$\begin{cases} \dfrac{1}{C_{sc}} = \dfrac{C + C_{fit}}{C \cdot C_{fit}} \\ C_{fit} = 10 \cdot C \end{cases} \qquad (4)$$

By replacing the C_{sc} value computed with (3) in (4), result the following values for C and C_{fit} parameters of the model:

$$\begin{cases} C = 385F \\ C_{fit} = 3850F \end{cases} \qquad (5)$$

If one considers that the supercapacitor was previously charged at 12.32V, when constant current charge process starts the initial voltage conditions of C and C_{fit} capacitors are:

$$\begin{cases} V_{IC\_C} \approx 11.2V \\ V_{IC\_C_{fit}} \approx 1.12V \end{cases} \qquad (6)$$

After 120 seconds of 3A constant current charging, the voltages across the two capacitors will be:

$$\begin{cases} V_{max\_C} \approx 11.2 + 3 \cdot 120 / 385 \approx 12.13V \\ V_{max\_C_{fit}} \approx 1.12 + 3 \cdot 120 / 3850 \approx 1.21V \end{cases} \qquad (7)$$

Since C_{fit} capacitance improves only the modelling of the dynamic processes, it can be considered fully discharged after 28 hours of self-discharge of the supercapacitor. Thus, one considers that the entire supercapacitor's voltage drop at the end of the self-discharge process is found on C capacitance. This voltage is, according to Fig. 4:

$$V_{min\_C} = 11.43V \qquad (8)$$

The value of the self-discharge resistance can be computed from the discharge equation of the C capacitor during the 28 hours (approximately 100000 seconds):

$$V_{min\_C} = V_{max\_C} \cdot e^{-100000 R_{sd} \cdot C} \qquad (9)$$

When replacing the known values in (9), results the value of the self-discharge resistence:

$$R_{sd} = -\dfrac{100000}{385} \Big/ ln\left(\dfrac{11.43}{12.13}\right) \approx 4.35 k\Omega \qquad (10)$$

For computing the equivalent series resistence R_{esr} of the PScap350, the method described in [9] is used. The EDLC is dynamically discharged with 15A current pulses (as in Fig. 6), monitoring the voltage drops when the current pulses occur.

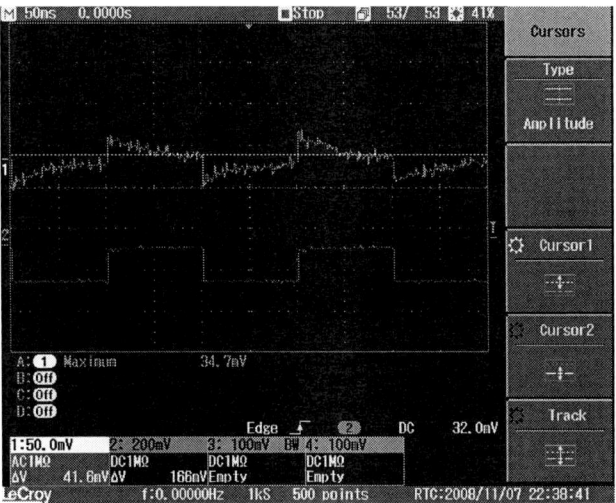

Fig. 6. Waveforms used for determinig the R_{esr} resistance of the PScap350.

As is can be seen in Fig. 6, the voltage drops on the supercapacitor at every transition moment are $\Delta V \approx 41.6mV$. Having the level of discharge current of 15A, the value of R_{esr} model parameter can be coputed with:

$$R_{esr} = \dfrac{\Delta V}{I_{dis}} \approx \dfrac{41.6V}{15A} \approx 2.77m\Omega \qquad (11)$$

where I_{dis} is the amplitude of the current pulses.

The value of the R_{fit} parameter was empirically obtained by adapting the simplified PSpice model to resemble, as much as possible, with the first measurements set from Fig. 4. For highlighting the R_{fit} identification method used, a comparison between the experimental self-discharge characteristic and different simulation results is presented in Fig. 7. As it can be seen, when the best curve fit was achieved, the recorded value of the R_{fit} parameter was:

$$R_{fit} = 1.5\Omega \qquad (12)$$

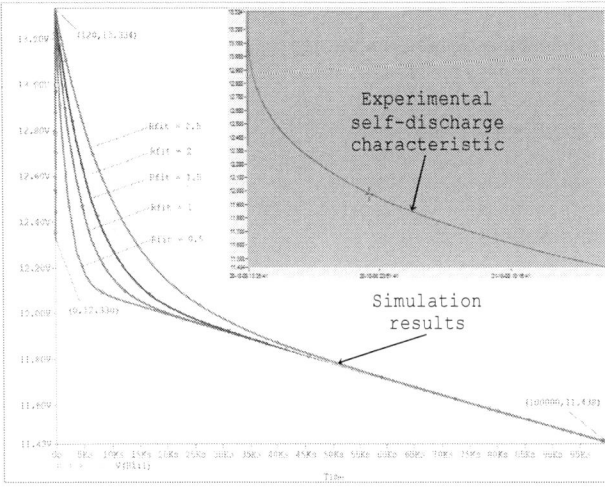

Fig. 7. Waveforms used for R_{fit} parameter indentification of the PScap350.

The proposed method for estimating the C_{fit} capacitance with (4) should be successfully used when the supercapacitor is allowed to self-discharge for a couple of hours (in this case, 28 hours have been taken into account). Thus, the C_{fit} identification method provides accuracy only when slow processes of charge/discharge are modelled. To assure the model accuracy when fast charging/discharging processes are happening, the C_{fit} parameter must be determined by using another set of measurements (in this case, the experimental data from Fig. 5 will be used). The rest of the model parameters will remain the same, but C_{fit} will be empirically determined by using a PSpice Parametric Sweep. A comparison between the experimental charge-discharge characteristic and different simulation results is presented in Fig. 8. As it can be seen, when the best curve fit was achieved, the recorded value of the C_{fit} parameter was:

$$C_{fit} = 1400F \qquad (13)$$

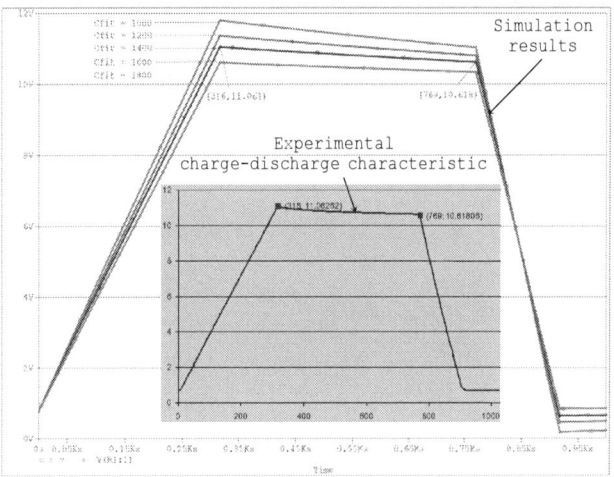

Fig. 8. Waveforms used for C_{fit} parameter indentification of the PScap350.

IV. CONCLUSIONS

In this paper an improved method for determining the parameters of a PSpice supercapacitor model is presented. The equivalent electrical circuit of the supercapacitor is a simplified one, because it has only 5 parameters. However, the parameters identification method can use both, slow and fast phenomena during the charge-discharge processes, and may be successfully used when the charge-discharge profiles of the supercapacitors are known. Three model parameters (C, R_{esr}, R_{sd}) are obtained by calculus, using the experimental data acheived from the two characteristics of the PScap350 supercapacitor, and the other two parameters (R_{fit}, C_{fit}) are empirically determined using PSpice simulations.

Table III presents the obtained results for PScap350 when modelling both types, self-discharge and charge-discharge

processes of the supercapacitor. The simplified PSpice model proposed and its parameters identification method can be used for any type of EDLC, when the experimental test conditions of the supercapacitor are known before.

TABLE III. PARAMETERS OF THE PSCAP350 SUPERCAPACITOR MODEL

Parameter / Processes	R_{esr} [mΩ]	R_{sd} [kΩ]	R_{fit} [Ω]	C_{fit} [F]	C [F]
Self-discharge	2.77	4.35	1.50	3850	385
Charge-discharge	2.77	4.35	1.50	1400	385

ACKNOWLEDGMENT

This paper was supported through the programme "Parteneriate în domenii prioritare – PN II", by UEFISCDI, project no. 53/01.07.2014.

REFERENCES

[1] R. Gălătuş, D. Petreuş, I. Ciocan, A. Grama, "Supercapacitors Study: Modeling and Sizing", The 14th International Symposium for Design and Technology of Electronic Packages (SIITME 2008), Braşov, România, pp. 55-59, September 2008.

[2] X. Li, B. Wei, "Supercapacitors based on nanostructured carbon", Nano Energy, vol. 2, pp. 159-173, 2013.

[3] A. Kuperman, I. Aharon, "Battery-ultracapacitor hybrids for pulsed current loads: A review", Renewable and Sustainable Energy Reviews, vol. 15, pp. 981-992, 2011.

[4] M. S. Halper, J. C. Ellenbogen, "Supercapacitors: A Brief Overview", MITRE Nanosystems Group from the MITRE Corp., McLean, Virginia, USA, March 2006.

[5] I. Ciocan, "Theoretical and experimental researches on energy generation and storage technologies in photovoltaic systems", PhD thesis, Technical University of Cluj-Napoca, pp. 57-58, 2014.

[6] J. Wohlgemuth, J. Miller, L. B. Sibley, "Investigation of Synergy between Electrochemical Capacitors, Flywheels and Batteries in Hybrid Energy Storage for PV Systems", Contractor report SAND 99-1477, Sandia National Laboratories, June 1999.

[7] L. Zubieta, R. Bonert, "Characterization of double-layer capacitors for power electronics applications", IEEE Transactions on Industry Applications, vol. 36, no. 1, pp. 199-205, 2000.

[8] A. Grama, D. Petreuş, R. Gălătuş, I. Ciocan, "Equivalent Models Study of Supercapacitors Behavior", The 14th International Symposium for Design and Technology of Electronic Packages (SIITME 2008), Braşov, România, pp. 50-54, September 2008.

[9] P. Johansson, B. Andersson, "Comparison of Simulation Programs for Supercapacitor Modelling. Model Creation and Verification", MSc thesis, Chalmers University of Technology, Sweden, 2008.

[10] C. Fărcaş, D. Petreuş, N. Palaghiţă and I. Ciocan, "Modeling and Simulation of Supercapacitors", The 15th International Symposium for Design and Technology of Electronic Packages (SIITME 2009), Gyula, Hungary, pp. 195-200, September 2009.

[11] A. Grama, D. Petreuş, P. Borza and L. Grama, "Experimental Determination of Equivalent Series Resistance of a Supercapacitor", The 32nd International Spring Seminar on Electronics Technology (ISSE 2009), Brno, Czech Republic, pp. 1-4, May 2009.

[12] I. Ciocan, D. Petreuş, C. Fărcaş, N. Palaghiţă, "A Study Review over Supercapacitors Modelling and Sizing Methods", The 7th Joint MmdE – IEEE ROMSC Integrnational Conference, Iaşi, România, June 2010.

2016 IEEE 22nd International Symposium for Design and Technology in Electronic Packaging (SIITME)

Developing a multi sensors system to detect sleepiness to drivers from transport systems

C. Dumitrescu, I. M. Costea, F. Nemtanu, I. Badescu. A. Banica

The Department of Telematics and Electronics for Transport
University "Politehnica" of Bucharest, Romania
ilonamoise@yahoo.com

Abstract— The increasing number of accidents in transportation (rail, road) recorded in the past few years have become a serious concern for the society. The accidents caused by driving under fatigue, drowsiness and disruptive (stress) factors have a high mortality rate due to decreasing of the abilities of perception, recognition and control which are necessary for the conductors in the field of road and railway. The paper proposes a multi-sensor system for predicting fatigue and drowsiness based on the analysis of the power spectrum density (PSD) of the electroencephalogram (EEG) and correlation with the cardiac cycle (ECG) or with the respiratory cycle. EEG data collecting uses a sensorial system, headset type (physical contact) or capacitive (no physical contact), and for ECG data collecting a new system of sensor is proposed, that does not involve physical contact against the capacitive system currently used. Our results demonstrated that it is possible to estimate with quantitative accuracy the driving performance in a 3D driving simulator, using the multi-sensor system proposed.

Keywords— *power spectrum density, sensor system, 3D driving simulator*

I. INTRODUCTION

In the past few years, the efforts in the engineering field were focused on developing methods for monitoring the cardiac or respiratory cycle, without involving the physical contact with the patients (subjects), but by means of sensors. The benefits of such an approach are obvious in situations where the investigated subject is the victim of a fire (presenting severe burns on his body), an avalanche or an earthquake, of an accident for testing the staff working in the Transportation and Critical Infrastructures field.

The EEG signals acquisition for drivers in the Transportation field can be achieved by using a contact system, helmet type (for example, EMOTION Kit) or a non-contact system, capacitive type. The capacitive sensor is installed in the driver seat backrest (headrest) and it monitors/acquires EEG signals from the occipital area. According to research in the field and to our studies, we found that alpha and beta waves are associated to the movements of the eyes open / closed, and this variation of alpha and beta waves can highlight the installing state of sleepiness and decreased ability of attention / concentration to drivers in transportation systems.

II. EXPERIMENTAL RESULTS

A. EEG Analysis using the method of power spectral density (PSD)

The EEG signals acquisition for drivers in the Transportation field can be achieved by using a contact system, helmet type (for example, EMOTION Kit) or a non-contact system, capacitive type. The capacitive sensor is installed in the driver seat backrest (headrest) and it monitors/acquires EEG signals from the occipital area. EEG signals that this system can monitor and identify are: Alfa, Beta, Theta and Delta.

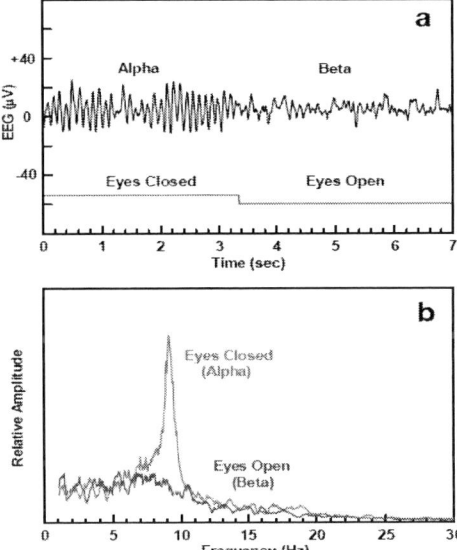

Fig. 1. Representation the the alpha and beta EEG signals in the time domain (a) and the frequency domain (b)

According to research in the field and to our studies, we found that alpha and beta waves are associated to the movements of the eyes open / closed, and this variation of alpha and beta waves can highlight the installing state of sleepiness and decreased ability of attention / concentration to drivers in transportation systems.[1][2][3]

By monitoring these EEG signals from different acquisition points Fz, Cz, Pz and Oz (international standard) and performing the PSD analysis using the Welch method, we obtain the distribution of the power spectrum for brain alpha,

978-1-5090-4446-7/16 $31.00 © 2016 IEEE 175 20-23 Oct 2016, Oradea, Romania

beta, theta and delta waves in every point. Achieving the correlation of the power spectra obtained for the specified points, we can determine with quantitative precision the performance of attention / relaxation of the conductors and determination / predict the development of fatigue and drowsiness. Figure 2 shows the theoretical simulation of the PSD Welch representation.

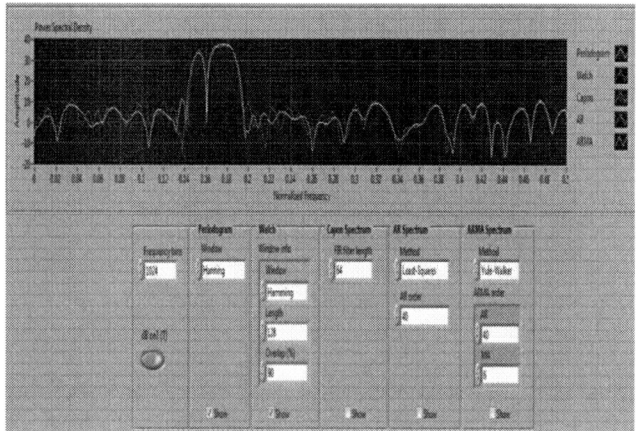

Fig. 2. PSD WELCH representation

B. EEG Analysis using the method of power spectral density (PSD)

To estimate / predict the development of fatigue / drowsiness to railway or car drivers, EEG recordings were made at points Fz, Cz, Pz and Oz. Using the Welch PSD method, the power spectra were determined for alpha, beta, theta and delta brain waves.

Figures 3 and 4 show the spectra of alpha, beta, theta, delta brain waves, purchased in point FZ, for the attention state and the moment of occurrence of somnolence.

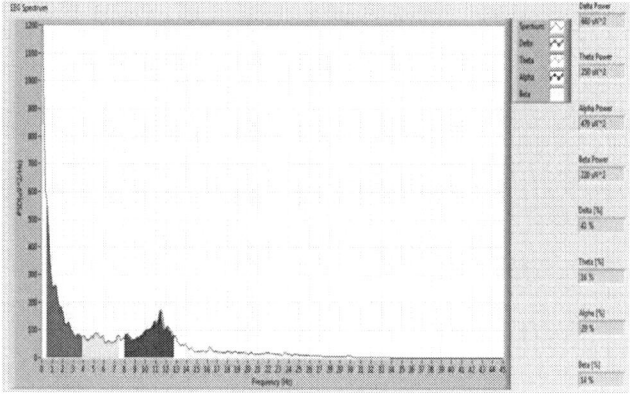

Fig. 3. The power spectrum for alpha, beta, theta and delta brain waves, purchased in point FZ, in the state of attention/concentration (eyes open)

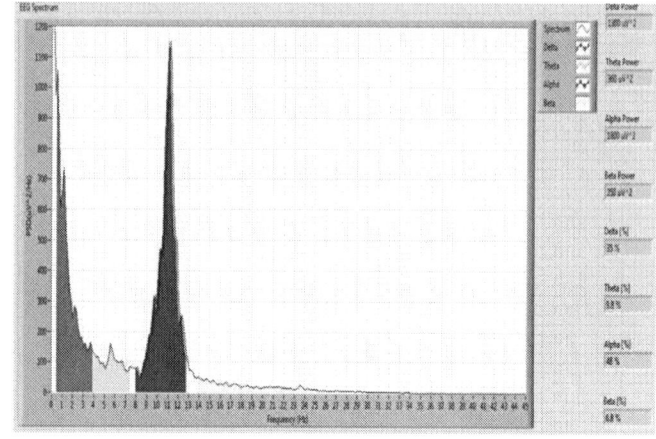

Fig. 4. The power spectrum for alpha, beta, theta and delta brain waves, purchased in point FZ, in the state of development of somnolence/fatigue (eyes closed)

C. ECG proposal with a new sensor which does not require physical contact

Our researches have focused on detecting very accurately the heart rate for very short periods of measuring and for a range of signal / low noise ratio by implementing an algorithm based on parametric spectral estimation of MUSIC type (Multiple Signal Classification).

For this study was used a continuous wave Doppler radar with frequency band 2.4 GHz. Parameter variations it was determined by: the reflected wave reception and the phase detection analysis. The acquired data were processed by: eliminating the DC component, filtering "high pass" and parametric estimation using MUSIC algorithm and non-parametric estimation using Welch algorithm. The result was evaluated for the frequency range of 0,15...3,7 Hz, in order to detect spectral peaks associated with heart rate or breathing.

III. THE MATHEMATICAL CONCEPT

The signal transmitted by a Doppler radar system with continuous broadcasting can be represented as shown in Fig 5.

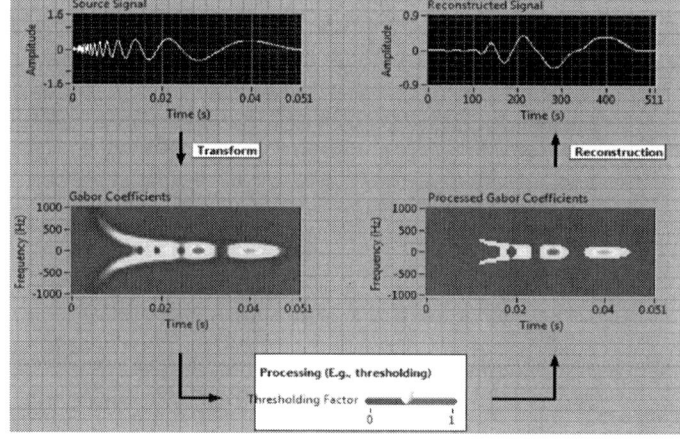

Fig. 5. The representation of the Doppler signal broadcasting and its reconstruction in broadcast using Gabor coefficients

*2016 IEEE 22nd International Symposium for Design and Technology in Electronic Packaging (**SIITME**)*

If the "target" presents a periodic oscillation x (t), according to the Dopper principle, the reflected and rebuilt signal will be have a variable phase, represented in Figure 5 (reconstructed signal). Thus, the received signal is a remissive version, delayed and modulated in frequency of the transmitted signal. In this case, the "target" is a human subject, and the signal processing obtained envisages the oscillations estimation x (t), associated to the periodic movement of the heart. In addition, the breathing causes a periodic movement of the surface of the human body, with a frequency of 0.2 ... 0.5 Hz; the amplitude given by this movement is at its highest on the chest level, where it can take between 3-15 mm. This is shown in Figure 6.

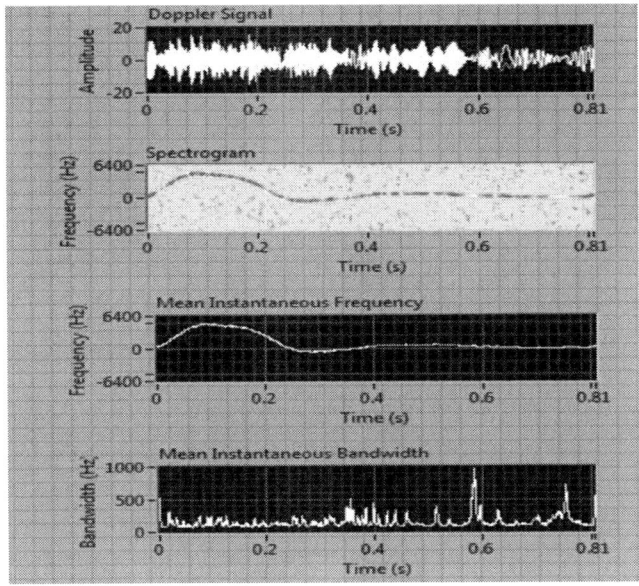

Fig. 6. Analysis of the EEG signal modulated with the Doppler signal, representation of the spectrogram and of the spectrum

According to data from the literature, the frequency of 2.4 GHz, the component determined by heart rate is very low compared with levels and the component due to breathing noises (which is why this signal is difficult to identify and recover); In addition, the signal level is in turn associated breathing very small compared with the given level of noise resulting from reflections. According squaring system architecture, the useful signal is decomposed in the two components (I and Q), which then is extracted phase. The useful sampled information with a frequency of 100 Hz was processed by the algorithm MUSIC and with spectral estimation algorithm was calculated heart rate (this values are lower than frequency of 3.5 Hz, because heart rate cannot normally exceed over the threshold of 210 beating / minute). The MUSIC algorithm uses the analyzed signal decomposition in a subspace of (useful) signal and in a subspace of noise.

This type of estimation is based on a model according to which the analyzed signal is composed of a finite number of sine waves of frequency, to which a Gaussian noise adds.

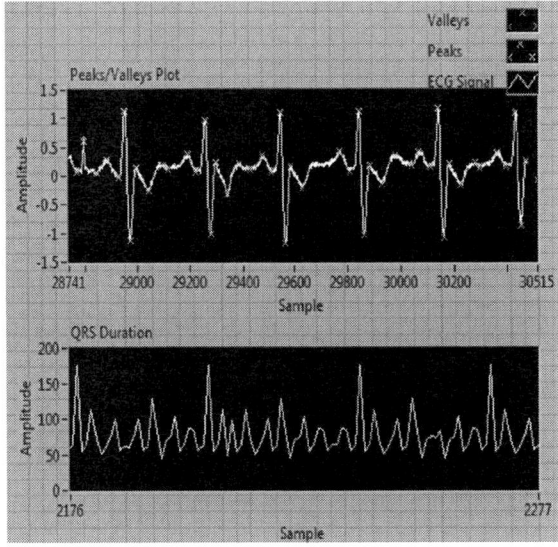

(a)

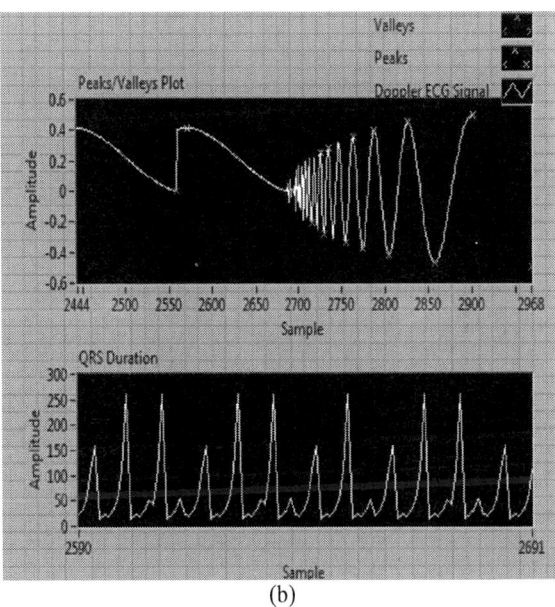

(b)

Fig. 7. The representation of the EEG signal obtained with the laboratory kit (a) and the Doppler signal received and the estimation of the QRS parameter using the MUSIC algorithm (b)

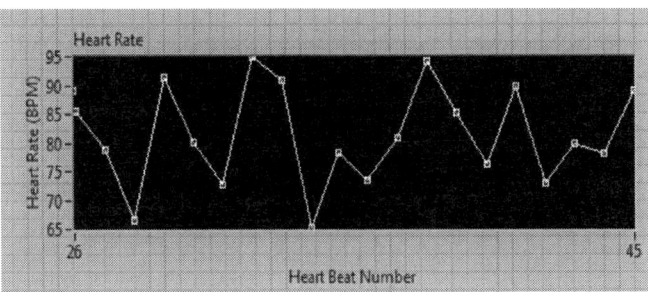

Fig. 8. The representation of the ~heart beat~ through MUSIC decomposition

According to Figure 7, the useful signal will contain at least two sinusoids (one appropriate to the heart rate and the other appropriate to the breathing), and the interferences will be represented by the thermal noise, the noise resulting from reflections caused by the environment and the noise caused by the radar system circuits.

The main parameter of the MUSIC algorithm is the size of the signal subspace; an oversized value of it can lead to false peaks and an undersized value leads to a smoothing of the spectrum, which means a loss of useful information (loss of spectral components). In implementing the algorithm, the values of size subspace were selected between 64 and 128. The result is represented in Figure 8.

When using MUSIC algorithm, only two relevant spectral components can be identified, and the component of maximum amplitude coincides with the frequency of the heart rate.

The results of the analysis highlight the capacity of the MUSIC algorithm in separating the useful signal noise and correctly identifying the heart rate, compared with the FFT algorithm, which has difficulty in evaluating this vital parameter, mostly for small segments of signal.

IV. CONCLUSIONS

Using the PSD Welch method to determine the density of the power spectrum associated to the EEG waves (alpha and beta) represents a great way to estimate the occurrence of fatigue and drowsiness.

The proposed system can use either physical contact sensors for EEG and ECG analysis, or a proposed solution with no physical contact sensors for the EEG/ECG analysis. Also, the system can be extended by introducing components of motion analysis using a kinetic type of sensor and the analysis of the facial expressions (emotional) from video images.

REFERENCES

[1] A. Sonnleitner, M. Simon, W. E. Kincses, A. Buchner, and M. Schrauf. Alpha spindles as neurophysiological correlates indicating attentional shift in a simulated driving task. International Journal of Psychophysiology, 83(1):110-118, 2012J. Clerk Maxwell, A Treatise on Electricity and Magnetism, 3rd ed., vol. 2. Oxford: Clarendon, 1892, pp.68-73.

[2] S. Wang, Y. Zhang, C. Wu, F. Darvas, and W. A. Chaovalitwongse. Online Prediction of Driver Distraction Based on Brain Activity Patterns. Transactions on Intelligent Transportation Systems, 16(1):136-150, 2015.

[3] Y.-K. Wang, T.-P. Jung, S.-A. Chen, C.-S. Huang, and C.-T. Lin. Tracking attention based on EEG spectrum. In HCI International 2013-Posters Extended Abstracts, pages 450-454. 2013.

[4] C. Wege, S. Will, and T. Victor. Eye movement and brake reactions to real world brake-capacity forward collision warnings - A naturalistic driving study. Accident Analysis & Prevention, 58:259-270, 2013.

[5] Y. Yorozu, M. Hirano, K. Oka, and Y. Tagawa, "Electron spectroscopy studies on magneto-optical media and plastic substrate interface," IEEE Transl. J. Magn. Japan, vol. 2, pp. 740-741, August 1987 [Digests 9th Annual Conf. Magnetics Japan, p. 301, 1982].

[6] Y. Zhang, E. Harris, M. Rogers, D. Kaber, J. Hummer, W. Rasdorf, and J. Hu. Driver distraction and performance effects of highway logo sign design. Applied Ergonomics, 44(3):472-479, 2013.

2016 IEEE 22nd International Symposium for Design and Technology in Electronic Packaging (SIITME)

Modelling and PSPICE simulation of a Photovoltaic/Thermoelectric system

P. A. Cotfas, D. T. Cotfas, and O.M. Machidon

Department of Electronics and Computers
Transilvania University of Brasov
Brasov, Romania
pcotfas@unitbv.ro, dtcotfas@unitbv.ro, octavian.machidon@unitbv.ro

Abstract— **This paper describes a model and its PSPICE simulation for a hybrid electric generator that is composed of a photovoltaic cell and a thermoelectric generator. The focus is to combine both models to obtain a hybrid system model. The electrothermal models of the photovoltaic cell and thermoelectric generator are described. The simulations are done using the NI Multisim electronic circuit simulators. The comparison between the results obtained through simulation and through real measurements on commercial PV and thermoelectric generators is also presented.**

Keywords— *hybrid system, photovoltaic cells, thermoelectric generator, SPICE simulation*

I. INTRODUCTION (*HEADING 1*)

The renewable energy domain is nowadays very dynamic and important for both research and market. A big effort is oriented towards the discovery of new and better systems to produce energy. Another direction is to improve the existing systems. An important part of renewable energy sources is photovoltaics. The photovoltaic panels PV based on different technologies, like monocrystalline silicon (mSi), polycrystalline silicon (pSi) or amorphous silicon (aSi) are mature. The efficiency of PVs is affected by different factors, one of the most important being temperature. The high PV temperature decreases its efficiency but also shortens its lifetime. Therefore, finding efficient solutions for decreasing the working temperature of the PV systems is an important direction for research.

One approach for increasing the efficiency of the PV systems is to use hybrid systems, such as: PV-TEG (photovoltaic-thermoelectric generator), PV-SC (photovoltaic - solar thermal collector) or PV-TEG-SC. By using the hybrid solutions, more solar energy is converted in useable energy (electric energy and thermal energy). At the same time the working temperature of the PV is smaller and thus the efficiency and lifetime increase.

In order to study, build and improve these hybrid systems it is useful to develop their models. This allows to simulate different configurations and to better understand the influence of each characteristic parameter.

In this paper we develop the thermoelectric models for PV and TEG and their connection in order to study the PV-TEG hybrid system.

II. HYBRID SYSTEM MODEL

A. TEG Model

The TEG is a solid-state device that converts the thermal energy into electrical energy based on the Seebeck effect. It consists of an array of groups of p and n semiconductors connected through a conductor. The groups are connected in series from the electrical point of view and in parallel from the thermal point of view.

The TEG model is developed based on analogies between the thermal and electrical systems. Based on manufacturer information offered through product datasheet the TEG parameters can be determined based on the following equations (which are obtained as is described in [1][2]):

$$R_i = \frac{V_{max}}{I_{max}} \frac{(T_h - \Delta T_{max})}{T_h} [\Omega] \tag{1}$$

$$R_t = \frac{\Delta T_{max}}{I_{max} V_{max}} \frac{2T_h}{(T_h - \Delta T_{max})} \left[\frac{K}{W}\right] \tag{2}$$

$$\alpha_{Sb} = \frac{V_{max}}{T_h} \left[\frac{V}{K}\right] \tag{3}$$

where R_i is the TEG electrical resistance, R_t is the TEG thermal resistance, α_{Sb} is the Seebeck coefficient, ΔT_{max} is the maximum temperature differences between the two TEG sides for a specified hot side temperature T_h, I_{max} is the impute current that will produce the largest temperature variation between the TEG two sides, V_{max} is the dc voltage that corresponds to the I_{max} when the largest temperature variation between the TEG two sides is obtained.

The implemented model is shown in Fig. 1 where used sources are described by the following equations [1]:

978-1-5090-4446-7/16 $31.00 © 2016 IEEE   179   20-23 Oct 2016, Oradea, Romania

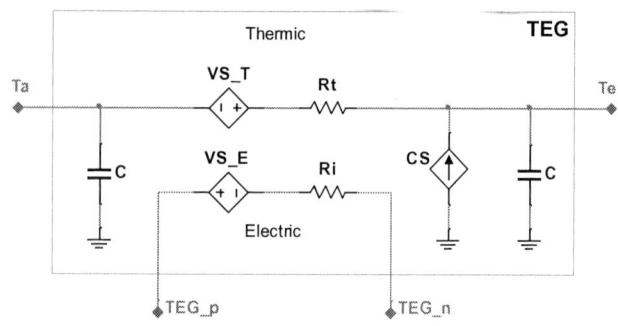

Fig. 1 TEG Model

$$VS\_T = \left(\alpha_{Sb}T_a - \frac{IR_i}{2}\right)IR_t \qquad (4)$$

$$VS\_E = \alpha_{Sb}(T_e - T_a) \qquad (5)$$

$$CS = V \cdot I \qquad (6)$$

where VS_T is the voltage source that corresponds to the Peltier cooling and heating on the heat-absorbing and heat-emitting sides, VS_E is the voltage source that corresponds to the Seebeck effect and CS is the current source that corresponds to the Joule heating of the TEG, I is the current on the electrical network of the TEG, V is the output voltage of the TEG, T_a and T_e are the temperatures of the heat-absorbing and heat-emitting sides of the TEG. The capacitors C represent the lumped heat capacitance of the ceramic plates and groups of the TEG.

B. PV Model

The photovoltaic cell (PV) is a solid-state system that converts the light radiation (electromagnetic energy) into electrical energy based on the photovoltaic effect. PV model is based on the one, two or three diode model function of the mechanism into PV. The most widely used model is the one diode model that is also used in this paper. The one diode model is described by the following equation:

$$I = I_{ph} - I_o\left(e^{\frac{V+IR_s}{V_T m}} - 1\right) - \frac{V+IR_s}{R_{sh}} \qquad (7)$$

where I_{ph} is the photogenerated current, I_0 is the reverse saturation current $V_T = kT/q$ is the thermal voltage, T is the temperature, k is the Boltzmann constant and q represents the elementary charge, R_s and R_{sh} are the series and shunt resistances of PV.

Since the temperature is a very important parameter that affects the operation of the PV, it should be taken into consideration as a dynamic parameter of the model and not as a constant one. The PV temperature can vary due to the irradiance level, due to self-heating or due to the environment temperature variation. Therefore, the electrothermal model of PV should be developed and used for the PV systems study. Such a model is presented in [3]. The model implemented in

this paper is shown in Fig. 2 and consists of two parts: electric and thermic. The electric part of the model is based on the following equations [3]:

$$I_{ph} = P_r S \frac{J_0}{1000}\left[1 + \alpha_T(T_j - T_0)\right] \qquad (8)$$

$$I_d = BST_j^3 \exp\left(-\frac{qV_j}{k_B T_j}\right)\left[\exp\left(\frac{qV_d}{nk_B T_j}\right) - 1\right] \qquad (9)$$

$$E\_Rs = V_{Rs}\alpha_{Rs}(T_j - T_0) \qquad (10)$$

where the P_r is the irradiance level (W/m²), S is the active area of the PV, J_0 is the current density of PV measured at 1000 W/m², α_T is the temperature coefficient of photocurrent, T_j is the internal temperature of the PV, and T_0 is the reference temperature, I_d is the current through the diode, B is a temperature independent constant, q is the elementary electrical charge, V_j is the built-in potential of PV, k_B is Boltzmann constant, V_d is the junction voltage, n is the ideality factor of diode, E_Rs represents the variation of the series resistance Rs due to the temperature variation, V_{Rs} is the voltage across the PV series resistance, and α_{Rs} is temperature coefficient of Rs. R_{sh} is the shunt resistance of the PV.

The thermic part is based on the following equations [3]:

$$P_{th} = P_r aS + (V_{out} - V_d)I_{out} + V_d I_d \qquad (11)$$

where P_{th} is the current source that model the fraction of the power of radiation absorbed by the PV, the power dissipated on series resistance and leakage resistances, a is the ratio of conversion of the radiation into heat, V_{out} and I_{out} are the output voltage and current of PV. R_{th} and C_{th} are the thermal resistance and the lumped heat capacitance of the PV. The T_{amb} is the ambient temperature and the T_{pv} is the PV temperature.

C. Hybrid System Model

The two models, TEG and PV, described in the previous two paragraphs can be combined in order to study the behaviour of the hybrid system.

The structure of a hybrid system that is modelled in this paper is shown in Fig. 5. The PV and the thermostat are

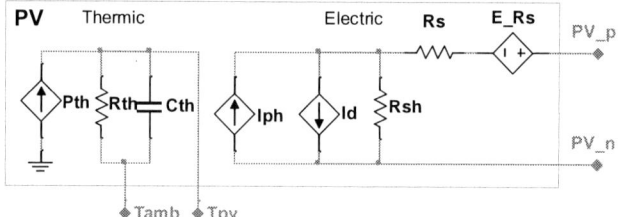

Fig. 2 PV Model

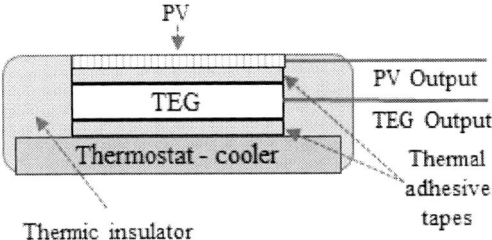

Fig. 5 Hybride system structure

thermally connected to the TEG through thermal adhesive tapes.

The system is placed in a thermal insulator, only the PV top side is let outside. The thermal adhesive tapes and the insulator are modelled based on a resistive-capacitive network, as is shown in Fig. 6 (PV to TEG and TEG to Thermostat).

The temperature of the thermostat is considered equal to ambient temperature. For measuring the I-V characteristics for PV and TEG two Voltage Controlled Voltage Sources are used. The both sources are controlled by the Vpol voltage source. In this manner the I-V characteristics can be measured in the same time.

III. RESULTS AND DISCUSSIONS

A. Testing the models

The models of PV and TEG were verified individually function of the temperature. The I-V characteristics for TEG and PV are shown in Fig. 3 a. and b. The values for the PV and TEG parameters used in this phase were taken from the datasheets of the PV and TEG modules or calculated as described in [1][2][3].

The influence of the thermal connection between TEG and PV upon the PV efficiency can be seen in Fig. 4. The I-V characteristics of PV obtained with TEG connected show that the V_{OC} of PV increases significantly while the current of PV has a small decrease in comparison with the case when the TEG is not connected. This means that the working

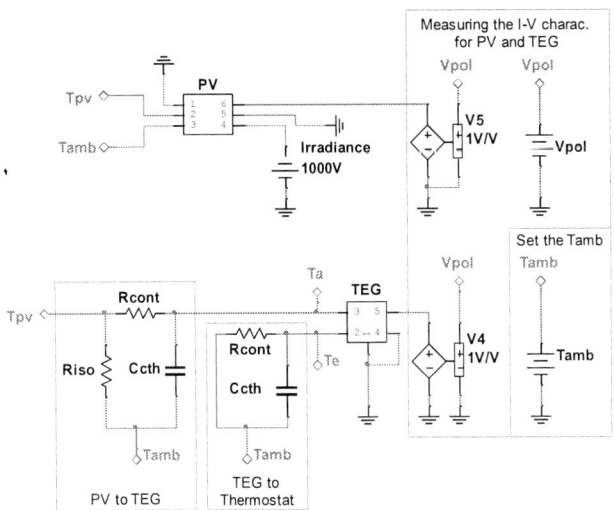

Fig. 6 Hybride system model

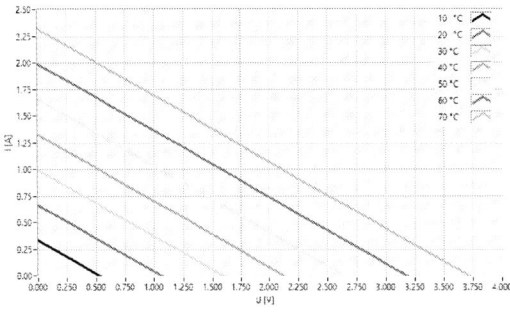

a.

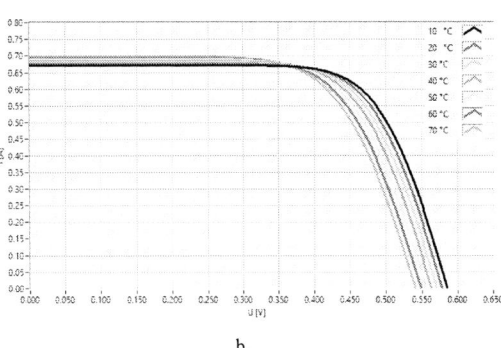

b.

Fig. 3 I-V characteristics function of temperature: a. for TEG, b. for PV

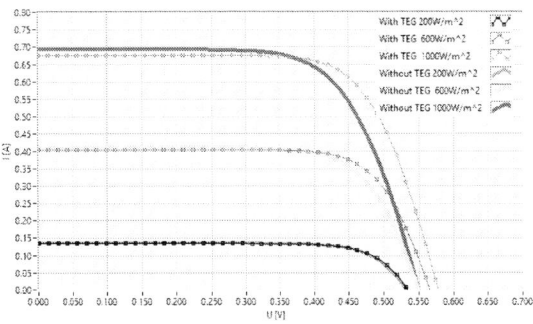

Fig. 4 I-V characteristics of PV with and without connection with TEG

temperature of PV with TEG connected is smaller than the temperature of the PV without TEG connected. Thus, the PV efficiency is increased and some extra electrical energy can also be generated by the TEG.

In the second phase the models were compared with real modules. The experimental setup consists of:

- a thermostat build from an aluminium block cooled with water. The temperature of the thermostat is quasi constant due to using a water tank.

- a thermoelectric generator Stonecold PM-62X62-267TEG made from Bi_2Te_3

- thermal adhesive tape with 1 mm thickness and its thermal conductivity is 6 W/m·K.

- monocrystalline silicon PV, with dimensions of 6 cm/6 cm.

- six thermocouples used for temperature measurements. Three between PV and TEG and three between TEG and Thermostat).

The PV and TEG parameters were calculated from the real measurements and were used into PSPICE models. For calculation of the PV parameters the five parameters method was used [5].

The comparison between the real measurements on the real hybrid system and the data obtained through simulation is shown in Fig. 8 a and b. The doted curves represent the simulated data and the solid line curves represent the measured data. There is a good correlation between the two sets of data. There are small differences between simulation and measured data. These differences are due to the imperfect thermal contacts between system components that can affect the heat transfer and also can introduce errors in temperature measurements.

The power generated by the hybrid system components is shown in Fig. 7 a and b. As it can be seen from Fig. 7 b the power generated by the TEG element, based on actual thermoelectric materials, is very small (4mW at 1000W/m^2), due to small temperature differences obtained along two sides of TEG. So the impact of the TEG in the power generation of the hybrid system is negligible. However the power generation of the hybrid system is higher than the power generation of the PV alone, as it is also observed in [6].

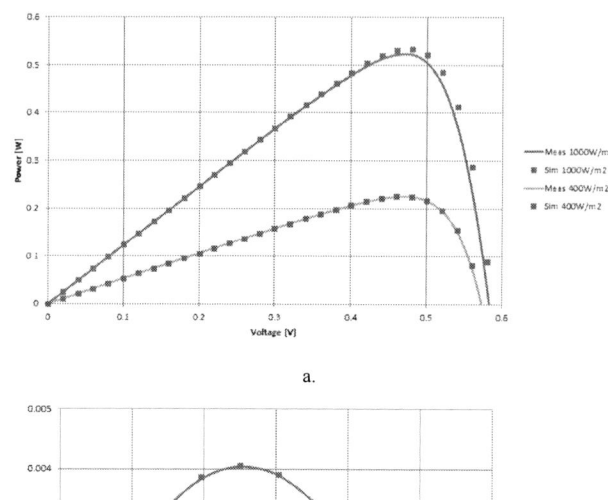

a.

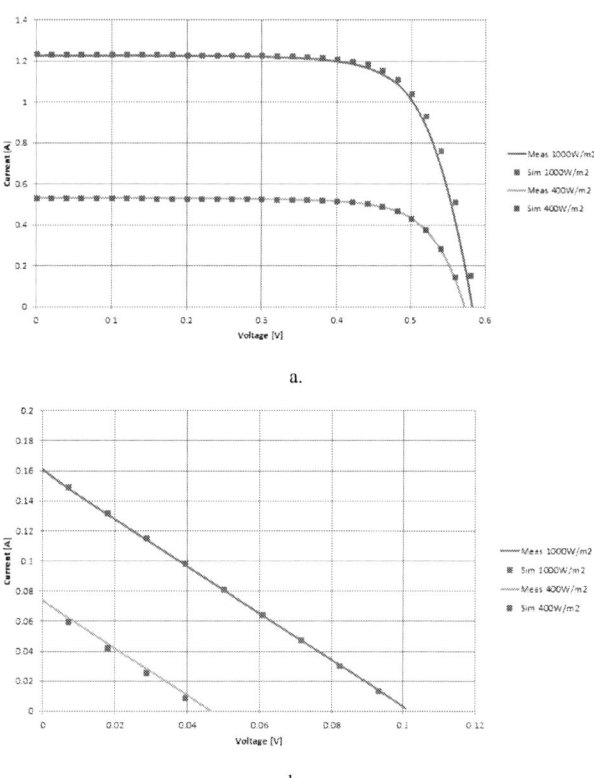

a.

Fig. 7 The power – voltage characteristics for the hybrid system at 400W/m2 and 1000W/m2: a. for PV, b. for TEG

IV. CONCLUSIONS

A PSPICE model for PV-TEG hybrid system was developed and described in this paper. The model allows varying the ambient temperature and the level of illumination. The I-V characteristics of the two components of the system can be obtained and analysed. The paper presents a comparison between the results obtained through the simulation based on the developed model and the results obtained on a real system. The obtained results are very close, thus validating the developed model.

Based on the analysis of the power generation it can be concluded that the contribution of the TEG is very small but the TEG has an effect on the PV efficiency through temperature decreasing. This leads to an increase in the PV power generation and also to prolonging the PV lifetime.

ACKNOWLEDGMENT (HEADING 5)

This work was supported by a grant of the Romanian National Authority for Scientific Research and Innovation, CNCS - UEFISCDI, project number PN-II-RU-TE-2014-4-1083.

REFERENCES

[1] S. Lineykin, S. Ben-Yaakov, " PSPICE-Compatible Equivalent Circuit of Thermoelectric Coolers ", IEEE 36th Power Electronics Specialists Conference, DOI: 10.1109/PESC.2005.1581688, 2005;

b.

Fig. 8 Comparison between real measurements and simulated data obtained for hybrid system at 400W/m^2 and 1000W/m^2: a. for PV, b. for TEG

[2] H. L. Tsai, J. M. Lin, "Model Building and Simulation of Thermoelectric Module Using Matlab/Simulink", Journal of Electronic Materials, 39, no. 9, 2105- 2111, 2010

[3] K Górecki, P Górecki, K Paduch, " Modelling Solar Cells with Thermal Phenomena Taken into Account", Journal of Physics: Conference Series, vol. 494, 2014;

[4] H. M. Bahaidarah, B. Tanweer, P. Gandhidasan, S. Rehman, " A Combined Optical, Thermal and Electrical Performance Study of a V-Trough PV System—Experimental and Analytical Investigations", Energies, vol. 8, pp. 2803-2827, 2015.

[5] Cotfas D.T, Cotfas P.A, and Kaplanis S, "Methods to determine the dc parameters of solar cells: A critical review", Renewable and Sustainable Energy Reviews 28, 588–596 (2013).

[6] A. Rezania, D. Sera, L.A. Rosendahl, "Coupled thermal model of photovoltaic-thermoelectric hybrid panel for sample cities in Europe", Renewable Energy 99, 127–135 (2016).

2016 IEEE 22nd International Symposium for Design and Technology in Electronic Packaging (SIITME)

Statistical Methods for Determining Components Non-liniarities, from Thermoluminescent Devices

NANORADDOS collaboration

E. Pajuste, J. Prikulis - University of Latvia, Latvia; M. Dima, D.D.-Rus - Institute for Nuclear Physics and Engineering, Romania; S. Sokovnin - Ural Institute of Electrophysics, Ekaterinburg, Russia; P. Krug - Köln University Applied Sciences, Germany   E-mail: *modima@nipne.ro*

C. Ionescu

Department of Electronics, Telecommunications and IT
Polytechnic University Bucharest
Bucharest - Romania

Thermoluminescent (TLD) dosimeters enjoy wide usage due to low cost and simplicity of use. They have however large errors at high doses in mixed-radiation fields, where non-linear effects occur. Algorithms based on the Akaike criterion [1] are presented for determining the maximal (physically meaningful) polynomial order with which the non-linearities are modeled. This depends on the number of points existing on a curve and on the points' errors.

Keywords – thermoluminescent devices, non-liniarity determination, Akaike criterion

I. INTRODUCTION

Thermoluminescent dosimeters offer an excellent example of how relevant non-linearities may, or may not, be with respect to the amount of information available about a device. In a context of high-errors and few measurement points it is improper to consider non-linearities when simply a flat line would do.

This seemingly simple principle is clear qualitatively, but what happens when it is important to know exactly "how much non-linearity" the available data conveys.

Firstly, without any loss of generality, it is possible to admit that any non-linearity is analytical – and can be expanded as a Taylor series. The above question then re-phrases itself as "what is the maximal polynomial order" to which the expansion makes physical sense.

TLD's are a good example because they have large errors at high doses in mixed-radiation fields – precisely where non-linear effects occur.

The statistical-informatics algorithm with which polynomial coefficients are extracted is presented, together with the Akaike criterion [1] prescribing the maximal order to which said algorithms are stopped.

Exemplifying results are shown, which confirm the aforementioned intuitive view – however, in a concrete, quantitative way.

II. NON-LINEARITY EXTRACTION

A. TLD non-linear response curves

Figure 1 shows a typical TLD-100 [2] response curve versus the radiation dose to which the TLD has been exposed.

In principle this is linear up to a very high dose. However, beyond this, additional defects are created by the radiation itself, defects which contribute to the chip's increased thermoluminescence.

Yet beyond such doses, another effect appears by which defect density distorts band structure and diminishes radiative recombination centers. This leads to a further decrease in thermoluminescence with respect to dose increase.

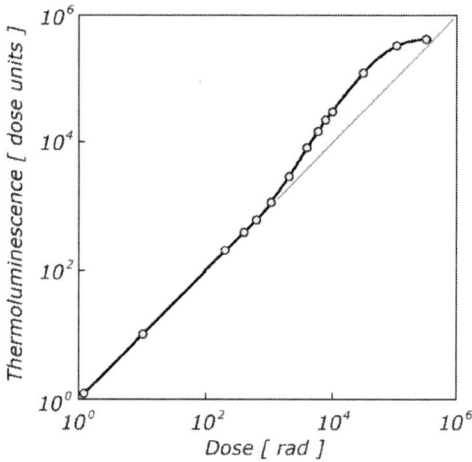

Fig. 1. Example of TLD thermoluminescence vs. dose non-linear curve. The material is TLD-100 [2].

The non-linearity is not in the typical regime for bio-doses, rather much higher, for instance for equipment, or materials' doses.

Yet the question remains: how many points are needed and how precise their errors, in order to be able to determine the curve's non-linearity ?

B. Polynomial coefficients' extraction method

Such a method has been reported before [3], for functions of 2 variables. In this case only one variable is needed. The method basically says that a change in dose (δx) is related to the change in thermoluminescence (δz) via a non-linear Taylor expansion with unknown coefficients:

$$\delta z = \Sigma \, C_k \, \delta(x^k) + \epsilon$$

978-1-5090-4446-7/16 $31.00 © 2016 IEEE

184

20-23 Oct 2016, Oradea, Romania

which are determined by minimising ε over the whole statistical sample: $\langle \varepsilon^2 \rangle$ = min., where the brackets denote statistical average.

By differentiating with respect to each C_k, the minimum condition ensues:

$$\langle \delta(z) \, \delta(x^n) \rangle = \Sigma \, \langle \delta(x^n) \, \delta(x^k) \rangle \, C_k$$

which, by inverting the $\langle \delta(x^n) \, \delta(x^k) \rangle$ matrix, gives the terms C_k.

The procedure can go up in polynomial order as high as demanded – however, it is evident that this would not be physical. It is also evident that this process can be iterative, adding one order at a time and for each iteration monitoring the residual $\langle \varepsilon^2 \rangle$.

The latter will register down-jumps, larger for terms with large coefficients, and small for the other. This can be observed in figure 2.

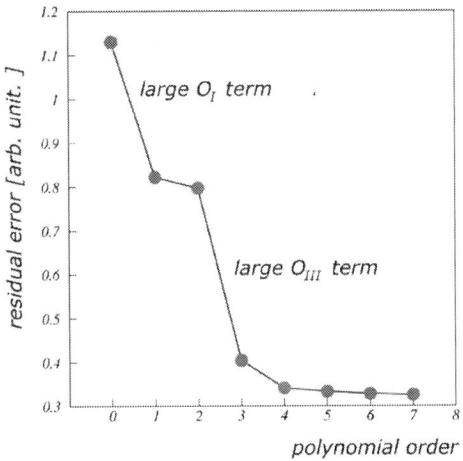

Fig. 2. Residual error decrease in the course of adding higher order polynomials to the model. The large jumps correspond to terms with high significance.

The behavior of the curve in figure 2 gives an intuitive idea, that where the curve saturates is probably the maximal physically meaningful polynomial order.

But how could this be discerned from a purely mathematical, steady slope decrease in the residual error, for curves with no model behind them ?

C. Akaike Criterion

For a statistical model of a given data, with L be the maximum value of the likelihood function for the model, n the number of parameters used in the model and N the number of observations (points), the AIC is:

$$AIC = 2n - 2L$$

For a set of candidate models to a data, the preferred one has the minimum AIC value.

Thus the AIC imposes a penalty on using more parameters than actually useful in minimising the residual error.

In practice however, there is almost always information loss by using the candidate model as "the" real-model. For this reason, in the univariate, linear model, with normally-distributed residuals (Gaussian probability) the formula becomes:

$$AIC'c = 2nN / (N\text{-}n\text{-}1) - 2L \approx 2nN / (N\text{-}n\text{-}1) - 2N \ln(\varepsilon/\sigma)$$

where σ is the effective standard deviation.

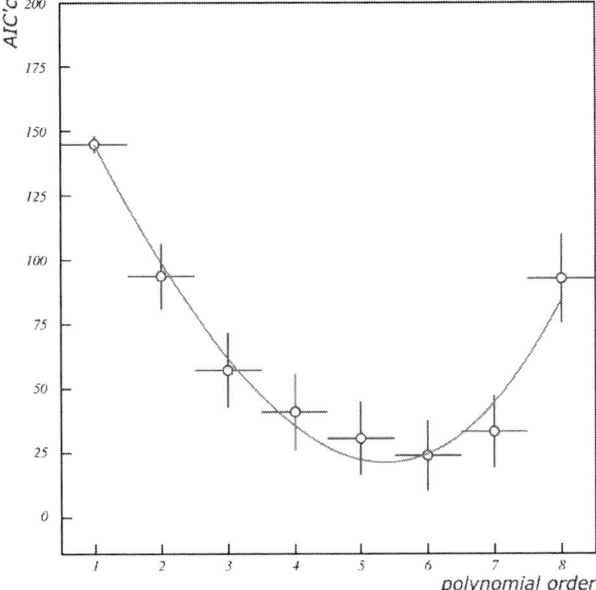

Fig. 3. AIC'c for simulated data on a curve generated with an O_V polynomial. The good performance of the AIC'c criterion can be seen, the minimum being very near to order 5.

III. RESULTS

A. TLD non-linear curve generation

Figure 4 shows the generated TLD curve, the points with errors added to them and the reconstructed curve (taking into account these errors).

The non-linearity in the case shown in figure 4 is very large and serves only to show the good performance of the reconstruction algorithm.

For the study performed, the non-linearity, as combined effect of the cubic and quadratic terms, was in the range 0 – 40% (i.e. – deviation from a straight line at the maximal dose).

The curve was generated with a quadratic and cubic term, thus the maximal polynomial order should be 3.

This was repeated producing statistical sets of 100 curves, for each pair of cubic and quadratic non-linear coefficients.

For all curves the reconstruction algorithm was applied up to polynomial order O_{VIII}. The AIC'c were calculated for each order and plots similar to the test in figure 3 were obtained.

These plots show the optimal order to be considered for each non-linearity case (as mentioned, statistical set of 100 curves with the same non-linearity).

The optimal polynomial should be of course 3, as present are cubic and quadratic terms.

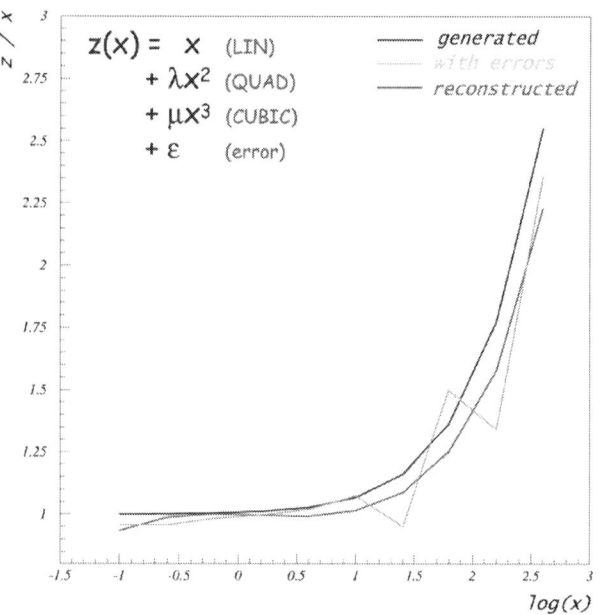

Fig. 4. Simulated data curve with a quadratic and a cubic term for non-linearity (blue). The said "perfect response" is altered by intrinsic errors (or reading errors) – green curve. This is fed as input to the software, the red curve being the reconstructed curve by the software.

It is shown in figure 5 that for $N_p = 10$ points on all curves, the maximal polynomial order decreases with increasing errors, which is what we would have expected intuitively: the larger the errors, the more the curve fit requires fewer polynomials.

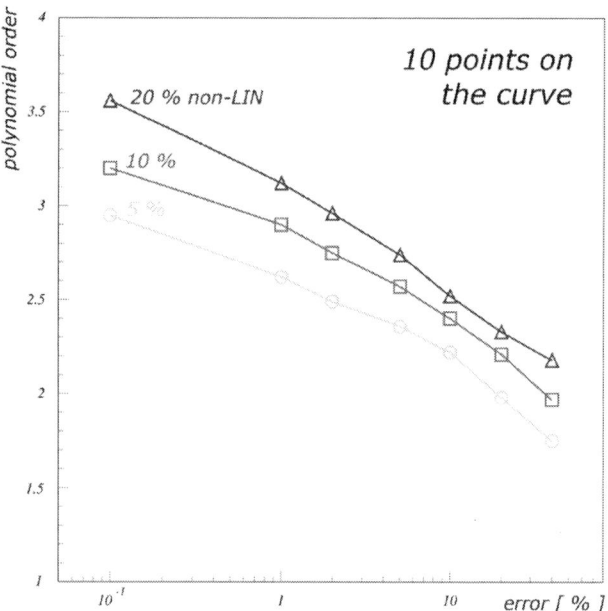

Fig. 5. Optimal polynomial order as predicted by the AIC'c criterion, versus errors on the curve's points. The number of points was kept constant to 10 in this case.

For all values of non-linearity the maximal order fell over 1 polynomial order by ranging the errors from 0.1% to 40%.

Basically this prescribes that instead of a cubic, for 40% errors on the points a quadratic is sufficient to describe the data.

Or corollary – for 40% errors the information available in the points cannot derive higher "bending" terms than quadratic ones.

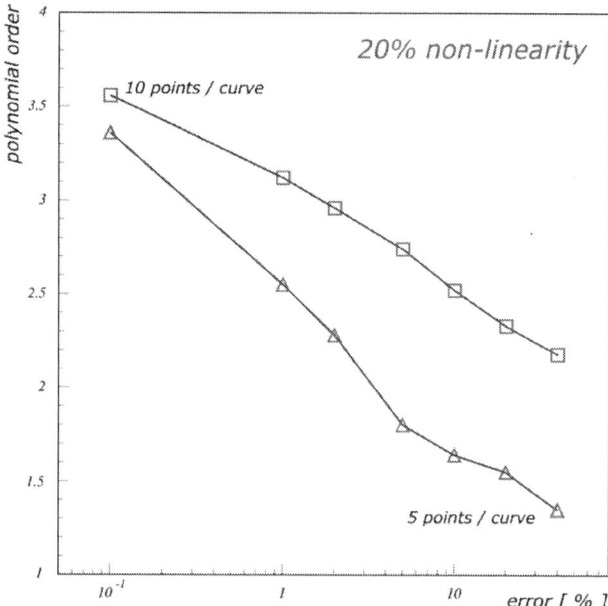

Fig. 6. Optimal polynomial order as predicted by the AIC'c criterion, versus errors on the curve's points. The non-linearity use was kept constant to 20% in this case.

The increase in maximal polynomial order with non-linearity is most noticeable for low errors, whereas for large errors there are small differences. This again is to be expected, as good precision allows perceiving the data more accurately in contrast to low precision.

Figure 6 shows how the number of points on the curve affects the maximal polynomial order. The fewer the number of points, the lower the maximal polynomial order that can (physically) meaningful be extracted from the data. It is obvious that for 2 points one could at most draw a line through them.

Notable (in contrast to polynomial order dependence on non-linearity magnitude) is that the dependence on the number of points present on the curve is much more significant at high errors. A low number of points at high errors quickly becomes equivalent to just 2 points on the curve (with a straight line through them).

Concluding remarks are that a quantitative way was developed to assert up to what maximal polynomial order is makes physical sense to consider a model for a non-linear component. Intuitive perception of the matter was reconfirmed, in a quantitative manner.

ACKNOWLEDGMENT

The NANORADDOS collaborators thank their national funding parties - respectively for M. Dima and D.D.-Rus this work was supported by a grant of the Romanian National Authority for Scientific Research and Innovation, CCCDI – UEFISCDI, project number EU-09/2016, within PNCDI III.

REFERENCES

[1] H. Akaike, "A new look at the statistical model identification", IEEE Trans. Autom. Control 19, 1974, 716.

[2] J. Zimmerman, "The radiation induced increaseof thermoluminescence sensitivity of the dosimetry phosphor LiF (TLD-100)", J. Phys. C: Sol. St. Phys. 4, 1971, 3277.

[3] M. Dima, Yu. Pepelyshev, "Extraction of Mechanical-Reactivity Influences from Neutron Noise Spectra at the IBR-2 Reactor", Chin. Phys. Lett. 30, 2013, 072801.

Bond-Graph Modeling of the Equivalent Circuit of an On-Chip Spiral Inductor

Adriana Grava, Cristian Grava
Department of Electrical Engineering, Department of Electronics and Telecommunications
University of Oradea
Oradea, Romania
agrava@uoradea.ro, cgrava@uoradea.ro

Abstract — **The aim of this paper is the bond-graph modeling and simulation of the equivalent electrical circuit of an on-chip spiral inductor with two ports and the validation of the proposed model comparing it with experimental results obtained by direct measuring. The bond-graphs allow a unitary analysis, modeling or design of any complex system containing any electric, magnetic, pneumatic and hydraulic or any other components independent of their physical nature. The experimental/simulation results are similar with the experimental results and prove that the bond-graph modeling allow to analyze and design an on-chip spiral inductor in an unitary environment taking into account even the mutual inductance and other physical phenomena. The bond-graph modeling and designing method could be applied in the case of microwave circuits, taking into account that the analysis could be made in a wide range of frequencies.**

Keywords — *bond-graphs; modeling; on-chip; spiral inductor.*

I. INTRODUCTION

A spiral inductor is one of the most common structures among the on-chip inductors and there are several versions of spiral inductors: rectangular, cylindrical, hexagonal, with 2, 3, 5 ports etc [3].

The devices using on-chip spiral inductors are widely used in biomedical industry, for example as pressure sensors that measures the arterial pressure, intraocular pressure or in angioplasty. Other medical applications using on-chip inductors are the defibrillators and the cardiac stimulators. The on-chip inductors are usually used as components of a complex network or as an inductive load. The main requirements for the on-chip inductors are: the resonance frequency and the quality factor that have to be high, and the occupied surface has to be small.

II. THE BOND-GRAPH MODEL OF A SYSTEM

The bond-graph model of a dynamic physical system is an intermediary representation between the physical model and the mathematical models associated to the system (state equations, functions or transfer matrixes etc) [6], [7]. The bond-graphs indicate not only the architecture of the system, but the organization of the equations as well, underlying the causes and effects that occur between the elements (through

causal lines). This characteristic is fundamental because it allows a systematic and organized approximation for writing the relations that characterize the system that leads to a combination of differential and algebraic equations given by the causal representation of the bond-graph. The user has often other very interesting possibilities as well, that come out from the graphical nature of a bond-graph, following the privileged paths, also named causal paths that are independent to the orientation of power between the bonds [8].

The bond-graph modeling and the 20-Sim simulation software [9] allow us to obtain parameters of a system from different physics fields as electrical, mechanical, thermal parameters or data concerning the interdependence of the parameters of the system. The bond-graph modeling represents a unique language, common for all the physics fields, based on the variables of power and energy corresponding to each field. Starting from the definition of the power (P) in any field of physics, two generic variables of effort (e) and flux (f) were defined:

$$P = e \cdot f \qquad (1)$$

Thus, in any physics field, the power represents the product between the effort variable and the flux variable as in (1). In the case of electrical domain, the effort variable is electrical voltage and the flux variable is the electrical intensity of the electrical current.

The bond-graph elements are called and written similarly to electrical domain. This is the reason why it has to be taken into account and avoid the possible confusion between the bond-graph elements when these elements are modeling systems (from magnetic, mechanic or other domains) that don't belong to the electrical domain. For the other domains, the bond-graph elements are determined taking into account the resistive, capacitive or inductive phenomena depending on the relations between the variables of power and energy, that is the phenomena of energy dissipation (resistive type) or energy storage (inductive or capacitive type). Using these rules, in any physics field they can be identified three types of passive elements of a bond-graph, namely the R, I and C elements.

The active elements of a bond-graph are the effort source S_e, and the flux source S_f. These elements are called active because they are modeling the energy supply in the system.

The power transmission and the power link between passive and active elements are realized through bonds (half-arrows) and junction elements that transmit and preserve the power/energy in the system. The junction elements of a bond-graph are the *0*, *1*, *TF* and *GY* elements. These elements link the passive to active elements and they transmit and preserve the energy independently on the connection between them and independently on the physics field. The junction elements preserve the power. The *0*-junction links the elements having the same effort and the *1*-junction links the elements having the same flux.

The transformer (*TF*) is a junction element that has the role to transform and preserve the power and usually is placed inside a certain domain of the system (e.g. electric, magnetic etc). The gyrator (*GY*) is a junction element that ensures the power transfer between different domains (e.g. electric to magnetic, mechanic to hydraulic etc). The bond-graph model of a system is the schematic structure that takes into account all the elements of the system and the physical phenomena that take place inside the system.

III. THE BOND-GRAPH MODEL OF AN ON-CHIP SPIRAL INDUCTOR

For the spiral inductor presented in Fig. 1(a) [1], [2] that has the equivalent circuit with two-ports, presented in Fig. 1(b) [2], we realized the bond-graph model presented in Fig. 2.

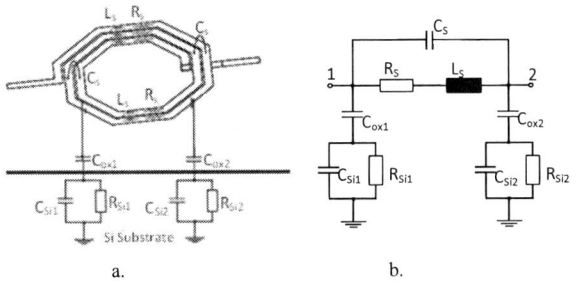

a. b.

Fig. 1. (a) On-chip spiral inductor; (b) Equivalent circuit with two-ports of the on-chip spiral inductor [1].

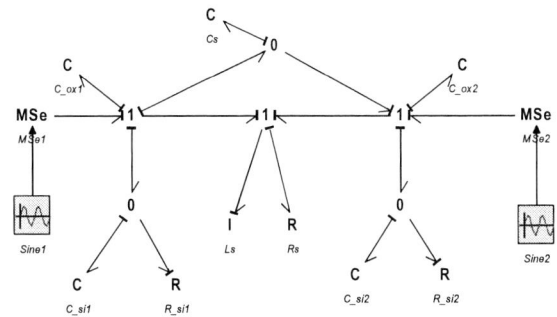

Fig. 2. The bond-graph model of the equivalent circuit with two-ports of the on-chip spiral inductor.

In order to obtain a more detailed model of the spiral inductor (from Fig. 1(a)) we also simulated the equivalent circuit with three ports presented in Fig. 3 [2].

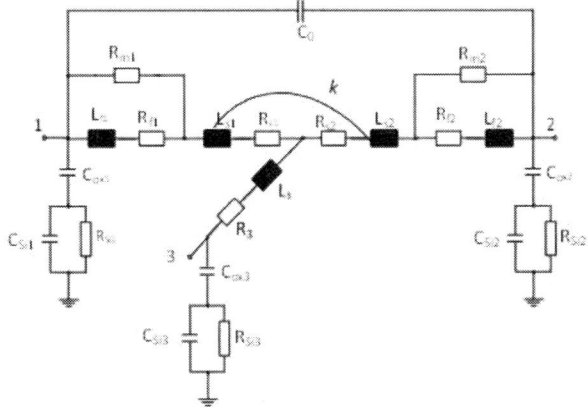

Fig. 3. Equivalent circuit with three-ports of the on-chip spiral inductor [2].

The bond-graph model of the equivalent circuit of the on-chip spiral inductor with three ports is presented in Fig. 4.

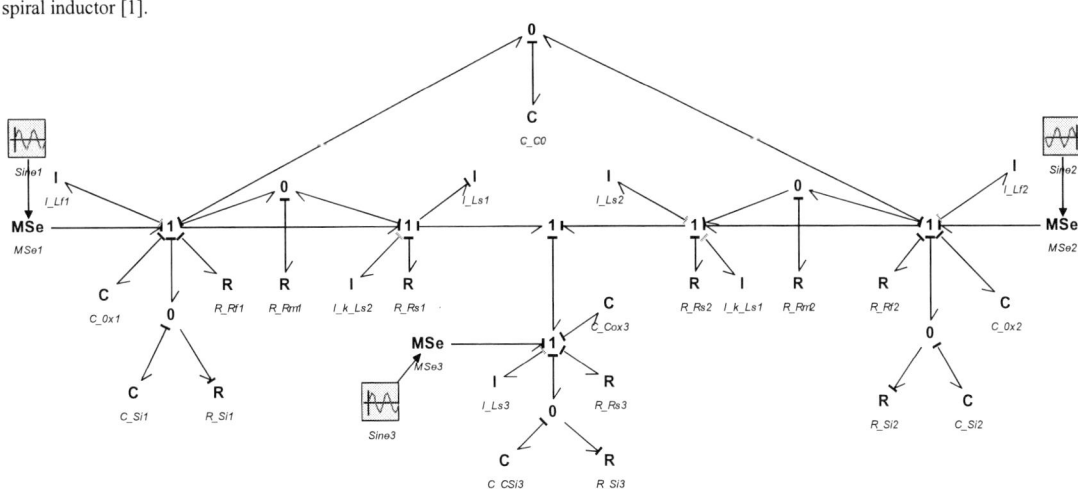

Fig. 4. The bond-graph model of the equivalent circuit with three ports of the on-chip spiral inductor.

In order to validate the simulation results, we experimentally realized a circuit that contains an on-chip spiral inductor (L_1 in Fig. 5).

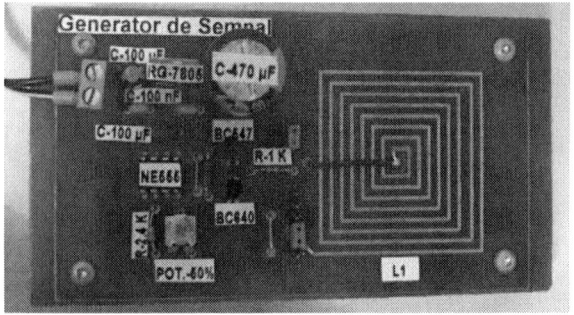

Fig. 5.   Experimental circuit containing an on-chip spiral inductor.

IV. EXPERIMENTAL RESULTS

The operating frequency range of the spiral inductor is $0...1$ GHz. The presented results were obtained for an operating frequency of 300 KHz and for a sinusoidal input voltage $u_1 = u_2 = 8,5\sqrt{2} \cdot \sin(\omega t)$. After the simulation of the bond-graph presented in Fig. 2 we obtained the results presented in Fig. 6.

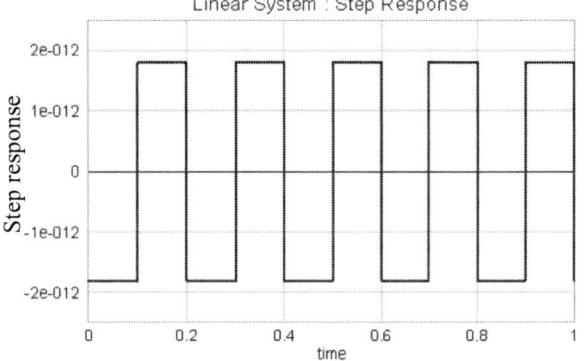

Fig. 6.   Simulated step response depending on time.

After measuring the output signal on the coil L_s, the obtained results are presented in Fig. 7.

Fig. 7.   Experimentally step response depending on time.

As it can be noticed, the simulation results are very close to experimental results, the wave-forms being very similar.

In Fig. 8 and 9 other simulation results obtained using the bond-graph model in Fig. 2, are presented. In Fig. 8 the voltages on C_{ox1} and L_s are presented.

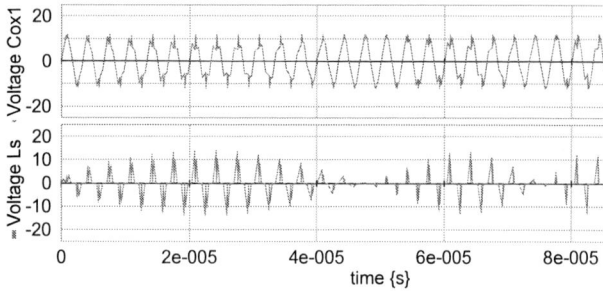

Fig. 8.   Voltages on C_{ox1} and L_s.

In Fig. 9, the time dependence of the voltages on C_{ox1} and C_{ox2} are presented, for an u_2 input applied on 2 (MS_{e2} in Fig. 2).

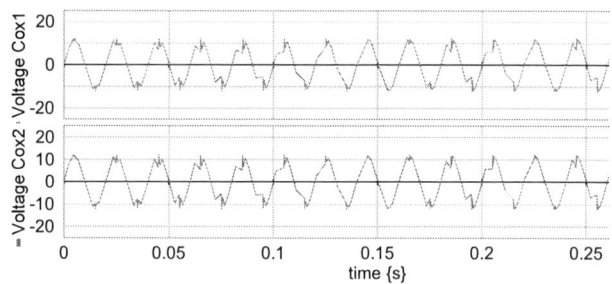

Fig. 9.   Time dependence of the voltages on C_{ox1} and C_{ox2}.

It can be noticed that the two voltages have the same amplitude and the same phase.

We also simulated the bond-graph model in Fig. 4. In Fig. 10...13 the simulation results obtained using the bond-graph model in Fig. 4, are presented. The presented results were obtained for an operating frequency of 300 KHz and for a sinusoidal input voltage $u_1 = u_2 = u_3 = 8,5\sqrt{2} \cdot \sin(\omega t)$.

In Fig. 10, the step response of the system, depending on time, is presented.

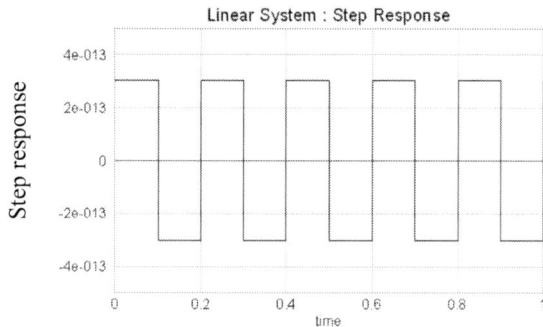

Fig. 10. Step response for the bond-graph model in Fig. 4.

In Fig. 11, the time-dependence of the voltages on C_{ox1} and L_{s1} are presented.

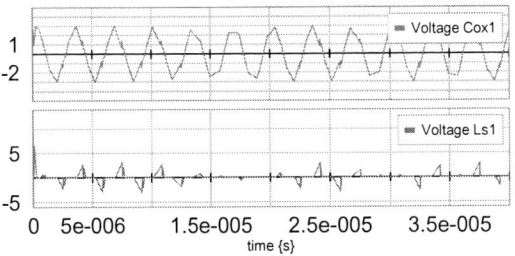

Fig. 11. Time-dependence of the voltages on C_{ox1} and L_{s1}.

In Fig. 12, the time dependence of the voltages on C_{ox1}, C_{ox2} and C_{ox3} are presented, for an u_2 input applied on 2 (MS_{e2} in Fig. 4).

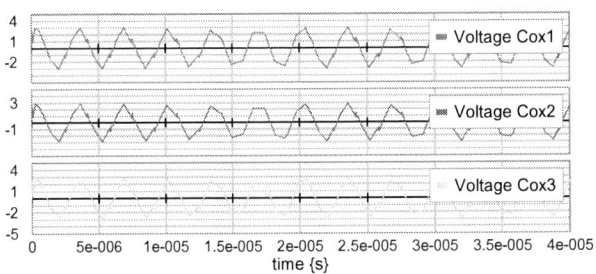

Fig. 12. Time dependence of the voltages on C_{ox1}, C_{ox2} and C_{ox3}.

It can be noticed that the three voltages have the same amplitude and the same phase. The phase of these voltages is the same as the phase of the input voltage u_2, as it can be noticed in Fig. 13.

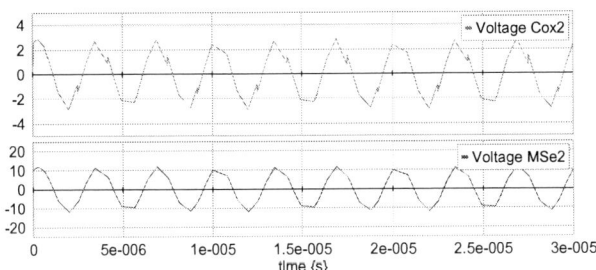

Fig. 13. Time dependence of the input voltage and the voltage on C_{ox2}.

The obtained results are useful in the design stage and also in the analysis of the designed circuits or systems. The bond-graph modeling and the simulation with 20-Sim software also allow the frequency analysis of the system. The results of the frequency analysis are not presented here only because of space limitation and will be presented in a future work that will better emphasize the possibilities and opportunities of bond-graph modeling in the analysis of an on-chip spiral inductor and of more complex systems containing parts of different physics domains.

V. CONCLUSIONS

The modeling/simulation results are similar to the experimental results and prove that the bond-graph modeling allow to analyze and design an on-chip spiral inductor in an unitary environment taking into account even the mutual inductance and other physical phenomena. The bond-graph modeling and designing method could be applied in the case of microwave circuits [4], [5], taking into account that the analysis could be made in a wide range of frequencies.

The differences between the simulation and experimental results are due to the fact that the 20-Sim simulation software uses the spiral inductor as an ideal component, but in the experimental implementation results losses through the deviation from the ideal case. Due to these differences the simulation has to be realized before the experimental implementation in order to avoid the supplementary costs. The bond-graph modeling and the simulation using the 20-Sim software could bring valuable benefits through the possible optimizations in the design stage, because the on-chip spiral inductor could be integrated in complex and expensive systems, especially in the biomedical field. The bond-graph modeling as well as the simulation using the 20-Sim software could emphasize simulation or implementation details even in the case of the complex systems where the use of an on-chip spiral inductor is needed. The advantage of the bond-graphs is that allows the modeling in a unitary environment of the components and phenomena from a wide range of physics fields as electrical, magnetic, hydraulic, pneumatic etc.

REFERENCES

[1] A. M. Niknejad, and R.G. Meyer, "Analysis, design, and optimization of spiral inductors and transformers for Si RF ICs", IEEE Journal of Solid State Circuits, Vol.33, Issue 10, pp. 1470-1481, 1998.

[2] V. Zhurbenko, Advanced microwave circuits and systems, Intech Publishing, 498 pages, April 2010.

[3] M. Danesh, and J. R. Long, Differentially driven symmetric microstrip inductors, IEEE Trans. on Microwave Theory Tech., Vol.50(1), pp. 332–341, 2002.

[4] D. Pozar, Scattered and Absorbed Powers in Receiving Antennas, IEEE Antennas and Propagation Magazine, Vol.46, No.1, pp.144-145, 2004.

[5] I. Bahl, Lumped Elements for RF and Microwave Circuits, Arthech House, Microwaves library, 2003.

[6] G. Dauphin-Tanguy, Les bond graphs, Hermes, 2000.

[7] W. Borutzky, Bond Graph Methodology. Development and Analysis of Multidisciplinary Dynamic System Models, Springer London, 2010.

[8] W. Borutszky, Bond graph model based fault diagnosis of hibryd systems, Springer International Publishing Switzerland, 2015.

[9] http://www.20sim.com/

2016 IEEE 22nd International Symposium for Design and Technology in Electronic Packaging (SIITME)

An Photovoltaic System Tester
with Three-Phase Off-Grid Supply

Simulation testing with real-equivalent parameters

or photovoltaic panels and electronic converter

Marius Ovidiu Neamțu, Nistor Daniel Trip
Department of Electronics and Telecommunications,
University of Oradea
Oradea, Romania
e-mail oneamtu@uoradea.ro

Abstract— **The system is designed dimensioning the area with photovoltaic panels, in conditions of available consumer in a house. The MPPT (Maximum Power Point Tracking) algorithm ensures efficiency and is included in Boost converter. At the output, there is used a multi-level inverter which forms a three-phase supply off-grid.**

Keywords— photovoltaic panel; boost converter; multilevel inverter

I. SYSTEM TESTER

Photovoltaic panels are dependent on the constructive structure that can be characterized by the intrinsic electrical parameters. Inputs are: irradiance and temperature. The electrical power required in a home will depend suitable dimensioning of the assembly in series and parallel, for such photovoltaic panels.

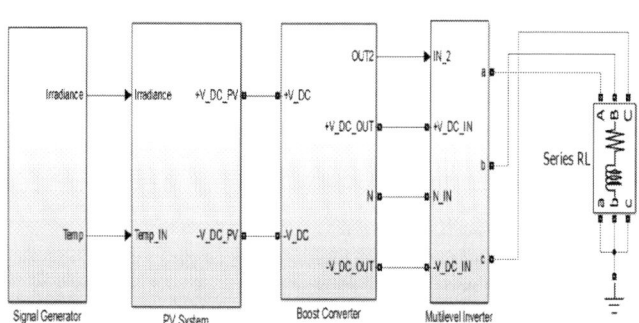

Fig. 1. System Tester in Matlab-Simulink.

The test system is capable of integrating various types of photovoltaic panels [1], from consecrated companies in the field. However simulation (figure1) contains the block PV (Photovoltaic) System and the method provides dimensioning based on the electrical parameters. Boost converter [2] contains an IGBT transistor, which is controlled on PWM

(Pulse-width modulation) provided by block MPPT. Multilevel inverter provides a constant power based on measurements of voltages from V_DC_OUT and output measurements: voltages and currents through [3]   the three-phase RL load.

II. PHOTOVOLTAIC SYSTEM

The photovoltaic system is composed of photovoltaic panels connected in series and in parallel. It can be analysis of the functionality and the power supply based on:

-   I_r Sun irradiance (W/m^2);
-   t^o Cell temperature (deg.C);
-   the number of photovoltaic panels and their interconnection.

In Fig.1 the first block is a nonlinear generator for continuous functions. Evolution of the values: I_r Sun irradiance and t^o Cells temperature are synthesized in the possible limits of variation present in the dynamically evolution in Simulink test scheme, and their visualization in Fig.2.

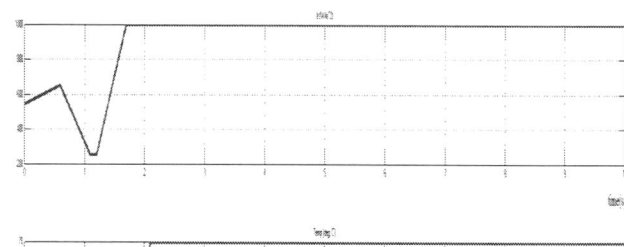

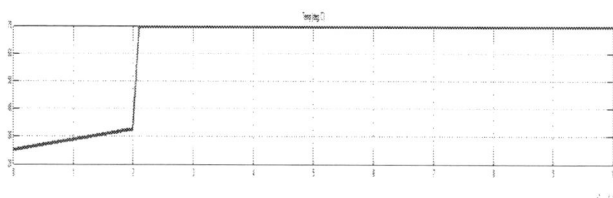

Fig. 2. Evolution of inputs at the photovoltaic pannels.

Simulation time is relatively short, but extensions applied testing can be carried for long periods in which the functions are dependent on the values measured for irradiance. The temperature can be set dynamically in real temperatures correlated with estimates of settlement areas of photovoltaic panels.

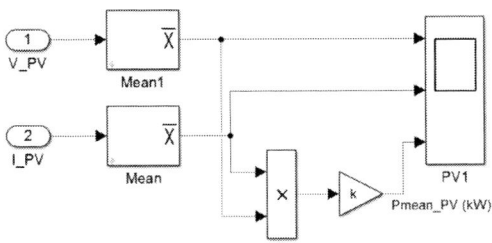

Fig. 3.   The outputs of the photovoltaic pannels .

The role of the test is to follow generating of electrical power when large variations mentioned inputs. Irradiance and temperature functions generated for photovoltaic cells can use real data from long-term measures.

Simulation is not restricted in time. In this work is an example evolution useful for sizing photovoltaic system designed for the supply of isolated farms.

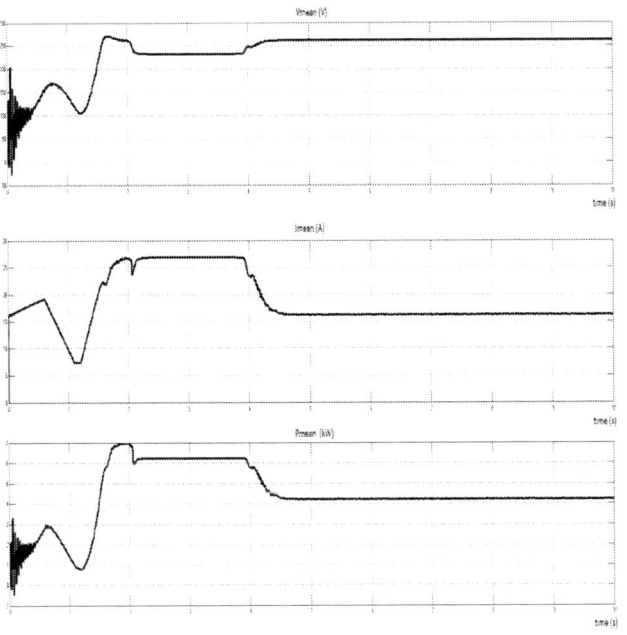

Fig. 4.   Signals at the photovoltaic pannels.

For photovoltaic generator, solar cells temperature increase leads to decreased electrical power delivered Fig. 4 even if I_r Sun irradiance is sufficiently large.

The medium voltage V_{mean} and the medium current I_{mean} are obtained Fig.3 with a soft block of computation existing in Matlab Simulink. The period for calculating the medium values was $T_{MPPT} = 0.2ms$

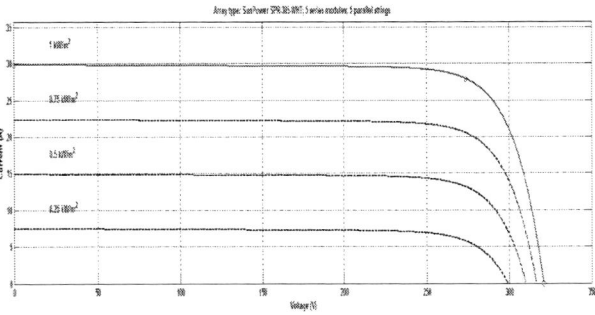

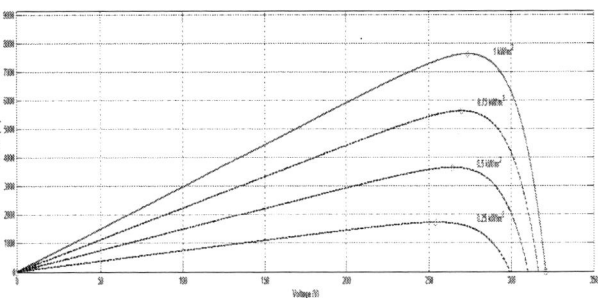

Fig. 5. Photovoltaic system depending on Sun irradiance at $25^o C$

It was chosen SunPower modules, for the photovoltaic system, connected in five parallel lines where each string contains 5 modules in series.

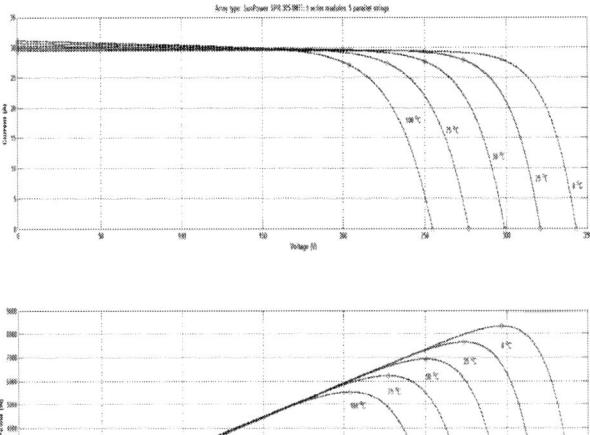

Fig.6. Photovoltaic system depending on temperature at $1000^o W/m^2$ irradiance

The default specifications for photovoltaic panels and systems used in the test are:
-   Open circuit voltage $V_{oc}(V)$;
-   Short-circuit current $I_{sc} = (A)$;

- Voltage at maximum power point $V_{mp} = (V)$;
- Current at maximum power point $I_{mp} = (A)$;
- Temperature coefficients of: $V_{oc}, I_{sc}, V_{mp}, I_{mp}$;
- Number of cells per module Ncells

Catalog values for the module are:

$$V_{oc} = 64.2V, I_{sc} = 5.96A$$
$$V_{mp} = 54.7V, I_{mp} = 5.58A$$
$$N_{cells} = 96$$

Characteristic functions Fig.5, Fig. 6 are achieved by simulation based on parameters characteristic of the ensemble of all solar panels. Voltages, current and power are dependent on electrical cabling panels. In Fig. 5 is obtained dependent on Sun irradiance (temperature is constant) and Fig.6 shows depending on temperature (Sun irradiance is constant).

In a similar manner are presented parameters for photovoltaic modules of other companies. The test is useful for photovoltaic panels are in series production or for future modules marketed.

III. CONVERTER WITH MAXIMUM POWER POINT TRAKING

Achieving maximum power is carried with a block MPPT but under dynamic conditions, simulated in maximum variation of the input values

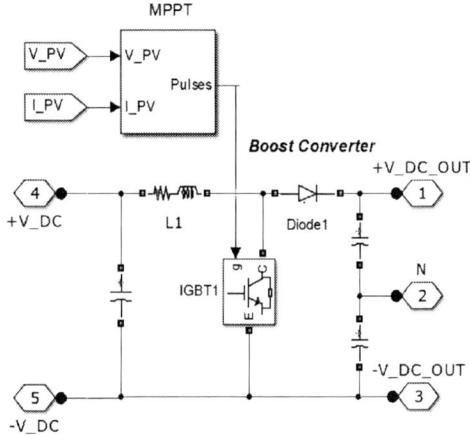

Fig. 7. The Boost Converter with MPPT control

MPPT is predefined module and ensures control of the Boost Converter by the PWM control pulses; MPPT output Fig.7 is connected directly to the grid transistor IGBT1.

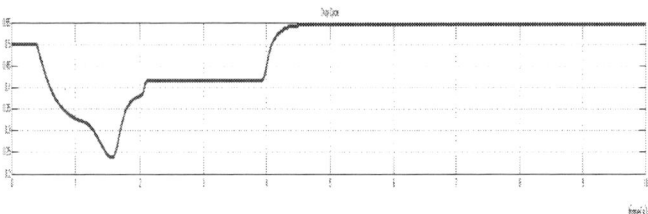

Fig. 8. The Duty Cycle for control IGBT driver in Boost Converter.

Is a DC-DC converter that provides constant output voltage symmetrical greater than the input voltage. In Fig.8 it can view the dynamic duty cycle of the PWM pulse

IV. POWER OUTPUT INVERTER

Bloc multilevel inverter Fig.1 has presented the internal structure in Fig.9. The solution [6] implemented is a classic 3-level inverter, His control is included based on measured data from the outputs. Power inputs in this block are coming from block Boost Converter.

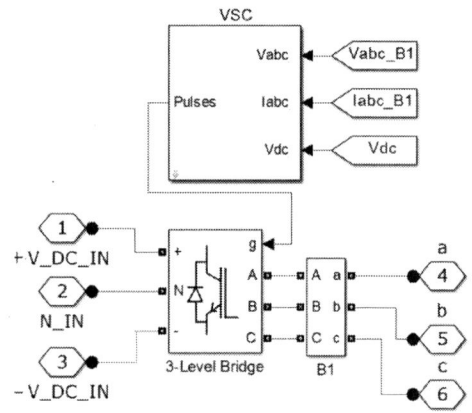

Fig. 9. The Inverter with VSC control

Bloc VSC (voltage-sourced converter) ensures the control for 3-Level Bridge. Control is PI (Proportional Integral) for both: three-phase voltages and for output currents.

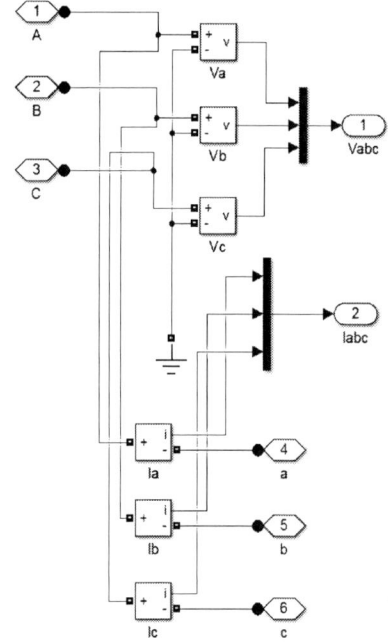

Fig. 10. The three-phase structure for electrical measuring

Block "B1" Fig.9 is a structure dedicated to measuring the output electrical values: Fig.10 voltages and currents for three-phase load

In Fig.9 is present in a single connection to input "g" at "3-Level Bridge", is practically represented a bus with 12 lines. They are controlled directly [6] in grids, a total of 12 IGBT.

Simulink scheme Fig.10 is included in block B1 Fig.9. Software blocks V_A, V_B, V_C, of the scheme are designed to measure the voltage inputs at A, B and C, towards a common ground.

To measure the currents, I_A, I_B, I_C, blocks are serially connected between the inputs A, B, C, and outputs: a, b, c which will connect the electrical load RL.

All these blocks have each output signal for providing a dynamic measurement that is carried out properly. The signal can be viewed on the oscilloscope can be stored or used in the calculations of other blocks. It is used multiplexers Fig.10 whose inputs are lines of measurement signals. The outputs of multiplexers will provide three-phase measurements: output 1 - Vabc and output 2 - Iabc. The control VSC inverter Fig.9 uses these signals.

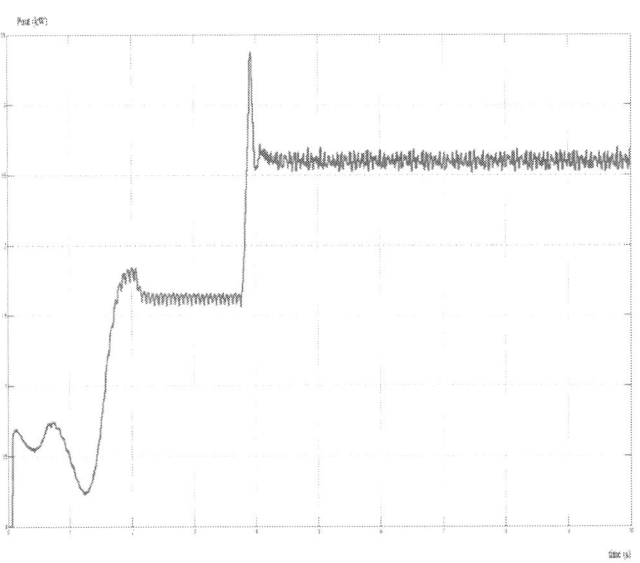

Fig. 11. The three-phase load Pout evolution.

Power output from the inverter Fig.11 is sufficient for an isolated farm, where [7] I_r Sun irradiance value Fig.2 is at least $1000 \, W/m^2$, even if the temperature of photovoltaic cells reaches high.

V. CONCLUSION

In this paper is defined functional scheme with simulation testing on real parameters, such as photovoltaic panels, Sun irradiance and cell temperature.

The project determines appropriate choice of the components for boost converter and 3- level bridge inverter.

All data obtained from the simulation are stored. The plots are eloquent for the functional schema validation. After the simulation the dates are saved for real system microcontroller. The controller will determine management algorithm based on the measured data or existing simulation results. It will be part of the project "Management system of energy from renewable sources in small isolated communities" *REMSIS*, UEFISCDI PN-II-PT-PCCA-2013-4, developed during 2014-2017.

ACKNOWLEDGMENT *(Heading 5)*

This paper is supported through the programme "Parteneriate in domenii prioritare – PN II", by MEN – UEFISCDI, project no. 53/01.07.2014

REFERENCES

[1] D.L. Talavera, G. Nofuentes, J. Aguilera, "The internal rate of return of photovoltaic grid-connected systems:A comprehensive sensitivity analysis", ELSEVIER, Renewable Energy 35 pp. 101–111, 2010.

[2] D. Panfilov, O. Husev, F. Blaabjerg, J. Zakis, K. Khandakji, "Comparison of three-phase three-level voltage source inverter with intermediate dc–dc boost converter and quasi-Z-source inverter", ISSN 1755-4535, 2015.

[3] S. Wensheng, F. Xiaoyun. S. Keyue, A Space-Vector PWM Method for Single-Phase Three-level Neutral-Point Clamped Converter, IEEE Applied Power Electronics Conference and Exposition,APEC, 978-1-4244-8085-2/11/, pp. 521-528, IEEE 2011.

[4] Li, Jun; Zhou, Xiaohu; Liang, Zhigang; Bhattacharya, Subhashish; "A Simplified Space Vector Based Current Controller For Any General N-level Converter", IEEE Energy Conversion Congress and Exposition (ECCE), 978-1-4244-5287-3/10/ - Atlanta, G, pp. 2156-2163, IEEE 2010.

[5] C. I. Odeh, "Improved three-phase, five-level pulse-width modulation switched voltage source inverter Power Electron"., 2015, Vol. 8, Iss. 4, pp. 524–535, The Institution of Engineering and Technology ISSN 1755-4535, 2015.

[6] N. Mohan, T.M. Undeland, W.P. Robbins, *Power Electronics, Converters, Applications, and Design*, John Wiley & Sons, Inc., New York, 1995.

[7] Gordan, Ioan Mircea, Cornelia Emilia Gordan, and Dorina Mioara Purcaru. "MODERN METHOD USED FOR SOLAR RADIATION INTENSITY MESUREMENT AND DATA PROCESSING." *Revista de tehnologii neconventionale" Nonconventional technologies review* 17.1 (2013).

Comparison between Zubieta Model of Supercapacitors and their Real Behavior

R. Negroiu, P. Svasta, Al. Vasile, C. Ionescu, C. Marghescu

1) University "Politehnica" of Bucharest, Romania, Centre of Technological Electronics and Interconnection Techniques, UPB-CETTI
E-mail: rodica.negroiu@cetti.ro

Abstract: The supercapacitor is a relatively young electronic component which bases its operation on the principle of the Helmholtz double layer of charge formed on the surface of an activated carbon (with an extremely high equivalent surface) and an electrolyte which allows both a short distance between the electric layers as well as a very large useful area.

For optimal use of an EDLC (Electrochemical Double Layer Capacitor) it is necessary to know the time behavior, specific electrical characteristics (discharge current, charge current, voltage variation at the terminals, equivalent series resistance, and especially leakage current (self-discharge current), as well as behavior over a long time – days or even months - under electric charge at different voltages, etc.) and the influence of environmental operating conditions (temperature, humidity, vibration, etc.). For modeling in more detail the behavior of an EDLC in time and voltage a comparison between the Zubieta model and actual behavior was performed. The research is focused on the accuracy of the Zubieta model related to the real behavior of supercapacitors.

Keywords: Supercapacitor, EDLC, Zubieta, comparison

I. INTRODUCTION

The approximation of the operation of a supercapacitor proposed by Zubieta (see Fig. 1) does not fully highlight its experimental behavior. As shown in Fig. 1 in the first branch, named immediate branch, the capacitor C_{i1} highlights the voltage dependence of the double-layer. Moreover, the leakage current is modeled using the R_P resistor.

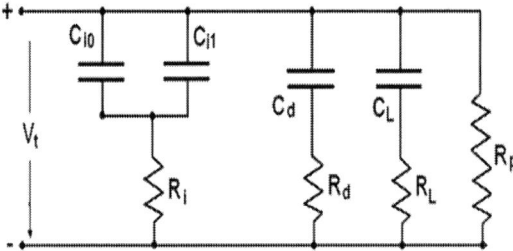

Fig. 1: The model with three branches (Zubieta)
C_{i0}-capacity of immediate branch; R_{i0}-resistance of immediate branch; C_{i1}-capacity that depend of the voltage of the double layer; C_d-capacity of delayed branch; R_d- resistance of delayed branch; C_L-capacity of long term branch; R_L- resistance of long term branch; R_P- resistance that modeling the leakage current.

The time constant for each branch of the Zubieta circuit has a different value, and each branch has a specific designation as follows:

- The branch that contains resistor R_i is called the immediate branch and this is defining the operation of the circuit in the first few seconds;
- The branch that comprises the resistor R_d is called the delayed branch and defines the circuit operation within a few minutes;
- The branch that includes resistor R_L is called long-term branch and characterizes the operation of the circuit after more than 10 minutes.

II. DETERMINIG EQUIVALENT PARAMETERS

Through this model with three branches, based on formulas, Zubieta determines the equivalent parameters of the circuit at different times and at different well-established voltage values. Zubieta's is mainly a mathematical model, and therefore a more comprehensive approach is required that takes into account the temporary behavior of EDLCs in order to correct and improve the model for each constructive type of EDLC.

The purpose of this work is to see if the model with three branches proposed by Zubieta differs from the real life behavior of a supercapacitor.

• For calculating the parameters of the immediate branch, the supercapacitor is charged with a constant current I_{ch}.

At t_1 (a few seconds after charging starts) we measure voltage V_1. Knowing the voltage value, and the current value respectively the resistance value for the immediate branch can be determined:

$$R_i = \frac{V_1}{I_{ch}} \qquad (1)$$

After t_2, the voltage that is measured at the terminals of the supercapacitor will increase by ΔV, and the capacity C_{i0}, will be calculated according to the formula below:

$$C_{i0} = I_{ch} \frac{t_2 - t_1}{\Delta V} \qquad (1)$$

Observation: t_2 was chosen so that $\Delta V = 50$ mV.

The time moment t_3 is when the voltage reaches the maximum value specified by the manufacturer in the supercapacitor's data sheet, and the power source is turned off. After this stage the EDLC begins the discharge process and that moment represents the time moment t_4 and the measured voltage will be V_4. Having all this information the value of C_{i1} can be computed:

$$C_{i1} = \frac{2}{V_4}\left(\frac{I_{ch}(t_4 - t_1)}{V_4} - C_{i0}\right) \tag{2}$$

With this information the total load of the supercapacitor can be calculated:

$$Q_{tot} = I_{ch}(t_4 - t_1) \tag{3}$$

- Having calculated all the parameters of the immediate branch, the parameters of the delayed branch can now be determined.

We will wait for a longer time t_5 until the voltage drops with ΔV_2 against to the value of voltage V_4. t_5 represents the time necessary to transfer the electrical current from the immediate branch to the timed branch. The resistance value R_d is calculated with:

$$R_d = \frac{(V_4 - \frac{\Delta V_2}{2})(t_5 - t_4)}{(C_{i0} + C_{i1}(V_4 - \frac{\Delta V_2}{2}))\Delta V_2} \tag{4}$$

For the calculation of the timed branch's capacity, voltage V_5 is measured at t_6, where $t_6 = 3t_5$:

$$C_d = \frac{Q_{tot}}{V_5} - (C_{i0} + C_{i1}\frac{V_5}{2}) \tag{5}$$

- To calculate the parameters of the long-term branch the following procedure should be followed: to determine resistance R_L we will wait for a longer time period noted with t_7 until the voltage drops with ΔV_3 against the value of voltage V_5:

$$R_L = \frac{(V_5 - \frac{\Delta V_3}{2})(t_7 - t_6)}{(C_{i0} + C_{i1}(V_5 - \frac{\Delta V_3}{2}))\Delta V_3} \tag{6}$$

Long-term capacity is calculated after 30 minutes, note t_8 since this is considered a long enough time for the charge to be fully distributed the long-term branch. This is when the voltage on each branch, will be balanced and equal to V_6.

The capacity's value C_L will be determined according to the formula:

$$C_L = \frac{Q_{tot}}{V_6} - \left(C_{i0} + C_{i1}\frac{V_6}{2}\right) - C_d \tag{7}$$

To perform simulations a circuit was developed allowing the charging and discharging of supercapacitors. Each supercapacitor must be charged up to maximum voltage without exceeding it. The electronic schematic from Fig. 2 realizes the voltage protection of the supercapacitors used. When the voltage reaches the maximum value specific to each supercapacitor (2.5 V, 5.4 V respectively), the transistors open and thus fulfill their role. The LED acts as a "witness" and specifies that the transistors were opened when the supercapacitos peaked at the maximum voltage value.

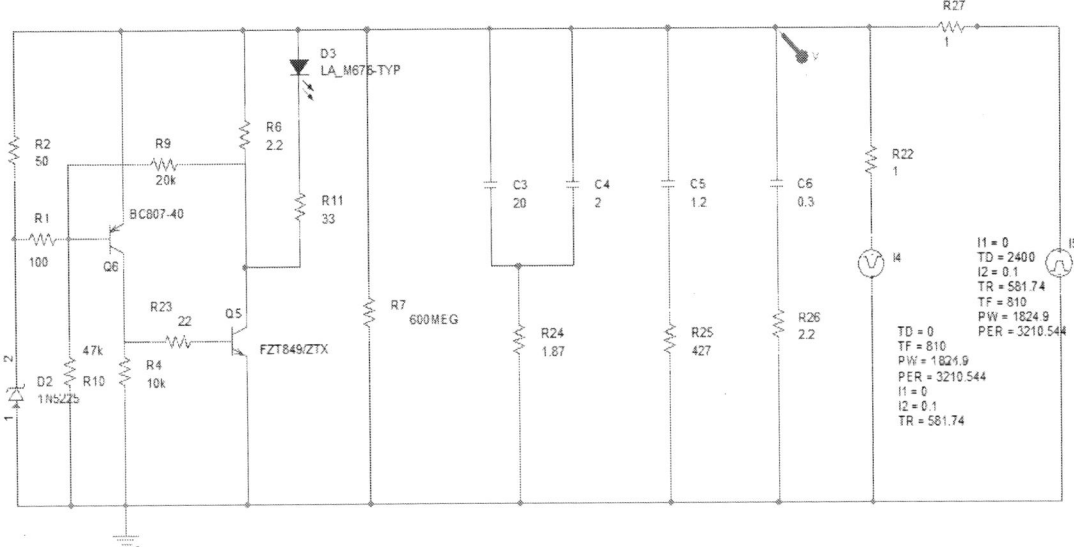

Fig. 2. The electronic scheme used for simulation at 2.5 V.

III. MEASUREMENTS AND RESULTS

In this work measurements were conducted for two type of supercapacitors with capacitances of 22F and 5F. The rated voltages for these supercapacitors are 2.5 V,

respectively 5.4 V. Initially the supercapacitors were charged with a current of 100 mA and the rules of the Zubieta model were applied. Subsequently, the supercapacitors were discharged with the same current using an active sink. All the data was saved on a PC using specialized software and with

2016 IEEE 22nd International Symposium for Design and Technology in Electronic Packaging (SIITME)

this data all the equivalent parameters of the Zubieta model for each branch were calculated. A model based on the obtained values was implemented and simulations conducted.

In Fig. 3 the measurements setup is presented and on the PC we can see the working software.

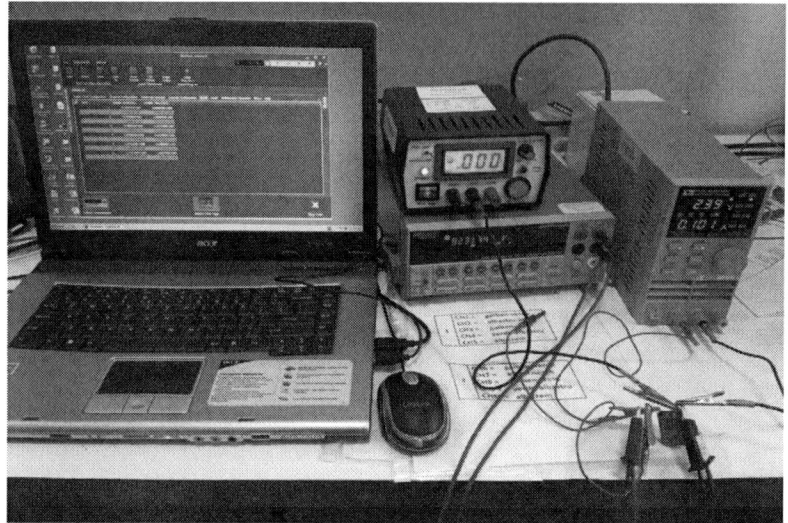

Fig. 3. The measurements setup.

The values obtained for each branch calculated with the formulas of Zubieta model for the used supercapacitors, are show in the table I.

The Zubieta model was implemented on the electronic scheme (see Fig. 2) with the values of equivalent parameters that were calculated and are presented in table1 and table 2. After this the simulation of electronic scheme for each supercapacitor was realized.

The comparison between the Zubieta model of supercapacitors (22 F, respectively 5 F) and their real behavior is presented in the figures below:

TABLE 1 THE VALUES OF THE PARAMETERS FOR 22F AND 5V

SUPERCAPACITORS

	22F/2.5V	5F/5.4V
Immediate branch		
R_i	1.87 Ω	2.3 Ω
C_{io}	20.032 F	4 F
C_{i1}	2.098 F	0.4 F
Q_{tot}	57,17 C	37.12 C
Delayed branch		
R_d	427,9 Ω	653 Ω
C_d	1.22 F	3.63 F
Long term branch		
R_L	2.2 kΩ	3.4 kΩ
C_L	0.32 F	1.3 F

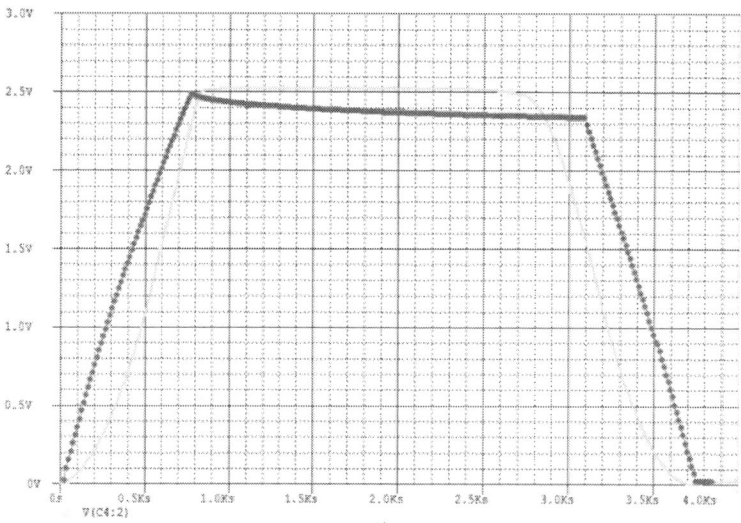

Fig. 4. The simulation (green) and the measurement (blue) for the 22 F/ 2.5 V supercapacitor.

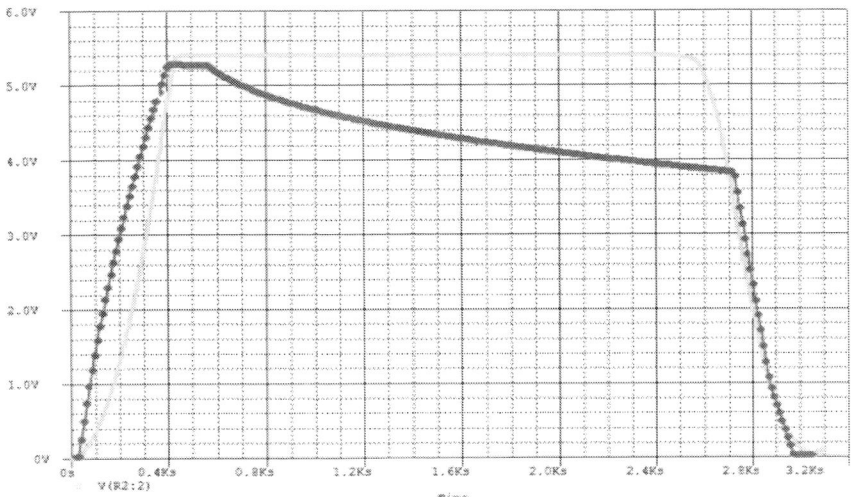

Fig. 5. The simulation (green) and the measurement (blue) for the 5 F/5.4 V supercapacitor.

We can see in the Fig. 4 and in Fig. 5 that the approximations of the operation of the monitored supercapacitors proposed by Zubieta does not fully highlight its experimental behavior.

IV. CONCLUSIONS

Zubieta assumes that in the model with three branches only the capacity of immediate branch C_{i0} is dependent on the voltage which can lead to errors for low voltages.

Because for the long-term branch is needed only 30 minutes that matter, this model can't be applied to the supercapacitors with higher capacities, tens or hundreds Farads. A solution to this problem would be extending the Zubieta model on multiple branches to monitor the EDLC's operation for a long time, tens of hours or even days.

For the simulation of each supercapacitor different voltage values were needed and for this we can use different electronic schemes.

After the comparison between the Zubieta model of supercapacitors and their real behavior we can see that differences arise because it does not monitories the phenomena of supercapacitors self-discharged.

ACKNOWLEDGMENTS

This paper was published under the frame of the "Partnerships in priority areas" (PN II) Romanian Research program, developed and supported by MEN-UEFISCDI, SIOPTEF project, PN-II-PT-PCCA-2011-3.2-899, no. 121/2012 and "MicroEIectronics Cloud Alliance" (MECA) project, no.: 562206-EPP-1-2015-1-BG-EPPKA2-KA, Grant Agreement Number: 2015 - 2967 / 001 – 001, program „Erasmus+".

REFERENCES

[1] P. Svasta, A. Vasile, C. Ionescu, "Condensatoare", Cavallioti, Bucharest 2010

[2] A. Vasile, Irina Bristena Bacîş „Optical solutions for unbundled access network" Optoelectronics, Microelectronics, and Nanotechnologies 2014 / 92580Q-92580Q-7; ISBN 9781628413250; International Society for Optics and Photonics; pag.382-38

[3] Plotog, Ioan," Investigations on electroluminescent tapes and foils in relation to their applications in automotive", 7th International Conference on Advanced Topics in Optoelectronics, Microelectronics, and Nanotechnologies (ATOM-N 2014), Constanta, ROMANIA, August, 21-24, 2014, WOS:000354179700077, ISBN:978-1-62841-325-0

[4] C. Ionescu, F. Draghici, D. Bonfert, "A SPICE model for electroluminescent foils", 38th International Spring Seminar on Electronics Technology (ISSE), 2015, pp. 526 - 531, DOI: 10.1109/ISSE.2015.7248057

[5] Andrei Drumea, Robert Alexandru Dobre, "Modelling, simulation and testing of an autonomous embedded system supplied by a photovoltaic panel", 20th International Symposium for Design and Technology in Electronic Packaging, SIITME2014, October 2014, pp.309-312

The framework of using models for comparative assessment of traffic sensors

F. C. Nemtanu (SM'08), I. M. Costea, I. Badescu and
V. Iordache
Telematics and Electronics Dept., Transport Faculty
Politehnica University of Bucharest
Bucharest, Romania
florin.nemtanu@ieee.org

J.Schlingensiepen
Technische Hochschule Ingolstadt
Ingolstadt, Germany

Abstract— The intelligent transport systems (ITS) are now one of the most dynamic sector in terms of finding solutions for transport domain. These systems are applications of electronics, IT, computer science and other connected disciplines which are installed in terms of increasing the efficiency of the transport activities and decreasing the negative impact of them. ITS are mainly based on a sensor network which is able to collect data from transport processes. The paper is focused on elaboration of a common model for traffic sensor which is able to support comparative assessment of various traffic sensors (the example in the paper is based on inductive loops and virtual loops based on CCTV cameras). The framework and the models are created in LabVIEW and an example of using the model in comparative assessment is provided. The authors presented the methodology for the elaboration of the model as well as the recommendations for comparative assessment based on the proposed framework.

Keywords— Traffic sensors, comparative assessment of sensors, LabVIEW models, ITS sensors

I. INTRODUCTION

The Intelligent Transport Systems are applications of electronics, IT, communications and computer science in transport domain in terms of increasing the efficiency and reducing the negative effects (accidents, pollution etc.). The effect of installing ITS has benefits, not only for transport systems, but also for mobility and urban area could be a good example (ITS as supporting systems for smart cities) [1].

The benefits of ITS are increasing if the systems are designed based on specific architecture [2], [3] and the main components are modelled and tested before the implementation of the system. The simulation of the ITS components, before the design and installation, is important to understand the behavior of the system as well as the relations of the system with other terminals in its environment [4], [5], [6].

One of the most important layer of intelligent transport systems is monitoring subsystem which is in charge with the collection of data as well as the local processing and transmission of data [7]. This monitoring subsystem is based on sensors and the network of sensors is able to collect data from transport processes and to send data and pre-processed data to other subsystems in terms of providing ITS services.

The network of sensors as part of monitoring systems has an important task to collect reliable and accurate data for all ITS processes and systems. For this reason, it is also important to understand the functionality of the sensors and what is the technical role of them in terms of finding the best way to replace them with new and improved sensors [8]. The functionality of the sensors, the role of the sensor in the network as well as the integration of the sensor into the monitoring subsystem has to be started with modelling and simulation activities and new techniques and approaches as well as the traditional methods in terms of simulation could be applied on ITS domain [9], [10].

The modelling and simulation phases are strong related to the software environment and a good solution could be a platform which has the software capabilities as well as the interfaces with hardware components. The authors selected as software environment LabVIEW and two main reasons are on the base of this selection: the power of the software in terms of implementing the algorithms and the integration with data acquisition board and other hardware components [11]–[13].

The traffic sensors are able to collect data from processes and phenomena which is characterizing the road traffic (any other transport modes could be considered as an extension of the application of this research) and various technologies are developed to be applied in transport field [8], [14]. The development of new sensing technologies and new methods for collecting, predicting and generating data pave the way for a new approach of integration various types of sensors based on fusion, data mining and big data technologies and techniques [15]–[18]. The replacement of a traditional sensor, the inductive loop cold be considered one of this traditional sensor, has to be done based on comparative assessment having as reference data collected from the existing and installed sensors. The new virtual loop sensors based on video processing is considered a better alternative for this type of traffic sensors in terms of operational and maintenance costs and the integration with other subsystems (automatic number plate recognition and so on) [19]. The replacement has to be done within the same limit of safety and functional reliability and the comparative assessment of the new sensors based on the performance of the exiting sensors has to be the main evidence of this action.

The authors present the framework, which is based on LabVIEW platform, of the comparative assessment of traffic sensors with a special focus on inductive loops and virtual loops based on CCTV cameras and video processing but this methodology and the framework could be applied to other types of sensors as well as to other domains.

The inductive loops are now accepted by all road administrators and operators and the main task is to find a solution to replace this type of sensors with new one (the example here is the virtual loop sensor based on image acquisition and video processing) in terms of reducing the cost and maintenance efforts. An important step in this task is to find a common set of key performance indicators (KPIs) for both technologies and to define a framework as well as the models for both technologies which are able to pave the way for comparative assessment. The main result of this comparative assessment is to validate the replacement of old technologies with new technologies using the same KPIs to measure the performance of the sensors.

II. METHODOLOGY

The main purpose of this paper is to set up a framework, which is a platform for comparative assessment and a set of procedures in terms of comparing different traffic sensors with

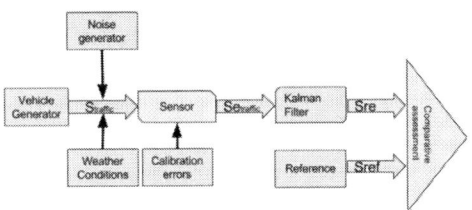

Figure 2 The architecture of the solution

some reference sensors.

The setting-up of the framework will be able to facilitate the comparison between a traffic sensor and reference traffic sensor. In Figure 2 is presented the architecture of the platform (which will be developed in LabVIEW environment) and the main components which are implemented in LabVIEW. The first step is to set-up the platform using the Vehicle Generator as sensor (this is a model of the sensor and will be able to model various types of traffic sensors). the second step is to replace the virtual sensor (Vehicle Generator) with a real traffic sensor which will be used as tested sensor.

The comparative assessment procedure has the following components and is shown in Figure 1:

- Setting-up the platform – at this stage the set of KPIs will be defined as well KPTs (key performance targets for these indicators), a model of the sensor will be defined to be used for calibration of the platform, a set of data from operational environment (noise, weather and calibration of sensor) will be determined and a testing scenario is created based on real traffic data.

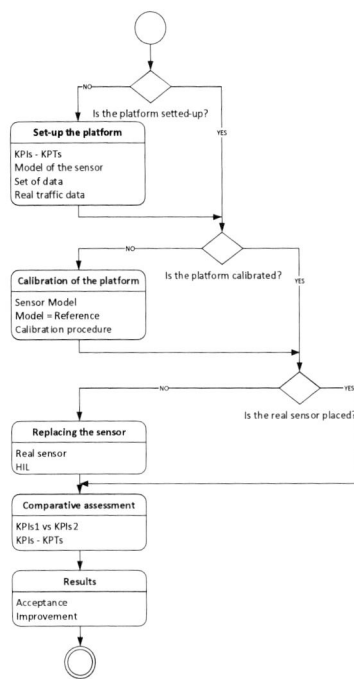

Figure 1 The comparative assessment procedure

- Calibration of the platform – is one of the most important phase of this procedure in terms of preparing the platform for testing. The calibration will be done using as sensor model the model of the reference traffic sensor which is the model of virtual loop. The calibration procedure will be the same with the final comparative assessment and all KPIs will tested and the KPTs will be defined.

- Replacing the sensor – after the calibration of the platform the replacement of the sensor will be done and the tested sensor will be integrated in the platform. This sensor could be a real sensor or a sensor model, similar with the one used in the calibration of the platform (this time the modelled sensor will be the tested sensor, in our case the virtual loop sensor based on video processing).

- Comparative assessment – the main tool for comparative assessment will be the set of KPIs. For these KPIs, the authors have already defined minimum and maximum accepted limits for variation of indicators (defined as key performance targets KPTs).

- Results – the results of comparative assessment are very important in term of taking the decision to replace an inductive loop sensor with alternative sensors. the behavior of tested sensor has to be at least as an installed sensor is and the KPIs have to be in line with defined KPTs.

III. THE PLATFORM

The platform was developed by the authors in LabVIEW environment and they selected this based on the capability to integrate hardware or software modules in a simple manner and to run simulation based on HIL (hardware in the loop) or hybrid simulation approaches.

The platform is created in LabVIEW and the VI was build using sub-VIs, one for every component described in the Methodology. The structure of the platform is presented in Figure 3.

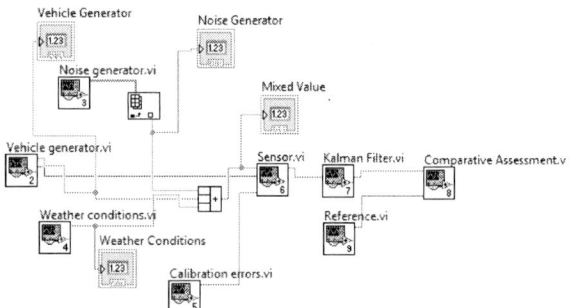

Figure 3 LabVIEW implementation of the framework (block diagram)

The Vehicle Generator will transmit the signal to Sensor. This signal will be mixed with Gaussian noise (this noise is added to create the real environment of the sensor) and with Weather Condition noise (this noise is the result of weather conditions on different sensors, the signal of inductive loop sensor is influenced by the level of water or ice on the surface of the road). The sub-VI Sensor will mix the signal with some additional noise generated by Calibration Errors and will transmit the signal to a Kalman Filter. Comparative Assessment is able to compare the signal/value from the tested sensor (the last component of it is the Kalman Filter) and the reference sensor (this reference sensor is an inductive loop sensor, but it could be every installed sensor).

The Vehicle Generator is shown in Figure 4 (a VI – virtual instrument in LabVIEW has two components: the front panel, which is a HMI, and block diagram, which is the logic of the programme). The user could select the type of the sensor, at this stage of the research only two sensors are implemented: an inductive loop and a virtual loop, based on video processing.

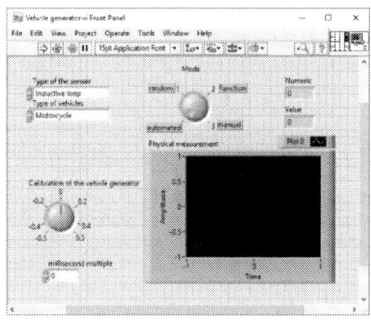

Figure 4 Vehicle generator (vi - front panel)

Another important control implemented in this VI is the type of vehicle. For this aspect the authors have defined 6 categories of vehicles: motorcycle, car, bus, van, truck, heavy truck (definition of the vehicles means the signal which is specific for every category of vehicles, in the case of inductive loop).

Calibration of the modelled sensor could be also done using the VI for vehicle generator. This control will add or subtract a value which is the error introduce by calibration. The role of this control is to make the model of the sensor as close as possible to the real sensor.

The Mode is a control which is able to select the functioning mode of the sensor: automatic (the sensor will generate the values without any intervention of the user and will use the values collected by other sensors), random (the same case as previous one but the application will generate values randomly), manual (the user is able to introduce values for this sensor) and function (the sensor will generate the values based on mathematical function).

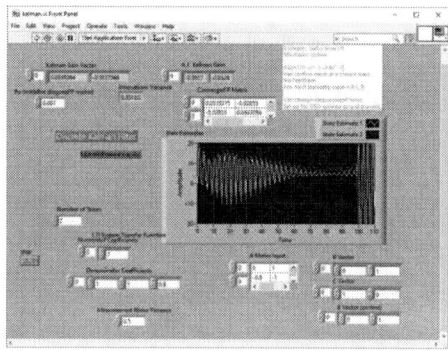

Figure 5 LabVIEW application – Kalman filter [20]

The Kalman Filter (this VI was developed in [20]) is used only for Random Mode of the Vehicle Generator and the main scope is to create a set of data, generated randomly, which is similar with real data collected by a reference sensor.

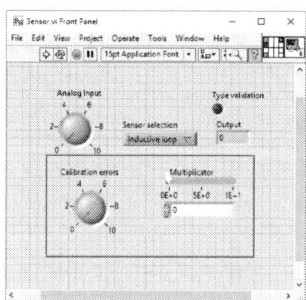

Figure 6 Sensor model - LabVIEW

The sensor VI has the role of integration different signals from different sub-Vis and the values generated by this sub-VI will be sent to Kalman Filter, if the sensor's values are generated randomly or directly, in any other cases. In this sub-VI the authors included calibration errors which are directly related to the tested sensor and a multiplier which will increase the value generated by the Vehicle Generator or real tested sensor.

IV. COMPARATIVE ASSESSMENT

The comparative assessment is done in two stages: the first one is a comparative assessment between a modelled sensor and the reference sensor and the second one, after the replacement of the modelled sensor with a real one, the comparative assessment between the tested sensor and the reference sensor.

The KPIs and associated KPTs are defined as following:

- Accuracy (Acc) (number of correct readings) – it is defined as the degree to which the result of the measurement conforms to the correct values collected by a reference system or sensor. In this case, the accuracy is defined related to the correct detection of every vehicle as part of the road traffic. In fact, it is the accuracy of vehicle's number. For this research, the KPT associated is 1 vehicle lost at 1000 readings. This indicator will be adapted to the real sensors which will set up the key performance target.

- False positive errors (FPE) (False vehicle is detected) – this indicator is defined as the number of false vehicles detected by the sensor. This indicator and the indicator related to the false negative errors are very important for sensors installed in an environment with a high level of noises. The KPT associated to this indicator is defined at the level of 1 vehicle out of 100 readings (if 100 vehicles are passing the sensor, 99 vehicles are correctly detected and 1 is a false vehicle, the sensor detected a vehicle but no one is passing the sensor). This value of KPT is set up for this research and it will be adjusted to the type of the sensor (the inductive loop has different values for this KPI and it is depending by the producer and the technology used).

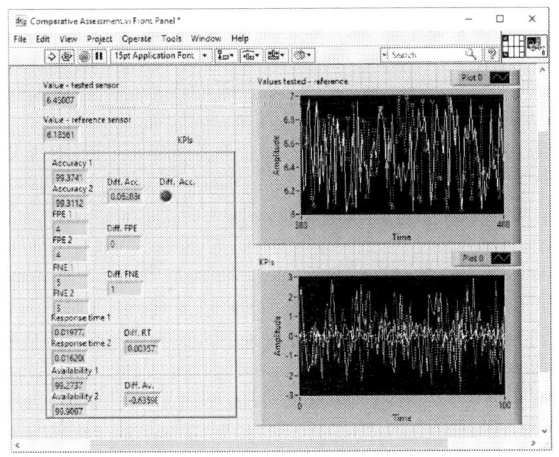

- False negative errors (FNE) (real vehicle is not detected) – this indicator is defined as the number of missing vehicles. The KPT defined by authors is 1 vehicle at 100 readings which could be translated in 1 missing vehicle out of 100 vehicles which are passing the sensor.

- Correctness (classification) – this indicator is defined only for the classification of vehicles. The authors defined 6 classes of vehicles detected by the sensor and the indicator is related to the ability of the sensor to classify correctly the vehicle. The key performance target defined in the paper is 1 vehicle out of 100 readings (99 vehicles out of 100 are classified well and 1 is outside of its class).

- Response time (the time needed to provide data) – this indicator is defined as the time which is needed to detect a vehicle passing the sensor. The KPT associated to this indicator is determined by the maximum speed of vehicles allowed to the road/motorway where the sensor is installed. If the maximum speed is 130 Km/h (36 m/s) and the length of the inductive loop is 2m, the response time has to be lower than approx. 56ms.

- Availability - is the degree to which the sensor is in operable state for a time mission. In this case the availability of a tested sensor is compared with the one of a reference sensor. The KPT of availability is determined by the reference sensor and it is a floating value.

The comparative assessment is started with the model of the sensor which is similar with the reference sensor and the platform is calibrated. The model of the sensor is replaced with a real sensor which is connected in the platform and the comparative assessment is done and the KPIs are calculated. The main point is to verify that the indicator is under the key performance target.

In Figure 1, the difference between the values of tested sensor and reference sensor is calculated and this difference is displayed for Accuracy, False Positive Errors, False Negative Errors, Response Time (RT) and a simulation for availability (this one was simulated because the time for the calculation of this indicator is longer than any other key performance indicator).

The final result is to obtain a set of KPIs of the tested sensor inside of the KPTs' limits and with better values than the reference sensor has.

V. CONCLUSIONS

The authors identified as the best solution for implementation the platform, the LabVIEW environment but other simulation/programming platform could be used (an example is Matlab with Simulink). The main advantage of this programming environment is the integration of hardware modules using data acquisition boards and the graphical user interface which is user friendly.

The authors tested only two types of sensors but the methodology as well as the framework (platform + procedures) could be applied for a generic set of two sensors. One of them has to be considered as reference sensor and the another one as tested sensor.

Figure 7 Comparative assessment and KPIs

The integration of a component which is a model of the tested sensor is able to open the platform for new sensors. In this case, the user has to create a model which will be used to calibrate the platform and after that the model will be replaced with a real sensor.

The KPIs are not limited at these 6 types defined by authors in this paper and the KPTs are defined by user related to the ITS application where the sensors will be installed.

A specific application of the results of the research described in the paper is to use comparative assessment for similar sensors, as example, the inductive loops produced by different manufacturers.

The platform is open for integration of new mathematical modules/nodes, especially for Kalman Filter implementation (the LabVIEW accepted Matlab nodes but the filters could be developed in LabVIEW, as the authors exampled in the paper) but also the noise generator could be implemented as mathematical models.

Hardware in the Loop (HIL) as well as the hybrid simulation is a very attractive simulation because it is placed between virtual environment and real one, having both components: real and virtual.

The framework, the platform and the associated procedures, presented by the authors could be used by technicians and could be considered as main tool for technical evaluation of different solutions provided by different manufacturers.

The next step of this research is to develop a platform which is able to be used in an autonomic manner: the real sensor will be integrated in the platform and the platform will select the best model for the first step and calibration, which will be automatically replaced by the real sensor; the KPIs will be defined and calculated without any human intervention (the user will define only the generic KPTs for different types of sensors).

REFERENCES

[1] J. Schlingensiepen, R. Mehmood, and F. C. Nemtanu, "Framework for an autonomic transport system in smart cities," *Cybern. Inf. Technol.*, vol. 15, no. 5, pp. 50–62, 2015.

[2] F. C. Nemtanu and M. Minea, "The development of its architecture for urban transport new components and new relations," *Zesz. Nauk. Transp. Śląska*, pp. 317–325, 2005.

[3] F. C. Nemtanu and D. Dumitrescu, "The National Architecture of Road Intelligent Transport Systems in Romania," *Proc. 13th ITS WORLD Congr. LONDON, 8-12 Oct. 2006*, 2006.

[4] F. C. Nemtanu, V. Iordache, and C. Cormos, "Modelling of intelligent sensors applied in intelligent transport systems," in *IEEE 21st International Symposium for Design and Technology in Electronic Packaging (SIITME)*, 2015.

[5] A. Fernández-Isabel and R. Fuentes-Fernández, "Analysis of Intelligent Transportation Systems Using Model-Driven Simulations," *Sensors*, vol. 15, no. 6, pp. 14116–14141, Jun. 2015.

[6] L. Verhoeff, D. J. Verburg, H. A. Lupker, and L. J. J. Kusters, "VEHIL: a full-scale test methodology for intelligent transport systems, vehicles and subsystems," in *Proceedings of the IEEE Intelligent Vehicles Symposium 2000 (Cat. No.00TH8511)*, pp. 369–375.

[7] I. M. Costea, F. C. Nemtanu, C. Dumitrescu, C. V. Banu, and G. S. Banu, "Monitoring system with applications in road transport," in *2014 IEEE 20th International Symposium for Design and Technology in Electronic Packaging (SIITME)*, 2014, pp. 145–148.

[8] M. Burtwell, "Assessment of Road Surface Freezing Point Sensors for the UK," Vol. 3. 1998.

[9] M. Bacic, "On hardware-in-the-loop simulation," in *Proceedings of the 44th IEEE Conference on Decision and Control*, pp. 3194–3198.

[10] H. D. Schwetman and H. D., "Hybrid simulation models of computer systems," *Commun. ACM*, vol. 21, no. 9, pp. 718–723, Sep. 1978.

[11] J. Y. Beyon and J. Y., *LabVIEW: programming, data acquisition and analysis*. Prentice Hall PTR, 2001.

[12] C. ELLIOTT, V. VIJAYAKUMAR, W. ZINK, and R. HANSEN, "National Instruments LabVIEW: A Programming Environment for Laboratory Automation and Measurement," *J. Assoc. Lab. Autom.*, vol. 12, no. 1, pp. 17–24, Feb. 2007.

[13] A. A. Maslov and T. A. Lepikhin, "Comparative characteristics and selection of optimal filtering algorithm signal using LabVIEW software package," 2015, vol. 1648, no. 1, p. 450013.

[14] L. A. Klein, *Sensor technologies and data requirements for ITS*. Artech House, 2001.

[15] R. Du, C. Chen, B. Yang, N. Lu, X. Guan, and X. Shen, "Effective Urban Traffic Monitoring by Vehicular Sensor Networks," *IEEE Trans. Veh. Technol.*, vol. 64, no. 1, pp. 273–286, Jan. 2015.

[16] Y. Lv, Y. Duan, W. Kang, Z. Li, and F.-Y. Wang, "Traffic Flow Prediction With Big Data: A Deep Learning Approach," *IEEE Trans. Intell. Transp. Syst.*, pp. 1–9, 2014.

[17] A. Abadi, T. Rajabioun, and P. A. Ioannou, "Traffic Flow Prediction for Road Transportation Networks With Limited Traffic Data," *IEEE Trans. Intell. Transp. Syst.*, pp. 1–10, 2014.

[18] F. Nemtanu, I. M. Costea, and C. Dumitrescu, "Spectral Analysis of Traffic Functions in Urban Areas," *PROMET - Traffic&Transportation*, vol. 27, no. 6, Dec. 2015.

[19] L. Calderoni, D. Maio, and S. Rovis, "Deploying a network of smart cameras for traffic monitoring on a 'city kernel,'" *Expert Syst. Appl.*, vol. 41, no. 2, pp. 502–507, 2014.

[20] T. Moir, "Community: Discrete Kalman Filter - National Instruments," 2016. [Online]. Available: https://decibel.ni.com/content/docs/DOC-1244.

Real-Time 3D Near-Field Visualization Using LED Field Sensors

Adrian Ioan PETRARIU and Eugen COCA

Department of Computers, Electronics and Automations
Stefan cel Mare University of Suceava
Suceava, Romania
apetrariu@eed.usv.ro, eugen.coca@usv.ro

Abstract—Measuring the radiated EM field generated by an antenna mounted on a wireless device can be a time consuming operation with traditional methods. In this paper, we present a real-time measuring technique in the near-field generated by a wireless device antenna, working at 13.56 MHz. The method implies LED Field Sensors placed in arrays, which provide fast real-time visual 3D results. The measurements can be compared, with good results, with the numerical simulations made with Ansoft HFSS.

Keywords—antenna, electromagnetic compatibility, near-field radiation pattern, numerical simulation, RFID

I. Introduction

Wireless communications are widely used for a large number of devices, causing the increase of electromagnetic (EM) pollution. Thus, transient signals can occur as unwanted electrical interferences affecting other communicating devices. Measuring the radiated EM fields generated by the antennas mounted on wireless devices can be made by using EM field probes. These probes are moved via a manual or automated mechanical system, step-by-step, into the near-field of the antenna defining the radiated volume around it. The procedure requires long acquisition time and specialized EM equipment tools to provide accurate results.

Reducing the acquisition time can be made using specialized field sensors [1-6]. These sensors are placed around the device under test (DUT) which in most of the cases are rotated to collect the EM field value, that are processed using specialized custom made applications [7,8].

One application used for determining the near-field radiation of an antenna is described in [9]. The device related requires an LED sensor for the field measurement and uses the so-called method "stop frame animation". The frames with the LED sensors, placed at different points around the DUT, are captured with a photo-camera and merged, having as a result a visual 3D radiation pattern of the near-field generated by the DUT. The disadvantage of this method is given by the fact the 3D radiation pattern isn't available until all frames are merged. This method is time consuming and requires image processing techniques to display the final result.

In [1] J. Rioult et all. presents a system where the DUT is placed in the center, and the 3D visualization of the near-field can be achieved using special field sensors. The sensors are placed along a dielectric loop, and depending on the radiation level transmitted by the DUT, some RGB LEDs will display the corresponding color shade to a different power level received. The device is a proof-of-concept for mobile terminals.

In the present paper, the authors propose a real-time measuring technique of the EM near-field that implies using LED field sensors (LFS). These sensors are placed in arrays and can display real-time 3D radiation patters for the DUT.

II. LED Field Sensor

The LFS are made using a standard double layer 0.8mm FR4 board as a support, with a relative permittivity of ε_r=4.4 and dielectric loss tangent of δ=0.02. The physical dimensions of the LFS are 20mm by 30mm, these dimensions being chosen randomly. On the PCB a resonant LC circuit is made with a PCB inductor and a 0603 package capacitor. The LC circuit resonance frequency is calculated using eq. (1). For the LC tank the resonance frequency is about 13.56MHz, which is a worldwide RFID unlicensed frequency used for most of today's NF (near-field) wireless devices. For optical visualization of the results 0603 package LEDs are used.

$$f = \frac{1}{2\pi\sqrt{LC}} \qquad (1)$$

The value of the inductance is determined after simulations performed using a full wave electromagnetic simulator based on the finite element method (Ansys HFSS) on the LFS circuit. Keeping the same dimensions of the board, we changed the track width, and the number of turns, in order to obtain the suitable capacitor value. In the Table I some simulated values of the PCB inductance are presented.

TABLE I. Simulated Inductor Value and Calculated Capacitance of the LC Circuit

Track width [mm]	Inductance [μH]	Capacitance [pF]
0.4	1.02	135
0.35	1.14	120.8
0.3	1.19	115.7
0.25	1.26	109.3
0.2	1.29	106.7

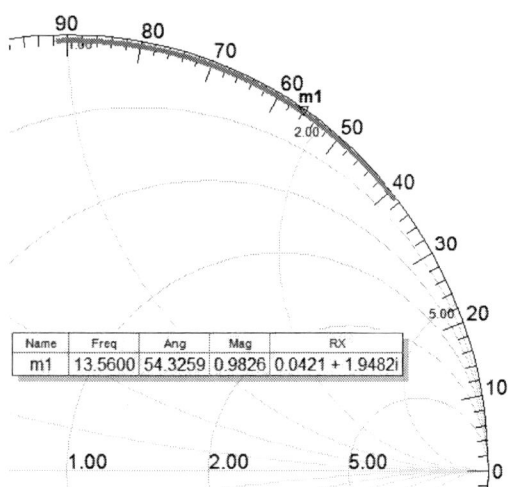

Fig. 1.   Simulated impedance of the PCB inductance

From the Smith Chart in Fig. 1, the impedance of the inductor is $X_L=1.9482\Omega$, with the reference point at 1Ω for the track width of 0.35mm. For the 50Ω reference impedance measurement, the impedance is 97.41Ω. From this value we can calculate the inductance value using eq. (2), this being the value used for matching the LC circuit at the resonance frequency.

$$L = \frac{X_L}{2\pi \cdot f} \qquad (2)$$

The inductance value is chosen to be 1.14µH because of the capacitance value of 120pF that can be easily found and mounted on the LFS.

The simulated LFS is prototyped and tested to ensure that the same results are obtained.

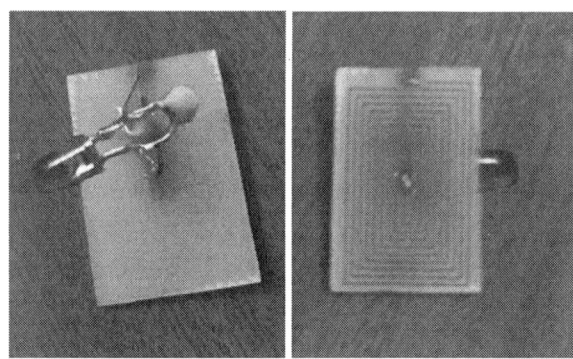

Fig. 2.   Prototype of the LFS

For the prototype, THT components and a standard single sided PCB have been chosen (Fig. 2). The calibration of the LC circuit is made using an additional capacitor, in order to ensure a perfect match for a maximum power transfer between the DUT antenna and the LFS. Thus, the total value of the matching capacitance is 126.8pF (using 120pF and 6.8pF capacitors).

In order to determine the coupling between two adjacent LFS, a set of consecutive, adjacent sensors is used in the simulation, changing the distance from 20mm to 50mm, and keeping the measuring point from the center of the LFS, on the long axis, like in Fig. 3.

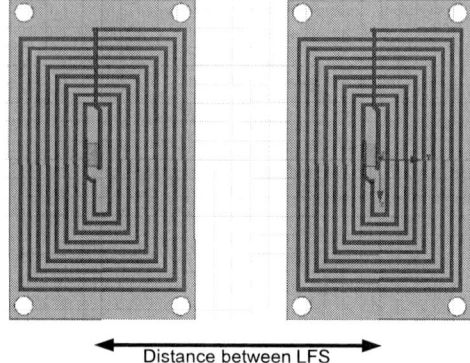

Fig. 3.   Configuration to determine the coupling between adjacent LFS

In the simulation, one LFS is fed by a source and the coupling parameter $|S_{12}|$ between the fed port and the adjacent LFS parasitic port is computed. The result in Fig. 4 shows that the coupling is decreasing with the decrease of the distance between LFSs (are take into account 3 situations with 30mm, 40mm and 50mm, respectively). The efficient distance can be considered from 40mm and up, where the coupling parameter is less than -50dB.

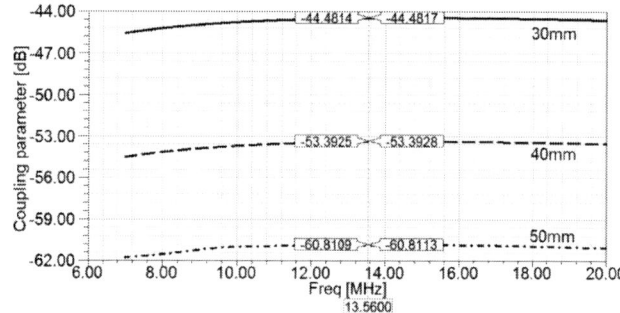

Fig. 4.   Coupling parameter for different distances between two LFSs

III. MEASUREMENT RESULTS

After all the simulations have been performed, keeping the obtained results as reference, the LFSs have been mass produced. For fine tuning, two capacitors have been selected for the final design (Fig. 5).

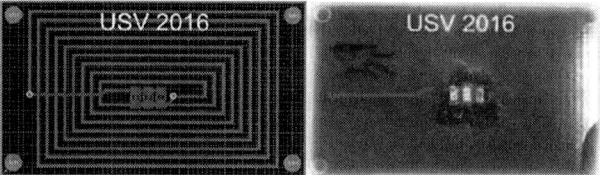

Fig. 5.   Coupling results after simulations performed

Another aspect that must be taken into account is choosing a proper LED for the sensor. Thus, are used several LED's with the same electrical parameters (power supply from 1.6V to 2.4V and current consumption between 2mA to 5mA). The only important parameter, on which the proper LED is chosen, is its luminous intensity. Several tests have been made by placing the LFS at the same distance from the DUT, in the same electromagnetic field, and measuring the luminous intensity using a light meter tester. For the tests were used 8 different LED's that can be observed in Fig. 6. For the final design is chosen the red LED, having a maximum luminous intensity of 170mcd.

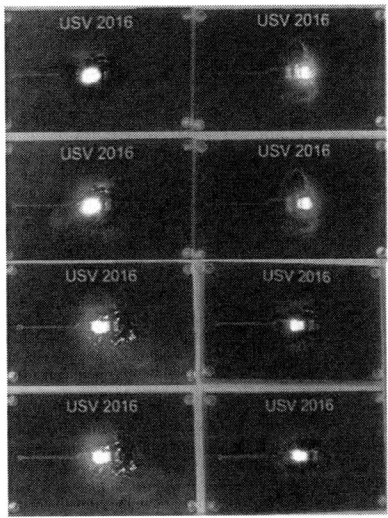

Fig. 6.   Testing the luminous intensity of different LEDs

Next, the LFS are placed in arrays, in a transparent cube with 50cm each side, in order to ensure coverage of the entire near-field that can be generated by a DUT. The distance between two adjacent LFS is kept at 40mm, the distance being measured between two adjacent LFSs centers. For the cube are used 16 columns with 9 rows and on each row being 13 LFS, being a total of 1872 sensors that can collect near-field signal from the DUT.

Fig. 7.   The LFS cube

The functionality of the LFS cube is tested using as DUT an RFID reader, working at 13.56MHz, having a maximum transmitting power of 4W. The reader has an external antenna attached. The RFID antenna must be in the middle of the LFS cube, for 3D visualization of the near-field to be observed. To demonstrate the validity of the proposed measurement method, the tested RFID antenna is simulated, in order to ensure the exact radiation pattern (Fig. 8). The scale of the simulation is made after the minimum power needed for activating the LFS LEDs, which is about 83dBµA or 0.014A/m.

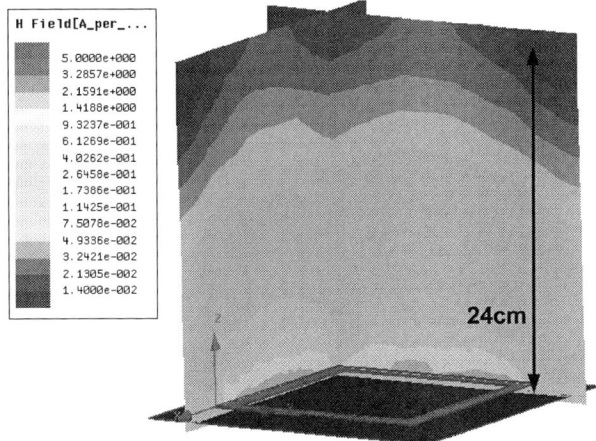

Fig. 8.   Simulation of the DUT antenna

The maximum distance obtained with a single LFS - for the LED to turn on, placed in the middle of the antenna, is about 24cm. If the DUT antenna is placed in the middle of the LFS cube, we obtain the 3D near-field visualization for a 22cm radius around tested antenna (Fig. 9), which will prove the simulation results.

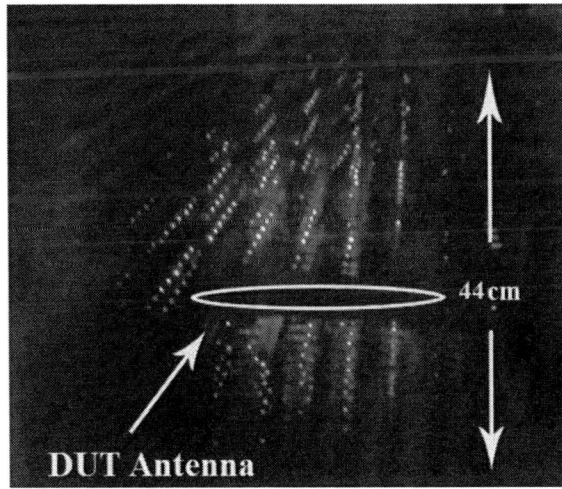

Fig. 9.   Measurements performed with the LFS cube

IV. CONCLUSION

This paper presents a method for visualization of the 3D near-filed generated by a 13.56MHz wireless device antenna, using simple LED Field Sensors placed in arrays. This kind of method is simple to implement and the results are satisfactory. Using Ansoft HFSS the performed simulations have been made to ensure the right components for the LC circuit and to determine the coupling distance between two adjacent LFSs. A prototype has been made and practically tested, confirming the simulation results. This method can be extended with success for different frequencies, obtaining real-time characteristics of the near-field.

REFERENCES

[1] J. Rioult, D. Seetharamdoo, M. Heddebaut, "Novel electromagnetic field measuring instrument with real-time visualization", 2009 IEEE International Symposium on Electromagnetic Compatibility, pp. 133-138, 2009; doi: 10.1109/ISEMC.2009.5284668

[2] J.-C. Bolomey, B.J. Cown, G. Fine, et all, "Rapid near-field antenna testing via arrays of modulated scattering probes", IEEE Transactions on Antennas and Propagation, vol. 36, no. 6, pp. 804-814, 1988, doi: 10.1109/8.1182

[3] P.O. Iversen, P. Garreau, D. Burrell, "Real-time spherical near-field handset antenna measurements", IEEE Antennas and Propagation Magazine, vol. 43, no. 3, pp. 90-94, 2001, doi: 10.1109/74.934906

[4] Y. Liu and B. Ravelo, "Fully Time-domain Scanning of EM Near-Field Radiated by RF Circuits", Progress In Electromagnetics Research B, vol. 57, pp. 21-46, 2014, doi: 10.2528/PIERB13072903

[5] G.-Y. Chen, J.-S. Sun Yd, C.-H. Lin, J.-Y. Yang, "The 3D Far-Field Antenna Measurement Technology for Radiation Efficiency, Mean Effective Gain and Diversity Antenna Operation", 7th International Symposium on Antennas, Propagation & EM Theory, pp. 1-4, 2006, doi: 10.1109/ISAPE.2006.353351

[6] D. Baudry, C. Arcambal, A. Louis, B. Mazari, P. Eudeline, "Applications of the Near-Field Techniques in EMC Investigations", IEEE Transactions on Electromagnetic Compatibility, vol. 49, no. 3, pp. 485-493, 2007, doi: 10.1109/TEMC.2007.902194

[7] A. Capozzoli, C. Curcio, A. Liseno, "Optimized Near Field Antenna Measurements in the Cylindrical Geometry", 9th European Conference on Antennas and Propagation (EuCAP), pp. 1-5, 2015

[8] P.-J. Chiu, D.-C. Tsai, Z.-M. Tsai, "Fast Near-Field Antenna Measurement Technique", European Microwave Conference (EuMC), pp. 594-597, 2015, doi: 10.1109/EuMC.2015.7345833;

[9] *Touch project* (n.d.). from www.nearfield.org, accesed on 23.09.2016.

2016 IEEE 22nd International Symposium for Design and Technology in Electronic Packaging (SIITME)

Electro-thermal Simulation Study of Different Core Shape Planar Transformer

Constantin Ropoteanu, Paul Svasta, Ciprian Ionescu

Center of Technological Electronics and Interconnection Techniques

Politehnica University of Bucharest

Romania

constantin.ropoteanu@cetti.ro

Abstract—**Using Finite Element Method this paper performs analysis over different type of magnetic core shapes from both electromagnetic and thermal perspectives. In this article,the planar transformer modeling and simulations intend to reveal the aspects that help the designer to choose between types of core shapes considering a more efficient use of copper, a better use of board space and an improved thermal dissipation. Using Solidworks application two planar transformer models have been developed taking into account both rectangular and a round centre post. For the FEM analysis 3F3 and 3F4 magnetic material properties were applied to models and the secondary top and bottom winding were used.**

Keywords—planar transformer; thermal simulation, electromagnetic.

I. INTRODUCTION

Planar transformers and inductors are designed in order to obtain significant performance of cost and size, but also optimal behavior for the magnetic cores when temperature is raised [1]. Taking into account that for the necessary work to perform calculations, in order to evaluate a particular core size, turns and copper losses [2] there is no substitute, modeling magnetic components is a plus in the designer approach. From the thermal point a view, simulation made on a previous work containing a virtual planar transformer model returned a non-uniform temperature distribution especially in the copper trace [3]. Coils having ferromagnetic core materials show a variable inductive behavior with the ambient temperature [4].

Although in power supply designs the flux path and distribution in the magnetic core is relative simple [5], using the Finite Element Method a model can be analyzed in order to predict the losses and the distribution of hot spots. The purpose of the present study is to estimate the magnetic components design challenges regarding distribution of losses in two different geometry models. For the two magnetic core models design and thermal simulation Solidworks application was used. Based on several steps of transformer calculations made considering constraints of power and dimensions, two types of magnetic core were modeled, meaning E32/6/20+PLT and ER32/6/25 [6], both types manufactured by Ferroxcube. The present study has two types of investigation, one considering the electromagnetic approach using Ansys, and the second

considering the thermal approach that uses as input data obtained in the first approach in Solidworks application.

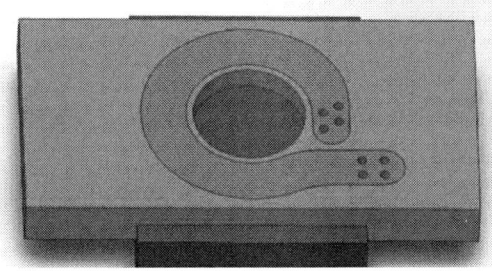

Fig.1: ER model planar transformer section [6]

An important aspect regarding the present analysis is that one of the cores has a round center post that forms the gap in the inner layers of the PCB stack (Figure 1). In the second case the gap closes the magnetic path near the copper that forms the winding (Figure 2).

For the present electro – thermal analysis the primary winding was suppressed due to the fact that the secondary winding, which was modeled from a single turn top and bottom parallel connected, carries the highest component current. In this case the copper thickness is 35μm. Both the electromagnetic and thermal analysis will be made over this secondary winding.

Considering that the skin effect may be an important source of heat in the winding, leading to a more complex thermal management the secondary winding layout was chosen to be as one turn.

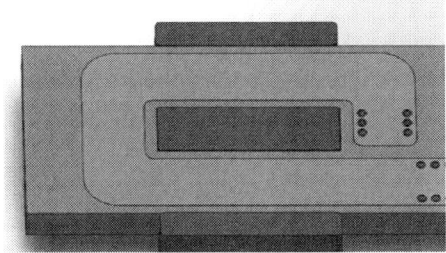

Fig.2: EI model planar transformer section [6]

978-1-5090-4446-7/16 $31.00 © 2016 IEEE   209   20-23 Oct 2016, Oradea, Romania

II. ELECTROMAGNETIC SIMULATION APPROACH

According to [7], for the calculation of maximum flux density it is necessary to admit that not only the copper losses will induce a temperature rise but even the core losses. Due to fact that properties will be applied to the virtual models, core loss density can be approximated by the following equation:

$$P_{core} = C_m \cdot C_T \cdot f^x \cdot B_{peak}^y \qquad (1)$$

where *Cm*, *x*, and *y* are parameters specific for a ferrite material and they have been found by curve fitting of the measured power loss data [7].

Table 1 describes the parameters indicated above in relation with the materials whose properties will be applied to the magnetic core.

TABLE I. CORE LOSS COEFFICIENTS FOR MAGNETIC MATERIAL

Magnetic material			
Properties		3F3	3F4
Relative permeability		4500	1000
Corel loss [W/m³]	Cm	$3.6 \cdot 10^{-9}$	$12 \cdot 10^{-4}$
	X	2.4	1.75
	Y	2.25	2.9

In order to proceed the electromagnetic analysis, the transformer model having ER32/6/25 core was imported from Solidworks application to Ansys. To the mentioned model 3F3 material properties were applied, being necessary for the Ansys solver to receive thermal dependence data that will estimate core losses. For the present study temperature difference is 35°C. Likewise, although in the manufacturer datasheet 3F3 material is recommended up to 500 KHz, data indicated in Table 1 fit for frequency operation between 500 KHz to 1 MHz. The solver adaptive frequency was set 700 KHz and the winding current for the eddy current solution data was 20A solid conductor.

Also an important aspect in transformer losses is the skin effect defined by equation 2:

$$\delta = \frac{1}{\sqrt{\pi \mu_r \mu_0 \sigma f}} \qquad (2)$$

where μ_r is the relative permeability of the conductor, μ_0 is the permeability of free space, σ is conductivity and f is the frequency.

For the second magnetic component model having the core E32/6/20+PLT, input data refers to 3F4 ferrite material. In this case the frequency domain recommended by the manufacturer is between 1 MHz and 2 MHz. It is to be mentioned that according to the specifications [6] improved loss values were

indicated between 50°C and 80°C in the material, for frequency range up to 2 MHz. For the Ansys present investigation purpose the solver frequency was chosen, as previous, 700 KHz. Material properties regarding core loss coefficients were those indicate in Table 1 3F4 column, for the same 20 A solid conductor.

Due to the fact that the relative permeability of the magnetic material varies as a function of temperature, in both cases was necessary to proceed to several iterations in order to determine the behavior of models.

III. THERMAL SIMULATION APPROACH

Thermal simulation tool, part of Solidworks application, is able to solve, based on a specific meshing operation, temperature distribution for different geometry volumes.

In a similar way with Ansys, meshing requirements for surfaces and volumes, Solidworks thermal solver needs to refine the dimension of the solid element in order to make an accurate prediction of surface or edge that will be analyzed. In this context, several simulation steps are needed to establish an appropriate solid element dimension that will describe the core edges or the copper thickness.

In this second step of the analysis, in a similar way with the previous run, to the planar transformer models were applied material properties in respect with thermal investigation.

TABLE II. MATERIAL PROPERTIES APPLIED TO THE THERMAL SOLVER

Material	Mass density (kg/m3)	Thermal Conductivity (W/m·K)	Specific Heat (J/kg·K)
Ferrite 3F3/3F4	4750/4700	4	800
Copper	890	390	390
FR4	1850	2.25	2.9

As it was mentioned before in this paper, electromagnetic simulation results will be data input for the thermal simulator. In this situation the application requires the total power that is converted into heat and also the virtual parts that represent thermal resistances or make bonded surfaces.

The relation between heat transfer leaving a surface at temperature T_s into a surrounding fluid temperature T_f is given by the Newton's law of cooling:

$$Q_{convection} = h\,A\,(T_s - T_f) \qquad (3)$$

where *h* is the heat transfer and has the unit W/m²K.

For both transformer models have been defined contact sets which allow the application to solve the thermal equations that describe the transfer between the copper trace and FR4 stack. Also, convection surfaces were defined not only as the exterior PCB but the exposed surfaces of the core. It is to be mentioned that the PCB structure was modeled as one piece, this choice being taken as a healing method for the meshed structure. Several meshing steps were made over the copper traces in order to relieve accurate details.

IV. RESULTS

Electromagnetic analysis performed over the two models relieved obvious differences not only on the magnetic flux dispersion on the core but in the current density distribution in the winding. The ER core transformer model was simulated with an adaptive frequency of 700 KHz for 20A current. Due to the eddy effects induced by the fringing flux when the magnetic path is closing in the gap, the solver returned in this case 20.85A.

In figure 3 can be observed that for the round core, current density takes significant values near the core center and its appearance is almost the same for both top and bottom layers.

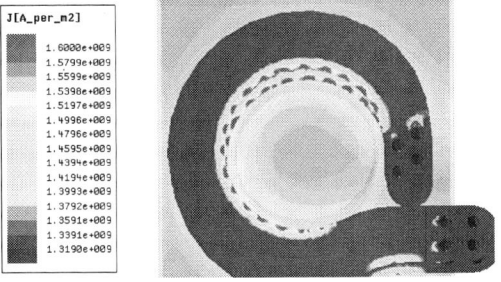

Fig.3: Current density distribution in the ER transformer secondary winding

Therewith, the opposite colors describe a current density domain between $1.32 \cdot 10^9$ and $1.6 \cdot 10^9$ A/m^2, this last value having red color in the corresponding chromatic plot. Also, it is possible that the non-uniform red may represent a result of the meshing phase. From Ansys solver 3.76 W is the winding power converted into heat which will represent de heat power in the thermal simulation tool.

In figure 4 the heat distribution in the PCB showed an average of 66ºC which corresponds partially to the area near center of the round core. The highest plotted temperature is an average of 78ºC and is located on the winding terminations and appears to suggest a better winding layout according to solver frequency.

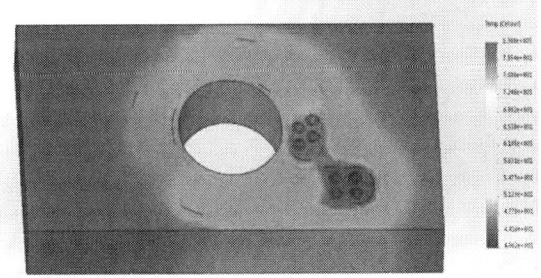

Fig.4: Heat distribution in the ER PCB model

In the case of the second transformer model, having rectangular center core, adaptive passes considering the same 700 KHz and 20 A winding current, went different due to the 3F4 magnetic material properties and copper layout. The electromagnetic analysis performed by achieving several simulation passes in order to observe the model behavior in terms of magnetic flux density during temperature variation. Likewise, the plotted density flux distribution in the core indicates for the rectangular core values near the saturation limits.

The field plotted by the solver in figure 4 illustrates the current density in a more uniform color distribution. It can be observed that a similar representation of density take place on the layout inner. For this case, the highest values are located in the inner corners of copper, current density values going from $1.16 \cdot 10^5$ and $3.04 \cdot 10^8$ A/m^2. The lowest value is located in the winding terminations.

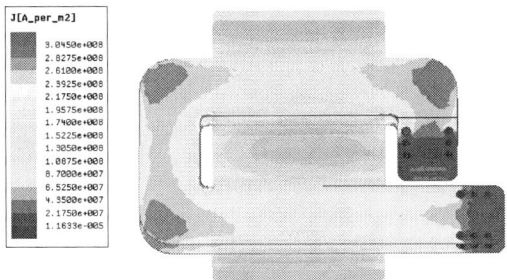

Fig.5: Current density distribution in the EI transformer secondary winding

For this type of core, E+PLT, the magnetic path closes in the gap near the copper turn. Ansys solver simulated 20.12A in the winding incuding eddy effects and 2.24 W power converted into heat.

Figure 6 represents the thermal distribution in the PCB regarding the 3F4 core, with all the material properties applied, simulation frequency passes and core losses coeficients, as they were previous indicated. Nevertheless, the highest temperature value can be observed in the area of winding terminations and is aproximately 70 ºC.

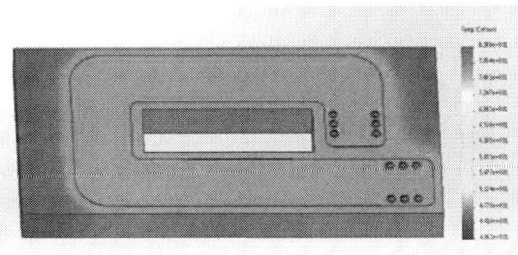

Fig.6: Heat distribution in the EI PCB model

Analysing the simulation results several aspects can be drawn from this approaches. Core models dimensions are partially similar and different from the center core point of view. The calculations conducted for models desing could not achieve identical geometry, power and material properties. Electromagetic and thermal approaches reveal that current density showed by the rounded central core converts more power into heat in comparison with the rectangulare model. One important aspect it is to be mentioned and that is the winding terminations in both situations are heat sources because of the skin efect.

By comparing the two models it can be observed that core and winding losses produce heat in both situations. Moreover, choosing one type of core and another is an aspect that in the design approach takes into account available space, power densities and heat management constraints.

V. CONCLUSIONS

As in the case of electromagnetic simulation, a transformer thermal analysis using Finite Element Analysis is intended to offer an accurate prediction about the thermal behavior of the magnetic component when connected to a load.

Simulations showed a non-uniform temperature distribution, no matter the winding geometry, but offers predictions about losses in one type of core geometry or another. From the two transformer models approach it was observed that magnetic material properties are an important factor that can reduce the design dimensions, considering the working frequency or can increase it by the need to ensure proper heat dissipation.

This is an important aspect when the PCB goes into high power densities. Losses in magnetic components are given by the skin effect, proximity effect but also by the core fabrication. An important aspect is that a finite element method cannot consider all of the physical and manufacturing core loss effects in powder or laminated cores, meaning that in the real PCB solution the designer needs to consider, for example the step gap fringing flux and magnetic material loss variations. The heat distribution in the secondary turn is affected by the current density due to the eddy effects and proper layout must

be taken into consideration in order to reduce the hot areas. Regarding this aspect, the thermal simulation offers important information about the secondary transformer turn termination layout, but the problem becomes more complex once the number of layers is increased, no matter primary or secondary winding. Regarding core losses aspect, due to the fact that in most cases designer does not have access to all measurements made by the manufacturer, laboratory measurements must be taken in order to a solution become validated.

Losses investigation in planar transformer will continue with experimental investigations in order to verify the virtual studies.

REFERENCES

[1] "Designing with Magnetic Cores at High Temperatures", Magnetics, Technical Bulletin No. CG-06, pp.1, 2008.

[2] R. Kankanala, "Full-Bridge Quarter Brick DC/DC Converter Reference Design Using a ds PIC DSC", Application Note AN1369, Microchip, pp.14, 2011.

[3] C.Ropoteanu, P.Svasta, I.Busu, "High-Frequency power loss investigation of a planar ferrite core transformer", 2015 IEEE 21st International Symposium for Design and Technology in Electronic Packaging (SIITME), pp.61-64, October 2015.

[4] B. Rall, H. Zenker, A. Gerfer ,T. Brander,"Trilogy of Magnetics",Wurth Elektronik, pp.45, 2010.

[5] M. Mu, "High Frequency Magnetic Core Loss Study", Ph.D. Thesis, Virginia Polytechnic Institute and State University, pp.30, Febryary 22, 2013.

[6] "Soft Ferrites and Accessories ", Ferroxcube, Data Handbook, pp.530-531 and 359-361, 2013.

[7] "Design of Planar Power Transformers", Application Note, Ferroxcube, pp.4.

Eye Blinking Detection to Perform Selection for an Eye Tracking System Used in Assistive Technology

Alexandru Păsărică, Radu Gabriel Bozomitu,
Vlad Cehan
"Gheorghe Asachi" Technical University,
Carol I No.11 Av., 700506,
Iaşi, Romania
bozomitu@etti.tuiasi.ro, vlcehan@etti.tuiasi.ro

Cristian Rotariu
"Grigore T. Popa" University of Medicine and Pharmacy,
Universitatii 16 Av., 700115,
Iaşi, Romania
cristian.rotariu@umfiasi.ro

Abstract— **The paper presents the analysis of methods used to implement eye blinking selection in an eye tracking system used in assistive technology for patients with neuro-motor disabilities. The system uses for selection the key words technology or ideograms presented on the user's screen. The method used for blinking detection is based on image segmentation using an adaptive local threshold determined using the integral sum image or Bradley method. Results obtained present that the method implemented for blinking detection can be used efficiently in a real time eye tracking system for assistive technology applications.**

Keywords—eye tracking, blinking, integral image sum.

I. INTRODUCTION

This paper presents a method used in order to determine voluntary eye blinking of patients with severe neuro-motor disabilities that utilize an eye tracking system in order to communicate with health care personnel [1]. This system is designed for those patients that cannot communicate by standard means such as oral or verbal communication, written communication, sign and gesture communication. The communication method relies on the key word principle that requires the patient to select an image shown on the user screen. These images are predefined and pertain to common necessities such as food or water, wellbeing or physiological requirements [2].

The main interest of this type of communication system is to use an efficient and robust method for the ideogram selection. This can be done through different methods.

The first method is implemented by maintaining the cursor position in the area corresponding to the ideogram for dwell time interval [3]. This type of selection raises the problem of false selection of ideograms due to the "Midas touch" effect [4] which consists of the random selection of each and every ideogram that is followed by the user's gaze. This is a problem when the user is not sure of the ideogram meaning and requires a longer time to understand and mentally process the action related to that ideogram. The solution is the learning process that the user has to undertake in order to efficiently utilize the system. Also, on the user screen there will be a resting zone where if the cursor is placed there will be no effect or selection.

The second method is done by identifying voluntary eye blinking for ideogram selection. This method relies on the capability to identify voluntary blinking that can be done by using the total number of consecutive frames that have blinking or using a dynamic threshold for selection [5].

Blinking detection is a technical solution with interest in different domains such as assistive technology or automotive industry [1-19]. This has led to a series of articles that present different methods to detect eye blinking, such as: Chau 2005 [6], Bhaskar et al. 2003 [7] or Mange et al. 2015 [8].

Article	Method	Accuracy
Chau 2005	Template matching	95.3%
Bhaskar et al. 2003	Frame differencing	97%
Mange et al. 2015	IR light reflection	-

II. MATERIALS AND METHODS

The hardware components of the system used for image acquisition are represented by a modified USB camera with the resolution 640x480 pixels mounted on a head support. The camera has an IR glass filter placed over the camera objective. This allows us to acquire IR eye images that have common use in eye tracking applications due to the dark pupil technique that represents the fact that the cornea has high reflective properties and this leads to the pupil being the darkest region of the image [9]. The eye image acquisition system used for this application is presented in Fig. 1.

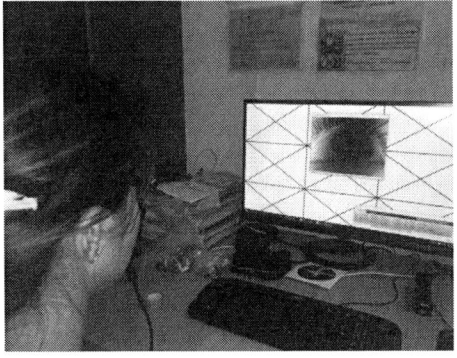

Fig. 1. Image acquisition system using infrared camera

The stages of the eye tracking system implemented and tested are presented in Fig. 2. These can be divided into three different categories [10-11]:

1. IR Image acquisition and filtering;
2. PDA – pupil detection algorithm;
3. Ideogram selection methods.

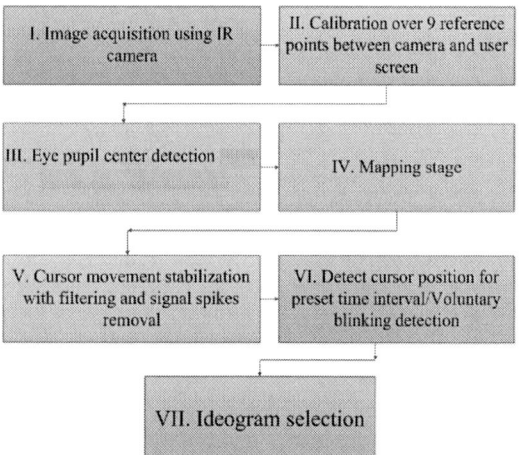

Fig. 2.   Eye tracking system architecture used in assistive technology

The ideogram selection was implemented using two methods. The first method was the selection by maintaining the cursor position a dwell time interval over the desired ideogram. The main disadvantages of this method are represented by false selection when the user is not focused. This type of characteristic is known as the Midas touch problem due to the fact that whichever zone the eye gaze is set upon can lead to an ideogram selection. This is avoided by creating a special zone on the user's screen where the cursor can be placed in order to not make any selections. This makes the system tiresome, meaning that the interval in which the system can be used highly depends on the subject's experience and focusing capability [12]. This makes the ideogram selection using this method difficult to use for patients with neuro-motor disabilities.

The second method uses the selection of ideograms by identifying voluntary blinking which is defined as a longer blinking interval than normal physiological one [13]. This requires us to identify consecutive images that present eye blinking and compare the interval to a threshold value that we determined experimentally. This method requires the patient to learn to use the system in order to determine the blinking interval necessary for ideogram selection. This implies that there will be a higher effort when they first start using the system but this will go down in time, as opposed to the first method that doesn't require a high learning curve but instead requires the subject's full concentration when using the system.

The blink detection algorithm is based on the method which essentially relies on the principle that during the frames that a person blinks the pupil is no longer present and cannot be detected [14]. This method is based on image binarization using an adaptive threshold determined using the integral sum

image or Bradley method. The integral sum I of the input image is the sum of all pixels intensities on the grey scale situated above and to the left of the analyzed pixel [15]. In the proposed algorithm, the computation of an integral image is used, given by following equation:

$$g(x, y) = \sum_{x=x_1}^{x_2} \sum_{y=y_1}^{y_2} I(i, j) \qquad (1)$$

where $g(x, y)$ is the sum of the function for any rectangle with upper left corner (x_1, y_1), and lower right corner (x_2, y_2) from the integral image I of the input image [16].

The method was adapted to eye image segmentation by using a moving window with the dimension that approximates the size of the pupil. We select this as a wxw window ($w=50$ pixels). We determine a local integral sum $S(x,y)$ for each window [17]. The local sum is determined as follows:

$$S(x, y) = \sum_{i=x-c}^{x+c} \sum_{j=y-c}^{y+c} I(i, j) \qquad (2)$$

where $c=(w-1)/2$. The open eye image segmentation compared to the blinking eye image segmentation is presented in Figs. 3 and 4.

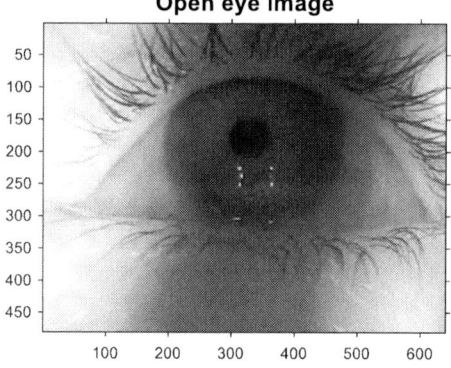

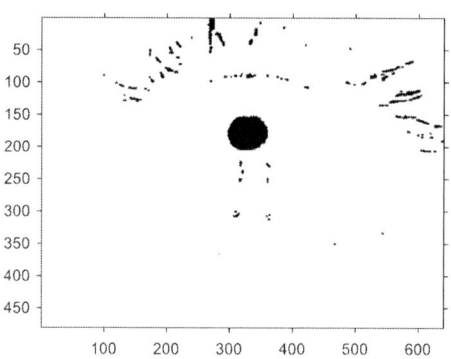

Fig. 3.   Segmentation of an open eye image

2016 IEEE 22nd International Symposium for Design and Technology in Electronic Packaging (SIITME)

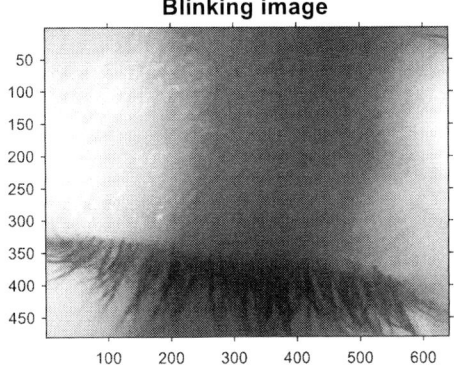

Blinking image

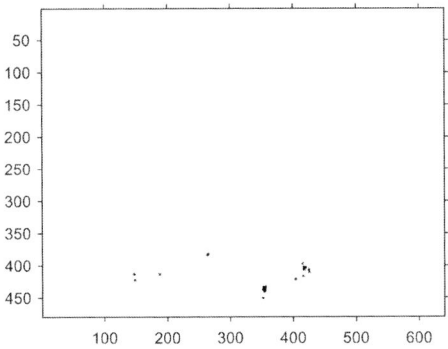

Fig. 4.   Segmentation of a blinking eye image

Figs. 3 and 4 show that the eye blinking image doesn't present a high amount of dark pixels in the segmented image. This can be used in to determine based on a threshold value of percentace of dark pixels in the entire image if the image is of a blinking eye or an open eye. The percentage threshold value was experimentally determined as t=0.3% of the total number of pixels in the image.

III.   EXPERIMENTAL RESULTS

Preliminary results indicated that the first method that relies on maintaining the gaze on an ideogram in order to perform a selection is difficult to use by patients with neuro-motor disabilities. The experimental results are determined only for the second method based on voluntary eye blinking detection. The algorithm's performance is determined by analyzing the number of correctly determined blinks over images acquired using the eye tracking system for a set of subjects in laboratory conditions. The results are determined for static images and the selection method is then tested in real time use conditions. For static images we determine the accuracy with which we identify blinking images from two sets of 500 IR eye images each. The difference between the two image sets is represented by the pupil position: for the

first set the pupil is positioned centrally as if the eye gaze is oriented straight forward; for the second set the eye gaze switches direction from forward to up/down or left/right.

In order to simplify the image type we assign open eye images the value 1 and blinking eye images the value 0, as presented in Figs. 5 and 6.

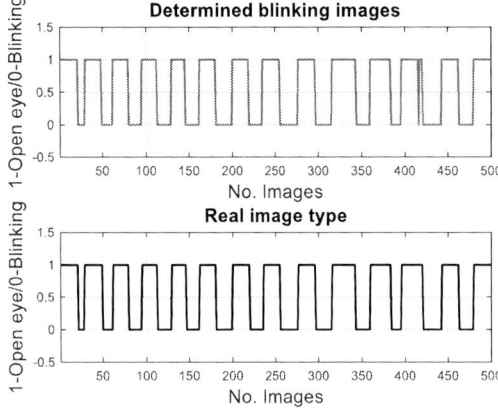

Fig. 5.   Open/blinking eye images identification for the first image set

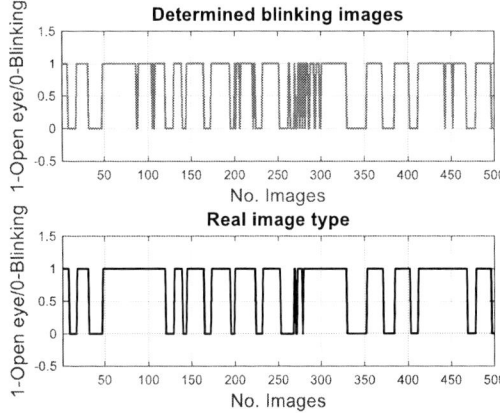

Fig. 6.   Open/blinking eye images identification for the second image set

Based on the comparison result between the determined blinking images and the real image type we classify the images as follows:

- CB – correctly identified blinking images;
- CO – correctly identified open eye images;
- FB – open eye images identified as blinking images;
- FO – blinking eye images identified as open eye images.

Based on the number of images in each class we determine the accuracy (Acc) with which we can determine blinking images by applying the following equation [19]:

$$Acc = \frac{CB}{CB + FB + FO} \times 100 \qquad (3)$$

978-1-5090-4446-7/16 $31.00 © 2016 IEEE   215   20-23 Oct 2016, Oradea, Romania

We obtain for the first image set the accuracy value *Acc=96.72%* and for the second image set the accuracy *Acc=84.04%*, leading to an overall accuracy value of 91.24%.

For real time use we determine if the system can identify voluntary blinking which have a longer duration than involuntary blinking. This aspect requires us to determine the threshold value for the number of frames required by voluntary blinking. This threshold value relies on the processing time for each frame that has an average duration of 0.04 s for each frame analyzed (approximately 25 frames/s). We establish the frame threshold value at 40 frames of consecutive eye blinking (over 1.5s duration) taking into consideration the fact that involuntary blinking is never longer than 0.4s (10 frames).

Based on the previous observation, we determine the ideogram selection time (*IST*) which is described by the following equation:

$$IST = RT + PF \times PT \qquad (4)$$

where *RT* is the response time that is specific to each user and represents the time required for the user to position the cursor over the desired ideogram, *PF* is the number of frames that have to be processed in order to determine voluntary blinking (40 frames as previously mentioned) and *PT* which is the average processing time for each frame. Based on the ideogram selection time (*IST*) we determine the maximum number of selections/minute. This value varies from user to user based on the degree of experience in using the system. We obtained a time interval for the *IST* between 1.68-1.72 s, which determines an average of 35 ideogram selections/minute. These results were obtained by a user that has a high experience level with using the eye tracking system.

IV. CONCLUSIONS

Ideogram selection is one of the software components of the eye tracking system implemented for people with neuro-motor disabilities. Based on the preliminary results we determined that ideogram selection implemented by maintaining the eye gaze over the desired ideogram for a dwell time interval is not a method that can be used for real time applications for patients with neuro-motor disabilities due to the level of focus that was required from the user. The second method implemented that is based on eye blinking detection is more user-friendly even though it requires a higher learning curve.

This method is only influenced by the system's capability to determine voluntary blinking. The results obtained for static images show an accuracy value of 91.04% of correctly identified eye blinking images and the results obtained for real time video analysis using the eye tracking system present a suitable number of selections/minute.

Blinking detection is a technical solution that can also be used to determine the degree of drowsiness of an automobile driver when incorporated in a gaze and percentage of eye closing (PERCLOS) detection system.

ACKNOWLEDGMENT

The work has been carried out within the program Joint Applied Research Projects, funded by the Romanian National Authority for Scientific Research (MEN – UEFISCDI), contract PN-II-PT-PCCA-2013-4-0761, no. 21/2014 (SIACT).

REFERENCES

[1] A.T. Duchowski, "Eye Tracking Methodology: Theory and Practice", Chapter 5: Eye Tracking Techniques, Springer Sciencea and Business Media, pp. 51-59, 2007;

[2] M. Cagliari, M. Godi, S. Guglielmetti, F. Franchignoni, A. Nardore, "Eye Tracking Communication Devices in Amyotrophic Lateral Sclerosis: impact on Disability and Quality of Life", Amyotrophic Lateral Sclerosis and Frontotemporal Degeneration, vol. 14, no.7, pp. 546-552, 2013;

[3] G. Andrienko, N. Andrienko, M. Burch, D. Weiskopf, "Visual analyrics methodology for eye movement studies", Transactions on Visualization and Computer Graphics, vol. 18, no. 12, pp. 2889-2898, 2012;

[4] B.B. Velichkovsky, M.A. Rumyantsev, M.A. Morozov, "New Solution to the Midas Touch Problem: Identification of Visual Commands Via Extraction of Focal Fixations", Procedia Computer Science, vol. 39, pp. 75-82, 2014;

[5] H. Sato, K. Abe, S. Ohi, M. Ohyama, "Automatic Classification Between Involuntary and Two Types of Voluntary Blinks based on an Image Analysis", International Conference on Human-Computer Interaction, Springer International Publishing, pp. 140-149, 2015;

[6] M. Chau, M. Betke, "Real time eye tracking and blink detection with USB Cameras, Boston University Computer Science Technical Reports", no. 12, pp. 1-10, 2005;

[7] T.N. Bhaskar, F.T. Keat, S. Ranganath, Y.V. Venkatesh, "Blink detection and eye tracking for eye localization", Convergent Technologies for the Asia-Pacific Region, vol. 2, pp. 821-824, 2003;

[8] A.A. Mange, A.V. Choudhari. S. Prasad, "Gaze and Blinking Human Machine Interaction System", IEEE International Conference on Computational Inteligence and Computing Research, pp. 1-4, 2015;

[9] N. Cherabit, F.Z. Chelali, A. Djeradi, "Circular Hough Transform for Iris localization", Science & Technology, vol. 2, no.5, pp.114-121, 2012;

[10] R. G. Bozomitu, A. Pasarica, V.Cehan, R.G. Lupu, C. Rotariu, E. Coca, "Eye Tracking System Implementation based on Circular Hough Transform Algorithm", IEEE International Conference on E-Health and Bioengineering (EHB), Iasi, Romania, 2015;

[11] Radu Gabriel Bozomitu, „Tehnici de comunicare prin detecția privirii utilizate în tehnologiile asistive", ISBN 978-973-621-458-5, Editura Politehnium, Iaşi, 2016;

[12] A. Pasarica, R.G. Bozomitu, V. Cehan, R.G. Lupu, C. Rotariu, "Pupil detection algorithms for eye tracking applications", 21st SIITME 2015, pp. 161-164, 2015;

[13] F. VanderWerf, P. Brassinga, D. Reits, M. Aramideh, B.O. de Visser, "Eyelid Movements: Behaviour Studies of Blinking in Humans under Different Stimulus Conditions", Journal of Neurophysiology, vol. 89, no. 5, pp. 2784-2796, 2003;

[14] J. Garry, K. Casey, T. Kling Cole, A. Rgensburg, C. McElroy, E. Schneider, D. Efron, A. Chi, "A Pilot Study of Eye Tracking Devices in Intensive Care", Surgery, vol. 159, no. 3, pp. 938-944, 2016;

[15] D. Bradley, G. Roth, „Adaptive Thresholding using the Integral Image". Journal of Graphics Tools, vol. 12, nr. 2, pp. 13-21, 2007;

[16] T.R. Singh, S. Roy, O. I. Singh, T. Sinam, Kh. M. Singh, "A New local Adaptive Thresholding Technique in Binarization" International Journal of Computer Science, vol. 8, Issue 6, no. 2, pp. 271-277, 2011;

[17] K.G. Derpanis, "Integral Image Based Representations", Department of Computer Science and Engineering, York University Paper, vol. 1, no. 2, pp. 1-6, 2007;

[18] A. Drumea, and R. Dobre, "Clicks counting methods for a scope knob", Hidraulica, vol. 4, pp. 79-84, October 2013;

[19] B. Rosner, "Fundamentals of Biostatistics", Nelson Education, 2015.

2016 IEEE 22nd International Symposium for Design and Technology in Electronic Packaging (SIITME)

VIBROMOD – An Electronic Equipment for Data Vibration Measurement and Analysis

I. Nacu, L. Luca, and N. Roman

Teamnet Engineering SRL
Automation Division
Bucharest, Romania

D. Aiordachioaie

Electronics and Telecommunications Department
Dunarea de Jos University of Galati
Galati, Romania

Abstract—**An electronic equipment was build, to measures and process vibration signals. The low level hardware contains Programmable Logic Controllers (PLC), and measures the vibration signals and makes some signal processing of low level complexity. The upper levels of the equipment use advanced signal processing hardware. The paper focuses on PLC part, hardware and software, and also describes the main signal processing methods. The advantages of the equipment, comparing with other available commercial solutions from market, are: the easy integration in the structure of other industrial control processes based on PLCs; the rapid implementation of advanced and - possible distributed - signal processing algorithms for change detection and diagnosis (CDD) purposes, with high level programming languages, e.g. Java; adaptability to various processes and CDD methods for testing.**

Keywords—*Electronic circuits, system, vibration, signal measurement, data analysis*

I. INTRODUCTION

There is a major activity for active maintenance of industrial processes and equipment, by detecting the change in the behavior of the processes, as effect of incipient faults in various elements of machineries. A concurrent engineering paradigm is necessary to solve change detection and diagnosis problem (CDD), with knowledge and experience from many engineering fields, [1]. Vibration signals are of great importance in solving CDD problem, [2].

In the framework of a research project, [3], an electronic equipment for vibration measurements and data processing was build. It is called VIBROMOD, with the structure presented in Fig.1. There are three processing levels: a low level, based on PLCs, for simple computation tasks, e.g. computation of some statistical moments; the level two for advanced signal processing algorithms; level III, for management, diagnosis and longtime data storage.

Paper [4] describes VIBROMOD from system engineering point of view. The present paper focuses on electronic and signal processing engineering solutions. The algorithms implemented in VIBROMOD must use a non-Matlab programming language. The solution based on PLCs for raw signal processing, i.e. mainly acquisition and filtering, was considered for an easy integration of VIBROMOD equipment in the electronic hardware of the industrial control units, which have many control unit based on PLC and fewer elements for fault detection and diagnosis. The following parameters are computed at this level: mean, median, variance, kurtosis and scenes coefficients. The computed functions are: scaling by

standard deviation, low pass filtering, high pass filtering, band–pass filtering, estimation of probability density function, discrete Fourier transform and power spectral density.

Section II presents the block structure of the equipment. Section III describes the first level, for data acquisition and raw processing mainly, which is made with PLCs. Section IV describes the signal processing tasks, mainly time-frequency analysis, information analysis based on Renyi entropy, and envelope detection based on Hilbert transform.

II. THE STRUCTURE OF THE VIBROMOD

The electronic equipment is organized as in Fig.1. There are three level of data processing. The vibration signals coming from the monitored process are collected and raw processed by the first level, based mainly on Programmable Logic Controllers (PLCs). The reason is that of rapid integration in the electronics of industrial processes.

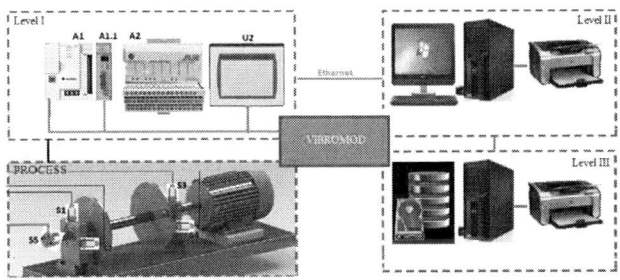

Fig. 1. The physical structure of the equipment

The second level is based on ordinary computer, as hardware, and Java programming language. Some algorithms for CDD based on time-frequency transforms and signal segmentation are implemented. The third level is dedicated for storage and statistic data processing, which is not considered here.

III. THE VIBROMOD – LEVEL I

Description of the PLCs, HMIs, server - hardware and software are presented now.

A. Hardware

With reference of Fig.1, the signals acquisition system is designed for a total of 4 to 6 vibration channels, with synchronous acquisition and conversion, plus an additional channel for rotation speed. The system uses a Programmable Logic Controller (A1) to perform some basic signal processing.

The PLC system provides sequential, process, motion, and drive control together with communications and state-of-the-art I/O. The system is modular, so you can design, build, and modify it efficiently - with significant savings in training and engineering. A simple PLC system consists of a standalone controller and I/O modules in a single chassis.

It is connected to a dedicated Vibration Measurement Module (A2), an intelligent 6-channel monitor that is designed for real time monitoring of direct vibration levels. The module measures and transmits the overall vibration level between selected high and low pass filters for each channel. The module can power and accept input from Integrated Electronics Piezo-Electric (IEPE) accelerometers. It can also accept signals from most standard voltage output measurement devices such as velocity transducers, making it well-suited for general machine monitoring and protection.

The Overall Vibration Modules are a series of intelligent 6-channel monitors designed to cost effectively serve applications for real time monitoring of overall (direct) vibration. Designed as a simple but complete monitoring system in a compact easily installed and easily maintained package, each module measures and reports the overall vibration level between selected high and low pass filters, as well as the bias (gap) voltage per channel.

The Vibration Module may be linked directly to a PLC or other control system via industry standard DeviceNet scanner cards. Host controllers can then scan the modules for data, alarm and relay status information in real-time. Prioritizing messaging insures that changes to any alarm or relay status is immediately communicated to network controllers.

A DeviceNet scanner module (A1.1) acts as an interface between Vibration Measurement Module and the PLC. Data received from the Measurement Module, input data is organized by the scanner module and made available to the controller. Data sent from controller, output data is organized in the scanner module and sent on to slave device. DeviceNet is a network system used in the automation industry, designed to provide a data exchange interface through a single cable from a programmable controller directly to "smart" devices such as sensors, push buttons, motor starters, operator interfaces and drives.

Four general purpose accelerometers (S1-S4) are connected to Vibration Measurement Module, having a 10.2 $mV/(m/s^2)$ sensitivity and 0.8…19 kHz frequency range. Sensor mounting magnets are included with the accelerometers, for easy installation in different measurement points. To make sure that installed sensors provides accurate measures of the signals, it is recommended that shaft speed is greater than 700 rpm.

For rotation speed measurement, an absolute multi-turn magnetic encoder (S5) is connected through a flexible coupling to the rotation shaft. The encoder uses a natural binary code format, with a total resolution of 26 bits and connects to the Measurement Module by DeviceNet network.

The encoder provides a bus interface and is configured as a DeviceNet slave according to DeviceNet specification.

Programmable functions:
- Electronic alignment (PRESET or Number-SET). Electronic alignment may be carried out via DeviceNet or via a pushbutton in the bus connector of the encoder;
- Clockwise or Counterclockwise (CW/CCW) counting direction;
- Scaling function - Setting the single-turn resolution as well as the number of turns. The single-turn resolution is any whole number from 1 to 8192. The number of turns must be 1, 2, 4, 8, 16, 32, 64... up to 8192;
- Up to eight programmable cams;
- Velocity Warning Flags.

Control and monitoring of acquisition process is done through a color display terminal (U2), touch input type, 6 inch display size, and is connected to PLC via Ethernet network cable. It contains different screens for:
- graphical display of signals from the sensors;
- network status view, using significant colors (red – Fault State devices, green – OK devices);
- command soft-buttons for the acquisition control (Start, Pause, Stop, etc.).

B. Programming

The architecture for this system has been described using the 4+1 architecture view model, developed by Philippe Kruchten for "describing the architecture of software-intensive systems, based on the use of multiple, concurrent views", [17].

The views are used to describe the system from the viewpoint of different stakeholders, such as end-users, developers and project managers.

The four views of the model are logical, development, process and physical view. In addition selected use cases or scenarios are used to illustrate the architecture serving as the 'plus one' view.

Hence the model contains 4+1 views:

- The Logical view is concerned with the functionality that the system provides to end-users.
- The Development view illustrates a system from a programmer's perspective and is concerned with software management.
- The Process view deals with the dynamic aspects of the system, explains the system processes and how they communicate, and focuses on the runtime behavior of the system.
- The Physical view depicts the system from a system engineer's point of view. It is concerned with the topology of software components on the physical layer, as well as the physical connections between these components.

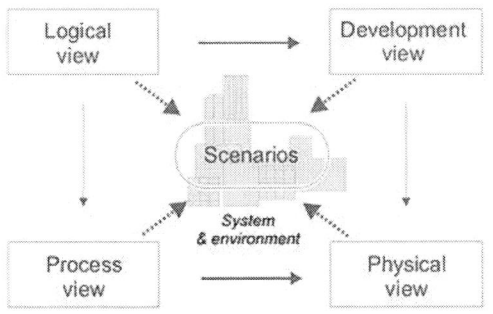

Fig. 2. Architectural View Mode (adapted form [17])

The description of the architecture is illustrated using a small set of use cases, or scenarios, which become a fifth view. The scenarios describe sequences of interactions between objects, and between processes, [17].

Modules represent a code-based way of considering the system. Modules are assigned areas of functional responsibility, and are assigned to teams for implementation. There is less emphasis on how the resulting software manifests itself at runtime. Module structures allow us to answer questions such as: What is the primary functional responsibility assigned to each module? What other software elements is a module allowed to use? What other software does it actually use? What modules are related to other modules by generalization or specialization (i.e., inheritance) relationships.

IV. THE VIBROMOD-LEVEL II

A. Graphic interface

It provides user interaction at level II.: Analysis Charts and Selection parameters. Following are a number of features of the application: Application window: On the Ox we have the time (which is relatively) and on the Oy the measured value for each size at a time. Each color on the graph corresponds to a size (vibration transducer 1, vibration transducer 2, vibration transducer 3, vibration transducer 4 and speed) evidenced the bottom where the legend. We can apply zoom out-in on a specific portion of the graphs to view more closely only a specific portion.

The stand/prototype is designed for accelerated testing by simulating mechanical and electrical defects, induction of disruptive forces or external stresses leading to reduced dynamic equipment reliability. On the stand can simulate various defects such as:

- static and dynamic imbalance
- misalignments
- weakening of the system stiffness
- eccentricity
- resonance phenomenon
- mechanical vibrations
- defects sliding bearings
- defects of rolling bearings – bearings

- bent shaft
- friction rotor
- electric motors faulty gear problems etc.

B. Examples

At level two, advanced signals processing algorithms, usually developed and tested on Matlab environment, are translated for implementation in VIBROMOD, by using Java. The main algorithms are for blind source separation, time-frequency transform, Hilbert transform, Renyi entropies and signal segmentation. All such advanced techniques use a FFT module only. As examples, details of Wigner-Ville time-frequency transform, Hilbert transform and Renyi entropy are presented.

The theory of time-frequency transform is well described in a series of books, e.g. [5], [6], and [7], and various articles, e.g. [8], [9] and [10]. If $x(t)$ is a continuous (possible complex) signal, the Wigner distribution (WD) of the signal $x(t)$ is defined as

$$W_x(t,f) = \int_{-\infty}^{\infty} x\left(t + \frac{\tau}{2}\right) x^*\left(t - \frac{\tau}{2}\right) \cdot e^{-j \cdot 2\pi \cdot f \cdot \tau} d\tau \quad (1)$$

where * denotes complex conjugation. If $X(f)$ is the Fourier transform of $x(t)$, an equivalent definition is

$$W_x(t,f) = \int_{-\infty}^{\infty} X\left(f + \frac{\eta}{2}\right) X^*\left(f - \frac{\eta}{2}\right) e^{j2\pi\tau} d\eta \quad (2)$$

For the case where $x(t)$ is an analytical signal, the Wigner distribution is termed the Wigner-Ville distribution (WVD), [11]. This distribution satisfies a large number of desirable mathematical properties, as described in the specialized literature. In particular, the WVD is always real-valued; it preserves time and frequency shifts and satisfies the marginal properties. The discrete WVD is defined as

$$W(n,m) = \frac{1}{2N} \sum_{k=0}^{N-1} x(kT) \cdot$$
$$x^*((n-k)T) \exp\left(-\frac{j\pi \cdot m \cdot (2k-n)}{N}\right) \quad (3)$$

and it is implemented by the below pseudo-code, which uses ordinary functions, compatibles with Java. Fig. 3 and 4 present a result of this analysis for an impulse modulated vibration signal.

Pseudocode WVT: // WV Transform
// **x** is the signal raw vector;
// only FFT and IFFT algorithms are used;
#1: Compute DFT of **x**;
    X = fft(**x**);
#2: Compute analytic signal of **x**:
    z = analytic(**x**);
#3: Compute WVT, by DFT of **z**;

$$\textbf{WV} = \text{wzt}(\textbf{z});$$

END.

The Renyi entropy is computed as

$$R_x^\alpha = \frac{1}{1-\alpha} \log \left[\int_{-\infty}^{\infty} f^\alpha(x)dx \right] \quad (4)$$

where α is the order of information and f is a probability density function. Third order Renyi information, applied to a time-frequency distribution $C_x(t,f)$, is defined as

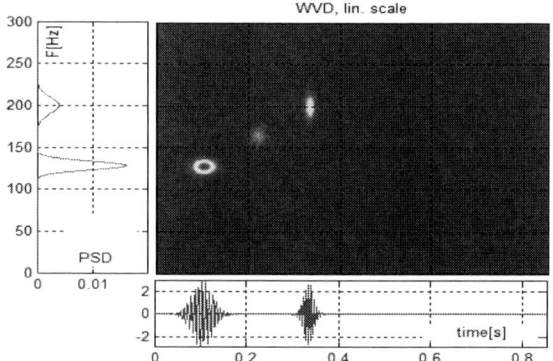

Fig. 3. An example of Wigner-Ville analysis for an impulse modulated vibration signal

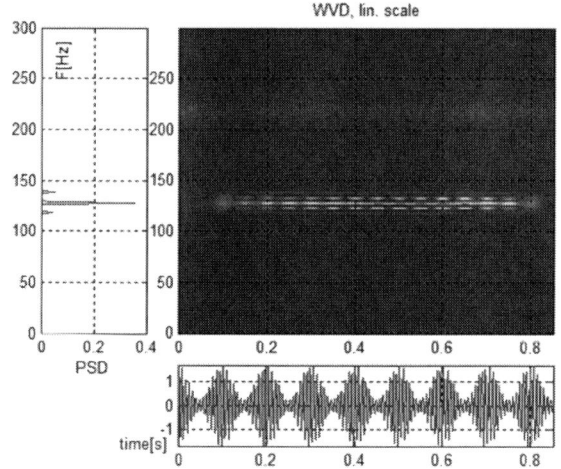

Fig. 4. An example of Wigner-Ville analysis for an amplitude modulated signal

$$R_C^\alpha = -\frac{1}{2} \log \left[\int_{-\infty}^{\infty} \int_{-\infty}^{\infty} c_x^3(t,f)dtdf \right] \quad (5)$$

with

$$c_x(t,f) = C_x(t,f) \Bigg/ \int_{-\infty}^{\infty} \int_{-\infty}^{\infty} C_x(t,f)dtdf \quad (6)$$

being the equivalent/associated probability density function. The discrete case, with dT sampling period and dF the resolution on frequency axis, is

$$R_C^\alpha = -\frac{1}{2} \log \left[dT \cdot dF \sum_{i=1}^{N} \sum_{j=1}^{N} c_x^3(i,j) \right] \quad (7)$$

and

$$c_x(i,j) = C_x(i,j) \Bigg/ \left[dT \cdot dF \sum_{i=1}^{N} \sum_{j=1}^{N} C_x(i,j) \right] \quad (8)$$

Some examples and more considerations are available in [12] and [13]. Third order Renyi entropy is used for the estimation of the number of components in multi-component signals.

The second order Renyi entropy of a data record of length N is computed by

$$H_2(\textbf{P}) = -\log \left(\sum_{j=1}^{N} P_j^2 \right) \quad (9)$$

where $P(j)$ are the probabilities of a finite set of independent events from a discrete information source. The computation of the entropy needs the availability of the exact or estimated pdf. (The probabilities set in the discrete case and the pdf in the continuous case). If we consider sets with N records and observation vectors of size $mx1$, the Renyi entropy estimator is based on Parzen window [14] with Gaussian kernel, [15]:

$$\hat{H}_2(\textbf{x}_m, \sigma) =$$
$$-\log \frac{1}{N} \sum_{n=1}^{N} \left(\frac{1}{N} \sum_{k=1}^{N} \left(G(x_m(n) - x_m(k), 2\sigma^2) \right) \right) \quad (10)$$

could be used for a conversion of the variance in average, very useful in change detection problems. An example is presented in Fig. 5. On the top the input signal and the changing moments for parameters are presented. On the bottom, the evolution of the Renyi entropy is introduced. The width of the processing window has 32 elements. In the output signal, the moments of changes are more clearly defined, and thus simple detections algorithm based on cumulative sum, see e.g. [18], [19] or [20], could run much better.

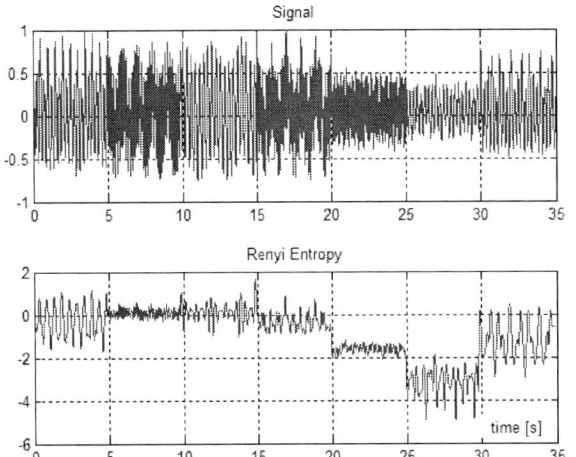

Fig. 5.   An example of second order Renyi entropy

The CUSUM criterion, which is sensitive to the changes of means, can be used also for change detection on processes with variances changes, see e.g. [21].

Renyi entropy could be used also for the estimation of the information content in various type of signal, mon-dimensional or bidimensinal (images). For example, the third order Renyi entropy of time-frequency transforms is used in the estimation of the number of components in the physical, recorded signal (time dependent). This is a very useful feature, especially for industrial applications, where many signals are involved. For details, see e.g. [11] or [22].

For some application, the computation time of this transform (Renyi entropy) could be a problem, but the information extracted by transform cover the above drawback.

The Hilbert transform is used in envelope detection for signal segmentation purposes. It is defined as

$$x_H(t) = \frac{1}{\pi} \int_{-\infty}^{\infty} \frac{x(s)}{t-s} ds \qquad (11)$$

The associated analytic signal is

$$x_A(t) = x(t) + j \cdot x_H(t) = |x_A(t)| \exp(j\phi(t)) \qquad (12)$$

The signal envelope is

$$|x_A(t)| = \sqrt{x^2(t) + x_H^2(t)} \qquad (13)$$

and the instantaneous frequency is

$$\omega(t) = \frac{d}{dt}\phi(t) \qquad (14)$$

The analytic signal is computed via FFT based on method described in [16]. There is a three-step algorithm: (1) calculates the FFT of the input sequence; (2) replaces those FFT coefficients that correspond to negative frequencies with zeros; (3) calculates the inverse FFT of the result. Based on this sequence, we may develop advanced processing algorithms with ordinary programming languages, e.g. Java, and without using special libraries for signal processing.

Fig.6 and Fig. 7 show two examples of envelope detection for an amplitude modulated harmonic signal. The results are very close to the theoretical shape of the positive envelope.

Fig. 8 shows an example of envelope detection for a sinus signal with additive Gaussian noise, plus a zoom from the instantaneous frequency estimation. The frequency of the input signal is 20 Hz and the variance of the noise is 0.01.

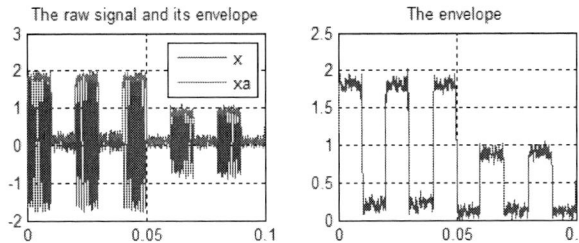

Fig. 6.   An example of envelope detection: rectangular signal

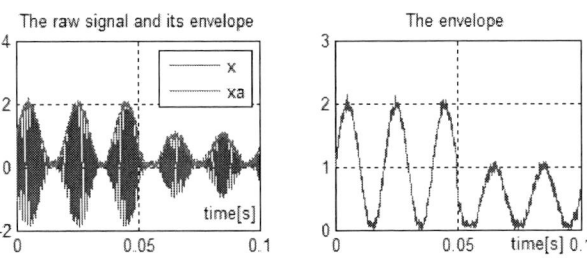

Fig. 7.   An example of envelope detection: exponential decay signal

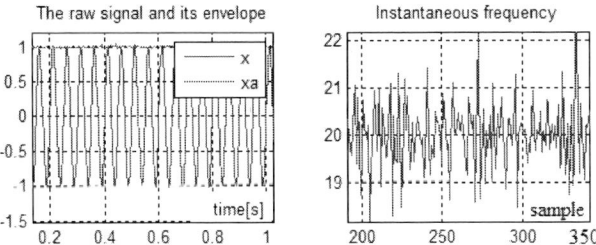

Fig. 8.   An example of frequency estimation

Fig. 9 presents a screen shoot from numerical results concerning the frequency estimation. The estimated value is an average of the last twenty estimated frequency values and depends on the signal to noise ratio.

Real frequency [Hz] = 20

Estimated values [Hz] = 19.9842

Fig. 9.   Numerical example of frequency estimation

CONCLUSION

The objective of the paper was to present some results obtained with new electronic equipment for data acquisition and analysis.

The specific feature of the equipment is the organization on three processing levels, each with modular and open structure.

The first level corresponds to acquisition and simple calculations, and is made with PLCs. This level is built for rapid integration in the electronics of industrial processes. The main modules were presented and briefly described. By this solution, the existent equipment might be upgraded for predictive maintenance and fault detection.

The second level is for advanced signal processing, and contains functions for time-frequency transforms, Hilbert transform, envelope detection, and Renyi entropy. All these functions are used in signal segmentation for change detection purposes. It use Java for computation and it is implemented on a separate computer.

The input data comes from the previous level. This level is in change with change detection and diagnosis. Some examples of time-frequency analysis and envelope detection were presented.

The last level, the number three, is for large time periods, for storage and statistics, and is not considered here.

ACKNOWLEDGMENT

The authors thank the Executive Agency for Higher Education, Research, Development and Innovation Funding (UEFISCDI) for its support under Grant PN-II-PT-PCCA-2013-4-0044 (VIBROCHANGE), and Contract 224/2014.

REFERENCES

[1] R. Isermann, "Supervision, fault-detection and fault-diagnosis methods - An introduction", Control Engineering Practice, pp. 639-652, 1997.

[2] H. Yang, J. Mathew, L. Ma, "Vibration feature extraction techniques for fault diagnosis of rotating machinery: a literature survey", Proc. of Asia-Pacific Vibration Conf., Gold Coast, Australia, 2003;

[3] The VIBROCHANGE project, "Experimental model for change det. and diag. of vibrational proc. using advanced measuring and analysis techn. model-based",at www.etc.ugal.ro/ VIBROCHANGE, 2015.

[4] D. Aiordachioaie and Th. D. Popescu, "VIBROMOD – An experimental model for change detection and diagnosis problems", (IMEKO-2016), Milan, Italy, pp. 317-322, June, 2016.

[5] Ronald L. Allen, Duncan W. Mills, Signal Analysis Time, Frequency, Scale, and Structure, IEEE Press, 2004.

[6] S.B. Damelin, and W. Miller , The Mathematics of Signal Processing, Cambridge University Press, 2012.

[7] R. Carmona, W.L. Hwang and B. Torresani, Practical Time-Frequency Analysis Wavelet And Gabor Transforms, Academic Press, 1998.

[8] L. Cohen, "Time-Frequency Distributions - A Review.", Proceedings of The IEEE, vol. 77, no 7 , 1989.

[9] W.J. Staszewski and A.N. Robertson, "Time-frequency and time-scale analyses for structural health monitoring", Philosophical Transactions of the Royal Society, vol. 365, 2007, pp. 449-477.

[10] M. Abed, A. Belouchrani, M. Cheriet, and B. Boashash, "Time-Frequency Distributions Based On Compact Support Kernels: Properties andPerformance Evaluation", IEEE Transactions on Signal Processing, Vol. 60, No. 6, June 2012, pp. 2814-2827.

[11] A.P. Flandrin, P. Gonçalvès, and O. Lemoine, "Time-Frequency Toolbox for Use with MATLAB", CNRS (France) / Rice Univ.(USA), 1995-1996.

[12] Popescu, Th. and D. Aiordachioaie, D., "Signal Segmentation in Time-Frequency Plane using Renyi Entropy - Application in Seismic Signal Processing", 2nd International Conference on Control and Fault-Tolerant Systems, (SysTol-2013), October 9-11, Nice, France, 2013, pp. 312-317.

[13] D. Aiordachioaie, "Signal Segmentation Based on Direct Use of Statistical Moments and Renyi Entropy", 10th International Conference on Electronics, Computer and Computation (ICECCO'13), Istanbul, Turkye, 2013, pp. 359-362.

[14] E. Parzen, "On estimation of a probability function and mode", The Annals of Mathematical Statistics., vol.33, no. 3, 1962, pp. 1065–1076.

[15] D. Erdogmus, et al., "Blind source separation using Renyi's alpha-marginal entropies", Neurocomputing, vol. 49, no. 1, 2002, pp. 25–38.

[16] S.L. Marple, "Computing the Discrete-Time Analytic Signal via FFT", IEEE Transactions on Signal Processing, Vol. 47, 1999, pp. 2600–2603.

[17] P. Kruchten, "Architectural Blueprints—The "4+1" View Model of Software Architecture", IEEE Software, vol. 12, no. 6, November 1995, pp. 42-50, available at https://www.cs.ubc.ca/~gregor /teaching/papers/4+1view-architecture.pdf.

[18] L. Shu, W. Jiang, Z. Wu, "Adaptive CUSUM proc. with Markovian mean estimation", Elsevier, Computational Statistics & Data Analysis (CSDA),vol.52,2008,pp. 4395–4409.

[19] L.Xin, P.L.H. Yu, K. Lam, "An Application of CUSUM Chart on Financial Trading", in Proceedings of the 2013 Ninth International Conference on Computational Intelligence and Security (CIS), 9th Int. Conf. on, pp.178-181.

[20] M. Pastell, H. Madsen, "Application of CUSUM charts to detect lameness in a milking robot", Elsevier, Expert Systems with Applications, vol. 35, 2008, pp. 2032-2040.

[21] D. Aiordachioaie and B. Dumitrascu, "On the Change Detection Methods with Sensitivity at Variance of The Processed Signal", The IEEE 39th International Conference on Telecommunications and Signal Processing (TSP-2016), in Vienna, Austria, June 27-29, 2016, pp. 417-420, 2016.

[22] D. Aiordachioaie, "On Time-Frequency Image Processing for Change Detection Purposes", The 7th Int. Workshop on Soft Computing Applications, 24-26 Aug. 2016 Arad, Romania (SOFA-2016), paper 18.

2016 IEEE 22nd International Symposium for Design and Technology in Electronic Packaging (SIITME)

Thermal Simulation of Traffic Lights in Extreme Weather Conditions

N. Bădălan (Drăghici), P. Svasta, C. Marghescu

University "Politehnica" of Bucharest, Romania, Centre of Technological Electronics and Interconnection Techniques, UPB-
CETTI
E-mail: niculina.badalan@cetti.ro

Abstract— With increasing use of LEDs as light sources for street lighting and traffic lights, snow and ice deposits on the external lens proved to be a problem since LED do not dissipate as much heat as incandescent lamps. There are solutions that address this problem based on thermistors or using dual mode heat transfer loops. This paper presents a traffic light model that uses a Peltier module to heat the external lens. The Peltier module can be used to heat the external lens as well as to cool the LED, thus assuring better operation during frost and snowstorms. The results are presented and analyzed.

Keywords— high power LED, traffic lights, thermal simulation, Peltier module.

I. INTRODUCTION

The lack of visibility of traffic lights and of vehicle headlights is the cause for many accidents that occur in traffic during snowstorms.

In extreme weather such as a blizzard or very low temperatures/frost it is possible for the external lens of traffic lights to be covered with snow or to be frozen so that the signaling element is not sufficiently visible and can no longer accomplish the function for which it was built. To avoid such situations, it is necessary to heat the external lens - to maintain it at a temperature high enough to prevent snow deposits or freezing.

LEDs have many advantages but in order to benefit from them it is necessary to ensure an adequate thermal management since a number of its parameters varies with temperature (e.g. forward voltage, color, luminous flux etc.).

A feature of power LEDs is that the dissipation occurs on a path opposite to the optical path, thus most of the studies on power LEDs endeavor to find solutions for removing dissipated heat from the LED. For traffic lights with high power LEDs the heat dissipated by the LED should be evacuated, but in some weather conditions it could be useful if a significant portion of the heat would be dissipated through the lens to heat through convection the external lens of the traffic light. Since this approach was not implemented, solutions were searched for heating the external lens [1]; one of this solutions uses thermistors – US patent US 20120255942 A1[2].

The solution presented in this paper uses as heat source the hot side of a Peltier module. The Peltier module fulfils a double role: it cools the LED that is mounted on its cold side but it

also heats the external lens by transferring heat from its hot side to the external lens.

This paper presents a model for a lamp that can be employed as a traffic light in harsh weather conditions. It is based on a RGBW type high power LED that can be used for all three traffic lights colors: red, yellow, and green and also for white light for special traffic lights. Depending on the command received it will be powered in a certain way to emit the desired light. Since this is a high power LED it will be necessary to implement a solution for cooling it. The LED used in the simulation is manufactured by Cree and can be seen below.

Fig.1. LED RGBW

If the RGBWs (Red,Green, Blue, White) LED power is P_e=9.5W, and assuming that the LED will have a 30% efficiency it follows that 70 % of the electric power will be converted into heat; thus the dissipated power will be P_d=9.5*70/100=6.65W. If we assume that the environmental temperature T_a is 50°C and the case temperature T_c should not exceed 85°C, then the temperature difference is $dT=T_c-T_a-85°C-50°C-35°C$. The thermal resistance shall be: $R_{th}=dT/P_d$=5.26°C/W. If we infer that the resistance of the interface material is 0.2°C/W it follows that we will need a cooler for the LED with $R_{th\ LED\ cooler}$=5.26°C/W-0.2°C/W=5.06°C/W if we use a passive heat sink [3-15].

In this paper a Peltier module (Fig. 2) was used to cool the LED. A Peltier module operates based on the Peltier effect. The Peltier effect consists in the removal and absorption of heat at the junction of two different conductors (metals or semiconductors), when an electric current flows through. They are generally used for cooling components.

Peltier modules have several advantages: they are silent, robust, can operate in any position for many hours, are reliable, and they can cool below ambient temperature. But the temperature difference between the two sides depends not only on the ambient temperature and humidity but also on the value of the current that flows through.

Fig. 2. Peltier module

The LED will be mounted on the cold side of it and on the hot side the head of a frustum will be mounted that will convey the generated heat to the external lens. That will have thin threads of a material with high thermal conductivity that will uniformly heat the lens but without affecting the lens transparency / role. (Fig.3 and Fig. 4).

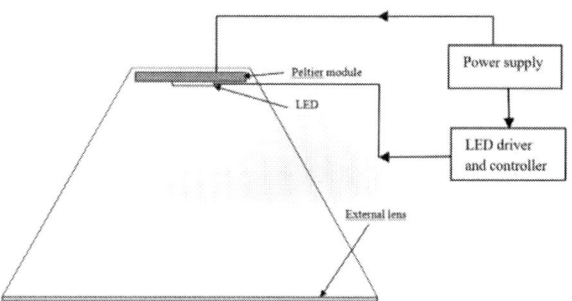

Fig. 3. Bloc diagram of the lamp.

Fig.4. Model of the lamp

II. RESULTS

Simulations were conducted in Solidworks.

Simulations were performed for different values of the ambient temperature and for different values of the supply current of the Peltier module.

Results obtained at an ambient temperature of -30° C, the power dissipated on the 10W LED, 1.2A current through the Peltier module, convective coefficient of 10 W / m²K are shown in Fig. 5.

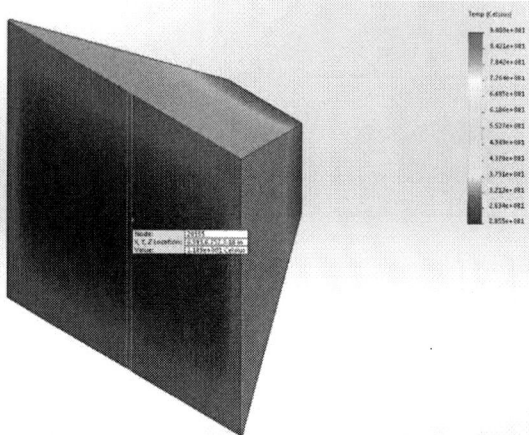

Fig. 5. Thermal map at -30°C and I_{EP}=1.2A.

When the ambient temperature reaches -30°C the lowest temperature on the lens is 23°C, enough to melt snow and ice.

The temperature was modified to -20°C; the other parameters remained unchanged. The results can be seen in Fig.6.

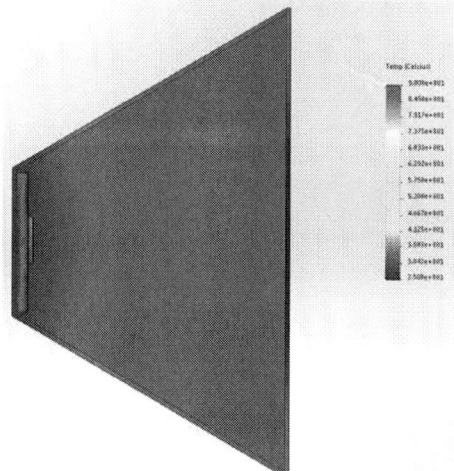

Fig. 6. Cross section on thermal map at -20°C and I_{EP}=1.2A.

When the ambient temperature reaches -20°C the temperature on the lens is 26°C.

The measurements were repeated for 0°C. The thermal map for this temperature can be observed in Fig. 7.

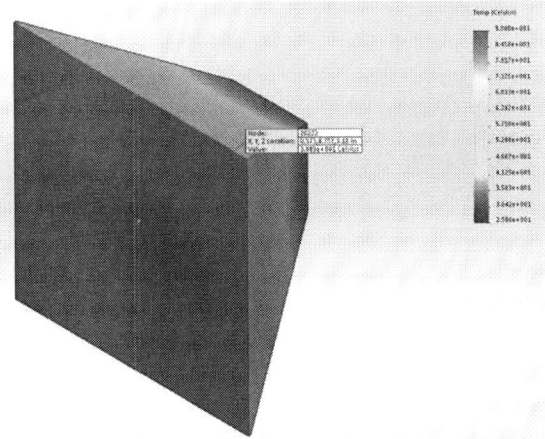

Fig. 7. Thermal map at 0°C and I_{EP}=1.2A.

When the ambient temperature is 0°C the temperature on the lens is 38°C.

The thermal map at 50°C is presented in Fig. 8.

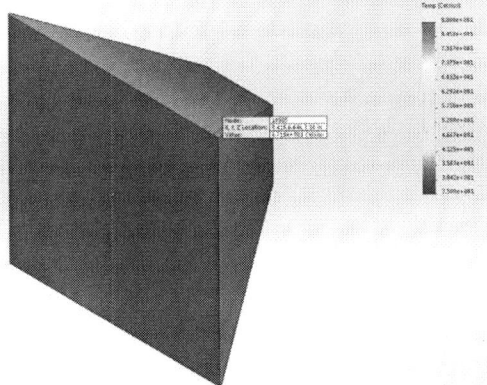

Fig. 8. Thermal map at 50°C and I_{EP}=1.2A.

When the ambient temperature reaches 50°C the temperature on the lens is 67°C - enough so that there are no mechanical problems.

The thermal map at 70°C is presented in Fig. 9.

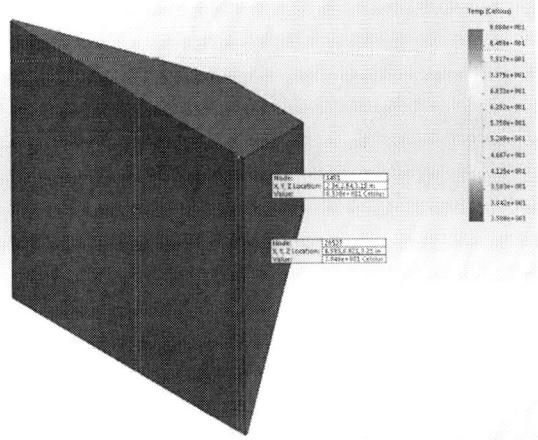

Fig. 9. Thermal map at 70°C and I_{EP}=1.2A.

Note that the lowest temperature on the lens is 78 ° C (in the middle of the lens) and the highest is 83 ° C (on the edge of the lens); this is not a high enough temperature to lead to mechanical problems for the lens. If the temperature obtained on the lens was too high, we might modify the value of the supply current of the Peltier module or we might add an extra radiator to remove this heat.

The value of the supply current of the Peltier module was modified to 1.5A. The simulation was conducted at the following ambient temperatures: 0° C, 70° C, and -30° C respectively. Simulation results are presented in Fig. 10, Fig. 11, and Fig. 12.

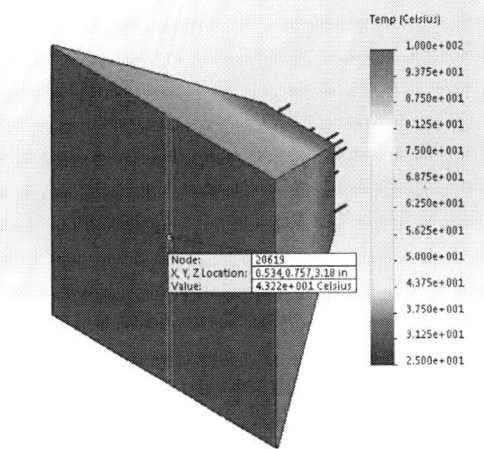

Fig.10. Thermal map at 0°C and I_{EP}=1.5A

At 0°C the temperature on the lens is 43°C.

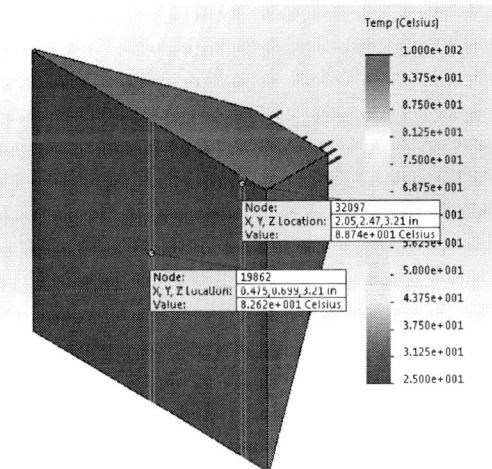

Fig.11. Thermal map at 70°C and I_{EP}=1.5A

At 70 ° C the temperature on the lens is 82 ° C, but this temperature is not high enough to cause mechanical or optical problems for the lens.

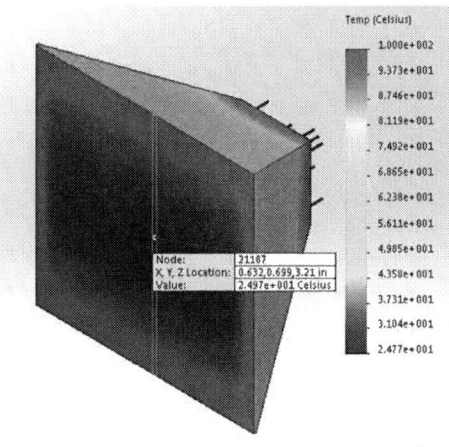

Fig.12. Thermal map at -30°C and I_{EP}=1.5A

At -30°C the temperature on the lens is 24°C.

The simulation did not take into account the temperature, the convection index and the radiation inside the lighting assembly. The simulation did not take into account the forced convection case (wind).

III. Conclusion

Traffic lights that use Peltier modules can solve the problem of lack of visibility during blizzards and frost.

If the ambient temperature is very low and the wind blows at high speed then we can use ITO (Indium tin oxide) (this solution is used for aircrafts) on the lens, but this would incur high costs in the manufacturing process as well as an additional energy expense.

Acknowledgments

This paper was published under the frame of the "Partnerships in priority areas" (PN II) Romanian Research program, developed and supported by MEN-UEFISCDI, SIOPTEF PN-II-PT-PCCA-2011-3.2-899, no. 121/2012 and BLCPL PN-II-PT-PCCA-2013-4-1546, no. 58/2014 projects.

References

[1] patents/US8262263;

[2] patents/US20120255942;

[3] http://www.osram-os.com/Graphics/XPic1/00165234_0.pdf/ Thermal %20Consideration %20of% 20Flash%20LEDs.pdf;

[4] Q. Shan, Q. Dai, "Analysis of thermal properties of GaInN light-emitting diodes and laser diodes," Journal of applied physics 108, 2010;

[5] N. C. Chen, Y. K. Yang, Y. N. Wang, and Y. C. Huang, "Heat flow in AlGaInP/GaAsAlGaInP/GaAs light-emitting diodes, " Appl. Phys. Lett. 90, 181104, 2007;

[6] C. Lasance, A. Poppe, "Thermal Management for LED Applications," Springer Publishers, Chapter 1, pp. 3-72, 2014;

[7] E. F. Schubert and J. K. Kim, "Solid-state light sources getting smart," Science 308(5726), pp. 1274–1278, 2005;

[8] M. Vidrascu and M. Vladescu, "Programmable pulsed current driver for high power LEDs applications," Design and Technology in Electronic Packaging (SIITME), 2014 IEEE 20th International Symposium for, Bucharest, 2014, pp.123-127;

[9] I. Plotog; M. Vladescu, "Power LED efficiency in relation to operating temperature," Proc. SPIE 9258, Advanced Topics in Optoelectronics, Microelectronics, and Nanotechnologies VII, 92582O, 2015;

[10] Y. Xi, E.F.Schubert, "Junction–temperature measurement in GaN ultraviolet light-emitting diodes using diode forward voltage method," Applied Physics Letters, Volume 85, Number 12, September 2004;

[11] M. Weilguni, J. Nicolics, R. Medek, M. Franz, G. Langer and F. Lutschounig "Characterization of the Thermal Impedance of High-Power LED Assembly based on Innovative Printed Circuit Board Technology," IEEE, Proc. of 33th International Spring Seminar on Electronics Technology - ISSE 2010, Warshaw, Poland, May 12th – 16th 2010, D14, pp. 1 – 6, 2010;

[12] http://www.osram-os.com/Graphics/XPic7/00165069_0.pdf/Details %20to%20the%20Assembly%20and%20Solder%20Pad%20Design%20 of%20the%20OSLON,%20OSLON%20SSL%20and%20OSLON%20S quare%20Family.pdf;

[13] P. Mashkov, B. Gyoch, S. Penchev and H. Beloev ,"Method for in-situ power LEDs' junction temperature measurements," Proc. of 33th International Spring Seminar on Electronics Technology - ISSE 2012, Bad Aussee, Austria, 9-13 May 2012, pp. 95-100;

[14] https://ledlight.osram-os.com/wp-content/uploads/2013/01/OSRAM-OS_LED-FUNDAMENTALS_Basics-of-LEDs_v1_09-01-10_SCRIPT.pdf;

[15] C. Ionescu, A. Drumea, N.D. Codreanu, and Al. Vasile, "Thermal Investigations on High Power LED's," Proc. of SPIE, vol. 8411, pp. 84112Y-84112Y-6, November 2012.

One Glass Solution Touch Panel Performance Variation over Temperature Exposure

Hunor Cutlac
Center of Competence Display,
Continental Automotive Romania,
Timisoara, Romania
Center of Technological Electronics and Interconnection
Techniques, "Politehnica" University of Bucharest
hunor.cutlac@continental-corporation.com

Paul Mugur Svasta
Center of Technological Electronics and Interconnection
Techniques, "Politehnica" University of Bucharest
Bucharest, Romania

Abstract—**Touch panel performance is very demanding problem in automotive market and it is requested to be kept during lifetime of the device with which it is sold. This paper presents a problem that may appear when touch panels with a specific structures are placed under temperature variation of automotive environment**

Keywords—OGS structure; ITO rezistivity change with temperature

I. INTRODUCTION (HEADING 1)

Touch controllers that are using capacitance sensing are detecting small changes in the capacitance formed between driving and sensing lines. Values detected on the touch panel by the controller are in range of femto-farads. The controlled capacitance between driving and sensing lines is influenced by any change that may appear in the material of the lines. This paper analyses the temperature influence that occurs with the conductive layer of the one glass solution (OGS) touch panel.[1]

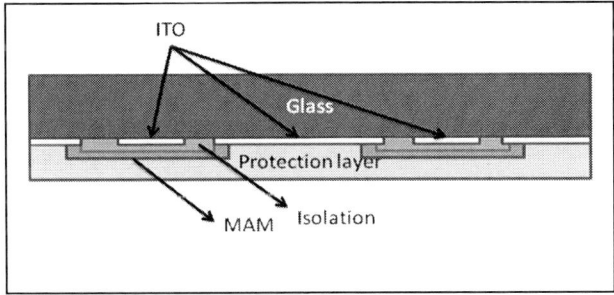

Figure : OGS touch panel stack-up[1]

II. TOUCH PANEL GENERAL DESCRIPTION

Capacitive touch panel is a uniformly distributed capacitor structure over a glass or plastic substrate. This paper takes into consideration only one specific type of touch panel which is One glass solution (OGS). The stack-up of for the OGS is represented in Figure 1. The OGS is type ITO(Indium

Tin Oxide) first. This means that in the process of manufacturing the touch panel the first layer placed is the ITO. Electrically the touch panel contains planar capacitors that are formed between the driving and the sensing lines. Each orthogonal intersection between sensing and driving lines is called node. This node is measured by the touch controller and the changes to its capacitors reported as a touch. A typical structure is the diamond pattern

Touch controller is made in such a way that it measures minute variations of capacitance. These capacitances can be approximated to femto Farads. A touch controller uses two types of measurements for obtaining the touch results. First measurement is a basis from which further calculus is made. This measurement is called a reference as it is used to express the touch panel state in a specific moment. [1]

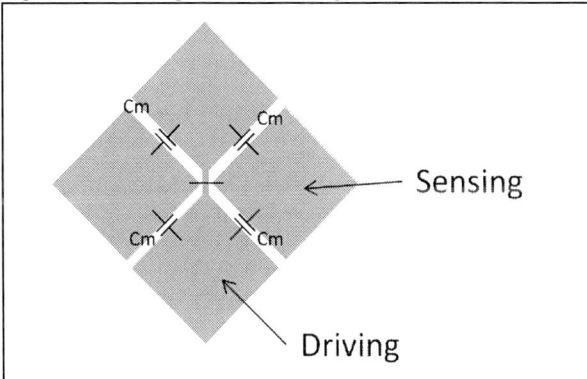

Figure 2 Capacitance formed in a node[1]

The reference measurement is considered as basis initial capacitance which is given by multiple factors like the sensor design, sensor structure, stack-up, distance to display. The variations that appear after initial capacitance measurement that exceed a certain threshold are called touch presses. These values are called delta variation between the initial and current state. The initial capacitance is used as a basis for measurement and it represents the charging time for each capacitor that is forms on the sensor layer.

Projected capacitive touch panel with a one glass solution (OGS) is a 1.5 layer structure that uses ITO as conductive surface. This 1.5 layer is due to the fact that, as presented in the Figure 1, besides the ITO there is structure made off bridges which doesn't cover the full active area. This structures are called MAM(Molybdenum Aluminum Molybdenum) and can be seen in Figure 1.[1]

The ITO is a semiconductor material that has a resistivity between 80Ohm/sq to 100Ohm/sq which is dependent on the thickness to which it is placed on the substrate. The higher the thickness of the layer of ITO the smaller the resistivity. Usually the ITO is sputtered with a 30 to 50 nm thickness in order to obtain minimal optical impact and also resistivity within the range specified.[1][2][3]

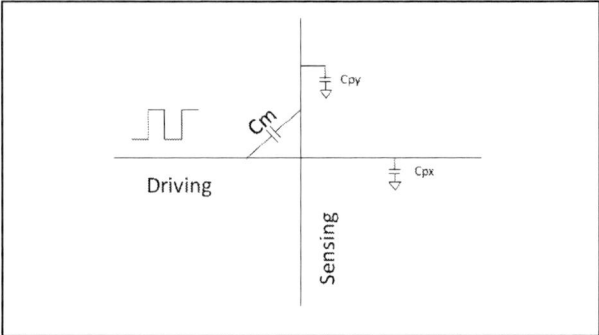

Figure 3 Equivalent circuit for a node

The charge time for each node of the touch panel is influenced by multiple factors such as the wiring structure and the resistance of the sensors conductive structure.

This paper examines only the delta variation that may occur after exposure to temperature changes. It concentrates on changes that consider alteration of charge time which in consequence lead to a performance alteration

III. TOUCH CONTROLLER DESCRIPTION

When the touch controller measures the finger position in mutual capacitance it uses one driving line and all the other sensing lines. This is done progressively for each driving line until all the driving lines are measured. The driving lines function is to induce a signal in the capacitance formed between the orthogonal driving and sensing line. This signal is measured by the sensing line and it estimates what is the charge time for the capacitance formed between the orthogonal lines to reach a certain threshold. This is also illustrated in the Figure 3Any small variation in this charging time, after initial measurement, can be seen as a touch if it exceeds a certain level. The charge time is greatly influenced by two main factors: factors related to design and factors related to electrical conductivity of the materials used.

An equivalent circuit of what the sensing and driving lines look like can be seen in Figure 3. In the same figure there are represented the parasitic capacitances formed between the driving and sensing lines and the ground. These capacitors are influencing the touch performance in obtaining good results and they are considered as noise. In Figure 3 there is also represented the mutual capacitance formed between driving

and sensing (Cm). This mutual capacitance is the useful information that the touch controller is measuring. Also this Cm is influenced by the users finger. In the circuit there is no representation of the resistance that the wires have.

IV. EXPERIMENTAL SET-UP

A. Data acquisition system

Structure of the testing environment can be found in Figure 4. Personal computer is connected to a PCB(Printed Circuit Board) via a USB cable. On the PCB there is a main controller that communicates with the touch controller. The touch controller is connected to touch panel via and FPC (film printed circuit).

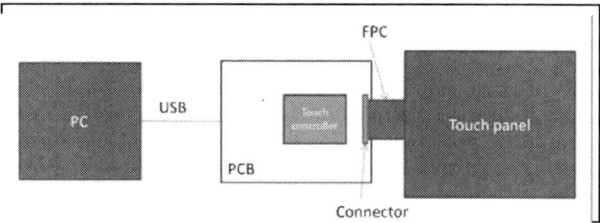

Figure 4 Test equipment

The FPC is on the left side of the touch panel and on the results drawings the position would be on the top. The touch panels including the PCB are inserted in the climatic chamber which controls the environmental temperature. The temperature inside the climatic chamber is measured using a temperature probe. Once the temperature becomes stable, the reference values are read and stored.

B. Measurements

The touch panel was measured at different temperatures that range from 25° to 85° Celsius. Two measurements were done at each temperature in order to check if the variation remains constant over time. The initial state of the touch panel can be seen in Fig.2. Variation of the charge time values is very small at 25° C. In Fig. 3 a degradation of reference values can be seen represented with higher values, hence the charge time of the capacitor. This changes in charge time may lead to variation in the performance of touch panel especially linearity and absolute accuracy errors may increase across the whole surface. The area most affected by this charge time changes is located near the FPC (flexible printed circuit) which connects the touch panel to the PCB (Printed Circuit Board)

C. Results analysis

In Figure 5 results are presented according to temperature progression. The value for 25 degrees Celsius is stored as a reference and that is why it is not represented. The values shown in each graphic are delta variation which are obtain by subtracting between initial charge time values , meaning at room temperature, the charge time values that represent the temperature written below the graph. According to the values there can be seen that at room temperature the impact of the resistivity change is smaller almost insignificant. At temperatures above 40 degrees the change begins on one side of the touch panel. At even higher temperatures above 65 degrees Celsius the differences extend on both left and top edges.

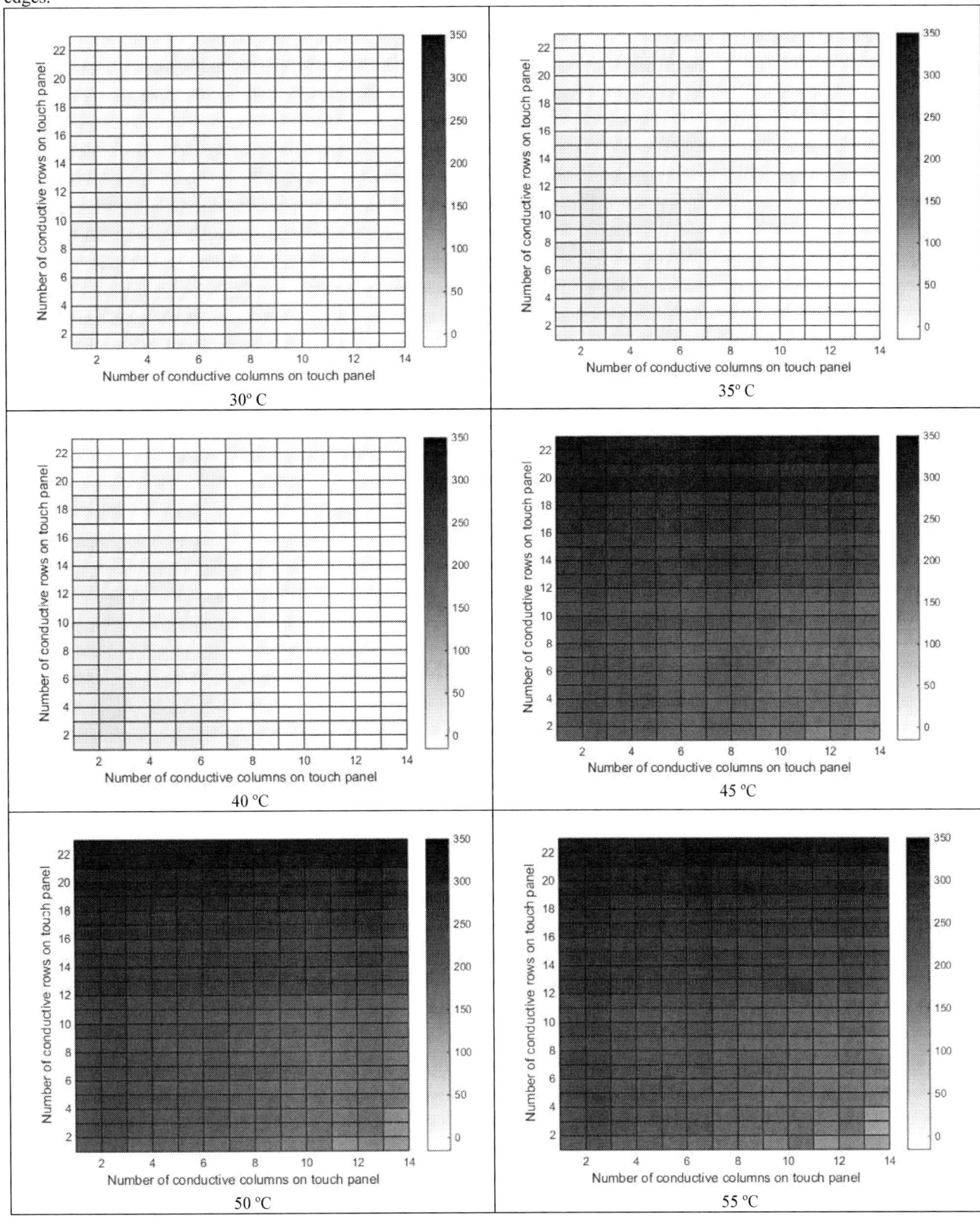

2016 IEEE 22nd International Symposium for Design and Technology in Electronic Packaging (SIITME)

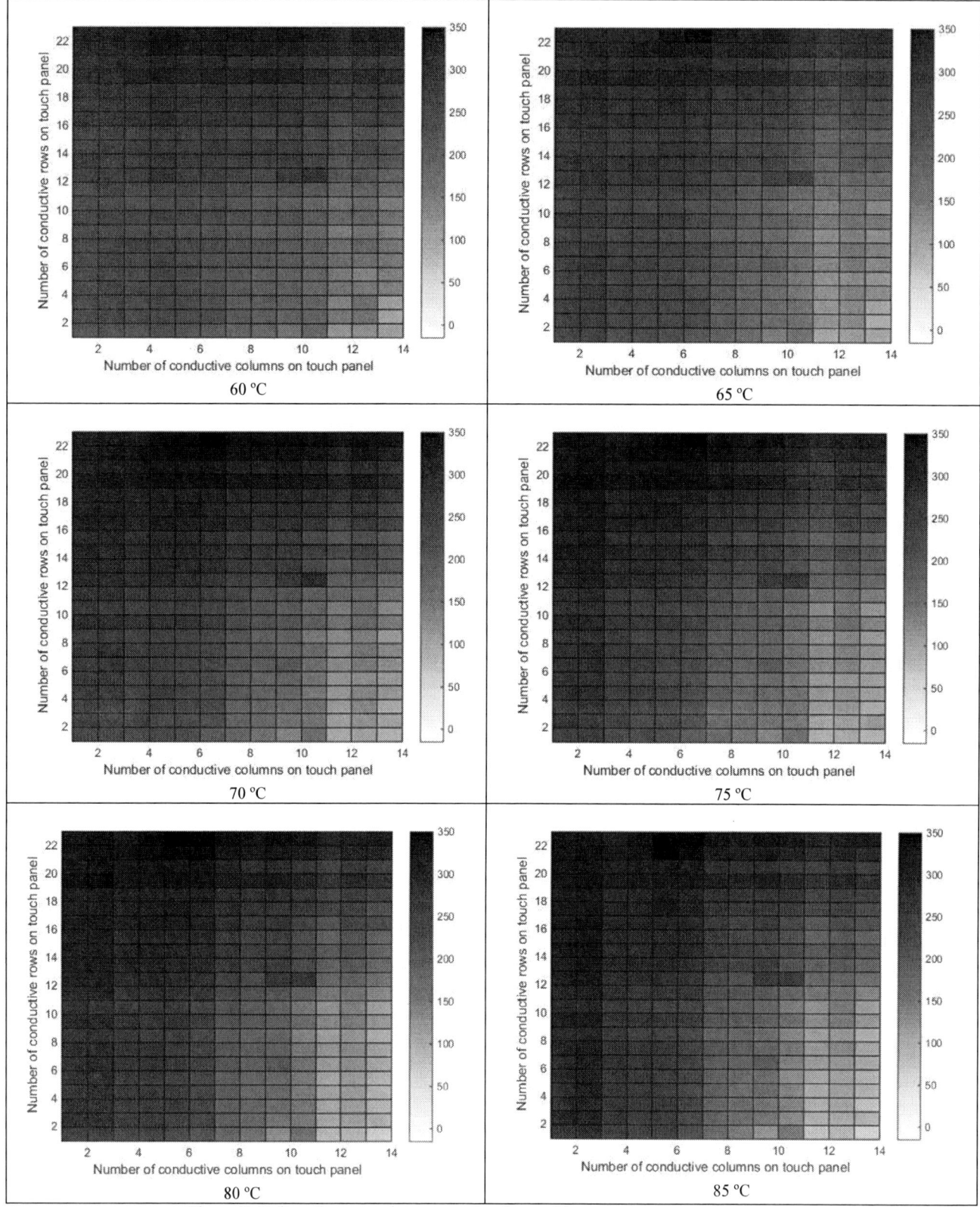

Figure 5: Delta reference values across touch panel in respect to temperature variation

V. Experimental Results

The results show an unequal variation of charge time distribution over the touch panel surface. This means that the capacitors at each node are not having enough time to completely charge. This in terms leads to change that is impacting the performance of finger detection which can lead to the touch panel performance alteration for linearity and accuracy.

VI. Conclusion

The results of this experiment show that the resistivity change of ITO affects the touch panels' basic capacitance. The capacitors closes to the FPC which have greater charging times are the nodes most affected. Problems that can determine these changes in performance are due to resistivity change of ITO, connectors' resistivity change, protection layer uniformity and wiring structure of the touch panel. Further study can be made in order to determine the root cause of this problem and how it can be improved.

Acknowledgment

The authors are grateful to Cosmin Moisa for continuous support. We also thank Bianca Mihăilescu for support regarding measurement data parsing

References

[1] G. Walker, "Fundamentals of Projected-Capacitive Touch Technology," in SID Display Week 2014 INTEL, 2014. K. Elissa, "Title of paper if known," unpublished.

[2] N. Straue, M. Rauscher, M. Dressler, A. Roosen and R. Moreno, "Tape Casting of ITO Green Tapes for Flexible Electroluminescent Lamps.," Journal of the American Ceramic Society, Vols. 684-689, p. 95, 2012

[3] Q. Luo, Indium tin oxide thin film strain gages for use at elevated temperatures, Rhode Island: University of Rhode Island, 2010.

Robust Audio Forensic Software for Recovering Speech Signals Drowned in Loud Music

Robert Alexandru Dobre, Cameila Elisei-Iliescu, Constantin Paleologu, Cristian Negrescu, and Dumitru Stanomir

Telecommunications Department
Politehnica University of Bucharest
Bucharest, Romania
rdobre@elcom.pub.ro

Abstract—**Audio evidence, when accepted by the court, can decide the final verdict in a trial. In order to be evaluated, these materials must be authenticated, but also the intelligibility of the message must be undoubtable. Two main categories of multimedia forensics solve these problems: content authentication and noise reduction. The application presented in this paper is part of the latter category. In order to conceal a conversation, the first action that comes into mind is also the easiest one: turn loud a nearby audio source. Since the most available audio sources play musical materials, if a microphone was placed in the room, it would record the speech signal heavily masked by music. A classical adaptive filtering method could be applied to recover the speech only if the speakers and the musical source remain perfectly still or, in other words, the acoustic environment does not change in time. This ideal situation is not to be found very often in real situations. This paper presents a method for recovering speech signals masked by loud music that is robust to acoustic environment variations. The method is thoroughly described, tested, and compared with a solution based on the recursive least-squares (RLS) adaptive algorithm using a variable forgetting factor.**

Keywords—multimedia forensics; adaptive filters; recursive least-squares (RLS) algorithm

I. INTRODUCTION

If a number of individuals would like to have a secret conversation and they suspect that there is a chance for a microphone to be placed in the room, the handiest way to assure themselves that their discussion will not be recorded would be to turn loud a noise source near them, which usually is a home audio system playing music. The microphone would record the speech and music mixture affected by the acoustic environment. Since the musical signal dominates the mixture, it can be easily identified using a song recognition software like Shazam or SoundHound. Given the recorded signal mixture and the identified song in studio quality, can the forensic engineer process them in such way to recover the speech signal? The scenario can be treated as an adaptive noise reduction problem illustrated in Fig. 1. Most available noise reduction solutions are oriented into extracting a noise profile and then subtracting it from the recording [1]. This paper presents a new approach on the problem. The studio quality signal is to be processed using an adaptive filter [2] in order to make it match its replica present in the mixture as explained forwards.

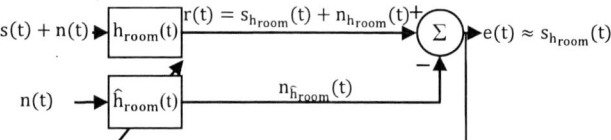

Fig. 1. The block diagram for adaptive noise reduction configuration.

In the previously described situation, the musical signal plays the noise's role and it is denoted with $n(t)$, $s(t)$ being the speech signal. The signal recorded by the microphone, denoted with $r(t)$, will be the speech and music mixture affected by the room's acoustic properties modelled as a finite impulse response (FIR) filter, $h_{room}(t)$ being its impulse response. $e(t)$ is known as an error signal and it is used by the adaptive algorithm to determine the adaptive filter's impulse response. Because the musical signal $n(t)$ and the speech $s(t)$ can be considered uncorrelated, by minimizing the error signal, the output of the adaptive filter can only converge to the recorded musical signal. So the adaptive filter's impulse response, after the convergence, will become an estimate for the room's acoustic properties. In these conditions the error signal will be the speech signal only affected by the room's reverberation, which represents a good estimate for $s(t)$, the signal of interest. A real situation is even worse because the acoustic properties of the room vary in time as the speakers change their positions or make natural movements that modify the path of the acoustic waves. It is shown in the paper that the proposed solution based on the variable forgetting factor recursive least squares (VFF-RLS) [3] adaptive algorithm succeeds in recovering the speech signal even if the acoustic environment changes in time, where other algorithms like classical RLS succeed only in the ideal situation with the acoustic parameters not changing in time.

II. ADAPTIVE FILTERS

Adaptive filters can be described as systems with variable impulse response that is updated in time according to an optimization algorithm [4] [5]. Four signals are characteristic to adaptive filtering: the input signal, the output signal, the desired signal, and the error signal. The output of the adaptive filter is subtracted from the desired signal resulting the error signal. This error signal is used further in order to update the adaptive filter's coefficients. The target is to minimize the power of the error signal, which is equivalent to making the output of the filter match the desired signal. In Fig. 1, the input signal's role will be played by the studio quality musical signal n(t), the desired signal will be the recorded signal r(t), and the error signal e(t) will be the recovered speech signal.

Adaptive filters' coefficients are updated according to an update equation. Many adaptive filtering algorithms exist, which provide different update equations. This paper uses the fast converging RLS and VFF-RLS algorithms for the implementation of the adaptive filter. These will be detailed using the standard notations: x – input signal, d – desired signal, w – the vector containing the adaptive filter's coefficients, e – error signal.

A. The RLS algorithm

Unlike the classical least mean squares (LMS) and normalized LMS (NLMS) algorithms, which use only one sample of the error signal in the update equation, the RLS algorithm takes into account the history of the error signal. Its cost function is

$$c_N(\mathbf{w}_N(n)) = \sum_{l=0}^{n} \lambda^{n-l} |e_N(l,n)|^2 \qquad (1)$$

where N is the length of the adaptive filter, λ is a constant known as forgetting factor and

$$0 < \lambda \leq 1 \qquad (2)$$

$$e_N(l,n) = d(l) - \mathbf{w}_N^T(n)\mathbf{x}_N(l) \qquad (3)$$

$$\mathbf{x}_N(n) = [x(n), x(n-1), \ldots, x(n-N+1)]^T \qquad (4)$$

where $()^T$ is the transposition operator since only real signals were considered.

As it was explained in Section I, the adaptive filter's coefficients are determined by minimizing the cost function with respect to w. The solution is

$$\mathbf{R}_N(n)\mathbf{w}_N(n) = \mathbf{D}_N(n) \qquad (5)$$

where $\mathbf{R}_N$ is the correlation matrix that can be computed with

$$\mathbf{R}_N(n) = \sum_{l=0}^{n} \lambda^{n-l} \mathbf{x}_N(l)\mathbf{x}_N^T(l) \qquad (6)$$

and $\mathbf{D}_N$ is the cross-correlation vector computed as

$$\mathbf{D}_N(n) = \sum_{l=0}^{n} \lambda^{n-l} \mathbf{x}_N(l)d(l) \qquad (7)$$

It can be shown that the vector w can be determined recursively. This characteristic gives the name of the RLS

algorithm. It has very fast convergence but it is much more computationally complex than the classic LMS or NLMS algorithms.

B. The VFF-RLS algorithm

The forgetting factor found in the RLS algorithm is a parameter that can take only the values shown in (2) and is chosen by making a compromise. Let $\mathbf{w}_o$ be the adaptive filter's coefficients that would assure the minimum power for the error signal, known as the optimal coefficients, and m(n)=|w(n)−w_o| be known as the misalignment. Obviously, a small value for the misalignment is desired. A small value for λ would assure fast convergence speed, but also a large misalignment, while a large λ would make the adaptive filter's impulse response to converge to the optimal one (small misalignment), but will make it not very able to follow variations that could occur in $\mathbf{w}_o$ (caused, for example, by changes in the acoustic environment in the presented application). The RLS algorithm gives very good results if the optimal coefficients do not vary in time. An algorithm similar to RLS, but with a variable forgetting factor that could adapt to the changes that could occur in the system would be preferred. It is shown in [3] and [4] that λ can be computed using (8)

$$\lambda(n) = \begin{cases} \min\left(\dfrac{\sigma_q(n)\sigma_v(n)}{\xi + |\sigma_e(n) - \sigma_v(n)|}, \lambda_{max} \right), & \sigma_e(n) \leq \gamma\sigma_v(n) \\[4mm] \lambda_{max}, & \sigma_e(n) > \gamma\sigma_v(n) \end{cases} \qquad (8)$$

where λ_{max} is a preset maximum value for the forgetting factor, ξ is a small positive constant to avoid division by zero and

$$0 < \gamma \leq 1 \qquad (9)$$

$$\sigma_e^2(n) = \left(1 - \frac{1}{KN}\right)\sigma_e^2(n-1) + \left(\frac{1}{KN}\right)e^2(n) \qquad (10)$$

$$\sigma_q^2(n) = \left(1 - \frac{1}{KN}\right)\sigma_q^2(n-1) + \left(\frac{1}{KN}\right)q^2(n) \qquad (11)$$

$$\sigma_v^2(n) = \left(1 - \frac{1}{K_\beta N}\right)\sigma_v^2(n-1) + \left(\frac{1}{K_\beta N}\right)e^2(n) \qquad (12)$$

$$K_\beta > K \geq 1 \qquad (13)$$

$$q(n) = \mathbf{x}^T(n)\mathbf{R}_N^{-1}(n)\mathbf{x}(n) \qquad (14)$$

III. THE DESCRIPTION OF THE PROPOSED SYSTEM

The problematic of the enhancement of speech signals masked by loud music was shortly presented in Section I. The details are presented onwards. The forensic engineer has at his disposal only the signal recorded using the microphone placed in the room. This signal is, in the mentioned speech concealment situation, a mixture of a very loud musical signal and an unnoticeable speech signal. Because a typical room is a reverberant space, the effect of acoustic reflections will be found in the recording. Since the reverberation time is usually short for an office or in spaces in which discussions usually occur and because it can be modeled as a FIR filter, its impulse response will not be very long. It is obvious from Fig. 1 that the room's impulse response must be estimated by the adaptive filter. The fact that the length of the filter to be estimated is generally not very large eases the work of the adaptive algorithm.

The recorded signal is analyzed using a song recognition software like SoundHound, Shazam, or Google Sound Search. These types of programs evolved to the point that they can recognize a song just if one hums it. The harder the people that are speaking try to drown their speech in music, the louder and clearer the recorded music is and the easier the song identification becomes.

After the song identification, the studio quality song must be made available to the engineer. Since many songs are available for free online, it is a great chance that it could be directly downloaded. In the other case it must be bought, but song prices will not be a notable obstacle in the process. The first preprocessing operations can now be applied.

Most voice audio recordings are done using a sample rate equal to 8 kHz, while studio quality music is usually available sampled at 44.1 kHz or 48 kHz. The studio quality song must be decimated accordingly.

Since the source for the musical signal is most likely to be a radio station, it is known that songs played on radio are usually crossfaded. So, the start of the song is most likely to miss from the recording. A pre-alignment of the studio quality song and the recording must be done in order to make the adaptive filter's task easier. Theoretically, the adaptive algorithm can take care of this aspect if the assigned adaptive filter's length is large enough, but this operation would greatly affect the

convergence speed and computational effort. The cross-correlation between the recorded signal and the decimated studio quality music signal can be a good tool to determine the time delay that must be added to one of them in order to obtain the temporal alignment of the musical parts. At this point, the two signals are ready to be processed with the speech enhancement system.

The speech enhancement system (implemented in Simulink) that uses RLS as the adaptive algorithm is presented in Fig. 2. It can be observed that the user interface is very easy to use, the parameters being set by turning virtual knobs. It contains two audio file readers ("From Multimedia File" and "From Multimedia File1"), a band-pass filtering stage that can be bypassed ("BPF1" and "BPF2"), the adaptive filter (in this case "RLS Filter"), an audio file writer ("To Multimedia File") and a block that sends the recovered speech signal to the Matlab's workspace if further processing is necessary.

The tunable band-pass filter stage's role is also to reduce the complexity of the task given to the adaptive filter by eliminating the spectral band that does not contain speech signal. In this way, the adaptive filter will have less useless data to process.

The adaptive filter is the block in which the speech recovery takes place. In Fig. 2, the RLS algorithm is used. The VFF-RLS algorithm was also used in a similar way.

IV. RESULTS

The software for recovering speech signals drowned in musical noise was developed using Simulink and Matlab. In order to determine its effectiveness in the presented situations, it was tested as follows: a reference vocal signal, illustrated in Fig. 3a), was mixed with a musical signal (considered noise) in a way that the signal-to-noise ratio was at most −40 dB. The mixture was filtered using an acoustic impulse response to simulate the room's acoustic properties. Fig. 3b) shows the performance of the two algorithms in terms of following the

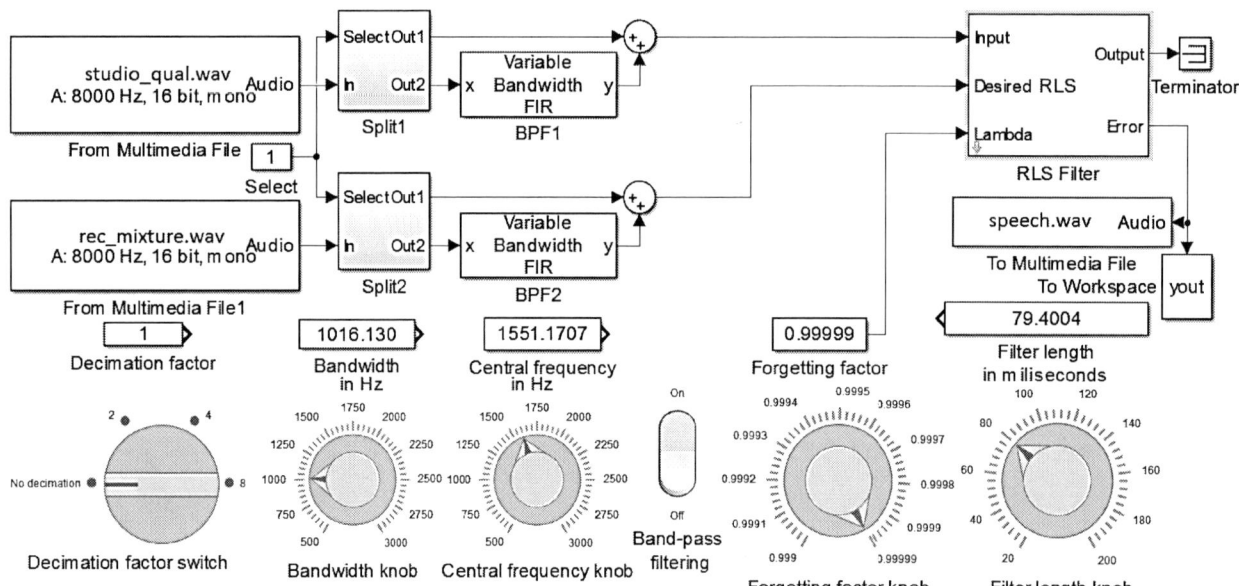

Fig. 2.   The developed speech enhancing system that uses the RLS algorithm.

*2016 IEEE 22nd International Symposium for Design and Technology in Electronic Packaging (**SIITME**)*

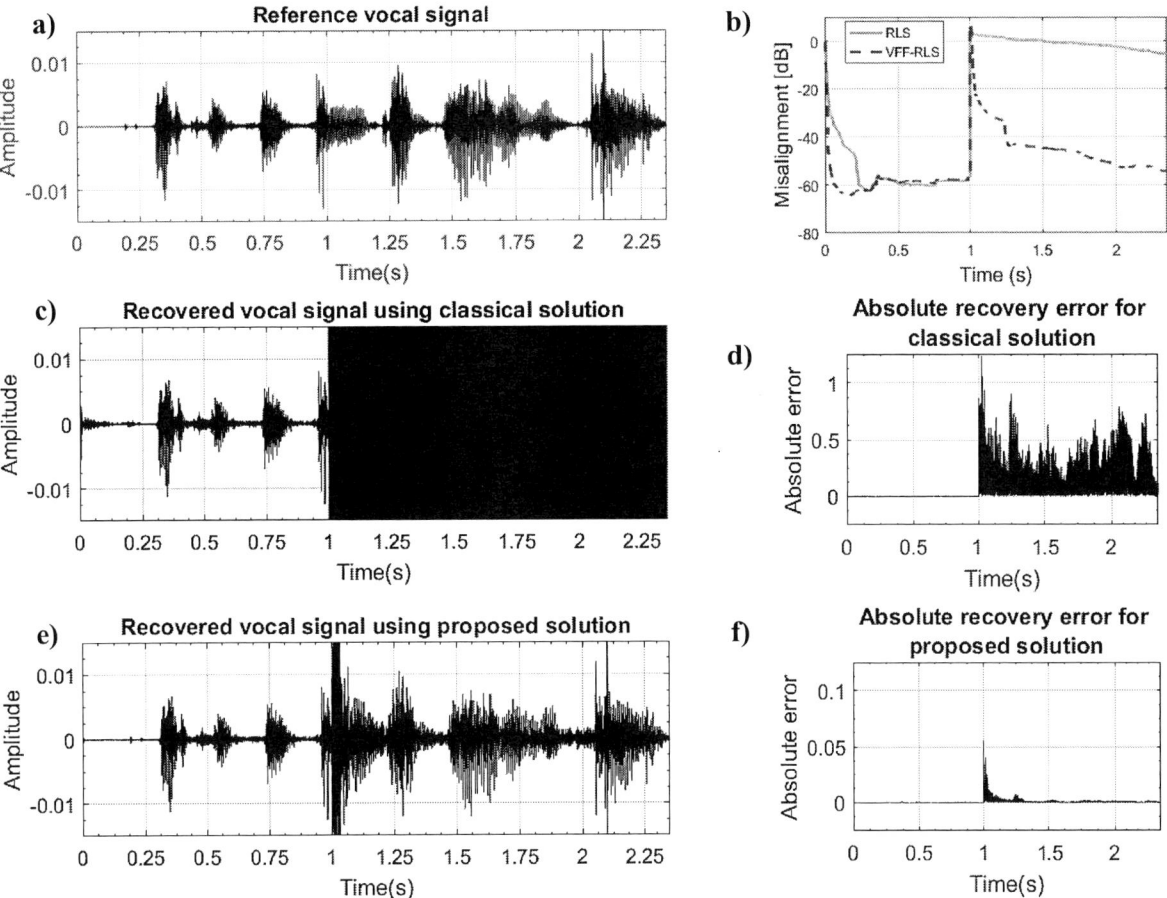

Fig. 3. a) The reference speech signal; b) The variation of the misalignment in time for the two analyzed algorithms; c) The recovered speech signal using the solution based on the RLS algorithm; d) The absolute recovery error for the solution based on the RLS algorithm; e) The recovered speech signal for the proposed solution based on the VFF-RLS; f) The absolute recovery error for the proposed solution based on the VFF-RLS algorithm.

changes of the acoustic parameters (lower values mean better tracking). After one second, the impulse response was changed to simulate a modification in the acoustic environment. It can be observed in Fig. 3c) and Fig. 3d) that the recovered signal using RLS is intelligible only for the part before the acoustic parameters changed (for one second), while by studying Fig. 3e) and Fig. 3f) it can be concluded that the speech signal recovered using the proposed system almost entirely match the reference one. Absolute recovery error means the absolute difference between the reference signal and the recovered signal. The change of acoustic parameters can still be observed, but its effect is annihilated very fast, in at most 40 ms, while the rest of the conversation can be understood. The proposed solution quickly reacts to acoustic environment changes.

ACKNOWLEDGMENT

This work was supported by the UEFISCDI under Grant PN-II-RU-TE-2014-4-1880.

REFERENCES

[1] A. Czyżewski, "Intelligent control of spectral subtraction algorithm for noise removal from audio," *Intelligent Tools for Building a Scientific Information Platform*, Springer Berlin Heidelberg, 2013.

[2] S. Haykin, *Adaptive Filter Theory*, Prentice Hall International, 4th edition, New Jersey, 2002.

[3] C. Paleologu, J. Benesty, and S. Ciochină, "A robust variable forgetting factor recursive least-squares algorithm for system identification," *IEEE Signal Processing Letters*, vol. 15, pp. 597–600, 2008.

[4] C. Paleologu, J. Benesty, and S. Ciochină, "A practical variable forgetting factor recursive least-squares algorithm," in *Proc. ISETC*, 2014, pp. 1–4.

[5] D. A. Visan, I. Lita, and I. B. Cioc, "Temperature control system based on adaptive PID algorithm implemented in FPAA," in *Proc. 34th International Spring Seminar on Electronics Technology (ISSE)*, 2011, pp. 501–504.

2016 IEEE 22nd International Symposium for Design and Technology in Electronic Packaging (SIITME)

Formula Student Single User Race Car Electronic Control

Vlad Lupu, Carmen Gerigan, Petre Lucian Ogrutan

Department of Electronics and Computers
Transilvania University of Brasov
Brasov, Romania
carmen.gerigan@unitbv.ro

Abstract—**The paper presents the implementation of the electronic control for the monopost BS16 built by BlueStreamline team at Transilvania University of Brasov to take part in Formula Student racings in 2016. Two electronic modules have been designed, implemented, tested and integrated in the system: the power module and the steering wheel electronic module. The electronic modules are communicating over CAN bus. The current implementation offers steering wheel interface for the pilot, increased protection functions and increased reliability through the design and verification technologies that have been used and by redesigning the software driving the electronic control.**

Keywords—automotive, electronic control, CAN bus

I.   INTRODUCTION

Nowadays car competitions have developed a lot, so that in addition to other branches of engineering, electronics has become indispensable for a competitive race car. Besides the electronic control of the engine, data acquired by the sensors and electronic actuation can result in both improving the control of the race car in competition and achieving higher reliability and performance. Unlike a normal car, the design of the cockpit of a race car aims to develop an environment in which the pilot can receive easy to see all the information he needs throughout the race (pressure, speed, gear position, etc.), as well as an ergonomic layout of the controls. Pilot position in BS16 monopost resembles a Formula 1 car, such as displays and controls are on the steering wheel. Commands are spread on the bottom and behind the steering wheel closer to the hands, depending on their relevance. The displays are located at the top of the steering wheel for a better view on them. In order to cope with the requirements and conditions of a Formula 1 racing, the design of the electronic modules represents a challenge.

The proposed control structure was conceived according to the principles given in [1]. The emergence and development of electric cars has led to increased complexity of the electronic part of a car. The appearance of electric cars at Formula Student contest [2] confirms the importance of the application presented in this paper. Driver comfort is essential in race car. Paper [3] discloses the concept of a cockpit for a pilot and a recent patent proposes electronic indicators directly on the steering wheel [4]. An electronic assistance system for a Formula Student racing car under the rules SAE (Society of

Automotive Engineers) is proposed in [5], but referred performances are poor. The importance attached for research by automotive companies is proved by the optional course organized by a company in the automotive field for the students at Faculty of Electrical Engineering and Computer Science from Brasov [6].

II.   DESCRIPTION OF THE APPLICATION

Student teams from around the world design, build, test, and race a small-scale formula style racing car. BlueStreamline is the first student team from Romania, which accepted the challenge to design and build a Formula Student single user race car. To face the challenges of a competition of this scale, the single user race car includes a major electronic control component which is improved year by year. Several options were analyzed as interconnection methods [7] but connecting via CAN was considered the best choice. The modules of the electronic control and their interconnection over CAN 2.0 Bus for the BS16 monopost are presented in Fig. 1.

The Power Module and the Steering Wheel Electronic Module have been redesigned and implemented to increase the functionality and to meet higher reliability of the electronic control for the 2016 race car.

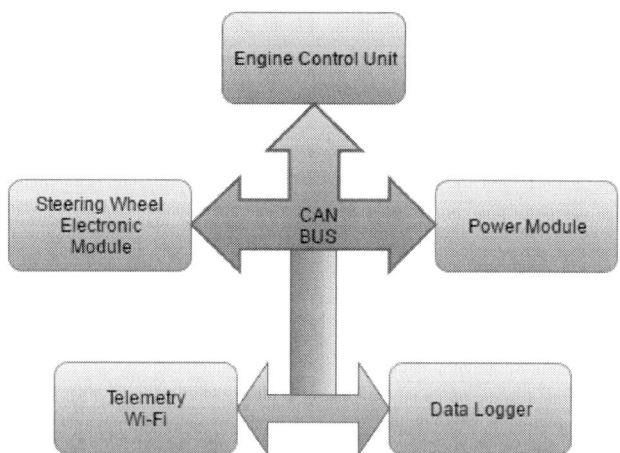

Fig. 1.   Electronic modules of BS16 monopost conntected over CAN 2.0 Bus.

Sponsor: Preh Romania S.R.L.

III. Module design and implementation

The Power Module and the Steering Wheel Electronic Module have been redesigned and implemented to increase the functionality and to meet higher reliability of the electronic control for the 2016 race car.

The Power Module includes monitoring for all consumers and a redesigned PCB to increase resistance to high current and high temperatures. The Steering Wheel Electronic Module is completely new for the current version of the car and it offers a wide range of monitoring and control functions for the driver.

Both modules are built around AT90CAN microcontroller. The electronic circuits, the software and the libraries to drive CAN communication have been redesigned [8], [9]. Both PCBs for the modules have been designed in two layers using Proteus [10].

A. Power Module

The Power module block diagram is presented in Fig. 2.

Power module is designed to supply the steering wheel electronic control module, the ECU, the telemetry module, lambda sensors and the brake lamp. Besides these, it can command the shifter, the clutch actuator and fuel pump automatically after pressing the "Ignition" switch and the cooling fan after reaching a critical temperature. The power module can send the command to the ECU to cut the fuel supply upon gear changes.

The communication with other functional modules is accomplished by the microcontroller over the CAN Bus. The communication with the controlled units is performed over the SPI interface (Serial Peripheral Interconnect) using a shift register to increase the number of units. The clutch actuator is driven using PWM.

B. Steering Wheel Electronic Module

Electronic steering wheel module receives data from the sensors in real time. Data from sensors either are processed, and based on the results certain decisions are taken, or are converted into a proper format to be displayed for the pilot to assist him to evaluate whether the monopost works in an optimal regime.

Fig. 3 presents the block diagram of the electronic steering wheel module.

The microcontroller serves to communicate with the other functional modules, to control the displays, to read and translate the status of switches and buttons, and to process the values indicated by the temperature sensor and by the Hall sensor. The communication with other functional modules is accomplished by the microcontroller over the CAN Bus. The display units (i.e. 7-segment devices, LEDs and the LCD display) are driven by the microcontroller using the SPI interface. In order to accommodate more display devices over a single SPI interface, a shift register array was used.

The information provided by the Hall sensor is used to drive the clutch actuator.

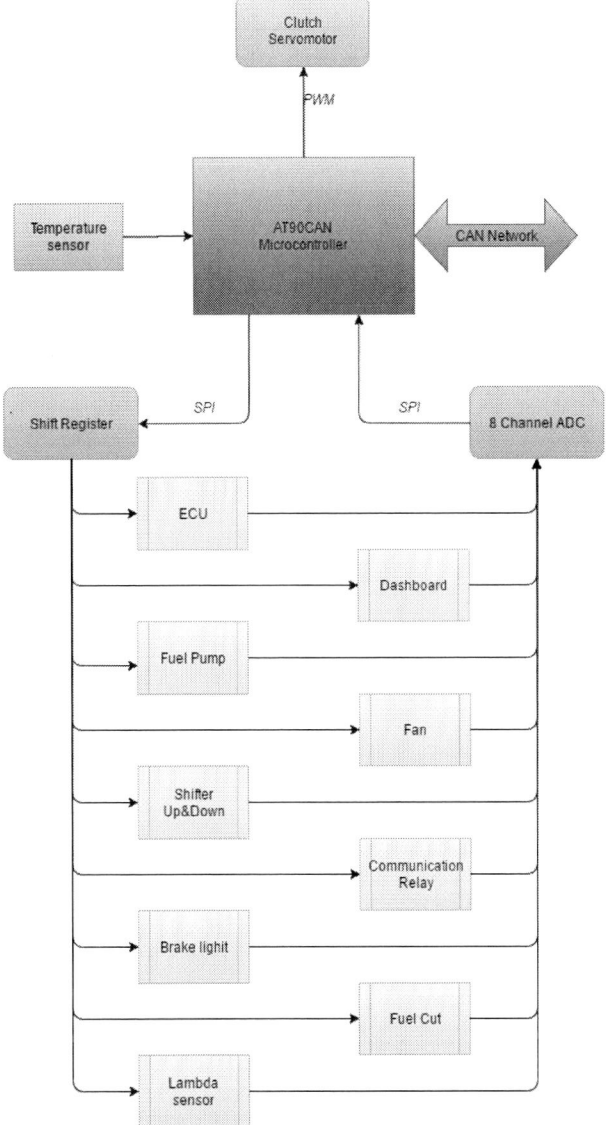

Fig. 2. Power module – block diagram.

The 7-segment displays are used to show the temperature of the engine, the speed and the gear position. The LEDs are representing graphically the engine speed. The LCD display is used to list a menu where the pilot can choose through rotary switches details of other information or is allowed to set options such as enable/disable the autopilot, adjust the volume of the communication, adjust the sensitivity of the paddles, etc.

All incoming information to the steering wheel electronic control module concerning the sensors on the engine come from ECU. ECU is providing together with the power module optimal operation of the internal combustion engine.

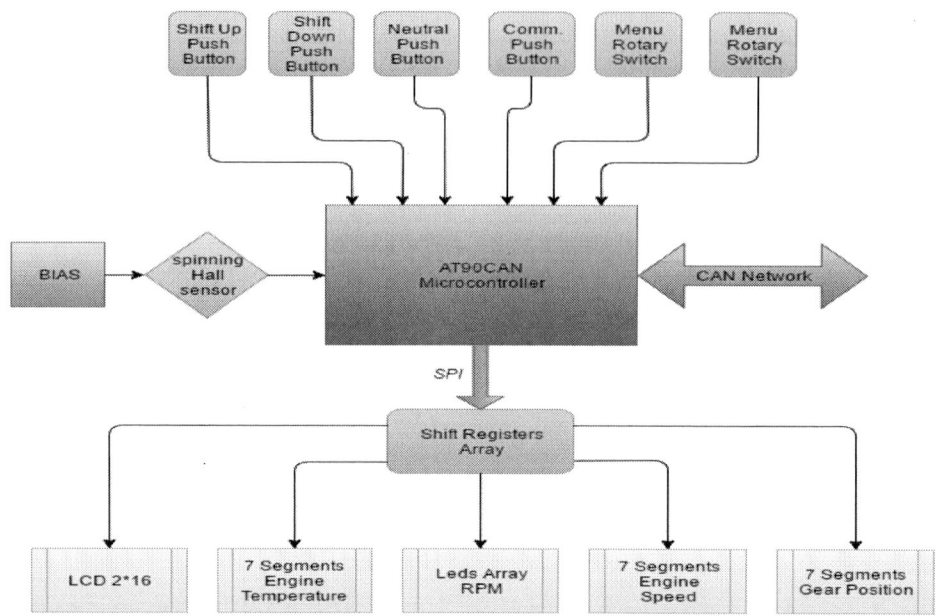

Fig. 3. Electronic steering wheel module – block diagram

C. PCB Design

The PCB modules have been designed entirely according to the requirements of the new race car dimensions and architecture alignment for enhanced ergonomics inside the race car. The PCB design was made using Proteus 8.2 with ARES package. The prints of all the components used have been defined. The height of the routes was set at 53 um, while their width was selected according to the voltage and current that crosses the wires. Thereby, the 12 and 5 volts supply routes width was set to 0.8mm and the signal routes width was set to 0.5mm. Aiming space-efficient PCB, the components were placed on both sides. On both sides of the PCB is one ground plane present with the role to eliminate disturbances occurred especially because of the presence of the microcontroller and the crystal quartz working at a frequency of 16 MHz. In addition to the role in electromagnetic compatibility, ground plans facilitate heat dissipation arising for example from voltage regulator.

D. Software implementation

The development of the software for the electronic modules is based on a number of libraries specially built for the BS16 monopost. Libraries for ADC, LCD, SPI, timers, CAN and function for the steering wheel were created. The main program runs a continuous loop to read data from the CAN bus, read buttons, process and display data. Software was developed using Atmel Studio 6.4 integrated environment.

IV. TESTING THE ELECTRONIC MODULES

Testing of the electronic modules designed for the BS16 monopost was performed using the tool CANoe from Vector

[11]. CANoe is a comprehensive software tool for development, test and analysis of entire networks and individual electronic control units. CANcaseXL module was used for simulations; it has two CAN channels, which can be linked to different networks with their modules.

The main nodes included in the simulation were ECU, steering wheel electronic control module and power module, as depicted in Fig. 4.

Signal conditioning and signal processing is carried out using programs written in the CAPL language, language developed by Vector. The physical steering wheel electronic module has been tested using CANoe, since CANcaseXL enables connection of a physical network in the simulation model. The setup used to test the steering wheel control module is presented in Fig. 5.

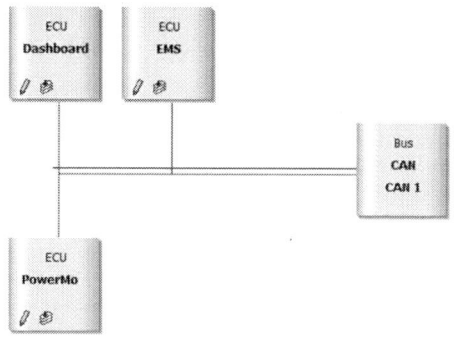

Fig. 4. CANoe simulation diagram.

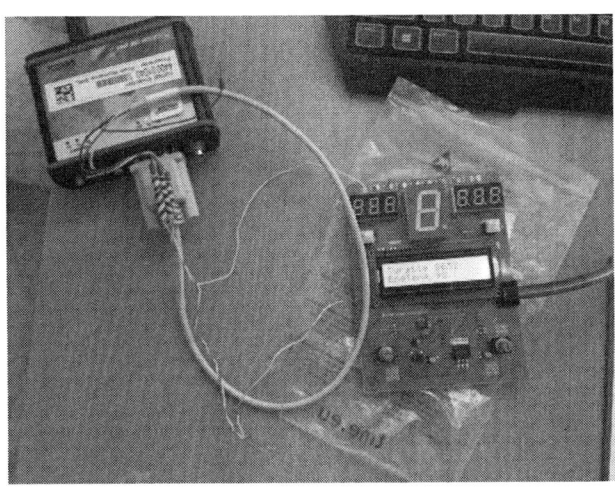

Fig. 5. Setup to test the steering wheel electronic module using CANcaseXL

V. RESULTS AND CONCLUSION

Electronic modules were designed, implemented, tested in the laboratory and then exercised in real competition in two races of the 2016 Formula Student competition: Silverstone in the UK and Barcelona Catalunya in Spain in July and in August 2016, respectively. Fig. 6 shows the steering wheel of the monopost BS16 at Catalunya.

The BlueStreamline team considers for the BS17 monopost upgrading the power module to accommodate turbo consumers, upgrading the steering wheel electronic module, testing all electronic modules using CANoe and wire harness design.

In future projects, data display on the steering wheel can be used in immersion (Virtual Reality) [12] and in studies on driver behavior [13]. In this perspective discussions with researchers in these fields of interest were initiated.

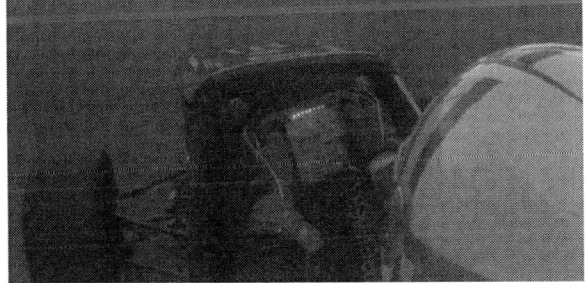

Fig. 6. BS 16 Steering Wheel – Catalunya-Barcelona, Spain

ACKNOWLEDGMENT

S.C. Preh Romania S.R.L supported development and testing of the electronic modules within this project.

REFERENCES

[1] Sommer, S., Camek, A., Becker, K., Buckl, C., Zirkler, A., Fiege, L., & Knoll, A. (2013, October). Race: A centralized platform computer based architecture for automotive applications. In *Electric Vehicle Conference (IEVC), 2013 IEEE International* (pp. 1-6). IEEE.

[2] Bauersachs, B., Mitrovic, D., & Schmidt, S. (2015). Development of a four-wheel drive electric racecar. In *15. Internationales Stuttgarter Symposium* (pp. 1131-1144). Springer Fachmedien Wiesbaden.

[3] Joehan, A. I. (2012). Design And Fabricate An Ergonomics Cockpit For Single Seater Formula Varsity Electric Racing Car., online at http://eprints.utem.edu.my/7200/

[4] Lisseman, J. C., Gardner, V. D., Van'tZelfde, D., Staszak, E. P., Mueller, N. H., & Andrews, D. W. (2016). *U.S. Patent No. 9,308,857.* Washington, DC: U.S. Patent and Trademark Office, online at https://www.google.com/patents/US9308857

[5] Wan, S., He, C., Xie, M., & Fan, Y. (2014). Design of Racing Electric Control System Based on AVR SCM. *Sensors & Transducers, 181*(10), 52.

[6] Stanca, A.C., Ogrutan, P., Kertesz, C.Z.: "Elective Courses with Support from Automotive Electronics Industry", Bulletin of the Transilvania University of Brasov • vol. 8 (57) no.1 – 2015, Series I - Engineering Sciences, ISSN 2065-2119.

[7] A. C. Stanca, F. Sandu, "Solutions to improve the features of local interconnected networks", Bulletin of the *Transilvania* University of Braşov, Series I: Engineering Sciences, vol. 6 (55) no.2 – 2013.

[8] F. Sandu, A.C. Stanca, A. Pasaroiu, G. Pana "Analog front-end and signal acquisition via sound channel", Proceedings of the 2012 IEEE 18th International Symposium for Design and Technology in Electronic Packaging (SIITME), October 25-28 2011, Alba Iulia, Romania, pp. 128-129, 2011.

[9] Cs.Z. Kertesz, C. Gerigan "CAN bus analyzer with general purpose microcontroller", Proceedings of the 10th International Conference on Optimization of Electrical and Electronic Equipment OPTIM 2006, Vol IV, May 18-19 2006, Brasov, Romania, pp. 61-64, 2006.

[10] David L. Jones, "PCB Design Tutorial", June 2004.

[11] F. Zhou, S. Li, X. Hou, "Development method of simulation and test system for vehicle body CAN bus based on CANoe", 7th World Congress on Intelligent Control and Automation, WCICA, pp.7515-7519, IEEE, 25-27 June 2008.

[12] Schmierbach, M., Limperos, A. M., & Woolley, J. K. (2012). Feeling the need for (personalized) speed: How natural controls and customization contribute to enjoyment of a racing game through enhanced immersion. *Cyberpsychology, Behavior, and Social Networking, 15*(7), 364-369

[13] Katzourakis, D. I., Velenis, E., Abbink, D., Happee, R., & Holweg, E. (2012). Race-car instrumentation for driving behavior studies. *IEEE Transactions on Instrumentation and Measurement, 61*(2), 462-474.

Wireless sensor networks as part of emergency situations management system

Claudiu Lung, Sebastian Sabou
Electrical, Electronics and Computer Engineering
Department
Technical University of Cluj Napoca - North University
Center of Baia Mare
Baia Mare, Romania
Claudiu.Lung@cunbm.utcluj.ro

Attila Buchman
Faculty of Informatics
Debrecen University
Debrecen, Hungary
buchman.attila@inf.unideb.hu

Abstract—This The new technology for sensors and wireless sensor networks has led to various fields of application. Environmental parameters monitoring, safety of persons and property, traffic control or some analyses into buildings to determine the cause of various problems, are some examples that highlights the ability of wireless sensors network to be deployed on a very short term basis. Wireless sensors can be placed on critical location or in critical pieces of building structure to detect, identify and diagnose faults . The local network sensor is generally an electronic embedded system, which is capable to process data received from a various type of sensors like: thermal, seismic, infrared, ultrasonic or magnetic. An advanced type of network sensor will not only store or send the acquired data, it will manage also the links to neighboring nodes, finding the best routes for data packets. Location for sensors can be obtained from a GPS or using a local positioning algorithm. All of this information's are vital to manage efficiently an emergency situation. (*Abstract*)

Keywords— Smart sensors, Smart city, Emergencies situations, FPGA, PicoBlaze. (key words)

I. INTRODUCTION

Many studies and nationally/internationally research programs are conducted on emergency situation management fields. The main targets for some of them are to design and implement different configurations of sensor networks in order to be able to predict and react efficiently on any type of dangerous situation and a better interaction with the environment. Generally the final purpose of those researches are only the data acquisition and database storage, for monitoring and statistics applications for reduced spaces which are associated with concept of "smart building"[1-4]. Starting from this idea, the research theme that will be carried out on this project and presented in this paper, is focused on design, implement and test several configurations of smart sensors networks, which are able to collect, transmit and store data from many type of sensors, with an extended coverage area: a city ("Smart city" concept) or regional/national. All of this integrated functionality's, are useful in the emergencies situations prevention and management system.

One part of this project is based on ZigBee communication modules. Main module implemented in FPGA and controlled by one PicoBlaze microprocessor, communicate serial with ZigBee Coordinator and sent data over wireless connection to

other control module witch receive the information and sent an ACK signal respond to ZigBee coordinator module.

On first start of the coordinator ZigBee module, it scan all available channels and initiate a network on the channel with the lowest noise. This network is identified by a 14 bit identifier (PAN ID). On network creation phase, coordinator chooses, by using a randomly number generator algorithm, a 128 bit key for data encryption. [5,10,14,15,19-22]

When a remote ZigBee device, will be started, it will perform also a scanning procedure upon all channels, to find a network to register.

When a Router or End Device, had been registered on the network, a Short address (16 bit) will be assigned to it. In this way, the data packets are sanded over the network to the particular module. The coordinator module will always have address 0x0000. The Short address for Routers module will be the same, unless a reset action will be performed.

The End Device module will change his Short address depending on the joined Router, which is named also "Father". If the End Device module is in outrange of its "Father", it will try to join to another Router, the Short address will be changed.

Also if it's "Father" it's turned off, the End Device will try to find another router and after connection to it, a new Short address will awarded.

When Router or End Device is registered to a network, and when an End Device change it's Father, the coordinator module will send through serial line some codes as follow described:

Device	Code
Coordinator	0x09
Router	0x07
End Device	0x06
Rejoin End Device	0x04

TABLE I :DEVICE REGISTRATION CODE

II. PROJECT IMPLEMENTATION

Designing and implementation of wireless sensor networks as part of emergency communication network by using multi-processor technology is the major goal of this project.[9,10]

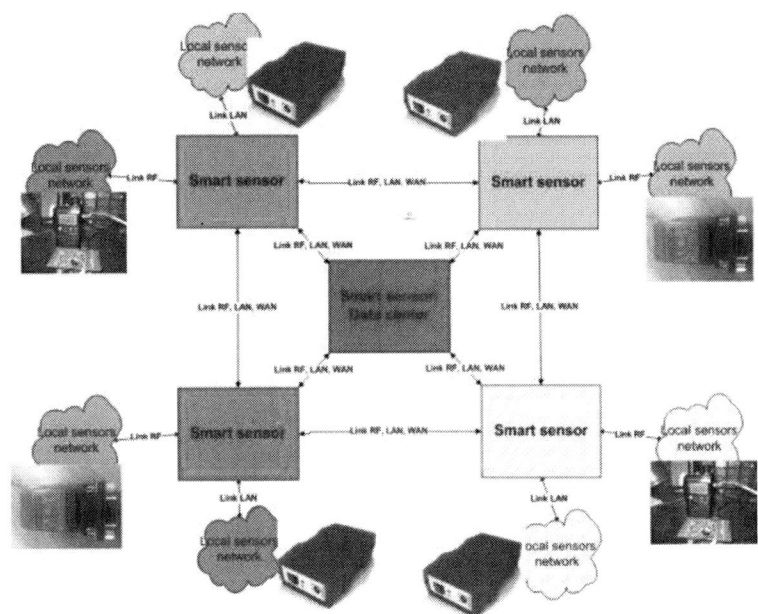

Fig. 1: Smart sensors network topology

The proposed design, implemented and presented in this paper use seven PicoBlaze microprocessors, which are free soft-core processors developed by Xilinx. [26,27]

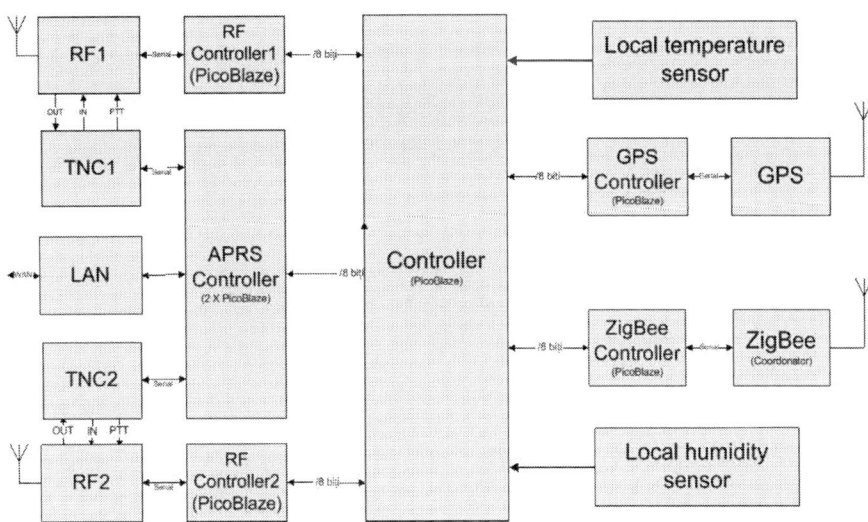

Fig. 2: Top level schematics of wireless sensor networks controller.

The ZigBee Controller module implementation in FPGA circuit is presented below.

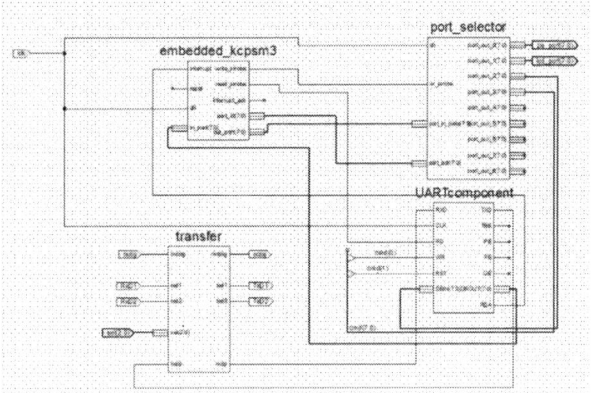

Fig. 3: ZigBee wireless network controller module

The UART component, writted in VHDL, includes a circuit for serial receiving data, and a circuit for serial transmitting data. The receiver takes serial byte data transmitted on RXD port and send out one byte of parallel data. This byte is avaible on the DBOUT port. The transmitter takes a parallel byte data avaible on the DBIN port, and send out a serial byte of data. This data are transmitted over the TXD port. The UART module is separated into two independent circuits as can be seen in Fig. 4.

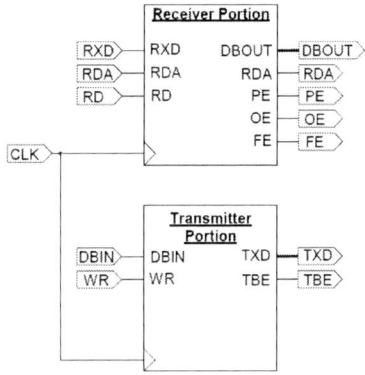

Figure 4. UART Rx/Tx circuits.

Port_selector block, also written in VHDL, permits the selection of one of the 256, 8 bit, ports that can be access by the PicoBlaze microcontroller. This block highlights the microcontroller's ability to use a large number of ports. The sincronization of data transfer from one module to another is done with WRITE_STROBE signal [8-10].

III. EXPERIMENTAL RESULTS

The wireless sensor network controller features are implemented in Digilent Spartan 3E FPGA development board and tested using the APRS network infrastructure.[25,28] Experimental stand is presented in Fig. 5 and Fig. 6.

Fig. 5: Local sensors network controller.

Fig. 6: ZigBee wireless network sensor

The presented wireless sensor offer various communication capabilities with other network sensors like: ZigBee, LAN, RF.[9,13,17,18]

Some results are presented in Fig. 7. and Fig. 8.

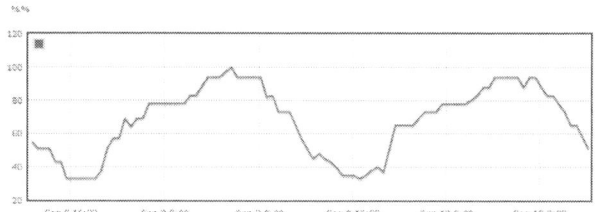

Fig. 7: Humidity data

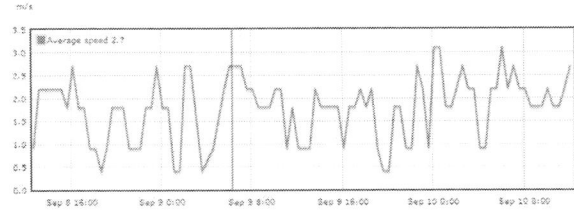

Fig. 8: Wind speed data

By using software tools developed by the authors of this paper it was possible to implement in FPGA circuit all the functionalities in order to obtain a fully operational local sensors network controller. [6]

IV. CONCLUSION

Proposed model of smart sensor wireless network is fully functional. Any functionality can be modified and improved by using PicoBlaze EDK/SDK tool, developed by the authors of this paper.

The implementation and results presented in this paper are part of a project which aims to design, test and develop a smart sensor controller implemented in FPGA circuits, using PicoBlaze multiprocessor technology.

REFERENCES

[1] Won-Suk Jang , William M. Healy, Mirosław J. Skibniewski, "Wireless sensor networks as part of a web-based building environmental monitoring system,"in Automation in Construction 17 , pg. 729–736, journal homepage: www. e lsevier.com/locate/autcon, 2008

[2] J.P. Lynch, An overview of wireless structural health monitoring for civil structures, Philosophical Transactions of the Royal Society of London. A 365, pg. 345–372. 2007

[3] K. Ducatel et al., "Scenarios for Ambient Intelligence in 2010", tech. report, Information Society Technologies Advisory Group (ISTAG), Inst. of Prospective Technological Studies (IPTS), 2001.

[4] C.Y. Chong, S. Mori, and K. C. Chang, "Distributed multitarget multisensortracking", Ed. Norwood, MA: Artech House, 1990, pp. 247–295.

[5] Khusvinder Gill, Shuang-Hua Yang, Fang Yao, and Xin Lu, "A ZigBee-Based Home Automation System", IEEE Transactions on Consumer Electronics, Vol. 55, No. 2,pg. 422-430, MAY 2009

[6] C. Lung, A. Buchman, „Software development tool for PicoBlaze multi-processor implementation", in Carpathian Journal of Electronic and Computer Engineering, Baia Mare, 2012, pg. 67-70

[7] M. Young, The Technical Writer's Handbook. Mill Valley, CA: University Science, 1989.

[8] C. Lung, S. Sabou, C. Barz „Smart sensor implemented with PicoBlaze multi-processors technology" – International Symposium for Design and Technology of Electronic Packages, SIITME 2012, 18th Edition, Alba Iulia, Romania, October 2012, p.241-245

[9] C. Lung, S. Sabou, I. Orha „Communication control system implemented in FPGA" – International Symposium for Design and Technology of

Electronic Packages, SIITME 2011, 17th Edition, Timişoara, Romania, October 2011, p. 245-248

[10] C. Lung, S. Sabou, I. Orha, A. Buchman, „ZigBee smart sensors networks" – International Symposium for Design and Technology of Electronic Packages, SIITME 2010, 16th Edition, Pitesti, Romania, September 2010, p. 309-312

[11] "21 ideas for the 21st century," Business Week, pp. 78–167, Aug. 30, 1999.

[12] Claudiu Lung „Intelligent thermometer with speech function implemented in FPGA" – Carpathian Journal of Electronic and Computer Engineering, North University Baia Mare, 2011, p. 65-68

[13] Sebastian Sabou, Claudiu Lung, Ioan Orha "Reference for Indoor Location Systems", Novice Insights in Electronics, Communications and Information Technology, Cluj Napoca, 2011, p

[14] Claudiu Lung, Sebastian Sabou, "ZigBee Smart Sensors Network", Novice Insights in Electronics, Communications and Information Technology, Cluj Napoca,2010,

[15] Claudiu Lung, „Embedded Ambient Assisted Living System" – Carpathian Journal of Electronic and Computer Engineering, North University Baia Mare, 2008, p. 37-40

[16] Daniel Mic, Ştefan Oniga, Emil Micu, Claudiu Lung, "Complete Hardware/Software Solution for Implementing the Control of the Electrical Machines with Programmable Logic Circuits" – International Conference on Optimization of Electrical and Electronic Equipment, OPTIM'08, Braşov 2008, Romania, May 22-24, 2008

[17] Jianliang Zheng, Myung J. Lee, "Will IEEE 802.15.4 make ubiquitous networking a reality?: A discussion on a potential low power, low bit rate standard", IEEE Communications Magazine, June 2004.

[18] C. Colonati, Radiocomunicaţii digitale, Articole, Note, Aplicaţii şi Software, Editura NErgo

[19] S. Oniga, J. Vegh, I. Orha, "Intelligent Human-Machine Interface Using Hand Gestures Recognition", Automation Quality and Testing Robotics (AQTR), 2012 IEEE International Conference on ,pp. 559 - 563,

[20] A. Alexan, A. Osan, S. Oniga, Personal assistant robot, 2012 IEEE 18th International Symposium for Design and Technology in Electronic Packaging, SIITME 2012, October 25-28 2012, Alba Iulia, Romania, pp. 69-72

[21] J. Suto, A. Mate, J. Vegh, I. Oniga, „Developing a general purpose data collector framework for robot", Carpathian Control Conference (ICCC), 2012 13th International , pp. 690 - 693, J

[22] Suto, S. Oniga, Remote controlled data collector robot, Carpathian Journal of Electronic and Computer Engineering, Volume 5, Number 1 - 2012, pp. 117-120

[23] I. Orha, S. Oniga, Assistance and telepresence robots: a solution for elderly people, Carpathian Journal of Electronic and Computer Engineering, Volume 5, Number 1 - 2012, pp.87-90

[24] A. Alexan, A. Osan, S. Oniga, AssistMe robot, an assistance robotic platform, Carpathian Journal of Electronic and Computer Engineering, Volume 5, Number 1 - 2012, pp.1-4

[25] ***, www.aprs.fi

[26] ***,http://www.xilinx.com/support/documentation/ip_documentation/ug129.pdf

[27] ***, KCPSM3_Manual.pdf , www.xilinx.com

[28] ***,http://stiintasitehnica.com/stiri/stratospherium-ii-zboara-alaturi-de-noi-in-spa-iul-cosmic

Investigation on Available Bandwidth in Visible-Light Communications

Alina-Elena Marcu
Center for Technological Electronics and Interconnection Techniques (UPB-CETTI)
"Politehnica" University of Bucharest
Bucharest, Romania
alina.marcu@cetti.ro

Robert-Alexandru Dobre
Telecommunications Department
"Politehnica" University of Bucharest
Bucharest, Romania
rdobre@elcom.pub.ro

Marian Vlădescu
Optoelectronics Research Center (UPB-CCO)
"Politehnica" University of Bucharest
Bucharest, Romania
marian.vladescu@gmail.com

Abstract—**Looking at the development rate of the radio communication systems, one conclusion can be clearly drawn: mobile data traffic is exponentially increasing and researchers do their best to keep up with the demand, while maintaining the power consumption as low as possible. In this context, new fast ways to transfer data are at their birth, like visible-light communication (VLC). Since LEDs are replacing conventional illumination solutions, they also proved to be useful as data transmitters simultaneously. This paper supports this new research field and studies the bandwidth that could be used for VLC by investigating four different kinds of LEDs: white, infrared, red, and green. The chosen types of LEDs play key roles in various applications not only from the comfort category, but from traffic safety also, which are detailed in this document. The dependency between the available bandwidth and the LED's bias current is presented along with the measuring procedure and technological setup details.**

Keywords—LED; visible light communication; bandwidth.

I. INTRODUCTION

Thanks to their power effectiveness white LEDs tend to replace conventional incandescent or fluorescent ambient lighting solutions. Since, historically, the illumination system was invented way before radio communications, its infrastructure is much more spread and also simpler. The latest trends in high speed communications is to use conventional lighting system to transfer data. Since most light sources are replaced with white LEDs, these are to play an important role in the downlink part of the communication, while infrared or ultraviolet LEDs are to be used for uplink in order to minimize interferences. Another top research field in which VLC [1] could bring breakthroughs is represented by the development of autonomous vehicles. It is known the crucial role that red and green light play in the safety of road traffic. If VLC could be added to the current traffic signalling system, the future's smart cars could receive a message about the state of the traffic lights, and, if it was red, automatic breaking could be engaged in order to force the driver not to enter the crossroads, avoiding accidents. Therefore, studying white, infrared, red and green LEDs is sustainable. Since communications are the main target, the available bandwidth is of utmost importance.

II. DESCRIPTION OF WORK

A test bench was developed in order to study the bandwidth that can be achieved in VLC. It consists of four LEDs produced by Led Engin, LZ4-00G108 green LED [2], LZ4-00R308 red LED [3], LZ4-00R708 infrared LED [4] and LZ9-00CW00 white LED [5], a constant current source, a signal generator, a DET410 high-speed InGaAs detector, a light meter and an oscilloscope, illustrated as a schematic diagram in Fig. 1, DUT meaning Device Under Test. A picture in which all the measurement devices and connections which form the developed assembly can be observed is presented in Fig. 2. The LEDs are biased at various currents and an alternative signal is added using a signal generator (denoted $V_{stimulus}$ in Fig. 1). This alternative signal will play the message's role. Power LEDs' DC bias currents are relatively large, from 200 mA to 800 mA for the ones available for the experiment. Because each power

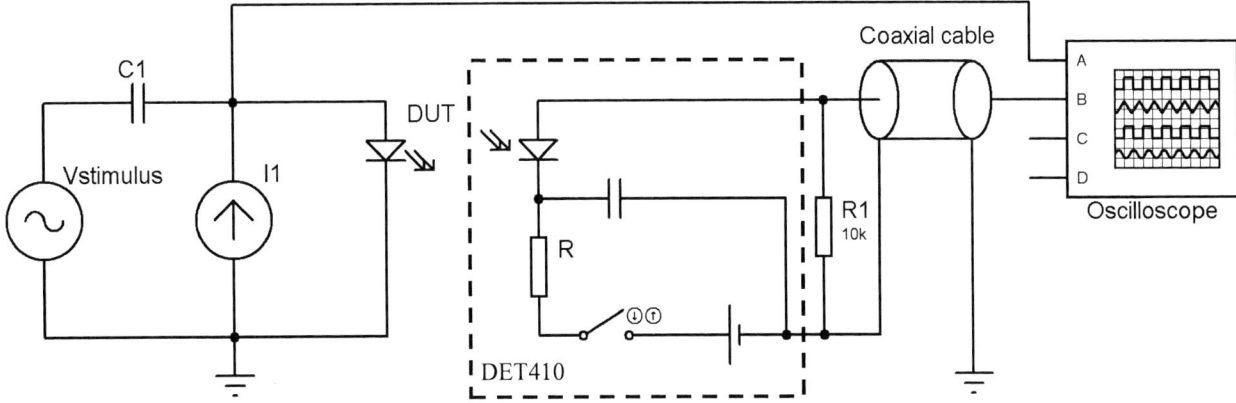

Fig. 1. The schematic diagram of the measurement setup.

LED structure is obtained by arranging single LEDs in a 2x2 or 3x3 matrix, forward voltages are also notable (as large as 8 V). A signal generator that could deliver the needed offset voltage to forward bias the LEDs and provide the needed amount of current is rare, so a separate power supply was used for obtaining the needed bias current (I1 in Fig. 1) and the AC signal was coupled using a 100 uF capacitor (C1 in Fig. 1). The waveform of the added signal can be a sine wave, for classical bandwidth measurement method or a square wave for a measurement based on the received signal's rise time using (1).

$$f_{BW} = \frac{0.35}{t_{RT}} \qquad (1)$$

where f_{BW} is the bandwidth and t_{RT} is the measured rise time.

The behavior of the VLC system is similar to a low-pass filter so the classical method for determining its cutoff frequency is explained onwards. The sine stimulus ($V_{stimulus}$ in Fig. 1) is maintained at a constant peak-to-peak amplitude and its frequency is swept in order to determine a suitable value for the start frequency that must be found well into the pass-band of the filter. By sweeping from 10 kHz to 20 kHz, the received signal's amplitude was also constant so 10 kHz was chosen as the start frequency. Then the received signal was measured at this start frequency and the measurement was kept as a reference, denoted V_{start}. The frequency was increased until the difference between the reference value and the received signal's amplitude was −3 dB. The frequency at which this

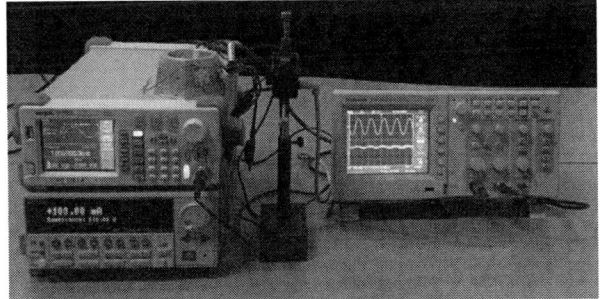

Fig. 2. The test bench used for measuring the bandwidth.

condition became true is the sought cutoff frequency. The condition to measure the stimulus signal was crucial because the coupling capacitor determines a high-pass filter behavior which means that the stimulus signal will have larger amplitudes as the frequency increases, this consequence compromising the whole measurement method. The amplitude of the AC signal was adjusted in order to maintain a constant peak-to-peak amplitude for the AC signal measured on the LED. These experiments were done in the darkest ambient illumination conditions that could be obtained in the laboratory (10 – 15 lux) at various bias current for each mentioned type of LED to determine their implications) on the available bandwidth.

$$\alpha + \beta = \chi . \qquad (1)$$

III. RESULTS

The results of the measurements are presented in Table 1. Peak-to-peak amplitude of the sine stimulus is the amplitude of the AC signal on the LED that was maintained constant during the bandwidth determination for each LED. It was observed that at the same generated amplitude (at the output of the signal generator), the smallest AC amplitude measured on the LED was at the largest DC bias current. So to be sure that this parameter can be held constant for all the four measurements, the measuring order was from the largest DC bias current to the smallest one. The smallest AC amplitude that could be imposed on the LED was for the green led, while the largest was in the case of the red LED.

The largest available bandwidth was found in the case of the infrared LED, while the smallest was determined for the white LED. Also, the infrared LED dominated the performances from this point of view. It must be mentioned that the infrared LED used in the experiment is a dual-junction LED.

All the LEDs involved in this experiment exhibit a very interesting behavior: the available bandwidth increases with the DC bias current of the LED, but again, the increase is the largest in the case of the infrared LED. This kind of behavior has beneficial consequences: LEDs that are part of a VLC system can be biased using a lower value DC current, needed for the primary application (ambient illumination, traffic control etc.) and increased only in the moment that a

TABLE I.   EXPERIMENT CONDITIONS AND MEASURED PARAMETERS

Green LED				
Peak-to-peak amplitude of the sine stimulus (mV)	LED's DC bias current (mA)	Received peak-to-peak amplitude (mV) at start frequency (10 kHz)	Cutoff frequency (Bandwidth) (kHz)	Received peak-to-peak amplitude (mV) at cutoff frequency
240	800	1.361	40	0.96
	600	1.68	38	1.201
	400	2.321	33	1.601
	200	3.601	28	2.561
Red LED				
Peak-to-peak amplitude of the sine stimulus (mV)	LED's DC bias current (mA)	Received peak-to-peak amplitude (mV) at start frequency (10 kHz)	Cutoff frequency (Bandwidth) (kHz)	Received peak-to-peak amplitude (mV) at cutoff frequency
330	800	2.081	47.2	1.441
	600	2.72	42	1.921
	400	3.921	38	2.801
	200	7.121	31.5	5.041
Infrared LED				
Peak-to-peak amplitude of the sine stimulus (mV)	LED's DC bias current (mA)	Received peak-to-peak amplitude (mV) at start frequency (10 kHz)	Cutoff frequency (Bandwidth) (kHz)	Received peak-to-peak amplitude (mV) at cutoff frequency
318	800	2.64	106	1.841
	600	3.36	87	2.401
	400	4.64	80	3.281
	200	8.16	67	5.761
White LED				
Peak-to-peak amplitude of the sine stimulus (mV)	LED's DC bias current (mA)	Received peak-to-peak amplitude (mV) at start frequency (10 kHz)	Cutoff frequency (Bandwidth) (kHz)	Received peak-to-peak amplitude (mV) at cutoff frequency
284	800	3.36	37	2.401
	600	4.16	33	2.96
	400	5.92	31	4.16
	200	10.64	26	7.44

communication is to take place. Further interpreting the results shows that the traffic control systems have a slightly advantage over the ambient illumination systems since the bandwidth that was determined for the red and green LEDs was larger than in the case of the white LED. Depending on the required bandwidth, the DC bias current can be accordingly set. The white LED is manufactured using the phosphor method: a short wavelength LED (UV or blue) is combined with a yellow phosphor coating resulting white light. The remanence caused by the phosphor coating is very likely to be responsible for the lowest bandwidth obtained for this kind of LED.

Leaving aside the special case of the infrared LED, for the LEDs radiating in the visible spectrum an average of 10 kHz bandwidth increase can be obtained by increasing the DC bias current from 200 mA to 800 mA. The largest bias DC current may assure the largest bandwidth, but care must be taken in applications that contain communications with large durations as this could lead to overheating [6] the LED. The bandwidth

for each LED as a function of the DC current is shown in Fig. 3.

It was also observed that the received signal's amplitude at the start frequency varied with the DC bias current. This variation is presented in Fig. 4 and offers the following conclusion: the received signal would be larger if the DC bias current of the LED would be smaller. So the design engineer must make a compromise between large bandwidth and large signal to noise ratio. The estimated third order polynomial equations for the four curves were also displayed in Fig. 4 near the legend to suggest the correspondence. Other functions like exponential, linear, logarithmic offered worse results at fitting. The red and infrared LEDs present very similar characteristics from this point of view, while the green LED shows the least variation. The derivative of the graphs in Fig. 4 would eliminate the effects determined by the placement or sensitivity of sensors.

2016 IEEE 22nd International Symposium for Design and Technology in Electronic Packaging (SIITME)

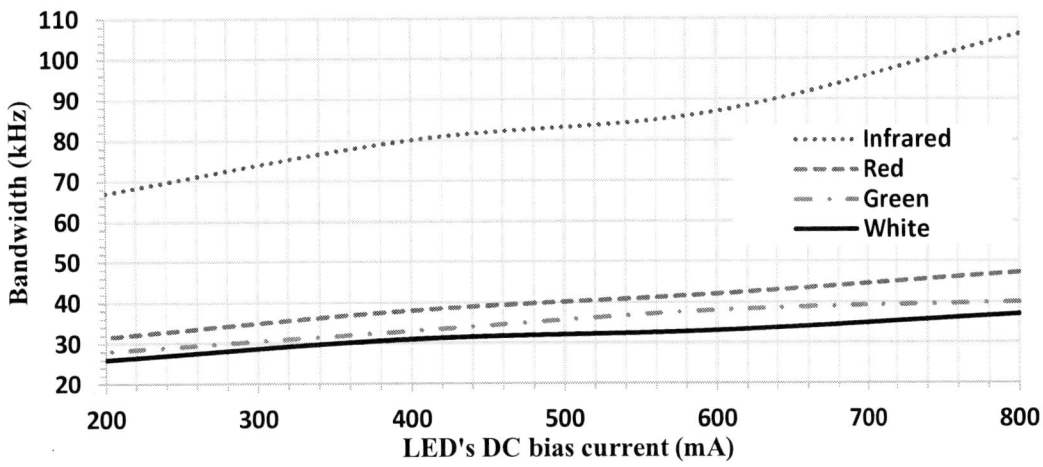

Fig. 3.   The variation of the bandwidth with the LED's DC bias current.

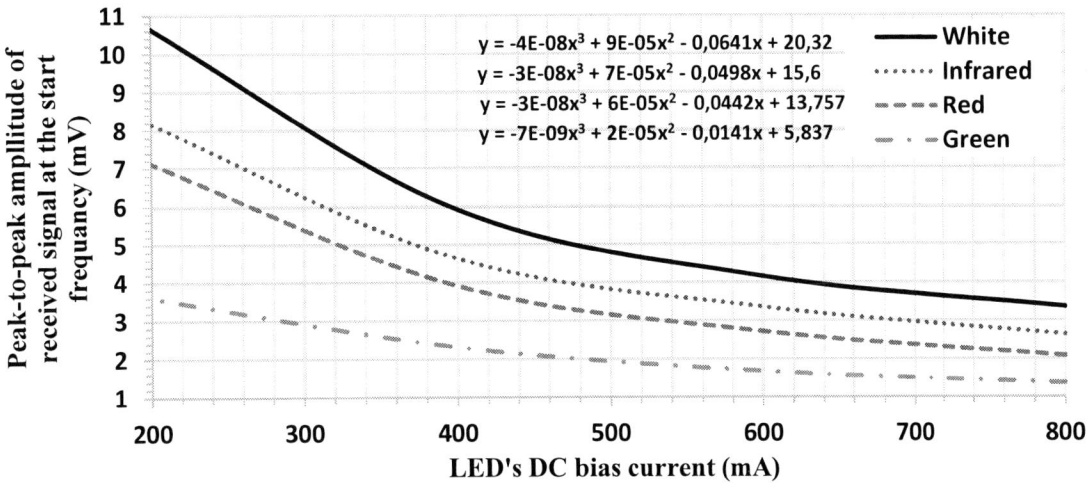

Fig. 4.   The variation of the peak-to-peak amplitude of the received signal at the start frequency (10 kHz) with the LED's DC bias current.

ACKNOWLEDGMENT

This paper is partially supported by Project PN-II-PT-PCCA-2013-4-0000, Contract no. 11/2014, "Fabrication technology of micro-biosensors arrays and development of a portable device for the diagnosis of acute myocardial infarction (AMI-DETECT)" and partially by Project PN-II-PT PCCA-2013-4-0967, contract no. 290/2014, "Instrument in the millimeter-wave spectroscopy (THz) for identifying hazardous substances (THz-DETECT)".

REFERENCES

[1] S. Haruyama, "Visible light communication," 17th International Display Workshops, IDW'10, 2010.

[2] LED Engin, "LZ4-00G108 Green LED Emitter Datasheet," Version 1.1, 2014.

[3] LED Engin, "LZ4-00R308 Far Red LED Emitter Datasheet," Version 1.1, 2014.

[4] LED Engin, "LZ4-00R708 940nm Dual Junction Infrared LED Emitter Datasheet," Version 1.0, 2014.

[5] LED Engin, "LZ9-00CW00 High Luminous Efficacy Cool White LED Emitter Datasheet," Version 1.8, 2014.

[6] C. Ionescu, A. Drumea, N. D. Codreanu, A. Vasile, "Thermal Investigations on High Power LED's," Proc. of SPIE Vol. 8411, 2012, pp. 84112Y-1.

E-bike electronic control unit

Florin Dumitrache, Marius Catalin Carp, and Gheorghe Pana

Electronics and Computers Department
Transilvania University of Brasov
Brasov, Romania
gheorghe.pana@unitbv.ro

Abstract — **The electric vehicles industry is continuously evolving. One type of such electric vehicle is the electric bicycle (e-bike). Electric bicycles, like other electric vehicles, use a BLDC motor (Brushless Direct Current Motor). This paper presents a way of designing and implementing an electronic module for an e-Bike. The paper shows how a low power, 8-bit microcontroller can be used to drive such a motor and also manage other useful functions on an e-Bike.**

Keywords —BLDC; e-Bike; controller

I. INTRODUCTION

Electric vehicles make use of BLDC motors as the propulsion method. Due to the fact that BLDC motors do not have brushes, they present some advantages over the DC brushed motors, from which we remember: (1) longer life span, (2) lower EMI (Electromagnetic interference) radiation, (3) noiseless operation, (4) grater torque to motor size ratio [1]. Due to the geometry of the windings in the motor, the BEMF (back electro-motive force) generated by the motor when in generator mode can be of two types: (1) trapezoidal and (2) sinusoidal. The latter can be of interest if the driven motor does not have Hall position sensors, and facilitates the calculation of the motor's rotor absolute angle.

The internal structure of a BLDC motor is presented in Fig. 1. BLDC motors are 3-phase motors, and to properly drive such a motor, a special control circuit must be used. The purpose of the control circuit is to energize the correct winding(s) at the right moment. This is achieved by reading information from certain rotor position sensors and generating PWM (pulse width modulation) signals.

According to [1], the main components of a system with BLDC motor are: (1) control logic, (2) power stage comprised of six switching devices (e.g. MOSFETs, IGBTs) and (3) sensors used for the closed-loop feedback.

The performance of a BLDC motor is dictated mainly by the motor structure and the control logic that is been used. By using different types of control logic, the torque ripple of the BLDC motor can be minimized [2].

II. BLOCK DIAGRAM

The block diagram of the designed electric system is presented in Fig. 2. The implementation of the proposed control unit is based on an 8-bit microcontroller.

The microcontroller receives information about the motor position (rotor angle), via signals generated by three Hall-effect sensors contained within the motor.

Using this data, the microcontroller uses a simple commutation table and switches the six power MOSFET transistors which drive the BLDC motor.

Internet-based literature (images only) suggests that BLDC controllers used in commercial e-Bikes contain linear power supplies. This is a main issue regarding power efficiency, due to the significate power loss as heat in the internal power supply.

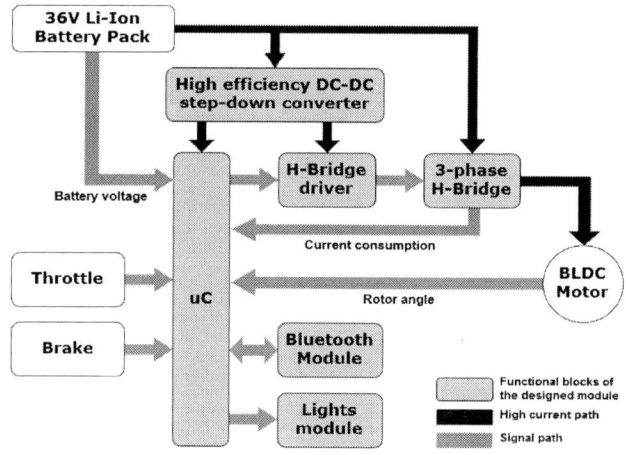

Fig. 2. Block diagram of the designed electric system

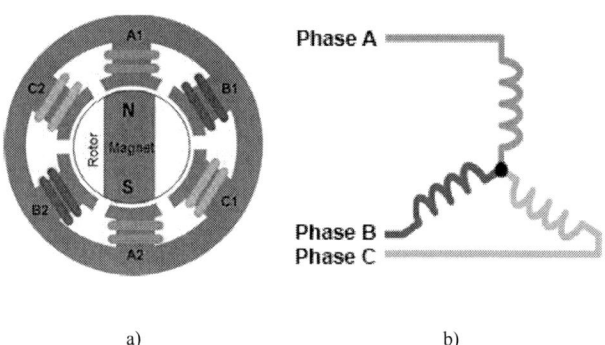

Fig. 1. BLDC motor: a) simplified internal structure; b) motor windings

Sponsor: S.C. PREH ROMANIA S.R.L.

One of the improvements this design brings is the use of a DC-DC step down converter, which greatly lowers the power consumption of the module and reduces the ambient temperature in the case of the module. The latter is an important issue, considering the fact that the electronics on an e-Bike must be housed in a water-proof enclosure.

The user of the e-Bike receives relevant data (e.g. instantaneous speed, battery state of charge) from the motor controller via Bluetooth protocol and can view the data on a GUI (graphical user interface), i.e. on a smartphone. By using a Bluetooth transmitter instead of a graphical display, the power consumption is furthermore reduced.

III. ELECTRONIC SCHEMATIC

The hardware implementation of the block diagram from Fig. 2 is presented in Fig. 3, without the power stage.

The power supply is designed using a dedicated DC-DC step down IC (integrated circuit) – U101 in Fig. 3 a. By using an IC with integrated power MOSFETs, EM radiations are reduced, due to the fact that there are no necessary external traces on the PCB.

The circuit requires the existence of a negative supply rail. This is achieved by using a 555 timer, U104, in a switched capacitors configuration.

Other supply voltages are obtained using LDO (Low dropout) regulators. The combination of a DC-DC pre-regulator followed by linear regulators provide a low output voltage ripple, while maintaining a good overall efficiency.

In Fig. 3 b the logic circuitry of the controller is presented. The heart of this block is an AVR ATmega128 microcontroller, U201. This solution has been chosen as a good compromise between processing power and low consumption.

In order to increase the EM radiation immunity of the microcontroller, the voltage supply level of the microcontroller is the maximum allowed, i.e. 5V.

The interface between the microcontroller and the Bluetooth module is made using a bidirectional level shifter, comprised of Q201 and Q202 transistors. The level shifters are necessary because the microcontroller and the BT module are powered from different supply voltages (5V and 3.3V respectively).

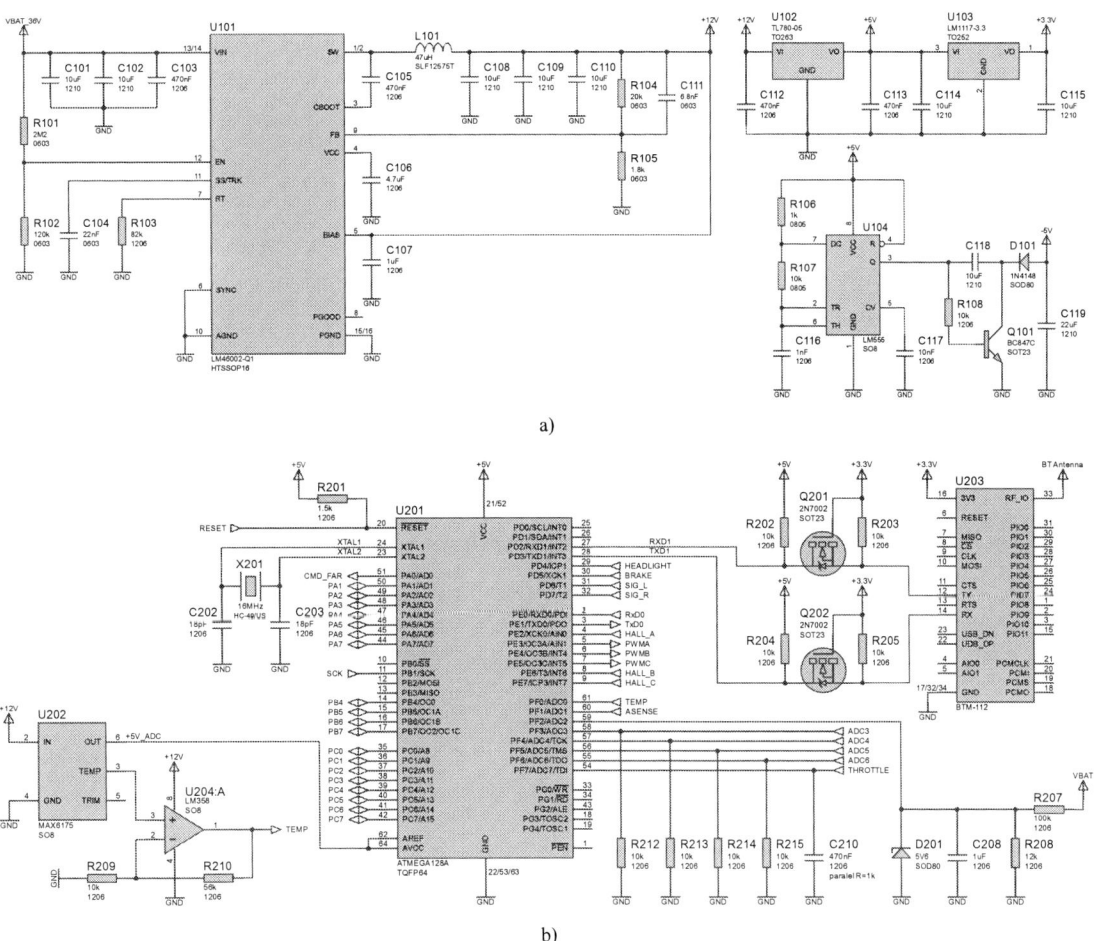

Fig. 3. BLDC electronic control unit schematic: a) power supply; b) logic circuitry

The BLDC motor is driven with six MOSFET power transistors, M1 to M6 in Fig. 4 a, arranged in a 3-phase H-bridge configuration.

The bridge current is measured using a 4-point type measurement using a differential amplifier. The differential amplifier is followed by a single diode precision rectifier [3]. The use of this measurement circuitry comprised of U301 and U302 makes possible the measurement of both the consumed power, when the motor acts as a load, and the recovered power, when the motor acts as a generator.

This novel method of current sensing can be further used to limit battery charge/discharge current, or calculate the battery state of charge more precisely.

The MOSFET driver ICs receive the PWM signals from the microcontroller and ensure the proper control voltage levels and timing for the power transistors. For the high side MOSFET, the control voltage must be positively offseted with a voltage equal to the voltage in the transistor's source connection. One of three MOSFET drivers is presented in Fig. 4 b. Using specialized MOSFET drivers rather than discrete level shifters with BJTs (bipolar transistors) assures fast, well-controlled rise and fall times of the power signals and thus, reducing power losses caused by the commutation of the transistors.

IV. SOFTWARE AND CONTROL LOGIC

Due to the fact that the software used by commercial e-Bikes is confidential, the software has been designed based on the fundamentals of BLDC motors presented in the introduction of this paper and other studies [2][4].

The commutation scheme used in this design is described in Fig. 5. By analysing the diagram in Fig. 5, the system of equations (1) has been deduced, through which the control logic has been modelled as a combinational system. The inputs to this combinational system are the Hall position sensors signals and a master PWM, and the outputs are the PWM signals.

$$
\begin{cases}
A_{HS} = PWM \cdot \left(HallA \cdot \overline{HallB}\right); \\
B_{HS} = PWM \cdot \left(HallB \cdot \overline{HallC}\right); \\
C_{HS} = PWM \cdot \left(HallC \cdot \overline{HallA}\right); \\
A_{LS} = HallB \cdot \overline{HallA}; \\
B_{LS} = HallC \cdot \overline{HallB}; \\
C_{LS} = HallA \cdot \overline{HallC};
\end{cases}
\tag{1}
$$

When choosing the commutation scheme the following factors have been taken into account: (1) simplicity of implementation, (2) delivered performances motor speed corresponding to an average speed of 20 to 25km/h and (3) amount of CPU load determined by the motor control function in conjunction with the secondary functions (i.e. data transmission, analogue signals processing, user commands processing).

The software flow diagram is presented in Fig. 6.

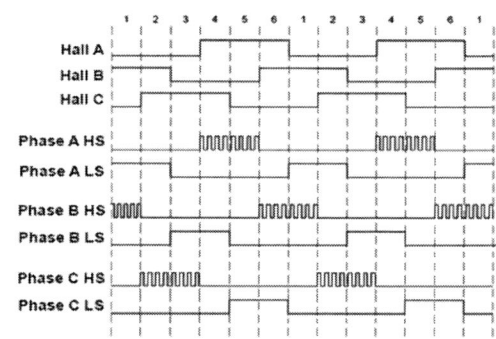

Fig. 5. PWM commutation scheme

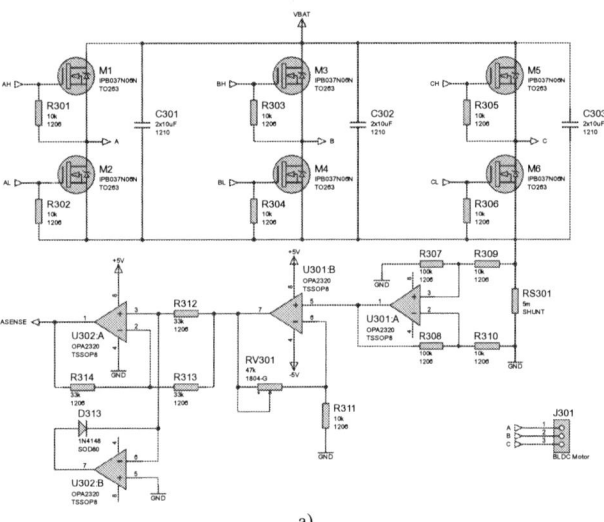

a)

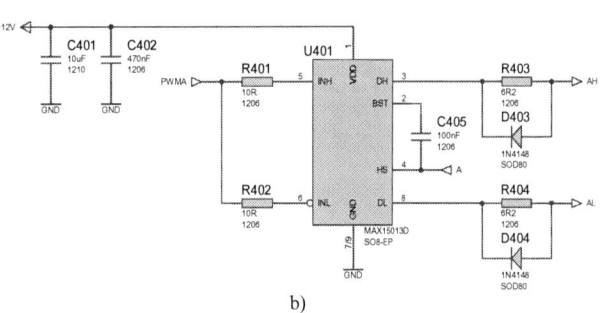

b)

Fig. 4. BLDC electronic control unit power circuitry: a) 3-phase H-Bridge and current sensing circuit; b) MOSFET driver circuit

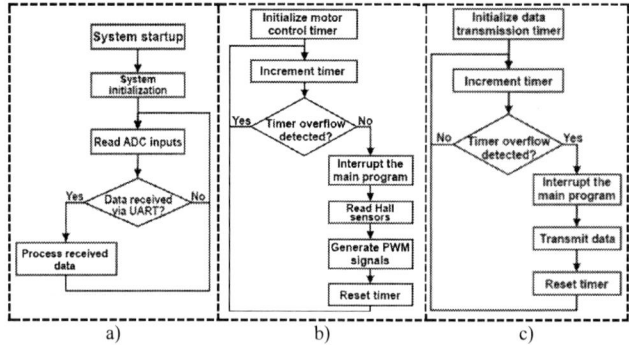

Fig. 6. Electronic control unit software flow diagram: a) main program; b) motor control thread; c) data transmission thread

The software presented in the flow diagram, in Fig. 6, is structured in two threads, one for each main function fulfilled. The threads are governed by two separate timers, the motor control thread having the highest priority.

V. SOFTWARE SIMULATION

The system has been simulated using Matlab Simulink simulation environment. The simulation took into account the commutation table presented in Fig. 5 and described by (1). The Simulink model is presented in Fig. 7. The results of the simulation are (1) the stator current waveform, (2) rotor speed and (3) electromagnetic torque ripple. These results are shown in the graphs in Fig. 8.

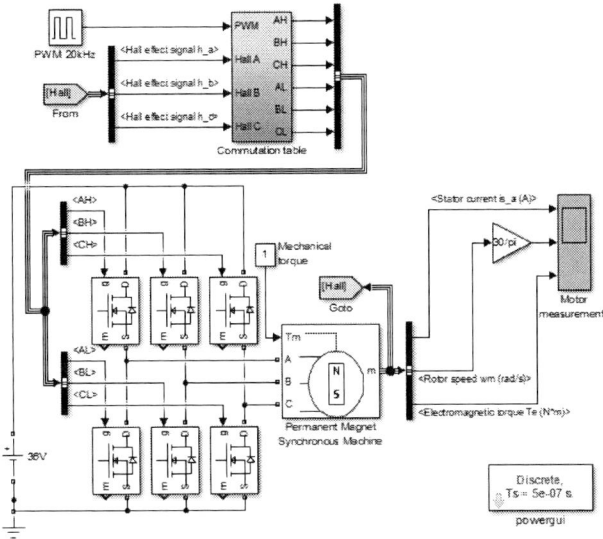

Fig. 7. Simulink model of the proposed system

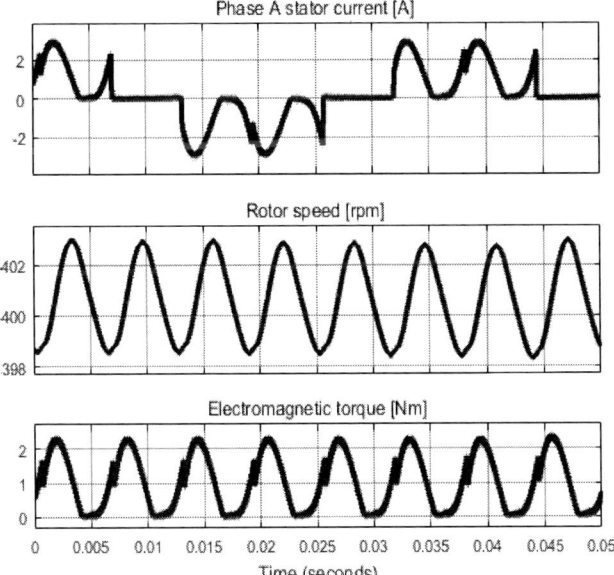

Fig. 8. Simulation results

The result shown in Fig. 8 suggests that the proposed commutation table will present good performances. This can be deduced from the rotor speed graph, in which a speed variation of around 0.5%

VI. CONCLUSIONS

The entire system has been implemented and tested. The results obtained are comparable with commercial e-Bikes, and the proposed implementation presents a higher power-autonomy ratio.

The results presented in Table I have been obtained using a 350W, 3-phase BLDC motor and a 36V / 5.5Ah Li-Ion battery pack, mounted on a 13kg bicycle.

The use of Li-Ion cells has been taken into consideration after studying the e-Bike market and observing that many commercial e-Bikes use Lead-acid batteries. The designed e-Bike is compared with a commercial e-Bike of about same cost and complexity.

TABLE I.   DESIGNED SYSTEM VS. COMMERCIAL SYSTEM

Criteria	Designed system	Commercial system
Maximum speed	33 km/h (software limited at 25 km/h)	25 km/h
Motor power	350 W nominal	250 W nominal
Assist type	Power-on-demand & pedal-assist	Power-on-demand OR pedal-assist
Battery	36V Li-Ion / 5.5 Ah	36V Lead–acid / 12 Ah
Autonomy	20-22 km	40-45 km
Total mass	19-20 kg	27 kg

An overall conclusion is that the hardware improvements (i.e. higher efficiency power supply, components with lower power consumption, reduced switching losses in the power stage), in conjunction with a simple yet effective software, yield a low-cost and reliable product.

ACKNOWLEDGMENT

This work has been developed in the frame of a partnership between *Transilvania* University of Brasov and the sponsor, S.C. PREH ROMANIA S.R.L.

REFERENCES

[1] P. Yedamale, "Brushless DC (BLDC) motor fundamentals", Application Note AN885, Microchip Technology Inc., 2003.

[2] H.S. Chuang, Yu-Lung Ke, and Y.C. Chuang, "Analysis of Commutation Torque Ripple Using Different PWM Modes in BLDC Motors", Conference Record 2009 IEEE Industrial & Commercial Power Systems Technical Conference, pp. 1-6, May, 2009.

[3] D. Jones, M. Stitt, "Precision absolute value circuits", Burr-Brown Corporation, December, 1997

[4] W. A. Salah, D. Ishak, K. J. Hammadi, "Minimization of torque ripples in BLDC motors due to phase commutation", Przeglad Elektrotechniczny, R. 87, pp 183-188, January 2011.

LabVIEW Simulator for Terrestrial MIMO Communications

Lucian Andrei Perişoară, Dragoş Ioan Săcăleanu, and Rodica Stoian

Research Center for Spatial Information (CEOSpaceTech), Department of Applied Electronics and Information Engineering
University Politehnica of Bucharest,
Romania
lucian.perisoara@upb.ro, dragos_sacaleanu@yahoo.com, rodicastoian2004@yahoo.com

Abstract—In the design phase of a communication system, computer simulations allow us to evaluate the system performances in order to choose the optimal techniques, algorithms, and parameters values for the constituent blocks, and to discover the design issues as soon as possible in order to correct them. This paper presents a software simulator developed in LabVIEW, which implements a Multiple-Input Multiple-Output (MIMO) communication system with two transmit and two receive antennas and Alamouti coding over terrestrial fading channels. Finally, the MIMO system performances are presented and compared for different channels, modulations and error correcting codes.

Keywords—*MIMO systems; fading channels; space-time block codes; virtual instruments.*

I. INTRODUCTION

In wireless communications, the use of multiple antennas at both sides of the link, known as MIMO (Multiple-Input Multiple-Output), can increase the capacity and reliability of the channel and combat the effects of multipath fading without any additional power or bandwidth expenditure compared to SISO (Single-Input Single-Output) systems [1], [2].

At transmitter side, the beamforming techniques can be used for known channels to assure both the diversity and array gains. If the channel is unknown, then the Space-Time Codes (STC) are used to assure only a diversity gain. STC are a more generally class of error correcting codes, the control symbols being inserted in both spatial and temporal domains. In [3], Alamouti has introduced a simple diversity scheme for two transmit antennas, which provides a diversity gain of 2 and no coding gain with minimum complexity. Later, the Alamouti code was generalized for an arbitrary number of transmit antennas in [4].

Computer simulation allows us to evaluate the performance of an information transmission system in the system design phase, by simulating the all transmission chain: transmitter, channel, and receiver. While the final objective is to build the physical system, we must discover the design issues as soon as possible in order to correct them. Some requirements for a software simulator should be: modular and highly flexible, the simulated blocks should model very well the models for the real world cases, the model parameters should be available to the user, easy to understand and implement, easy to extend to include other blocks and modules in the future (different modulations, different error correcting codes, different channel models, etc.), low computational complexity [5].

The LabVIEW development environment from National Instruments is a visual programming language based on virtual instruments (VI's), which is usually used for data acquisition, instrument control, and industrial automation. The major advantages of using LabVIEW are the possibilities of connecting, communication and control with measurement instruments, the design and development of performant, flexible and scalable systems, and it is easy to learn, use, maintain and update. With the aid of NI Modulation Toolkit, custom applications can be rapidly developed for research, design, characterization, validation, and testing of communications systems [6].

Section II presents the signal model of Alamouti encoder / decoder for a 2x2 MIMO system. Section III describes the MIMO Simulator developed in LabVIEW and Section IV shows the simulation results and performance comparison for different channels, modulations and error correcting codes. Section V draws the conclusions of this paper.

II. THE MIMO SYSTEM

We suppose a 2x2 MIMO system, with two transmit antennas, Tx1, Tx2, and two receive antennas, Rx1, Rx2. The transmission of the same bit stream from the two antennas can be made using Alamouti code described by the transmission matrix or the codeword [3]:

$$\mathbf{G} = \begin{bmatrix} x_1 & x_2 \\ -x_2^* & x_1^* \end{bmatrix}. \quad (1)$$

The transmitter picks two complex symbols from a Quadrature Amplitude Modulation (QAM) or Phase-Shift Keying (PSK) constellation, for example x_1 and x_2. In the first time slot, the two antennas send the symbols x_1 and x_2, simultaneously. Then, in the second time slot, the symbols $-x_2^*$ and x_1^* are send. In this way, both symbols x_1 and x_2 are spatially spread over the two antennas and temporally spread over the two time slots.

For a 2x2 MIMO flat fading channel, the channel matrix can be written as $\mathbf{H} = \begin{bmatrix} h_{ij} \end{bmatrix}$, where the complex fading coefficients h_{ij} correspond to propagation paths from the transmit antenna j to the receive antenna i, and are independently and identically Rayleigh distributed (the real and imaginary parts are Gaussian random variables with zero-mean and variance 1/2 per dimension) [7].

This work was supported by Romanian Space Agency, Space Technology and Advanced Research (STAR) Programme, project no. 75/2013. The work of L. A. Perişoară was funded by the Sectoral Operational Programme, Human Resources Development, financed from the European Social Fund and the Romanian Government under the contract POSDRU/159/1.5/S/137390.

For two receive antennas, the symbols at the output of the MIMO channel, at two time slots, are:

$$y_{11} = h_{11}x_1 + h_{12}x_2 + n_{11}, \qquad y_{12} = -h_{11}x_2^* + h_{12}x_1^* + n_{12},$$
$$y_{21} = h_{21}x_1 + h_{22}x_2 + n_{21}, \qquad y_{22} = -h_{21}x_2^* + h_{22}x_1^* + n_{22}, \qquad (2)$$

where n_{ij} are the additive noise samples for the receive antenna i and time slot j, which are modeled as independent complex Gaussian random variables with zero-mean and variance σ_n^2. In matrix form, we can write:

$$\mathbf{y}_1 = \mathbf{H}\begin{bmatrix} x_1 \\ x_2 \end{bmatrix} + \mathbf{n}_1, \qquad \mathbf{y}_2 = \mathbf{H}\begin{bmatrix} -x_2^* \\ x_1^* \end{bmatrix} + \mathbf{n}_2. \qquad (3)$$

The relation between the input of the Alamouti encoder and the output of the MIMO channel can be derived as:

$$\mathbf{y} = \begin{bmatrix} \mathbf{y}_1 \\ \mathbf{y}_2^* \end{bmatrix} = \begin{bmatrix} h_{11} & h_{12} \\ h_{21} & h_{22} \\ h_{12}^* & -h_{11}^* \\ h_{22}^* & -h_{21}^* \end{bmatrix} \begin{bmatrix} x_1 \\ x_2 \end{bmatrix} + \begin{bmatrix} \mathbf{n}_1 \\ \mathbf{n}_2^* \end{bmatrix} = \mathbf{H}_{ef}\mathbf{x} + \mathbf{n}, \qquad (4)$$

where $\mathbf{H}_{ef}$ is the matrix of equivalent channel formed by the Alamouti encoder and the 2x2 MIMO channel.

Supposing that we always know the channel at receiver (we have Channel State Information at Receiver, CSIR), we can use the Maximal Ratio Combining (MRC) technique to combine the two streams from the two antennas in one single stream. The coefficients of the linear combination can be optimally chosen equal with the Hermitian of equivalent channel matrix, so the combined symbols are [7], [8]:

$$\tilde{\mathbf{x}} = \mathbf{H}_{ef}^H \mathbf{y}, \qquad (5)$$

which can be detailed as:

$$\tilde{x}_1 = h_{11}^* y_{11} + h_{12}y_{12}^* + h_{21}^* y_{21} + h_{22}y_{22}^*,$$
$$\tilde{x}_2 = h_{12}^* y_{11} - h_{11}y_{12}^* + h_{22}^* y_{21} - h_{21}y_{22}^*. \qquad (6)$$

Finally, the combined symbols $\tilde{x}_1$ and $\tilde{x}_2$ from the two time slots are applied to a classical Maximum Likelihood (ML) decoder in order to obtain the transmitted symbols $\hat{x}_1$ and $\hat{x}_2$, the decision rules being:

$$\hat{x}_1 = \underset{x_1}{\arg\min} |x_1 - \tilde{x}_1|^2, \qquad \hat{x}_2 = \underset{x_2}{\arg\min} |x_2 - \tilde{x}_2|^2. \qquad (7)$$

III. THE MIMO SIMULATOR

The software simulator for a MIMO communication system over terrestrial fading channels using Alamouti coding is implemented in LabVIEW 2014 using the Modulation Toolkit.

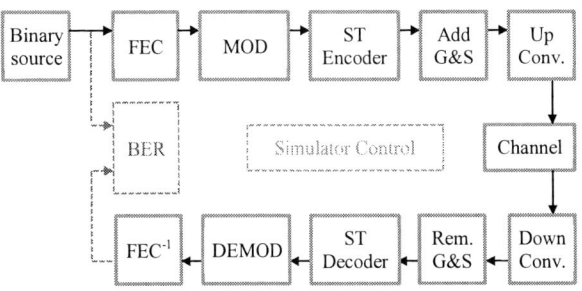

Fig. 1. The block diagram of the developed MIMO simulator.

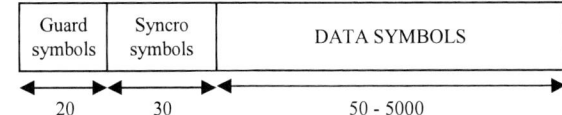

Fig. 2. The structure of the transmitted packet.

The block diagram of the MIMO simulator from Fig. 1 contains following main blocks with relevant parameters:

- random binary information source: number of frames (1000), number of bits per frame (1000), total number of generated bits indicator;

- channel encoder / decoder: type of Forward Error Correcting (FEC) code (none, block or convolutional), rate of block code (1/3, 4/7, 11/15, 26/31 or 57/63), systematic or nonsystematic block code, rate of convolutional code (1/2, 1/3, 1/4, 2/3 or 3/4), constraint length of convolutional code (between 3 and 17), trellis decoding depth, channel interleaver (none, block or convolutional). Also, we have the option not to use any code in simulations, so the channel is uncoded;

- modulator / demodulator (MOD/DEMOD): modulation type (PSK or QAM), modulation order (2 – 256 symbols) and filter type for pulse shaping (none, raised cosine, root raised cosine or Gaussian). For PSK we can choose further particular modulations (shifted PSK, Offset QPSK or Differential PSK);

- space-time (ST) encoder / decoder to use multiple antennas: type of space-time block code (Alamouti), type of combining technique (Maximum Ratio Combining, MRC) and Maximum Likelihood (ML) decoding. These options are activated only if the MIMO 2x2 channel is chosen;

- guard and synchronization (G&S) symbols insertion/ extraction: number of guard symbols and number of synchronization symbols inserted at the beginning of a transmitted packet, see Fig. 2. These symbols are BPSK modulated, whatever would be the modulation for data. To check the correct synchronization between transmitter and receiver, the start position of recovered synchronization sequence is displayed, which must be equal with the number of guard bits;

- baseband to passband converter (Up Conv./Down Conv.): the frequency of the carrier signal, and the power of the transmitted signal. We can deactivate this step, because the simulation of complex baseband transmission system is sometimes enough, the up/down converters having reached perfection of the hardware design and do not contribute to the bit error rate;

- communication channel: channel type (SISO, MIMO 2x2), signal-to-noise ratio (SNR) for each receive antenna, fading type (none, flat fading, selective fading). For no fading case, an Additive White Gaussian Noise (AWGN) channel is simulated. For flat or selective fading cases, we can choose the distribution and the model of fading generator (Rayleigh-Gans, Rayleigh-Jakes, Rician-Gans or Rician-Jakes), the fading variance and Doppler spread. For Rician distribution we can modify the K parameter. If we choose the MIMO 2x2 channel, then the options of the ST encoder & decoder will be activated;

- simulator control: start and stop buttons, indicator and progress bar for running a simulation, current frame indicator, total transmitted bits indicator, the number of bits per symbol indicator. To control the speed of simulation we have a pointer slide, which introduces a simulation delay between two loops. Other system parameters are the number of samples per symbol (16), and the transmitted symbol rate (1 MHz);

- performance evaluation of simulated communication system: the number of errors per transmitted frame, number of accumulated errors and cumulative Bit Error Rate (BER) from the beginning of a simulation and till the current moment.

Based on the block diagram of MIMO system from Fig. 1, we have implemented in LabVIEW a MIMO simulator having the front panel shown in Fig. 3. Following areas are marked:

- A contains all numeric controls and indicators, combo boxes, buttons, etc., used for setting and visualization of parameters for all blocks of the simulator;

- B contains 6 waveform graphs to visualize the information messages (left), the coded messages (center) and the coded messages with the guard and synchronization symbols inserted (right), for the transmitter (up) and for the receiver (down);

- C contains 4 waveform graphs to visualize the I and Q signals after modulation or at the input of the channel (up) and the noisy signals from the output of the channel (down), which are affected by noise and fading. For the SISO channel, there are activated only the left graphs (Tx1, Rx1), and for the MIMO channel are activated also the right graphs (Tx2, Rx2);

- D contains 4 constellation graphs to display the symbols constellations from the transmitter (up) and the receiver (down), and for Tx1, Rx1 antennas (left) and Tx2, Rx2 antennas (right);

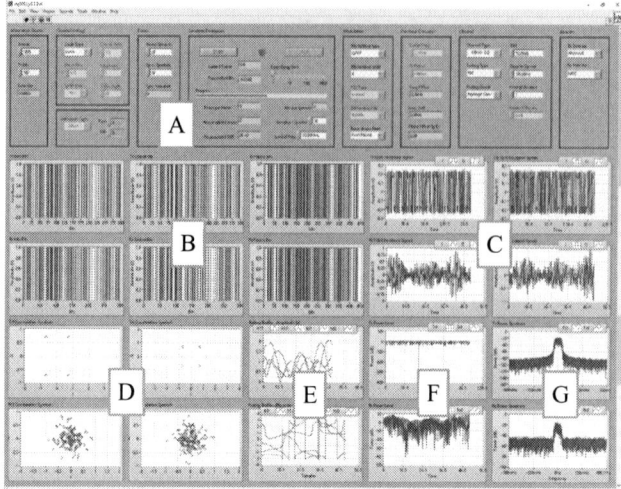

Fig. 3. The LabVIEW front panel (user interface) of the MIMO simulator.

- E contains two waveform graphs to visualize the amplitude and phase characteristics of the fading channel;

- F contains two waveform graphs to visualize the instantaneous power for transmitter (up) and for receiver (down);

- G contains two power spectrum densities graphs for transmitter (up) and for receiver (down).

IV. PERFORMANCE RESULTS

To evaluate the MIMO system performances, the Bit Error Rate (BER) is measured and plotted versus the channel Signal to Noise Ratio (SNR) using Monte-Carlo simulations. The simulation ends after 500 errors are found for low SNR and 1000 errors for high SNR.

Firstly, the BER(SNR) performances for a SISO uncoded system with PSK and QAM modulations over AWGN channels are shown in Fig. 4. To assure a certain BER for transmission, for noisy channels we must use low order modulations, like BPSK. For high order modulations, the performances decrease (the BER curves are shifted to right) because the constellations are more dense and are more sensitive to errors. Also, high order QAM modulations assures a higher bitrate than PSK modulations for the same BER performances, e.g. 256-QAM vs. 64-PSK. For 8-QAM, 32-QAM and 128-QAM, the BER(SNR) curves have different slopes, because the constellation maps are not perfectly squared like the other modulations.

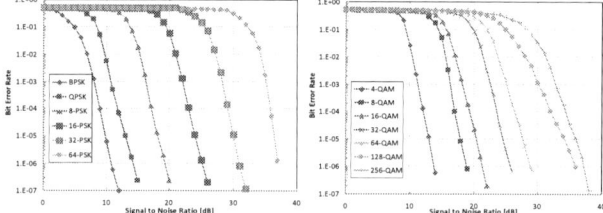

Fig. 4. BER(SNR) for an uncoded SISO AWGN system with PSK and QAM.

2016 IEEE 22nd International Symposium for Design and Technology in Electronic Packaging (SIITME)

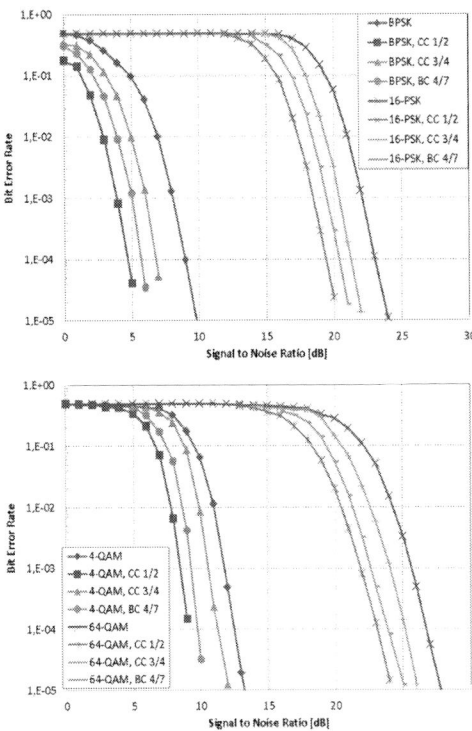

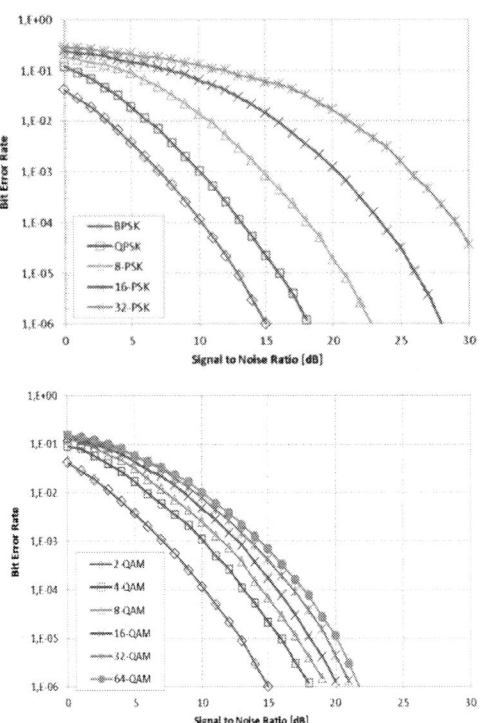

Fig. 5. BER(SNR) for SISO AWGN coded system with PSK and QAM.

Fig. 6. BER(SNR) for MIMO 2x2 Alamouti system with PSK and QAM.

Secondly, the BER(SNR) performances for a SISO coded system over AWGN channels are shown in Fig. 5 only for BPSK, 16-PSK, 4-QAM and 64-QAM modulations. The error correcting codes are convolutional codes (CC) of rates 1/2 and 3/4 and block code (BC) of rate 4/7. For all modulations, the best performances are obtained for the convolutional code of rate 1/2 (BER = $4.1*10^{-5}$ at SNR = 5 dB for BPSK), because it has the smallest rate and inserts the biggest redundancy in the transmitted message. For BPSK, at BER=10^{-4}, the coding gains are about 2.2 dB (CC 3/4), 3.2 dB (BC 4/7) and 4.2 dB (CC1/2) compared to uncoded channel. Also, the coding gains remain almost the same for other modulations.

Thirdly, the BER(SNR) performances for the MIMO 2x2 system using Alamouti coding are presented for PSK and QAM modulations in Fig. 6. The fading channel model is Rayleigh-Gans. The error correcting codes are not used. If we increase the data rate (high order modulations), the performances decrease (higher BER) for a fixed SNR. Between all modulations, the best performance is obtained for BPSK (BER = 10^{-6} at SNR = 15 dB). The performances of BPSK and 2-QAM are the same; also for QPSK and 4-QAM (constellations are identical). But, for high order modulations, 64-QAM gives better performance than 8-PSK.

V. CONCLUSIONS

In this paper, we have implemented in LabVIEW a software simulator for a MIMO communication system with two transmit and two receive antennas using Alamouti coding. The obtained performance results are according with those from the literature, which confirms the correct implementation of the simulator.

As future work, the system performances may be improved by using more powerful error correcting codes, like parallel concatenated codes (turbo codes). For wideband channel models, the Orthogonal Frequency-Division Multiplexing (OFDM) technique is necessary to be used.

This work is a first step in implementing a real-time MIMO communication system using FlexRIO modules for PXI instruments from National Instruments.

REFERENCES

[1] G. J. Foschini, M. J. Gans, "On the limits of wireless communication in fading environment when using multiple antennas", Wireless Personal Communications, vol. 6, no. 3, pp. 311–335, Mar. 1998.

[2] L. A. Perişoară, M. Neghină, R. Stoian, "Performances of FEC Coded MIMO OFDM Systems," 4th European Conference on Circuits and Systems for Communications (ECCSC 2008), Bucharest, Romania, pp. 295-300, July 10-11, 2008.

[3] S. M. Alamouti, "A simple transmit diversity technique for wireless communications," IEEE Journal on Selected Areas in Communications, vol. 16, no. 8, pp. 1451-1458, Oct. 1998.

[4] V. Tarokh, H. Jafarkhani, A. R. Calderbank, "Space–time block codes from orthogonal designs," IEEE Trans. Inform. Theory, vol. 45, no. 5, pp. 1456-1467, July 1999.

[5] C. E. D. Sterian, Space-Time Trellis Coding. A Primer, Politehnica Press, Bucharest, 2016.

[6] http://sine.ni.com/nips/cds/view/p/lang/en/nid/210568

[7] H. Jafarkhani, Space-Time Coding. Theory and Practice, Cambridge University Press, New York, 2005.

[8] L. A. Perişoară, "BER Analysis of STBC Codes for MIMO Rayleigh Flat Fading Channels", Telfor Journal, Vol. 4, No. 2, pp. 78-82, 2012.

978-1-5090-4446-7/16 $31.00 © 2016 IEEE   255   20-23 Oct 2016, Oradea, Romania

2016 IEEE 22nd International Symposium for Design and Technology in Electronic Packaging (SIITME)

Single-Phase Inverter for Solar Energy Conversion Controlled with DSpace DS1104

Dorin Petreus, Toma Patarau, Radu Truta, Cristian Orian and Radu Etz

Department of Applied Electronics
Technical University of Cluj-Napoca
Cluj-Napoca, Romania
toma.patarau@ael.utcluj.ro

Abstract— **The present paper presents a system intended to be used in the development of photovoltaic inverters with maximum power point tracking capability. A simulation model was implemented in MATLAB/SIMULINK and an experimental model composed of a DSpace 1104 controller board, a Semikron inverter, a coupling transformer, an LC filter and two 75W solar panels was developed. The system facilitates the development and testing of digital control and MPPT algorithms for single-phase inverters connected to the utility grid.**

Keywords— Solar inverter, DQ control, DSpace, MPPT.

I. INTRODUCTION

Exhausting fossil fuels lead to increased interest in harvesting and utilizing renewable energy resources. PV is one of the top eco-friendly energy sources being pollution-free, maintenance-free and having long life span. Moreover, this type of energy is inexhaustible [1, 2]. Compared with other types of renewable energies like wind, geothermal or biomass, the PV energy source is a DC power source and works in conjunction with an inverter to convert the energy to AC [3]. The majority of PV systems require inverters as interfacing units.

In spite of their advantages, solar systems also have some drawbacks. Some of the main issues posed by PV systems are: the impact of different loads on the output voltage of the inverter, nonlinearity, low efficiency of PV panels, intermittent nature of solar energy, electromagnetic interference, harmonic

content of the generated energy and many more. It was observed that the control strategy of the inverter is a key component in the proper and efficient operation of the entire PV system [4].

It was also noted that the main cause of failure in a PV system is the failure of the inverter [5, 3]. Decreasing solar panel prices lead designers to increase their project loading ratio of their inverters. The loading ratio is defined as the ratio of DC module capacity to AC inverter capacity. It was observed that for a loading ratio higher than 2.0 more than 16% potential energy generation is lost [6]. Hence, good operation of PV systems depends heavily on the design and operation of the DC-AC inverters.

As follows opportunities still exist for power electronics engineers to improve the operation of PV inverters and controllers. The present paper presents a rapid prototyping system intended to be used in the development of photovoltaic inverters with maximum power point tracking capability. The bock diagram of the system is represented in Fig. 1. It comprises of two solar modules connected in series, a Semikron inverter and a dSpace DS1104 controller board. The control algorithm for the inverter is implemented in Matlab/Simulink compiled and loaded to dSpace controller board. A simulation model for the entire system is also developed in Simulink to facilitate the development of the control and maximum power point tracking (MPPT) algorithms.

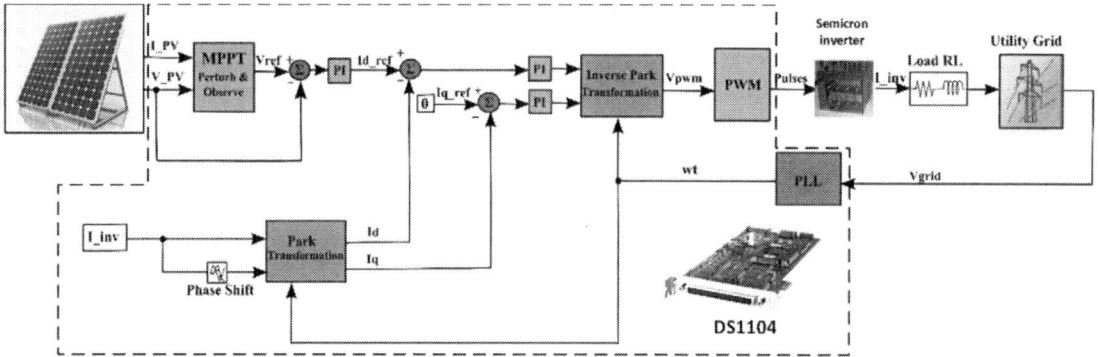

Fig. 1. Bloc diagram of the inverter

978-1-5090-4446-7/16 $31.00 © 2016 IEEE
256
20-23 Oct 2016, Oradea, Romania

The remainder of the paper is organized in six main sections. First, in Section 2, the simulation model is presented. Section 3 presents the simulation results. Section 4 describes the experimental model and section 5 presents the obtained results. The final section will draw the conclusions.

II. SIMULATION MODEL

The following section describes each main bloc of the photovoltaic system simulation model and its implementation in Simulink leading to the description of the complete model at the end of the section.

A. Solar panel

Photovoltaic cell is basically a semiconductor p-n junction based photodiode which generates electrical power when it is exposed to sunlight. The power produced by a single photovoltaic cell is not enough for general use therefore, by connecting them in series and parallel, higher voltages currents and power can be obtained. Figure 1 illustrates a simple equivalent circuit of a photovoltaic cell [7].

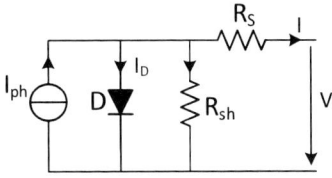

Fig. 2.   Equivalent circuit of the photovoltaic cell

This model consists of a current source, I_{ph}, which represents the generated current from photovoltaic cell, a diode D in parallel with the current source, a shunt resistor, R_{sh} expressing the leakage current, and a series resistor R_s describing the internal resistance to the current flow.

B. Inverter model

The inverter schematic is represented in Fig. 3. It is an H-bridge inverter composed of 4 IGBT transistors IGBT1 to IGBT4 and an input capacitor C.

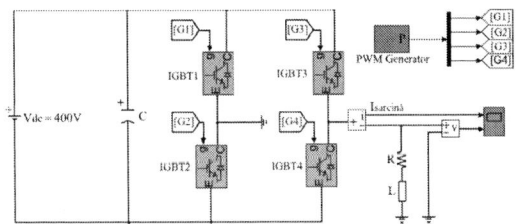

Fig. 3.   Inverter

The input capacitor can be determined based on the DC voltage ripple.

$$C = \frac{S}{2w_g V_{DC} \Delta V_{DC}} \tag{1}$$

where S is the apparent power, w_g is the grid frequency, V_{DC} is the nominal input voltage and ΔV_{DC} is the input voltage ripple.

C. Maximum power point tracking algorithm (MPPT)

A photovoltaic array has an optimum operating point, known as the maximum power point, which varies according to cell's temperature and irradiation level. The usual approach for maximizing the power drawn from solar panels under varying atmospheric conditions is to use a MPPT algorithm. Perturb and Observe algorithm is widely used in photovoltaic systems because of its simplicity and ease of implementation. The operating voltage of the photovoltaic array is perturbed by changing the quantity into a given direction and the power drawn from the photovoltaic array is measured. If it increases, then the operating voltage is further perturbed in the same direction whereas, if it decreases, the direction of operating voltage perturbation is reversed. The drawbacks of this method are that the operating point oscillates around the maximum power point, slow response speed, and even tracking in wrong direction under rapidly changing atmospheric conditions [8]. The sequence of operations performed is shown in the flowchart given below in Fig.4.

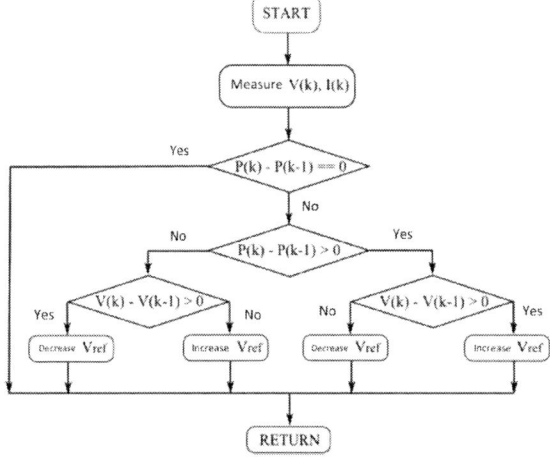

Fig. 4.   Perturb and Observe

D. Grid syncronization

In the synchronous frame control method, the amplitude and phase of the grid voltage needs to be known for the control system. This information is essential for the current and voltage control loops to work properly. Therefore, a synchronization method is employed to synchronize the inverter output and utility grid. There are various methods to extract the phase information from a given signal [9]. Because of its advantages and wide practical usage, phase locked loop (PLL) will be employed in the simulation model. A basic PLL circuit consists of three essential components: a phase detector, a loop filter, and a voltage controlled oscillator (VCO). Using a negative feedback loop, PLL minimizes the phase and frequency errors between the input and output signals [10, 11, 12]. The structure of PLL is shown in Fig. 5. As it can be noticed, this structure uses the coordinate Park transformation and the lock is realized by setting the Vd to zero. A proportional integral regulator can be used to control this variable and the output of this regulator is the grid frequency. After the integration of the grid frequency, the utility voltage angle is obtained, which is fed back into the Park transformation module in order to transform into the synchronous rotating reference frame.

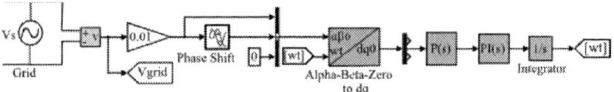

Fig. 5. Grid syncronization PLL

E. Control loops

The system uses two control loops: one external voltage loop that gives the reference values for the inner current control loop of the inverter. There are two current PI controllers for direct, Id, and quadrature, Iq, components of inverter output current and one for the voltage loop. As can be seen in the Fig. 6, the voltage set-point for voltage control loop is made through the maximum power point controller. The set-point for current component, Id, is made by the voltage control loop and the reference value for the quadrature current component, Iq, is forced to zero to minimize the injection of reactive power to the grid [4 truta].

To simplify the control design process of a three-phase grid-connected system usually two fundamental transformations are used in order to reduce the dimensions of the mathematical model of the system and decouple the differential equations. These transformations are Clarke transformation, and Park transformation. Clarke's transformation converts a three-dimensional system to a two-dimensional stationary system, $\alpha\beta$. The two coordinate orthogonal $\alpha\beta$ components are then transformed into dq rotating frame quantities using Park's transformations [8 truta]. If the dq frame is rotating with the same frequency as the grid voltage, then the output components of the Park transformation are stationary. Because the inverter is single phase only the Park transformation is used. For the α axis component the grid current is used and the β component is created artificially delaying the α component with $\pi/2$, Fig. 6. The equations describing the direct and inverse park transformation are presented below:

$$\begin{bmatrix} I_d \\ I_q \end{bmatrix} = \begin{bmatrix} \cos(\theta) & \sin(\theta) \\ -\sin(\theta) & \cos(\theta) \end{bmatrix} \begin{bmatrix} I_\alpha \\ I_\beta \end{bmatrix} \quad (2)$$

$$\begin{bmatrix} I_\alpha \\ I_\beta \end{bmatrix} = \begin{bmatrix} \cos(\theta) & -\sin(\theta) \\ \sin(\theta) & \cos(\theta) \end{bmatrix} \begin{bmatrix} I_d \\ I_q \end{bmatrix} \quad (3)$$

where θ is the angle supplied by the PLL loop. The transformation allows the use of classical PI controllers to be used to control the inverter.

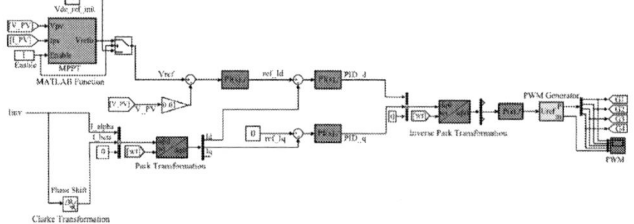

Fig. 6. Control loops

F. Complete system description

The detailed simulation model of the inverter and its controller is shown in Fig. 7. The solar panel model, the inverter, the control loops and the PLL are represented.

The model shown in Fig. 7 represents only the Simulink implementation of a solar inverter control technique. The next step was to modify it in order to use it with dSpace DS1104 platform. The DS1104 is a so called rapid prototyping real time development board, which means that generation of a working prototype is involving very small procedural delay. Real time interface provides some new Simulink blocks as: analog to digital converters, digital to analog converters, and PWM generator block.

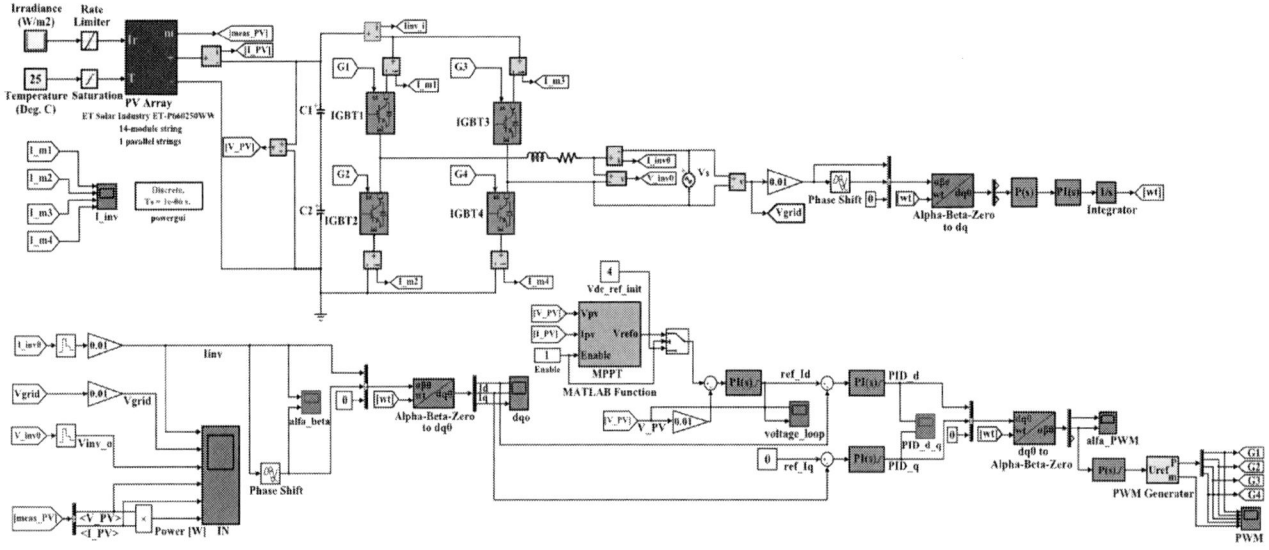

Fig. 7. Complete simulation model

2016 IEEE 22nd International Symposium for Design and Technology in Electronic Packaging (SIITME)

III. SIMULATION RESULTS

The effect of irradiance on current-voltage and power-voltage characteristic curve is shown in Fig. 8. From the figure, it can be concluded that when the irradiance is 1000W/m², the output power of the solar panel is 150W. As the solar radiation decreases, the output power decreases also.

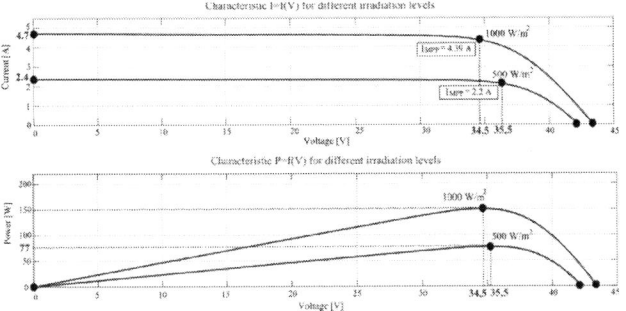

Fig. 8. The solar panel characteristics at different values of irradiance

Fig 9 represents the waveforms that describe the operation of the system when a change in irradiance occurs. The irradiance [W/m²], power [W], inverter output current [A] attenuated by a factor of 100, grid voltage [V] attenuated by a factor of 1800, solar panel voltage [V] and the solar panel current [A] are shown.

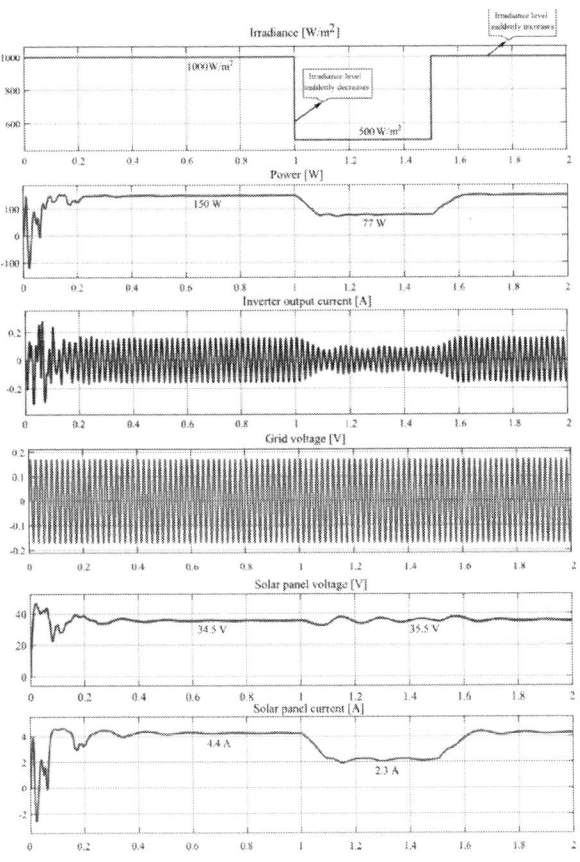

Fig. 9. Waveforms for the system being modeled.

It can be observed that once the maximum operating point is found the system starts oscillating around it. Analyzing Fig. 8 and 9 it can be concluded that perfect match exists between the solar panel values obtained from its I-V and P-V characteristics and simulation results of the proposed model.

The response of the MPPT algorithm are represented in Fig. 10. A zoom is shown in Fig 10 that shows the oscillation of the MPPT around the MPP point.

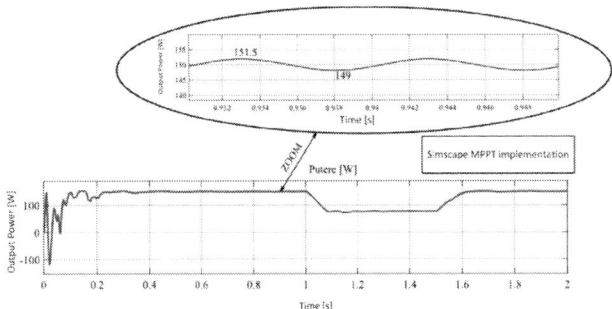

Fig. 10. MPPT response

IV. EXPERIMENTAL SETUP

A. DSpace controller

DS1104 is a controller board suited for inverter control. It enables the link with Matlab/Simulink. Using Simulink, the algorithms are implemented, compiled and converted to C-code which is then loaded automatically to the real time dSPACE processor. Linking process implies using dSPACE input-output library components like: DS104ADC, DS104DAC, etc. A block diagram of dSPACE is represented in Fig.10

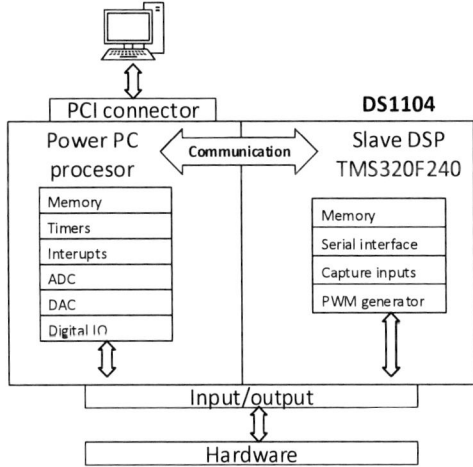

Fig. 11. Simplified diagram of dSPAEC DS1104 controler board

B. DSpace implementation

The algorithm implemented in dSPACE is represented in Fig. 11. Three DS1104ADC blocks are used to read the PV voltage, the grid voltage and the grid current. Four DS1104DAC are used for testing and a DS1104SL_DSP block is used to generate the PWM signals to control the inverter.

2016 IEEE 22nd International Symposium for Design and Technology in Electronic Packaging (SIITME)

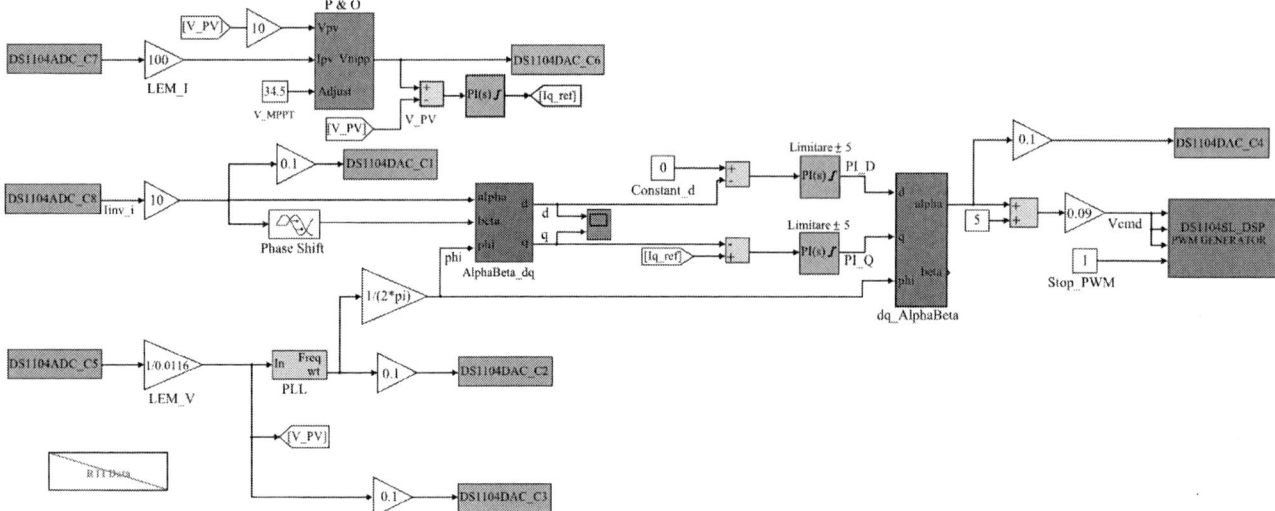

Fig. 12. Control algorithm implemented in DSpace

V. EXPERIMENTAL RESULTS

The experimental results that describe the operation of the inverter are represented in Fig. 13. The first waveform presents the phase of the grid voltage detected by the PLL circuit, second waveform represents the output current of the inverter injected into the grid and the last waveform represents the output voltage of the inverter filtered.

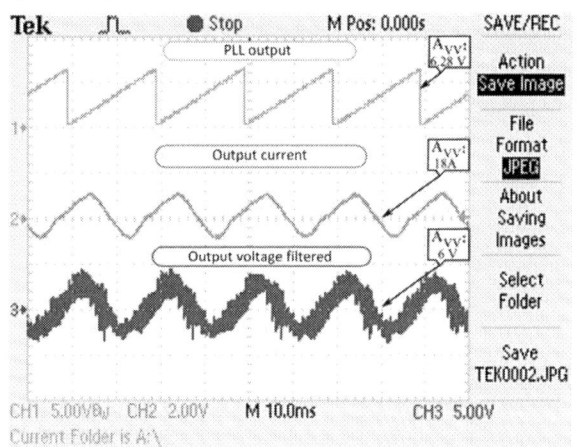

Fig. 13. Experimental results

Fig 14 represent the operation of the MPPT algorithm. The step frequency is 1kHz and the amplitude of the step voltage is 30mV.

The experimental setup is represented in Fig. 15. It consists of: 1 – oscilloscope, 2- current probe, 3 - DSpace DS1104 control board, 4 – inverter driver, 5 – voltage and current sensors, 6 – differential voltage probe, 7 – grid coupling transformer, 8 – laboratory power supply, 9 – single phase H-bridge inverter, 10 – LC filter.

Reference voltage supplied by the MPPT algorithm

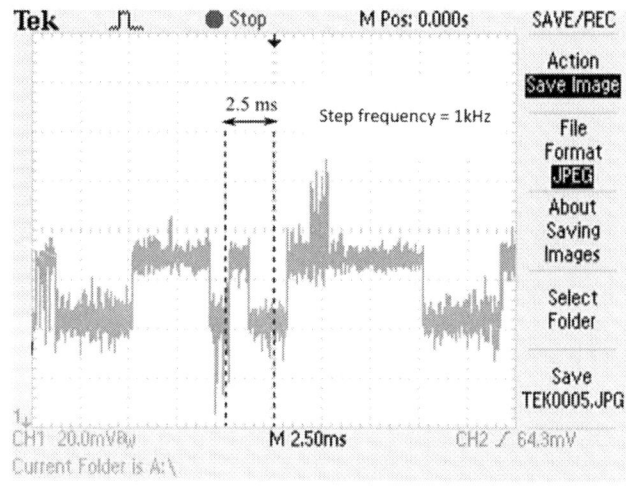

Fig. 14. MPPT operation

Fig. 15. Experimental setup

VI. CONCLUSIONS

The paper presented the development of a system intended to be used in the development of photovoltaic inverters with maximum power point tracking capability. A simulation model was implemented in MATLAB/SIMULINK and an experimental model composed of a DSpace 1104 controller board, a Semikron inverter coupled to the utility grid through a coupling transformer, an LC filter and two 75W solar panels was developed. It can be concluded that the proposed model presented in this paper achieves good results and can be used successfully in rapid prototyping development, digital control and MPPT algorithm development and test of solar inverters.

ACKNOWLEDGMENT

This paper is supported through the program "Parteneriate in domenii prioritare – PN II", by MEN – UEFISCDI, project no. 53/01.07.2014.

REFERENCES

[1] G.E. Ahmad, H.M.S. Hussein, H.H. El-Ghetany "Theoretical analysis and experimental verification of PV modules" Renewable Energy, vol 28, pp. 1159–1168, 2003.

[2] A. Chel, G.N. Tiwari, A. Chandra "Simplified method of sizing and life cycle cost assessment of building integrated photovoltaic system", Energy and Buildings, vol. 4, pp. 1172–1180, 2009.

[3] F. Blaabjerg, Z. Chen, S. Kjaer "Power electronics as efficient interface in dispersed power generation systems", IEEE Transactions on Power Electronics, vol. 19, pp. 1184–1194, 2004.

[4] Z.A. Ghania, M.A. Hannana, A. Mohameda, "Simulation model linked PV inverter implementation utilizing dSPACE DS1104 controller", Energy and Buildings, Vol. 57, Pages 65–73, February 2013.

[5] R.A. Messenger, J. Ventre "Photovoltaic System Engineering" CRC Press, Boca Raton, FL 2004

[6] Jeremy Good, Jeremiah X. Johnson, "Impact of inverter loading ratio on solar photovoltaic system performance", Applied Energy, Vol. 177, , Pages 475–486, 1 September 2016.

[7] A.A. Hassan, F.H. Fahmy, A.A. Nafeh, M.A. El-Sayed, "Modeling and simulation of a single phase grid connected photovoltaic system", WSEAS Transactions on Systems and Control, vol. 5, pp. 16-25, 2010.

[8] R. Kiranmayi, K. Vijaya Kumar Reddy and M.Vijaya Kumar, "Modeling and a MPPT Method for Solar Cells", Journal of Engineering and Applied Sciences, Vol. 3, N1, pp. 128 - 133, 2008.

[9] Remus Teodorescu, Marco Liserre and Pedro Rodríguez "Grid converters for photovoltaic and wind power systems" 2011 John Wiley & Sons, Ltd. ISBN: 978-0-470-05751-3

[10] Arti Gadekar, V. B. Virulkar, "Effective dspace Inverter Controller for PV Application", IEEE Students' Conference on Electrical, Electronics and Computer Science, Amravati, India, 2014.

[11] Michael E. Ropp, Member IEEE, and Sigifredo Gonzalez, "Development of a MATLAB/Simulink Model of a Single-Phase Grid-Connected Photovoltaic System", IEEE Transactions on Energy Conversion, Vol. 24 no. 1, pp 195 - 202 · April 2009; DOI: 10.1109/TEC.2008.2003206

[12] A.A. Hassan, F.H. Fahmy, A.A. Nafeh, M.A. El-Sayed, "Modeling and simulation of a single phase grid connected photovoltaic system", WSEAS Transactions on Systems and Control, Vol. 5, pp. 16-25, 2010.

Obstacle avoidace algorithm

Sabou Sebastian

Electrical, Electronic and Computer Engineering
Department
Technical University of Cluj Napoca - North University
Center of Baia Mare
Baia Mare, Romania
sebastian.sabou@cunbm.utcluj.ro

Lung Claudiu

Electrical, Electronic and Computer Engineering
Department
Technical University of Cluj Napoca - North University
Center of Baia Mare
Baia Mare, Romania
claudiu.lung@cunbm.utcluj.ro

Abstract— **Algorithm for autonomous navigation of robots are the most diverse, given the wide range of use of robots and their applications. Robot's environment is crucial for navigation and its orientation, and therefore all approaches to this problem begin from this start point. One of the most important aspects is to avoid obstacles found on the path the robot. This paper presents an obstacle avoidance algorithm with increased performance, starting from Newton method for calculating the direction, adding an additional priority function. The results showed increased efficiency of this type of algorithm.**

Keywords— avoid obstacle algorithm, autonomous navigation

I. INTRODUCTION

There are certain specifications for autonomous system to consider when designing the avoid obstacle algorithm, for example updating position is at a frequency of (approximately) 30 Hz, for faster calculations it was assumed that the robot and all obstacles have a circular shape and have a diameter of 50cm . The position updating frequency is given by the speed of obtaining data from the sensors and the time required for calculations, therefore the type of microprocessor used and the algorithm used.

The reason for using Newton direction is the use 2 order derivative, which means that the trajectory is evaluated after the first derivative, it is possible to observe the trend to move towards left or right. This information is used to decide the moving direction of the mobile sistem.

In figure 1 can be seen two methods for determining the direction, the green line in first order derivative and the red line is second order derivative or Newton direction methods. The difference between the two methods is visible, the optimal was

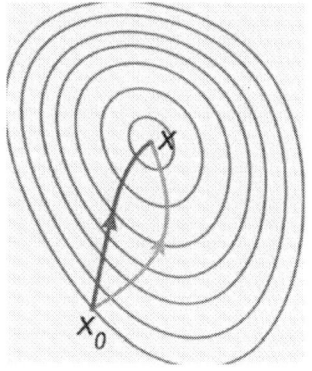

Fig. 1 Direction determining

the Newton direction (red line path).

In figure 1 can be seen two methods for determining the direction, the green line in first order derivative and the red line is second order derivative or Newton direction methods. The difference between the two methods is visible, the optimal method is the Newton direction (red line path).

II. DESCRIPTION OF ALGORITHM

A. Destination function

The autonomous system have a start point and a destination point. The function for destination is

$$f(\vec{p}, \overrightarrow{p_T}) = (x_T - x)^2 + (y_T - y)^2 \qquad (1)$$

Where $\vec{p} = [x, y]$ is current position of sistem and $\overrightarrow{p_T} = [x_T, y_T]$ is position for the destination point. The minimum of this function is reached where $\vec{p} = \overrightarrow{p_T}$.

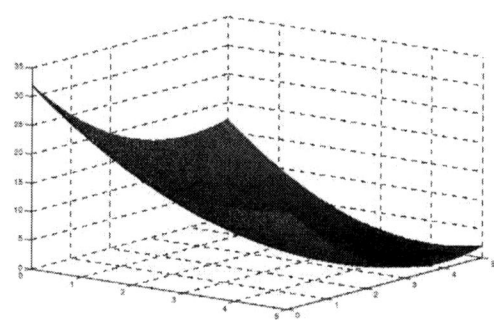

Fig. 2 Evaluation of destination function

B. Limit function

All surfaces where the sistem is allowed have a limit. The limit function offer information about localization, shape and dimension of objects. In this study was assumed that all objects have a circular shape, all have a center $\overrightarrow{p_b}$ and radius R of 25 cm. The mobile system have no dimension and no shape, it is modeled as a point. This constrains makes fast data processing and to compensate this fact (the system model is a point) all limits need to be bigger then 25cm.

The limit function for a circular shape object j:

$$g_j\left(\vec{p}, \vec{p}_{B_j}, R\right) = -\left(x_{B_j} - x\right)^2 - \left(y_{B_j} - y\right)^2 + R^2 \quad (2)$$

The interior area of this object will be allways positive and the exterior area of this circular object will be always negative.

C. Barrier Function

Finding the minimum of function is easy, more difficult is to generate the path for the mobile system without passing over any limit.

The barrier function generate a protection area, showed in Fig. 3.

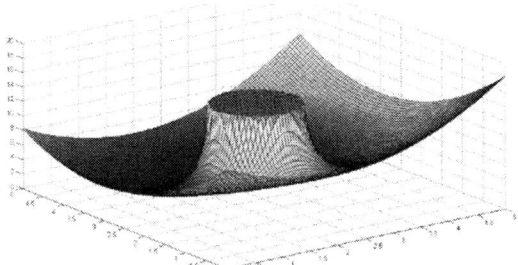

Fig. 3 Barrier function

The equation for barrier function for each object from the map is presented in (3).

$$B_j\left(\vec{p}, \vec{p}_{B_j}, R\right) = \frac{1}{-g_j\left(\vec{p}, \vec{p}_{B_j}, R\right)} \quad (3)$$

D. The priority function

The priority function it aims to highlight the importance of an obstacle. Current position of the system $\vec{p}$ determine whether an object should have a greater or lesser importance in the destination function.

The importance of priority function decreases as the distance to the obstacle grows so only the most important obstacles that are closest to the mobile system will be considered when calculating the trajectory to the point of destination. This priority is calculated determining the distance between those objects and the mobile system.

The distance between the mobile system and obstacles change continuously and the result of the priority function will increase or decrease the importance of the item leading to lower or increased weight in barrier function.

$$r_{B_j}\left(\vec{p}, \vec{p}_{B_j}\right) = r + e^{-a \cdot d} \qquad d = \sqrt{\left(x - x_{B_j}\right)^2 + \left(y - y_{B_j}\right)^2} < 0 \quad (4)$$

$$r_{B_j}\left(\vec{p}, \vec{p}_{B_j}\right) = r - e^{-a \cdot d} \qquad d = \sqrt{\left(x - x_{B_j}\right)^2 + \left(y - y_{B_j}\right)^2} > 0 \quad (5)$$

a it is a constant value, equal with 2.0. This value was determined empirically, after several experiments.

Distance between mobile system and objects is calculated with (6).

$$d = \sqrt{\left(x - x_{B_j}\right)^2 + \left(y - y_{B_j}\right)^2} \quad (6)$$

The range for priority function is 0.05 to 1, value is always positive and have low value. If this value is greater than 1 then the robot will always keep a big distance to objects, which is not always recommended because it can lead to the impossibility of calculating a route.

If the value returned by the function is less than 0.05 will be approximated to 0.05, where the value is greater than 1 shall be approximated to 1.

The final equation is:

$$T\left(\vec{p}, \vec{p}_T, \vec{p}_{B_j}, R\right) = f(\vec{p}, \vec{p}_T) + \sum r_{B_j} \cdot B_j\left(\vec{p}, \vec{p}_{B_j}, R\right) \quad (7)$$

E. Direction finding

This algorith compute the path finding the best direction for mobile system from the current position. This algoritm for finding direction have the Newton direction name.

The equation for finding Newton direction is show in (8),

$$\vec{s}(\vec{p}_k) = -\underline{H}_T^{-1}(\vec{p}_k) \cdot \nabla T(\vec{p}_k) \quad (8)$$

The reason for using this algorithm is the ability to determine the trend of movement.

III. IMPLEMENTATION

The base algorithm is presented in the following part.

1. Set the step value k=1. Set the tolerance factor δ. Set default value.

2. Loop

   Evaluate Hessian matrix $\underline{H}_T^{'}(\overrightarrow{p_k})$, gradient $\nabla T(\overrightarrow{p_k})$. Check if the result is positive or negative, if negative then changes according to become positive.

   Find Newton direction $\vec{s}(\vec{p}_k) = -\underline{H}_T^{-1}(\vec{p}_k) \cdot \nabla T(\vec{p}_k)$

   Normalize direction $|\vec{s}| = \dfrac{\vec{s}}{\sqrt{s_x^2 + s_y^2}}$

   Determining the step size $\overrightarrow{\Delta q_k} = \vec{s}(\overrightarrow{p_k}) * 0.125$

   Determining the new point $\overrightarrow{p_{k+1}} = \overrightarrow{p_k} + \overrightarrow{\Delta q_k}$

3. If $\|\nabla T\| < \delta$, k=k+1 and repeat step2, else the algorithm is end.

In the simulation scenario to simplify the calculation, it was considered that the mobile system has a constant speed of 4m/s and data update frequency of approximately 32 Hz will give an update every 12,5cm driven, so the minimum step will be 12,5cm.

The algorithm described above, in addition to the advantages have several disadvantages, one of the largest being shaped trajectory in certain situations, a serrated form, as shown below.

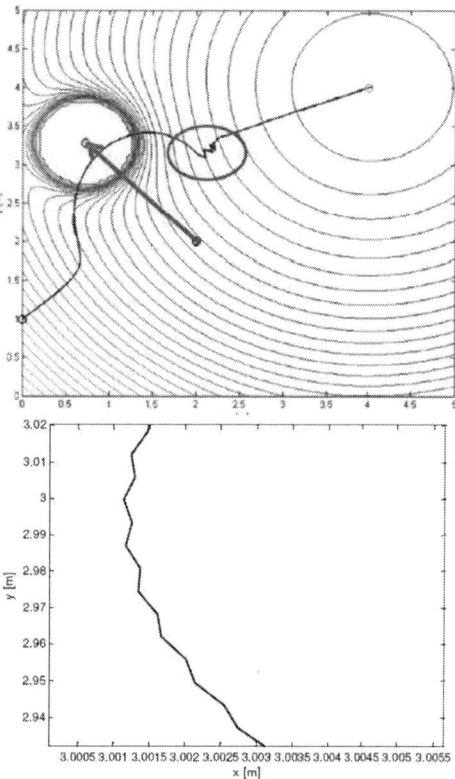

Fig. 4 The shape of path, up- the simulated path, down – zoom for path

This irregularly shaped trajectory is the result of the step size, decrease this step result in smoothing the path but involves a decrease in speed mobile system. Reduction the step size does not eliminate completely the trajectory deviations. This occurs when the next point is not the optimal choice, is due to step fixed distance that can pick the next point further than the optimum.

IV. RESULTS

The simulations were made using Mathlab software, were simulated from 1 to 3 obstacle, in each case with diferent value

for priority function. Somne of the result is shown Fig. 5, Fig 6 and Fig. 7.

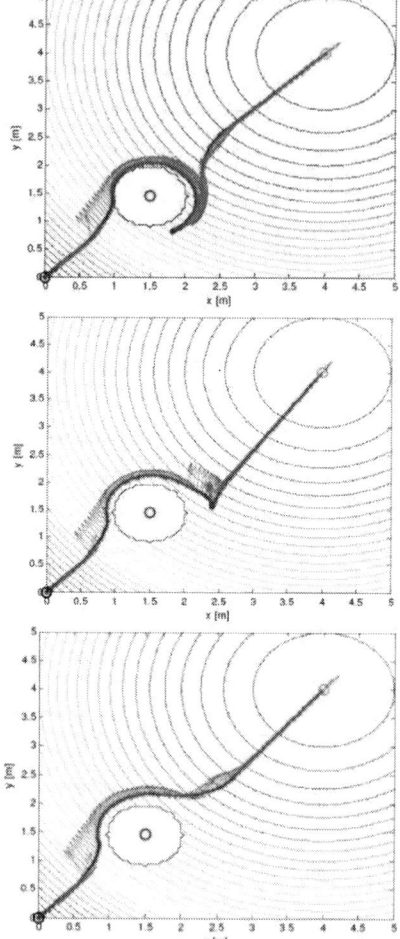

Fig. 5 Simulation with one fixed object and with 3 diferent priority value: 0.1, 0.5 and 1.0

In Fig. 5, was used a single fixed object and regard the priority value used, path does not collide with the object. In the case of low value for priority, the path follow the shape of the object, for higher priority value the path is more liniar.

In Fig. 6, is simulated the scenario with 2 fixed objects. In this case each object have their own priority factor, in figure is show the situation with 0.1 priority factor for both objects. In some situation, with diferent priority factor for objects, path may collide with objects, as in Fig. 7 where one object have 1.0 priority factor.

V. Conclusion

The importance of prioritizing function decreases as the distance to the obstacle grows, so only the most important obstacles that are closest to the mobile system will be considered when calculating the trajectory to the point of destination.

Benefits:

• Algorithm implementing requires a very small volume of calculations, which we recommend for embedded applications.

• Works with cheap sensors.

• Can be easily adapted for different sensors.

• Keep in mind the geometry of robot.

• Works well on narrow corridors.

• Can detect and avoid any obstacles, including some moving obstacles, with the restrition of the obstacles speed to be less then sistem speed.

Disadvantages:

• Movement is not linear, has a trajectory with some swings.

• Failure is possible even if there is a path to the target.

• The movement is initiated even though there isn't a path to the target.

The results obtained show an improvement over the initial algorithm (Newton direction), locals minimum situations are avoided (robot blocked between obstacles), this avoidance is achieved by choosing a direction and compliance constraints. Also by changing the algorithm the trajectory approximate a liniar path, eliminating the "saw tooth" trajectory of the original algorithm.

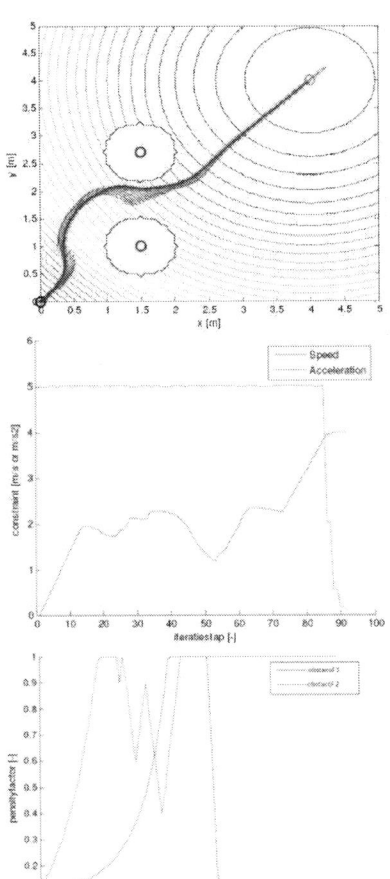

Fig. 6 Simulation with 2 fixed objects

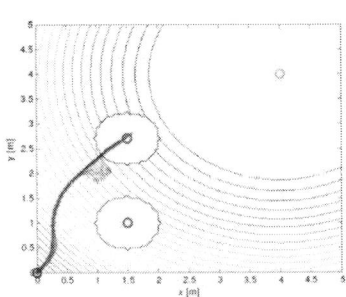

Fig. 7 Path collision

References

[1] Susnea I., Filipescu A., Vasiliu G., Coman G., Radaschin A., "The buble rebound obstacle avoidance algorithm for mobile robots", in Proceedings. 8th IEEE International Conference on Control and Automation, Xiamen, pag. 540-545, 2010;

[2] Bakker T., Kees van Asselt, Jan Bontsema, Müller J., Gerrit van Straten, A path following algorithm for mobile robots, 21 April 2010 This article is published with open access at Springerlink.com.

[3] Carlo L. Bottasso, Leonello D., and Savini B., Path planning for autonomous vehicles by trajectory smoothing using motion primitives, Paper presented at the AHS International Specialists' Meeting on Unmanned Rotorcraft, Phoenix, AZ, USA, January 23–25, 2007..

Wireless sensor node
for fruit growing monitoring

Dragoș Ioan Săcăleanu, Lucian Andrei Perişoară, Rodica Stoian and Lucian Şucu

Research Center for Spatial Information (CEOSpaceTech), Department of Applied Electronics and Information Engineering

University Politehnica of Bucharest

Romania

dragos_sacaleanu@yahoo.com, lucian.perisoara@upb.com, rodicastoian2004@yahoo.com, lucian_sucu@yahoo.com

Abstract—This paper presents a wireless sensor node designed and developed for fruit growing monitoring. The data acquisition node, the sensors signals conditioning interface and the low cost diameter sensor are described in detail. The proposed solution offers several advantages like minimal human interaction, low costs of implementation and monitoring of many fruits. The results for the node energy consumption measured in laboratory are presented and an acquisition for many days is performed for following environmental parameters: air temperature, air humidity, light intensity and rainfall. Finally, the developed sensor node is characterized and its performances are evaluated.

Keywords—wireless sensor node, sensors, environmental parameters, fruit growing,

I. INTRODUCTION

The food production has become a global priority nowadays. The technology implemented in agriculture brings multiple advantages by monitoring and controlling different vital parameters for plants growing: quantity of water, temperature, humidity, fertilization, light, etc. In this way, the resources are used more efficiently, waste of water is minimized and the quality of the fruits and vegetables is improved. More information could be obtained by correlating environmental factors with fruit development. Therefore, the data permit the direct assignment of plant responses to environmental influences.

The field of Wireless Sensor Networks (WSN) has been characterized by constant growth in the last years. Due to the flexibility of the system and the diversity of applications that can be implemented, these intelligent data acquisition systems are used successfully in agriculture to monitor environmental parameters [1]. WSN can be considered as a distributed, precise and low cost measurement tool that eliminates the human presence and gather field data continuously.

The base component of a WSN is the wireless sensor node, which includes sensors (transducers), a microcontroller, a communication module and a power source. Since in most cases the modules are placed in locations where a wireless system is imposed, the goal is to make a wireless sensor network in a manner that the energy consumption is minimized, to increase the network lifetime.

One of the first and well-known wireless sensor node for data acquisition is the Mica node, developed by Crossbow

Technology at University of California, Berkeley, followed by a series of nodes including Mica2, Mica2Dot and MicaZ [2]. For MicaZ node, an expansion interface provides data acquisition from humidity, temperature, pressure, light, acceleration, acoustic or sounder sensors.

Memsic brought further improvements by developing TelosB node [2]. Equipped with temperature, humidity and light sensors and two expansion connectors, it is focused mainly on research and development.

For research and final users applications, Libelium released a wireless sensor node based on Arduino programming language with implementations in large areas through specific expansions boards [3]. The data can be transmitted through different protocols (ZigBee, Bluetooth, WiFi, 3G/GPRS) by replacing the transceivers. The expansion boards makes it suitable to monitor parameters like temperature, humidity, pressure, light, radiation, gas, sound, presence, etc.

For fruit growing monitoring, present solutions involves dendrometers, custom sensors connected at data-loggers or manual sensors that need human presence and offline data analysis. The Ecomatik fruit and vegetable dendrometer [4] offers an accuracy of ±2 μm and a resolution till 0.3 μm but at a very high price (>500€). The parameters are measured for long or short term with data loggers. A fruit gauge sensor based on a high performance linear potentiometer is described in [5]. It has a custom-built stainless steel frame and is connected to a data logger. If offers less precision than the Ecomatik dendrometer but at a much lower cost.

This paper presents a new solution to monitor fruit growing and discuss the influences of the environmental parameters in order to implement the proper recipe to obtain better products. The WSN consists of wireless sensor nodes specially designed and developed to assure three main requirements: low energy consumption, flexibility and high adaptability to a wide range of applications in many domains. The proposed fruit growing sensor offers good performances at very low costs. Data of the wireless sensor node document the reactions of plants to environment in high temporal resolution.

The paper is organized as follows: Section II presents the hardware architecture of the wireless sensor node, being detailed the data acquisition node, the signal conditioning interface and the environment sensors, focusing on diameter fruit sensor. Section III presents the current consumption

---

This work was supported by Romanian Space Agency, Space Technology and Advanced Research (STAR) Programme, project no. 75/2013.

measurements performed in laboratory and the functionality simulations in outdoor environment. Section IV draws the conclusions of this paper and offers the perspectives for further developments.

II. THE NODE'S ARCHITECTURE

To monitor fruit growing and some of the influential environmental parameters, we propose a wireless sensor node with the block diagram from Fig. 1. It consists from DASMote node [6], a signal conditioning interface and sensors, all powered from a battery.

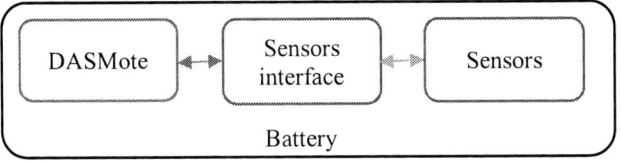

Fig. 1. The block diagram of the wireless sensor node.

A. The Data Acquisition Node

The node used for data acquisition is DASMote (Mote for Data Acquisition Systems), see Fig. 2, developed by the authors following three main requirements: low energy consumption, flexibility and high adaptability to a wide range of applications. It consists of low power components in order to minimize the energy consumption. The main component of the DASMote node is the MSP430F2618 microcontroller from Texas Instruments that provides 8 analog to digital (ADC) inputs with 12-bit resolution, digital to analog converter with 3 outputs, 4 universal serial communication interfaces, SPI, I2C, 116 kB flash memory and 8 kB RAM memory to perform complex data processing.

The node transmits the acquired data through a low power transceiver, XBee ZB Pro S2 with ZigBee communication protocol and RP-SMA connector for antenna. The transceiver is active only in transmission, in the rest of the time being introduced in power down mode. It assures a data transmission range over 100 m line of sight.

The power supply is connected to the node through a 3.3V voltage regulator with 6 V maximum input voltage. This permits to use different types of batteries, including three common 1.5 V AAA batteries.

DASMote node includes a temperature sensor for self-monitoring and offers two 16 contacts sockets for connection with external sensors boards.

Fig. 2. The DASMote node.

B. The Signal Conditioning Interface

The connection to external sensors is achieved through signal processing interfaces that could offer 22 digital inputs/outputs (I/O) and 8 analog inputs to microcontroller. The interface designed for environmental parameters monitoring permits the connection of 4 analog output sensors and 20 digital I/O, see Fig. 3.

Fig. 3. The signal conditioning interface.

C. The Sensors

1) Environmental sensors

To correlate the fruits growing with the environmental factors, a series of environmental sensors are used to acquire data, see Table 1. The node temperature sensor is attached to the node printed circuit board and the other sensors are connected through the signal conditioning interface.

TABLE I. SENSORS LIST

Parameter	Sensor
Node Temperature	TEXAS INSTRUMENTS TMP102
Air Temperature	SENSIRION SHT75
Air Humidity	SENSIRION SHT75
Light level	SHARP GA1A2S100LY
Rain presence	Libelium Leaf wetness sensor (LWS)

2) Fruit growing sensor

The low cost fruit diameter sensor relies on a slide potentiometer embedded in a case, see Fig. 4. For this prototype sensor it was used a 10 kOhms linear potentiometer with a 45 mm slider length and 100 mV maximum noise.

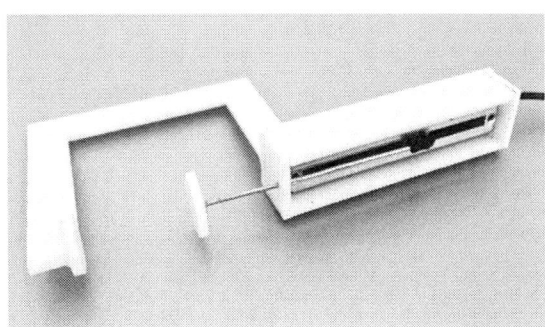

Fig. 4. The fruit diameter sensor.

The potentiometer is connected to a 3.3 V voltage regulator and the output voltage signal from the wiper terminal is applied at the ADC input. The slide is connected to a rod, which is moved by the fruit growing. The case is built from 3 mm white perspex material, to make the sensor light and to minimize the expansion due to sun heat.

The measurements performed in laboratory are presented in Fig. 5, where we verified the linear dependence between the acquired voltage and the potentiometer slide displacement. The developed sensor has a linear dependency between 3 mm and 35 mm, making it suitable for small fruits, like walnuts.

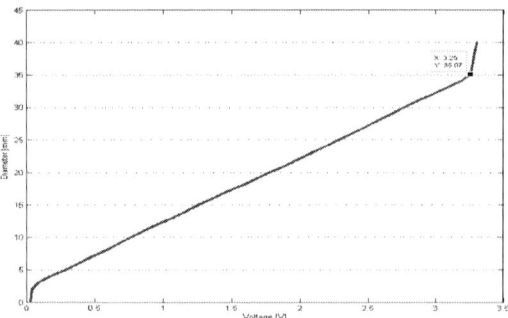

Fig. 5. Voltage vs. Diameter dependency.

From the experimental data, we derived the equation:

$$y = 9.96 \times x + 2.38. \qquad (1)$$

The fruit diameter y can be computed from the voltage x measured at the potentiometer wiper.

The maximum error for the linear dependency can be calculated as the difference between the real diameter measured with a 0.01 mm precision caliper and the diameter calculated using the equation (1). The obtained error is 0.11 mm. If we add the quantization error introduced by ADC, approximately 0.02 mm, the resulting measuring error is 0.13 mm. Considering this two types of errors, for a minimum 5 mm diameter fruit the relative error is roughly 0.4%. Based on cheap components, the price of the sensor does not exceed 10€, making it suitable for monitoring multiple fruits from a plant.

III. RESULTS

To evaluate the lifetime of the wireless sensor node, tests were performed in laboratory and in real environmental conditions. We considered a time span of 100 days for fruit growing. The sensor node should acquire and transmit one measurement per hour, considering the characteristics of the monitored environmental parameters, resulting 2400 sets of data acquired. In the rest of the time, the node is introduced in the sleep mode, both the microcontroller and the transceiver.

Firstly, the energy consumption during data gathering and transmission was monitored. In Fig. 6, we present the sensor node current waveform during a data acquisition and transmission interval. The current average value is about 27 mA. In sleep mode, the manufacturers declare a current consumption of 0.5 µA for the microcontroller and less than 10 µA for the transceiver. The sensors are shut down during this period.

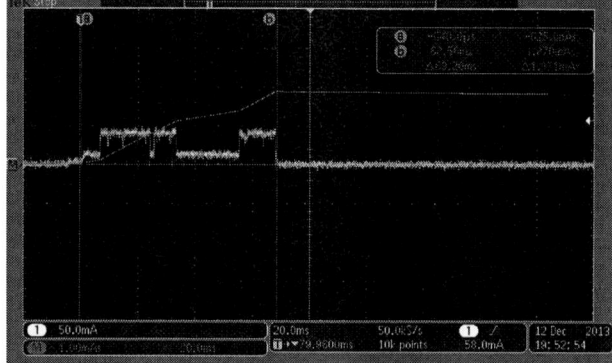

Fig. 6. The node current consumption waveform in data acquisition and transmission.

Secondly, to simulate the functionality of the wireless sensor node in real life environmental conditions for the same number of data acquisitions as in a full fruit growing period, an outdoor data acquisition was made for 40 hours with 1 minute interval between two acquisitions, see Fig. 7. The measurements were performed at the end of September 2016.

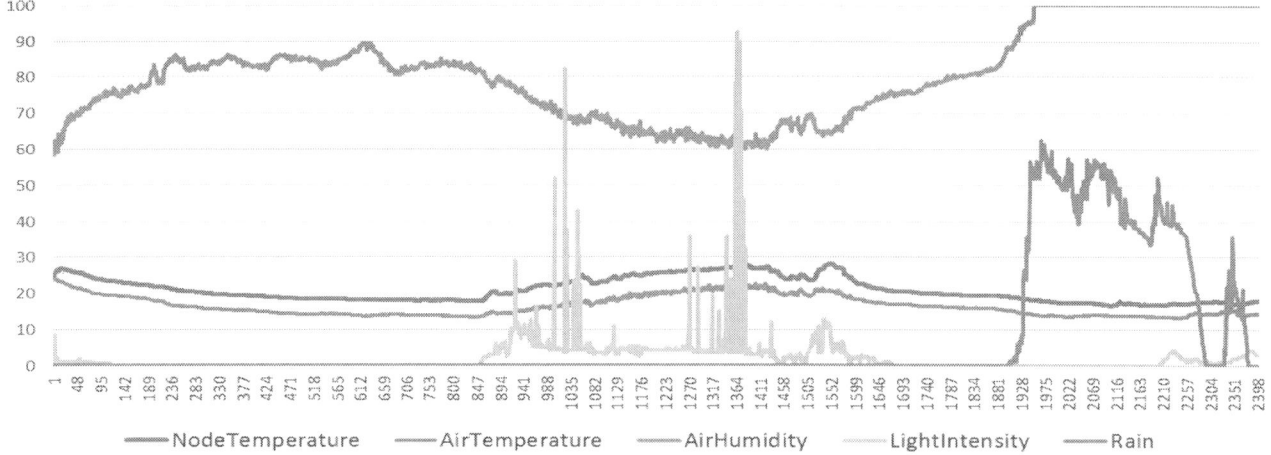

Fig. 7. Data acquisition for environmental parameters.

The wireless sensor node sends the acquired data to a base station connected to a computer. To store the values in a database and further interpret it, a data acquisition interface was developed in LabVIEW. The application acquires information from the USB port where the base station is connected, decodes the received packet and saves the resulted data in an Access Database. The values for the fruit diameter sensor were also acquired, but they are not shown in the chart, not being significant during 40 hours.

In Fig. 7, the rainfall is represented with green. In the last quarter of the acquisition it started to rain, influencing also the air humidity (represented in grey). The light intensity is represented in yellow. The light sensor has a linear output and it is scaled to 100 units for full sun and zero units for night. The cloudy weather characteristic can be observed from the light curve, with small widows (spikes) of sunny time. The air temperature (with orange) and the node temperature (with blue) have a common starting point, with slightly increase of node temperature due to circuit heating. The maximum air temperature during the 40 hours was 24°C, while the minimum was 13°C. The average difference between the air temperature and node temperature was approximately 5 °C.

During data acquisition, the battery voltage level was also monitored. It was used a 3.7 V Li-Ion rechargeable battery with a 2300 mAh capacity. In Fig. 8 is presented the voltage drain, with values between 3.61 V and 3.56V.

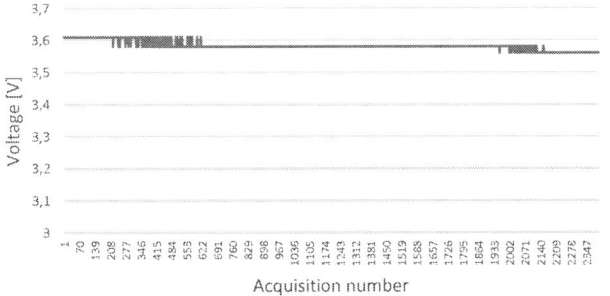

Fig. 8.   The voltage battery drain for the data acquisition from Fig. 7.

For testing purposes and for short term data acquisition, the mounting of the wireless sensor node on the tree and the fruit growing sensor on the fruit are presented in Fig. 9. The node is attached without a cover to a 3 years old walnut. The diameter sensor is attached to a full grown fruit.

Fig. 9.   Wireless sensor node (left) and fruit growing sensor (right).

IV. CONCLUSIONS

In this paper, we presented a wireless measurement system for fruit growing monitoring. The development of the three main components was detailed: data acquisition node, signal conditioning interface and diameter sensor. Current consumption tests and simulation of data acquisition in outdoor conditions were performed in order to demonstrate the functionality for the entire fruit growing period. The diameter sensor offers acceptable performances at a very low cost.

This work is a first step in implementation of a practical solution with multiple wireless sensor nodes to gather valuable information about the fruits development. As future work, the diameter sensor error may be analyzed in detail for different potentiometer sizes. Also, each node may be equipped with multiple diameter sensors and a very long distance communication may be implemented.

REFERENCES

[1] L. M. Oliveira, J. J. Rodrigues, "Wireless sensor networks: a survey on environmental monitoring", Journal of Communications, vol. 6, no. 2, pp. 143-151, April 2011.

[2] http://www.memsic.com/wireless-sensor-networks/

[3] http://www.libelium.com

[4] http://ecomatik.de/wp-content/uploads/2016/08/catalog.pdf

[5] B. Morandi, L. Manfrini, M. Zibordi, M. Noferini, G. Fiori, L.C. Grappadelli, "A low-cost device for accurate and continuous measurements of fruit diameter", HortScience, 42(6), 1380-1382, 2007.

[6] D. I. Săcăleanu, L. A. Perişoară, V. Lăzărescu, R. Stoian, „A New Multipurpose Wireless Sensor Node for Data Acquisition Systems", 6th International Conference on Electronics, Computers and Artificial Intelligence (ECAI 2014), Bucharest, Romania, Vol. 6, No. 2, pp. 35-38, Oct. 23-25, 2014.

Testing Immunity to Portable Transmitters with Helical Antennas: Key concepts

Andrei-Marius Silaghi / Aldo De Sabata

Faculty of Electronics and Telecommunications
"Politehnica" University of Timisoara
Timisoara, Romania
andrei.silaghi@student.upt.ro, aldo.de-sabata@upt.ro

Alexandru-Marius Silaghi

Faculty of Electrical Engineering
University of Oradea
Oradea, Romania
masilaghi@uoradea.ro

Abstract— **Radiated immunity testing in the automotive industry is an activity with ever increasing importance. Immunity tests are being performed in the semi-anechoic chamber of the EMC laboratory, within Continental Automotive Romania. Testing results rely on measurements performed by operators and nominal parameters of the equipment involved, which are input to the testing software. In this paper, we show how deviations of the equipment parameters from nominal values influence the testing outcomes. We performed measurements of impedance and VSWR's of antennas and directional couplers involved in EMC radiated immunity to portable transmitters testing. The results can be used for devising correction procedures for the testing software in order to improve the reliability of the tests.**

Keywords— automotive, immunity testing, helical antennas, directional coupler, vector network analyzer

I. INTRODUCTION

Modern automotive vehicles depend increasingly on electronic devices. Radiated immunity testing is an important activity required by safety related issues [1]. The main purpose of EMC testing is to make sure that a device is able to function properly in its intended electromagnetic environment [2].

Various authors tackle the problem of radiated immunity. In [2] the authors report investigations of different EMC test facilities from immunity point of view. In [1], the directives that have been issued in the field in the last 40 years are discussed. Test field calibration is described and analyzed in [3].

Tests are usually performed by operators and measurement results are input to specially devised software, which automatically outputs the decision on pass or failure. Important data for the operation of the software are parameters of the test equipment. Nominal values of these parameters are used in general.

This paper reports an assessment of some of the existing equipment in Continental Automotive Timisoara, used for immunity testing. It is shown how differences between nominal and measured values of the parameters influence testing results.

Immunity to portable transmitters is one of the tests that are currently conducted within the EMC Laboratory. Other tests include: radiated immunity, radiated emissions,

conducted emissions, immunity to bulk current injection, immunity to electrostatic discharge. For the measurement of immunity to portable transmitters some key components need to be analyzed. These components are: the semi-anechoic chamber, the signal generator, the amplifier, the directional coupler and the antenna. In this paper, an analysis of the directional coupler and antenna parameter effects is presented. In Section II, the test procedure is discussed. Measurements on two directional couplers used in the test procedure are reported in Section III. Test examples using three different antennas in various frequency ranges are presented in Section IV. Conclusions are drawn in the last Section.

II. TEST PROCEDURE

The considered test is intended to check the immunity of equipment to electromagnetic field generated by handy transmitters in close proximity or in contact [4], [5].

For the frequency range 28 MHz – 360 MHz, normal mode helical antennas (NMHA) with counterpoise tuned at each test frequency are used. The length (L) of antenna element including connector is in the interval 100 mm $\leq$ L $\leq$ 250 mm.

Measurements are performed following a test plan. The general procedure for performing immunity to portable transmitter test is described in the Standard ISO 11452-9 [6].

Each DUT power supply lead is connected to the power supply through an artificial network 50Ω /5μH. The battery voltage is maintained at 13.5±0.5V for 12V systems.

The DUT is placed at a 50 mm height on a non-conductive material ($\varepsilon_r \leq$ 1.4): polystyrene.

The distance between the antenna and the surface of the DUT has to be between 5 mm and 50 mm. The test antenna is positioned in step sizes of 100mm. No supplemental attenuators (10dB) are used at forward (FWD) and reflection (RFL) ports. The test is performed in two phases:

- Calibration (test level setting);

- Test of the DUT with wiring harness and peripheral devices connected.

A. Calibration

Calibration is performed before every test. The simulated handy transmitter antenna is placed on a dielectric support at a position no closer than 1000 mm to the floor and ground plane and no closer than 500 mm to any absorbing material and no closer than 1500 mm to the wall of the shielded enclosure. The DUT is unpowered. The calibration is made only with CW modulation. The measured VSWR has to be less than 3:1 for normal mode helical antennas.

After performing the test, we have introduced the test power level from the test plan in the calibration field and compared it with the obtained net power (forward-reflected). Finally, the obtained forward power has been recorded.

B. Test of the DUT

The limit is set equal to the forward power obtained in the calibration step [6].

C. Equipments used for measurements

For the measurement of radiated immunity, the main measuring components involved are: a semi-anechoic chamber, a signal generator, an amplifier, a directional coupler and an antenna [6], Fig. 1.

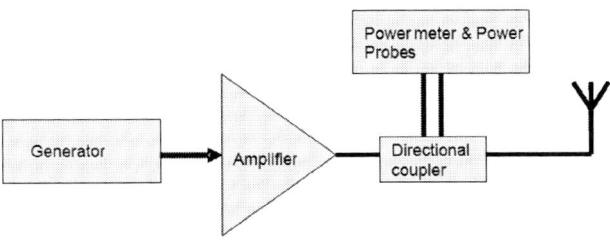

Fig. 1. Equipment used for immunity testing

III. COMPARISON BETWEEN TWO DIRECTIONAL COUPLERS

We measured the VSWR obtained with two different directional couplers (AR and Werlatone). Werlatone works in the range 1-400 MHz and AR works in the range 80-1000 MHz [7]. The test conditions were identical: same antenna position (pointing towards the ceiling and in the middle of the chamber), antenna connected directly to the output of the directional coupler and 1 W net power used for testing. VSWR's provided by the software used for immunity testing are reported in Table I.

TABLE 1. MEASURED VSWR FOR DIRECTIONAL COUPLERS

Frequency [MHz]	VSWR	
	AR	Werlatone
125	2.44	2.36
145	2.16	1.99
155	1.82	1.82
165	2.51	2.38
174	2.08	1.99
190	1.23	1.11
223	2.71	2.53
350	2.39	2.38
385	3.74	3.02

Results in Table 1 reveal that the Werlatone directional coupler provides a better VSWR for all considered frequencies.

We have also measured the directivity of the two couplers, using an R&S Vector Network Analyzer. Results presented in Fig. 2 clearly show that the directivity of Werlatone is closer to -40 dB.

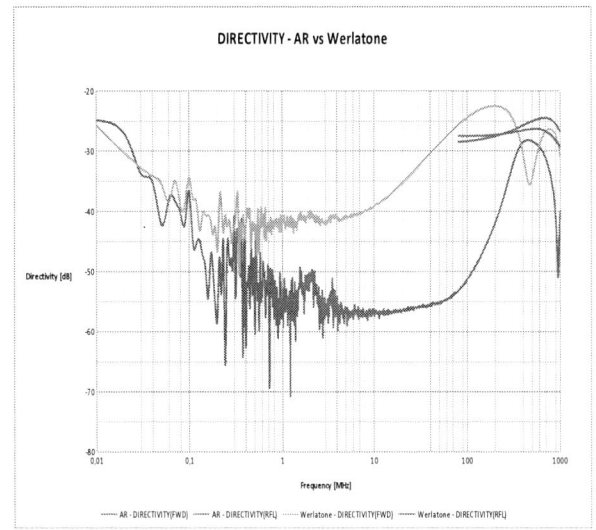

Fig. 1 Comparison between the directivity of the two directional couplers

IV. TEST DATA

We used an R&S Vector Network Analyzer for measuring the VSWR of helical antennas working on 52 MHz, 125 MHz and 145 MHz respectively, Figs. 3-5. We compared the measured VSWR's with the ones provided by the producer in the calibration certificates [8].

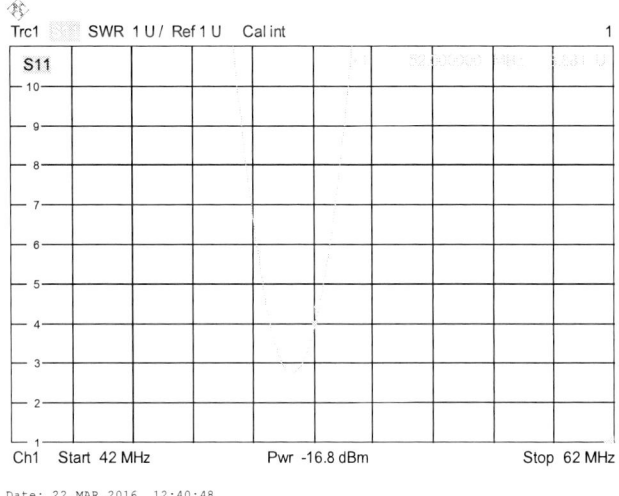

Fig. 2 VSWR of the NMHA 52 MHz antenna

2016 IEEE 22nd International Symposium for Design and Technology in Electronic Packaging (SIITME)

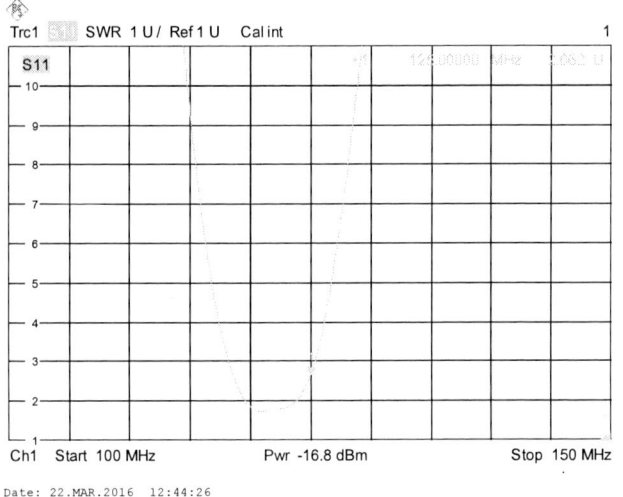

Fig. 3 VSWR of the NMHA 125 MHz antenna

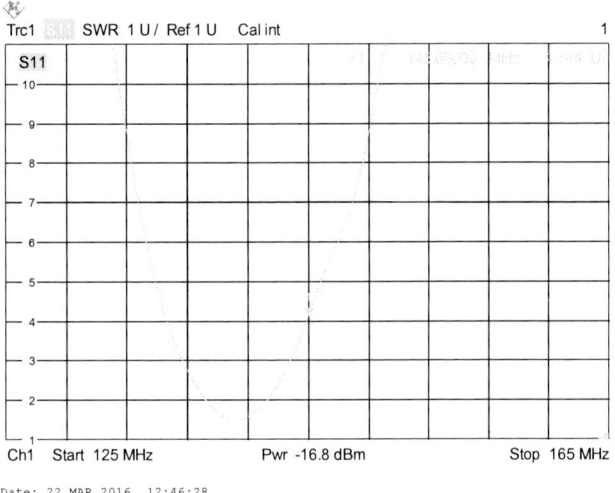

Fig. 4 VSWR of the NMHA 145 MHz antenna

By comparing results reported in Figs. 3-5, with the nominal specifications, deviations can be observed as follows. For 52 MHz, a VSWR of 3 is claimed, but with our equipment we obtained a value of 3.8. For 125 MHz, a VSWR of 2 is claimed, but with our equipment we obtained a value of 2.66. For 145 MHz, a VSWR of 3 is claimed, but with our equipment we obtained a value of 4.1. Frequencies for which the VSWR's present minima are also deviated.

In Fig. 6-8 we present plotted Smith charts for the same antennas. The antennas impedances reported in these figures can be used for devising matching networks.

We used the three antennas for immunity to portable transmitters testing, each combined with the two measured directional couplers, Werlatone and AR. We relied on specialized software for immunity testing. This software is used for testing in Continental Automotive Timisoara. The results of the tests are reported in Figs. 9-10.

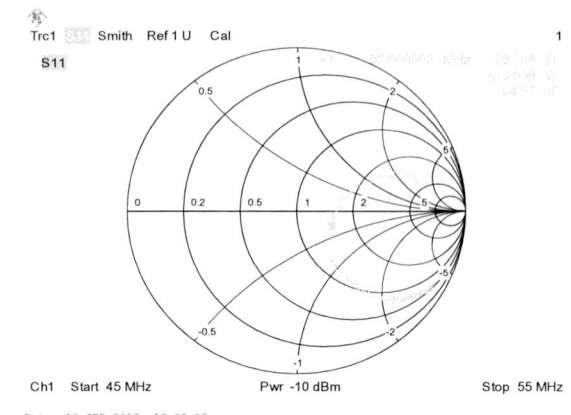

Fig. 5 Smith Chart for NMHA 52 MHz

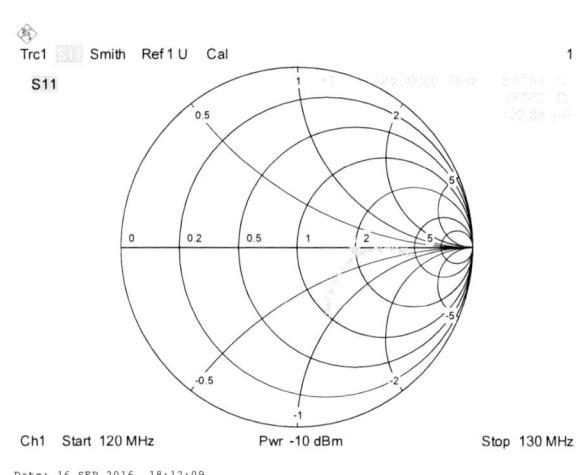

Fig. 6 Smith Chart for NMHA 125 MHz

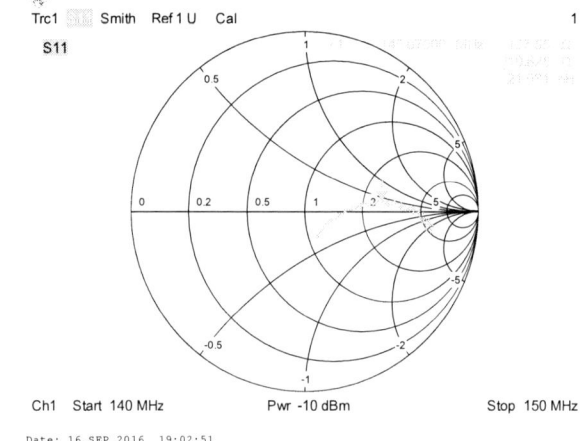

Fig. 7 Smith Chart for NMHA 145 MHz

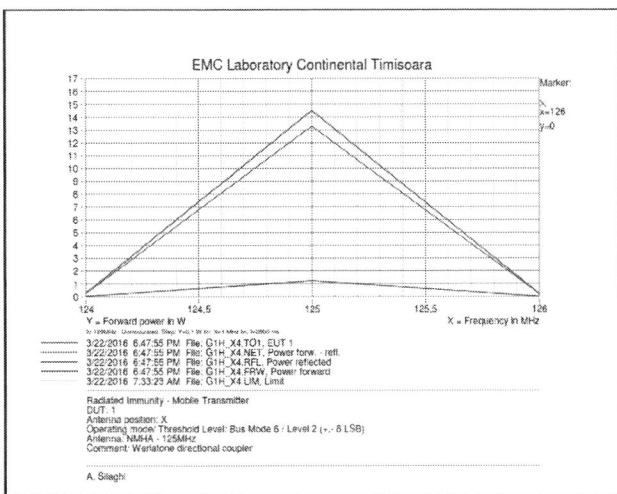

Fig. 8 Test using Werlatone coupler (125 MHz helical antenna)

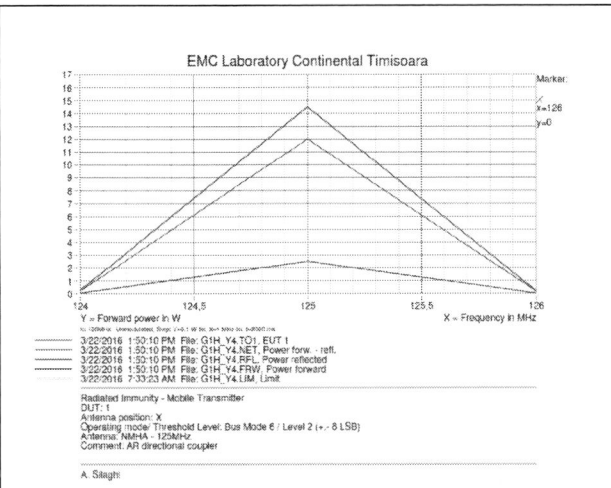

Fig. 9 Test using AR coupler (125 MHz helical antenna)

As we can see in Fig.9-10, by using Werlatone coupler we obtained a smaller reflected power as reported by the software used for immunity testing. We obtained similar results by using the other antennas: 52MHz and 145 MHz. This variability in results demonstrates the necessity to use measured parameters for the used equipment. Situations might occur when DUT's that fulfill emission regulations could be subject to rejection after failing to pass the test.

V. CONCLUSIONS

In this paper we have presented some key concepts for immunity to portable transmitters testing. We started with the test procedure used for testing: ISO-11452-9. After that, we made a comparison between two directional couplers, to see which one suits better this application.

We also measured the VSWR and Smith charts of three helical antennas that are common for immunity testing, to see if they match the nominal values provided by the producer.

Finally, we made tests with the two directional couplers and the three antennas, and reached the conclusion that we need to use Werlatone coupler in order to obtain more realistic results. We demonstrated the necessity to rely on measured parameters of equipment instead of nominal values in order to avoid potential false results of immunity tests.

In future work, we will use the reported results for devising and introducing corrective procedures allowing for the tests to be performed correctly.

References

[1] I.E. Noble, *Electromagnetic compatibility in the automotive environment*, IEEE Proceedings - Science, Measurement and Technology, Vol. 141, No. 4, pp. 252 – 258, Jul 1994.

[2] Holger Streitwolf, Ralf Heinrich, Heinz-Günther Behnke, Lothar Dallwitz, Uwe Karsten, Comparison of radiated immunity tests in different EMC test facilities, Electromagnetic Compatibility, 2007. EMC Zurich 2007. 18th International Zurich Symposium on, 24-28 Sept. 2007, Munich, Germany, pp. 229 – 232, 2007.

[3] E. L. Bronaugh, J. D. M. Osburn, Whole-vehicle radiated EMI immunity tests in automotive EMC: establishing and calibrating the test field, Electromagnetic Compatibility, 1992., Eighth International Conference on, 21-24 Sep 1992, Edinburgh, pp. 39-42, 1992.

[4] C. R. Paul, *Introduction to Electromagnetic Compatibility: Second Edition.*, JOHN WILEY & SONS, INC., Hoboken, New Jersey 2006.

[5] DH. W. Ott, Electromagnetic Capability Engineering, John Wiley & Sons, Inc., Hoboken, New Jersey, 2009.

[6] ISO 11452-9: 2012 Road vehicles - Component test methods for electrical disturbances by narrowband radiated electromagnetic energy: International Standardization Organisation, Part 9 Portable Transmitters;

[7] C. A. Balanis, *Antenna Theory: Analysis and Design*, Third Edition, JOHN WILEY & SONS, INC., Hoboken, New Jersey, 2012.

[8] J. D. Kraus and R. J. Marhefka, *Antennas for all applications,* Third Edition, New York: McGraw-Hill, c2002., vol. 1, 2002

Electrochemical migration of Sn and Ag in NaCl environment

Bálint Medgyes, Dániel Szivós, Sándor Ádám, Lajos Tar, Patrik Tamási, László Gál, Richárd Berényi and Gábor Harsányi

Department of Electronics Technology
Budapest University of Technology and Economics
Budapest, Hungary
medgyes@ett.bme.hu

Abstract— The impact of chloride ion concentration on electrochemical migration (ECM) of tin and silver was studied by using an in-situ optical and electrical inspection system. It was found, that in both cases, dendrites grow not only in an electrolyte solution at low chloride concentration but also in an electrolyte at medium and high or even saturated chloride concentrations as well. According to the results, the migration susceptibility has decreased at low and medium concentration levels in both cases. However, the ECM susceptibility of Ag has increased, while the migration susceptibility of Sn was decreased at the saturated concentrations.

Keywords— electrochemical migration; water drop test; NaCl; Sn; Ag.

I. INTRODUCTION

Many reliability issues of electronics have to be solved nowadays [1-3]. One of the most dangerous ones is the electrochemical migration (ECM) failure phenomenon.

ECM may cause a relative high reliability risk regarding electronics applications [4]. This failure phenomenon occurs at the presence of humidity in the case of operating circuits: the dissolution of metal traces begins at the anode and produces metal ions, which are directed by the applied electric field and move to the cathode, where they can reduce as pure metals – dendrites (See Fig. 1). Dendrite growth can lead to bridging. Many types of the metals, which are widely used in the electronics industry have huge ability for ECM, such as Sn, Ag or Cu [5]. Silver is also widely investigated related to ECM [5-10].

Usually, migration tests were tested by three main methods; water drop (WD) test [11], environmental tests under different thermal-humidity conditions [12] and by various electroanalytical methods, like voltammetry or polarization tests [13]. On the one hand the electrochemical corrosion and migration investigations of Ag were usually tested in bulk solution. On the other hand, migration tests in thin electrolyte layers (TEL) is also important, since ECM ability depends on the thickness of water layer as well [14]. Furthermore, the concentrations of the used electrolytes had a wide range as well: deionized water [5], acidic or alkaline electrolytes [15] and also salt electrolytes [16], which usually simulate the possible contamination effects. One of the most common contaminant is the Cl, which can strongly modify the

electrochemical corrosion mechanism and therefore, the migration ability of tin and silver as well. Tin and silver ions can combine with hydroxide groups and also with Cl⁻ ions and form complexes in aqueous solutions [4, 17]. The impact of chloride ion on Sn and Ag ECM behavior was already investigated at low concentrations [17, 18].

Although the impact of NaCl on ECM was already investigated at low concentration, the effect of other concentrations of chloride ion was not reported related to the ECM of tin and silver. Therefore, the migration behavior of Sn and Ag modified by chloride ion is not well understood. So, in this study, different concentrations of NaCl solution were applied in order to investigate the impact of Cl⁻ concentration on migration in terms of galvanic tin and immersion silver as widely used surface finishes in the electronics.

Fig. 1. Scanning electron microscope (SEM) image of dendrites.

II. EXPERIMENTAL

In order to simulate the impact of chloride ion on ECM in the case of silver, a pure copper layer was immersed with silver (iAg). In the case of tin, electroplated method was used on copper and both of the surface finishes were fabricated according to the conventional PCB technology. The dimensions of the patterns were 2×5 mm, which structure was prepared on a widely used fiberglass epoxy resin substrate (FR4). The gap size between the conductor, lines on the sample was 0.5 mm (See Fig. 2).

2016 IEEE 22nd International Symposium for Design and Technology in Electronic Packaging (SIITME)

The schematic of the measuring platform can be seen in Fig. 3. The system can follow the various electrochemical migration mechanisms by visual means and by real-time voltage measurements as well. Simultaneously, the ECM mechanisms were in-situ observed by using a USB microscope (type: XCAM MAN1001-SA) for visual inspection and real-time voltage measurement (Novus Automation MyPCLab 6014) was also applied to better understand (verify) the optical results. The real-time voltage measurement is characteristic for the dendrite growth (Since the dendrite grows as a function of time, the voltage increases also as a function of time, simultaneously). The voltage was measured on a serially connected 10 MΩ resistor, which is located on the negative pole of the test board (See Fig. 3.), while the optical part provides qualitative information, for example about the dendrite structures, growing directions, different residue formations, (out)gassing, etc.

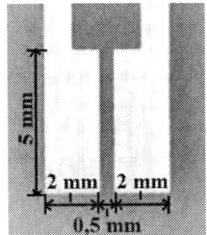

Fig. 2. An iAg test board for ECM test.

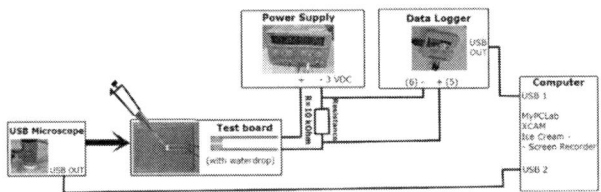

Fig. 3. Schematic of the measuring platform for ECM investigation.

Before every water drop (WD) test, the Sn and Ag samples were washed out with deionized water (2 µS/cm), degreased with IPA, and dried in room temperature. The test solution was NaCl with different concentrations (0.1 mM, 1 mM, 10 mM, 500 mM and saturated NaCl), which was made from deionized water and from analytical grade reagents. The droplets (volume was 25 µl in case of Sn and 15 µl in case of Ag due to the different surface wetting behaviour) were placed by a micropipette onto and between the Sn and Ag electrode surfaces. A bias of 3 VDC was used between anode and cathode. At least eight times were all WD test repeated to check reproducibility. The Mean-Time-To-Failure (MTTF) was also calculated from the real-time voltage measurements, where the failure criterion was the first "voltage jump" (see Fig 4). To investigate the chemical compositions of the dendrites and residues, (SEM)-EDS method was also applied.

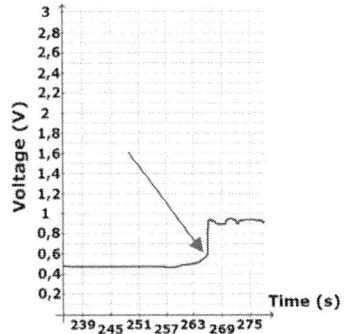

Fig. 4. Failure criterion (Voltage jump) for MTTF calculation.

III. RESULTS AND DISCUSSION

A. ECM of Ag and Sn in 0.1 mM NaCl solution

In this case, different colored and shaped dendrite formation occurred during the WD test. Furthermore, the dendrite formation was very rapid in both cases; MTTF was about 28 sec in case Ag and 72 sec in case of Sn. A typical optical result can be seen in Fig. 5 for silver, where mainly Cu was the dominant migration element. Only the red type of dendrite showed some Ag composition according to EDS results (See Fig 5).

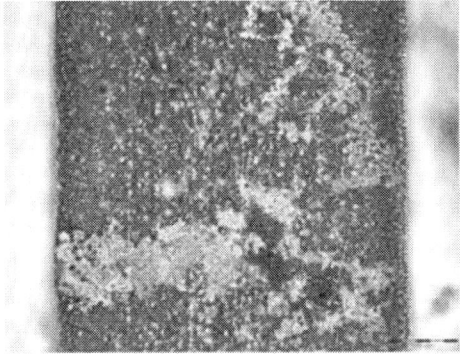

Fig. 5. Different colored dendrites (red, yellow, green) were formed in 0.1 mM NaCl solution during WD test in case of iAg sample.

In Figure 6, a typical needle-like dendrites and blue colored residues can be seen in case of Sn.

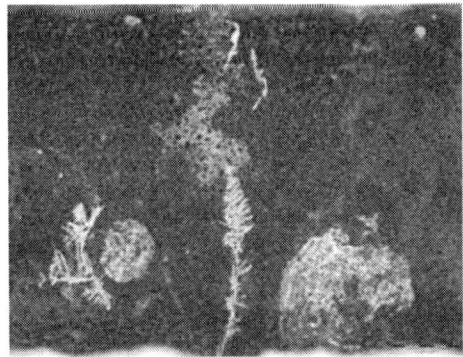

Fig. 6. Dendrites and blue colored resideus were formed in 0.1 mM NaCl solution during WD test in case of Sn sample.

978-1-5090-4446-7/16 $31.00 © 2016 IEEE   275   20-23 Oct 2016, Oradea, Romania

B. ECM of Ag and Sn in 1 mM NaCl solution

At this concentration level very similar results were obtained in case of Ag compare to 0.1 mM level. That means the MTTF, the colors of the formed dendrites and residues did not show any significant differences. However, in case of Sn, the dendrite growth was more dense (see Fig. 7) and the MTTF (around 113 sec) was also different compare to 0.1 mM level.

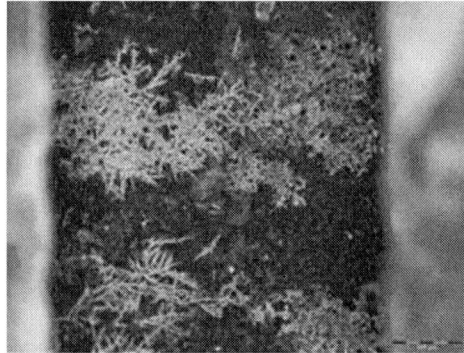

Fig. 7. Dendrite and blue colored residues formed in 1 mM NaCl solution during WD test in case of Sn sample.

C. ECM of Ag and Sn in 10 mM NaCl solution

In both cases the ECM susceptibility was significantly decreased. In case of silver, a pale-blue residue wall was formed (Fig. 8), while in case of tin, white colored residue wall was occurred (Fig. 9). However, in all cases dendrite formation was also detected (MTTF$_{Ag}$ ~ 783 sec; MTTF$_{Sn}$ ~ 195 sec).

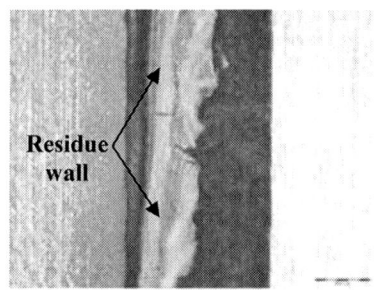

Fig. 8. Blue colored residue was formed in 10mM NaCl solution during WD test in case of immersion Ag.

Fig. 9. White colored residue and dendrites were formed in 10mM NaCl solution during WD test in case of galvanic Sn.

In case of immersion silver the dominant migration element was copper in the dendrites as well as in the residues. According to the EDS results and the literature [20], the formed blue residue is $Cu(OH)_2$ with all likelihood.

Despite of the huge white residue wall some dendrite formation can be detected in case of galvanic tin (Fig. 9), which result is similar to the literature [21]. Namely, in [21] only a huge residue wall was occurred with no dendrites. The difference between the current results and [21] can lead back to the different electrolyte thicknesses. According to the EDS results the dominant migration element is Sn in the dendrites as well as in the residues. The residue is probably $Sn(OH)_2$ or $Sn(OH)_4$.

D. ECM of Ag and Sn in 500 mM NaCl solution

In this case of Ag the MTTF value (about 2133 sec) was significantly increased. In the most cases after evaporation of the bulk solution a thin electrolyte layer (TEL) was remained, where and when the short was occurred. So, this is also a confirmation for the impact of electrolyte layer thickness on the ECM. In Fig. 10 a short can be seen, formed under TEL after evaporation of the bulk solution.

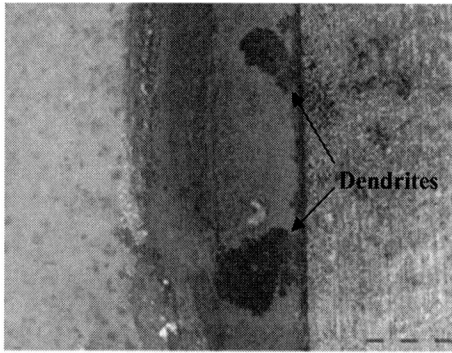

Fig. 10. Dendrites formed in 500mM NaCl TEL during WD test in case of immersion silver.

After applying the test voltage (3 VDC) gas evolution (H_2) was occurred, simultaneous with the red-orange colored product formation at the cathode side. This product formed parallel with the electrodes (like a wall) and migrated towards anode. Later, dendrite formation was started from the cathode, which was probably hindered by a green-yellow colored residue wall on the anode (Fig. 10). Finally, after evaporation of the bulk solution either new dendrites were grown or the existing ones found ways to form short across the medium (Fig. 10). According to the EDS results copper was the dominant element in the dendrites as well as in the different colored (red-orange, green) products. The red-orange product is probably Cu_xO (x = 1, 2) and the green-yellow residue is probably $CuCl_2 - 3Cu(OH)_2$, [22]. The supposed chemical states are based on the color, EDS and literature information [22].

In case of tin, there was no significant change compare to the 10 mM case. It means, the value of MTTF, the density and the structure of the dendrites were similar as well as the white residue wall, which is probably $Sn(OH)_2$ or $SnCl_2$.

E. ECM of Ag and Sn in saturated NaCl solution

On the one hand, the migration susceptibility of silver was dramatically increased (MTTF ~ 218 sec), which was an unexpected result. During the electrochemical processes very intensive H_2 gassing on the cathode and metal dissolution (discoloration) on the anode was observed (See Fig. 11).

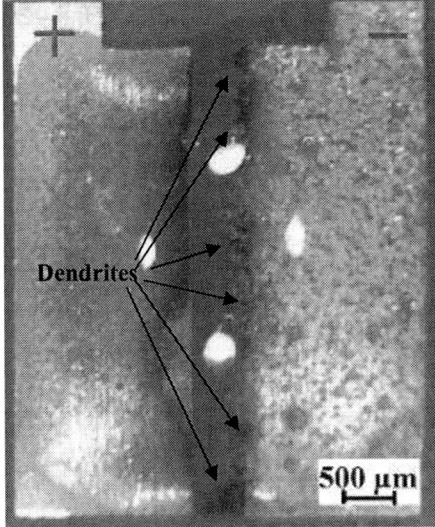

Fig. 11. Short in case of Ag during WD test using saturated NaCl solution.

It can be also observed that the dendrite growth showed a relative homogeny distribution along the cathode border line (Fig. 11). There was no or less residue formation, which can confirm the low MTTF values (high susceptibility of ECM). It should be also noted, that the shape of the dendrites was very similar compare to the ones, which grown at low NaCl levels. The summarized MTTF values of Ag can be seen in Fig. 12.

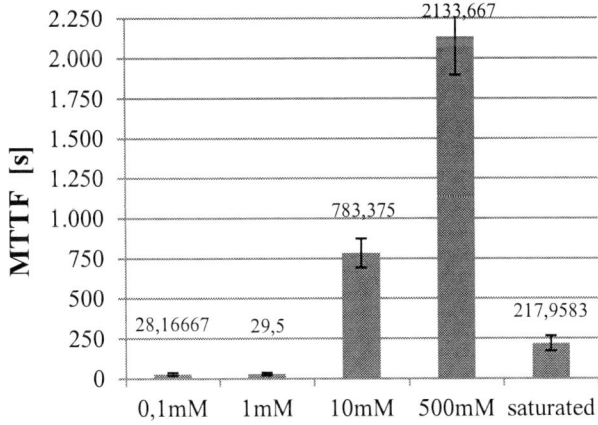

Fig. 12. MTTF values of Ag in different NaCl concantrations after WD test.

In case of tin the tendency of ECM ability was not changed (MTTF ~ 363 sec). The summarized MTTF values of Sn are presented in Fig. 13.

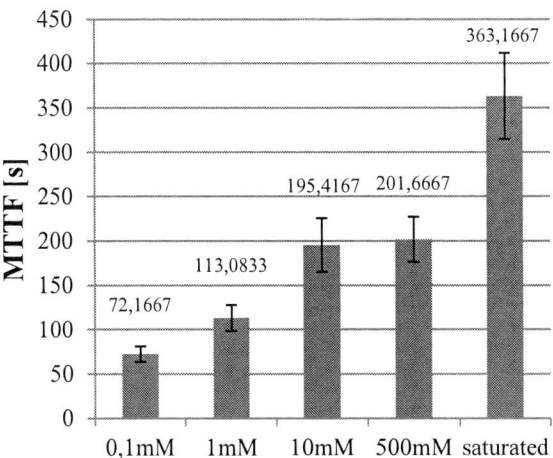

Fig. 13. MTTF values of Sn in different NaCl concantrations after WD test.

IV. CONCLUSIONS

The impact of chloride ion concentration on the electrochemical migration of Ag and Sn was tested by using an in-situ visual and electrical inspection platform under various NaCl concentrations: 0.1 mM; 1 mM; 10 mM; 500 mM and saturated solution.

In case of immersion silver the followings were found:

- MTTF ranking: 0.1 mM ≤ 1 mM < saturated NaCl solution < 10 mM < 500 mM

- According to the EDS results the dominant migration element was the copper in the dendrites as well as in the residues.

- At the low and at the saturated concentration levels no or less residues were formed, while at the 10 mM and 500 mM NaCl levels different colored products and residues were observed {probably $Cu(OH)_2$ at 10mM and CuO / [$CuCl_2 - 3Cu(OH)_2$)] at 500 mM NaCl}.

- The relative low MTTF value in case of saturated NaCl solution was an unexpected result. It can be explained by the relative less residuc formation.

In case of galvanic tin the followings were found:

- MTTF ranking: 0.1 mM < 1 mM < 10 mM ≤ 500 mM < saturated NaCl solution

- According to the EDS results the dominant migration element was the tin in the dendrites as well as in the residues.

- At low concentration levels blue and white residues were formed (probably $Sn(OH)_2$ was formed at 10mM NaCl). White residues were mainly formed at 500mM and at saturated NaCl levels (Probably $SnCl_2$ was occurred).

Acknowledgment

The work reported in this paper was supported by the János Bolyai Research Scholarship of the Hungarian Academy of Sciences. The authors would like to thank to the Pro Progressio Fundation (Hungary) for the financial support as well.

References

[1] O. Krammer, "Comparing the Reliability and Intermetallic Layer of Solder Joints prepared with Infrared and Vapour Phase soldering", Soldering and Surface Mount Technology, Vol. 26, Issue 4., pp. 214–222, 2014.

[2] A. Geczy, M. Fejos and L. Tersztyánszky, "Investigating and compensating printed circuit board shrinkage induced failures during reflow soldering", Soldering & Surface Mount Technology, Vol 27. issue 2. pp. 61-68, 2015.

[3] B. Horváth, B. Illés, T. Shinohara, G. Harsányi, "Copper-Oxide Whisker Growth on Tin-Copper Alloy Coatings Caused by the Corrosion of Cu6Sn5 Intermetallics", Journal of Materiaé Science, Vol. 48 pp. 8052-8059, 2013.

[4] D. Minzari, F. B. Grumsen, M. S. Jellesen, P. Møller, R. Ambat, "Electrochemical migration of tin in electronics and microstructure of the dendrites", Corrosion Science, Vol. 53, pp. 1659–1669, 2011.

[5] B. Medgyes, B. Illés, G. Harsányi, "Electrochemical migration behaviour of Cu, Sn, Ag and Sn63/Pb37. ", Journal of Materials Science: Materials in Electronics, Vol. 23, pp. 551-556, 2012.

[6] B. I. Noh, J-B Lee and S-B Jung, "Effect of surface finish material on printed circuit board for electrochemical migration", Microeletronics Reliability, Vol. 48, pp. 652–656, 2008.

[7] S. Yang and A. Christou, "Failure Model for Silver Electrochemical Migration", IEEE TRANSACTIONS ON DEVICE AND MATERIALS RELIABILITY, Vol. 7, No. 1, pp. 188-196, 2007.

[8] S. Yanga, J. Wua and A. Christou, "Initial Stage of Silver Electrochemical Migration Degradation", Microelectronics Reliability, Vol. 46, pp. 1915–1921, 2006.

[9] B.I. Noh, J.W. Yoon, K.S. Kim, Y.C. Lee and S.B. Jung, "Microstructure, Electrical Properties and Electrochemical Migration of a Directly Printed Ag Pattern", Journal of Electronic Materials, Vol. 40, No. 1, pp. 35-41, 2011.

[10] K. S. Kim, Y. T. Kwon, Y. H. Choa and S. B. Jung, "Electrochemical migration of Ag nanoink patterns controlled by atmospheric-pressure plasma", Microelectronic Engineering, Vol. 106, pp. 27–32, 2013.

[11] B. I. Noh, J. W. Yoon, W.S. Hong, S.B. Jung, "Evaluation of Electrochemical Migration on Flexible Printed Circuit Boards with Different Surface Finishes", Journal of Electronic Materials, Vol. 38, No. 6, pp. 902-907, 2009.

[12] B. Medgyes, B. Illés, Richárd Berényi and G. Harsányi, "In situ optical inspection of electrochemical migration during THB tests", Journal of Materials Science: Materials in Electronics, Vol. 22, pp. 694-700, 2011.

[13] G. Harsányi, G. Inzelt, "Comparing migratory resistive short formation abilities of conductor systems applied in advanced interconnection system", Microelectronics Reliability, Vol. 41, pp. 229-237, 2001.

[14] X. Zhong, G. Zhang, X. Guo, "The effect of electrolyte layer thickness on electrochemical migration of tin", Corrosion Science, Vol. 96, pp. 1-5, 2015.

[15] O. Devos, C. Gabrielli, L. Beitone, C. Mace, E. Ostermann, H. Perrot, "Growth of electrolytic copper dendrites. II: Oxalic acid medium", Journal of Electroanalytical Chemistry, Vol. 606, pp. 85-94, 2007.

[16] B. Medgyes, L. Gál, D. Szivós, "The effect of NaCl on water condensation and electrochemical migration", In: IEEE, 20th International Symposium for Design and Technology in Electronic Packaging (SIITME). Bucarest, Romania, pp. 259-262, 2014.

[17] K-K. Ding, X-G. Li, K. Xiao, C-F. Dong, K. Zhang, R-T. Zhao, "Electrochemical migration behavior and mechanism of PCB-ImAg and PCB-HASL under adsorbed thin liquid films", Trans. Nonferrous Met. Soc. China, Vol. 25, pp. 2446-2457, 2015.

[18] V. Verdingovas, M. S. Jellesen and R. Ambat, "Influence of sodium chloride and weak organic acids (flux residues) on electrochemical migration of tin on surface mount chip components", Corrosion Engineering, Science and Technology, Vol. 48. pp. 426-435, 2013.

[19] M. Pourbaix, Atlas of Electrochemical Equilibria in Aqueous Solution, Pergamon Press, Oxford, 1966.

[20] B. Medgyes, X. Zhong, G. Harsányi, "The effect of chloride ion concentration on electrochemical migration of copper", J Mater Sci: Mater Electron, Vol. 26, pp. 2010-2015, 2015.

[21] X. Zhong, G. Zhang, Y. Qiu, Z. Chen, W. Zou, X. Guo, "In situ study the dependence of electrochemical migration of tin on chloride", Electrochemistry Communications, Vol. 27, pp. 63–68, 2013.

[22] A.G. Masey, N.R. Thompson, B.F.G. Johnson, R. Davis, "The Chemistry of COPPER, SILVER and GOLD", Pergamon Press, New York, 1975.

A Low-Cost Pavement Image Acquisition System

Claudiu Chiculiţă and Laurenţiu Frangu

Dept. of Electronics and Telecommunications,
University „Dunărea de Jos”
Galaţi, Romania
Claudiu.Chiculita@ugal.ro, Laurentiu.Frangu@ugal.ro

Abstract—The paper presents an acquisition system, intended to be used for recording the images of the pavement in front of a vehicle and the geographic position where the images were taken. The main advantages of this system are: easy to deploy in a regular car, low-cost, easy to use acquisition and recording software. The system is based on a cheap and widespread board, Raspberry Pi. It can record images up to 8Mpix, at an average rate of 4 fps, or lower resolution at higher speeds. The position provided by a GPS receiver is also recorded. Supplementary, this system can support a glueless interface with a car vibration acquisition subsystem. The images and the vibration signals are useful for the inspection of the pavement. Alternative solutions (such as a smart phone or image recorders) are not suitable, as they can not record double camera images, nor do they record multiple vibration signals. The presented system was built and tested on real vehicles, in order to validate its properties.

Keywords—image acquisition, pavement inspection, Raspberry Pi, data storage

I. Introduction

Diagnosing the quality of the roadway pavement is a major concern for the companies who build or maintain the roads ([1]). They need systematic detection of the cracks or holes of the pavement, as well as evaluation of its roughness, profile and wearing. Multiple sensory systems provide the data for this purpose. The extensive set of sensors includes cameras, laser reflectometer, accelerometers, gyro, ultrasonic sensors and GPS receivers (see [2]). Because the surface to be inspected is huge, the sensors are installed on a dedicated vehicle and the data are acquired at the vehicle cruise speed, as in Fig. 1 ([2], [3]). Mainly, the useful information relates to the images of the pavement faults and to the vibration spectrum ([2], [3], [4], [5], [6]). The image information is used to detect cracks, pavement distress ([2]) or even the microtexture of the asphalt ([5]). Both the set of sensors and the vehicle are expensive components of the diagnose equipment, so the frequent evaluation of the roads becomes difficult to accomplish. Replacing them with a cheap system of sensors, installed on a regular car, is a convenient alternative solution. The expensive systems also perform online image processing ([3], [4]), but this part is not relevant in our paper, as the results of the processing are useful later, when the maintenance starts.

This paper deals with the system performing image acquisition, for diagnosis purposes. Unlike the dedicated solutions mentioned above, the proposed system is able to work on a running regular car, it is cheap, it allows 2 simultaneous cameras and it allows the integration of the vibration data.

Fig. 1. Vehicle dedicated to the pavement inspection ([3])

The images are taken at speeds of up to 70km/h, then recorded on an external memory device. The vibration data (provided by the subsystem described in [6]) are recorded on the same memory device and the geographic position is attached to the image and vibration files. The field of view and the data transfer speed were chosen so as to allow a spatial resolution of 2mm/pixel, in the region of interest. The image acquisition is triggered by the program running on the microcomputer which handles the data. The system works properly during the daylight (the image becomes poor during the low light conditions). In order to make this system work, some problems had to be solved: acquisition speed, transfer speed, position acquisition and storage of image and vibration data.

The rest of the paper is organized as follows: section II presents the requirements of the image acquisition system, section III presents the hardware design and the last section presents the experimental results and concludes the paper.

II. Requirements of the Data Acquisition System

The main purpose of the system is to store images of the pavement, while the vehicle is running at a reasonable speed: up to 70km/h. The required spatial resolution is 2mm/pixel. According to these objectives, the proper camera and its position should be chosen. More, the data flow and storage speed are influenced by these parameters. We assume here a camera having the following features (they match the features of Omnivision OV5647): resolution 2592 x 1944 (i.e. 5Mp camera), horizontal field of view 53,5°, vertical field of view 41,41°, fixed focus from 1m to infinity, rolling shutter. We also assume a reasonable position of the camera, as presented in Fig. 2, which contains the longitudinal cross section of its field of view. The vertical position h is 1.5m (top of a regular car), and the tilt angle of the central axis θ is 65°. The angle α is the vertical field of view. The length of the area covered by the camera, in longitudinal section, is 18.5m, according to:

$$l_y = h \cdot (\operatorname{tg}(\theta + \alpha/2) - \operatorname{tg}(\theta - \alpha/2)) \quad (1)$$

The spatial resolution (in mm/pixel) ranges between the limits presented in (2) and (3), for the upper and lower limits of the cross section. The constant N_y is the number of pixels on the vertical axis.

$$y_p = \frac{2 \cdot h}{N_y} \cdot \frac{\operatorname{tg}(\alpha/2)}{\cos(\theta + \alpha/2)} \quad (2)$$

$$y_p = \frac{2 \cdot h}{N_y} \cdot \frac{\operatorname{tg}(\alpha/2)}{\cos(\theta - \alpha/2)} \quad (3)$$

Now, considering the inferior part of the field of view (right side in Fig. 2.), having the length of 5m, the spatial resolution is equal or better than 2mm/pixel. The resolution degrades on the sides (left or right), but these parts of the image have lower influence on the diagnosis. Assuming the speed of the vehicle is 70km/h, i.e. 19.4 m/s, the sampling rate should be about 4 frames/s, in order to have a record of the part observing the minimum spatial resolution, at any time. The values determined above were considered as main requirements of the image acquisition. However, because the length of the covered area is 18.5m and the speed is 19.4m/s, the recorded images overlap, which allows reducing the covered area, in order to gain resolution and reduce the sampling rate. A special mention should be added about the image sampling time, inside the camera, as the camera is installed on a running car, so the image can be blurred or smeared. A shutter is a possible answer to this question (OV5647 presents a rolling shutter).

The position of the recorded pavement images should be recorded too, in order to decide where the detected pavement distress lies. Assuming the necessary precision of the position information does not exceed 20m, a regular GPS receiver is enough, as its sampling period (NMEA protocol) is 1s.

For a 5Mp camera, compressed images may take 3-4 Mb, which requires less than 60GB of external memory, per recording hour. The space doubles for two cameras of the same resolution. The data transfer speed has to allow 4 frames/s, i.e. 16MB/s.

A third data flow comes from the vibration measuring subsystem, which requires a considerable lower speed of 10kB/s.

The main board connecting these peripheral devices has to provide the appropriate communication interfaces for the cameras, GPS receiver, external memory device and vibration subsystem. Its processing performance is useful, but not compulsory.

The vibrations of the vehicle can affect some parts of the electronic devices. In this case, an HDD and the image itself are sensitive to the vibrations. However, the position of the device inside the car can be chosen out of large accelerations (less than 0.2g), so the memory device does not require special precaution. On the contrary, the quality of the image may be affected by the car vibrations; this aspect will be detailed in the following section.

Finally, the price of the system is also important, it should be comparable to other subsystems, such as the car vibration acquisition (less than 400euros).

III. HARDWARE AND SOFTWARE DESIGN

For designing the image acquisition system, two versions were analyzed. They can use respectively a smart phone and a platform Raspberry Pi. Other solutions are also possible, adding more connectivity and more processing power, but they become considerably more expensive.

The smart phone can be a convenient solution. It contains its own camera and its own GPS receiver, it allows image recording at a high rate, it allows vibration data to be transferred on one of its ports (USB), and the sampling moments can be imposed by a program. The resolution of the camera observes the requirement set in the previous section. However, it does not allow two cameras and the price is higher than the alternative. So, the Raspberry Pi version will be presented here.

The main features of this board, linked to the purpose of the paper, are: single board computer, powerful processor (BCM2837 64bit Quad Core), 1.2GHz, 1GB RAM, 4 USB connectors, CSI connector (camera), SD card connector, LAN connector and Bluetooth module. It runs a free OS (Linux), loaded on the SD card. It also has the advantage of its huge popularity, meaning a lot of tested hardware and software projects available. The peripheral devices can be connected as follows: one camera on CSI, one camera on USB, GPS receiver on USB, memory device on USB (or SD card on its dedicated connector), vibration data subsystem on USB and LAN connection on the dedicated connector (or Bluetooth module). The structure of the system is presented in Fig. 3. The recent version Raspberry Pi 3 fits better the requirements of the system, when compared to previous versions (memory access and power availability).

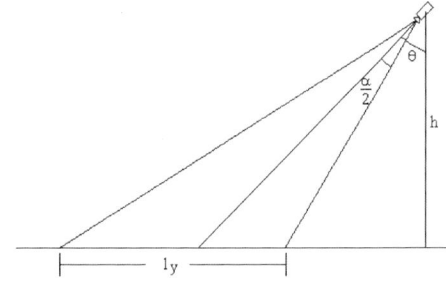

Fig. 2. The longitudinal section of the field of view of the camera

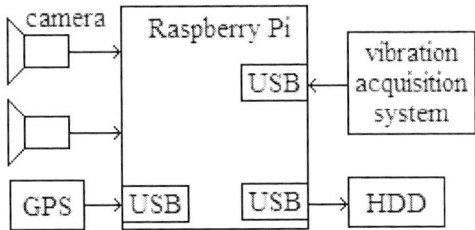

Fig. 3. The structure of the system

One of the cameras is the Raspberry Pi camera module. The design based on OV5647 (5Mp) was recently replaced by another one, based on Sony IMX219. Its features cover all the requirements previously set: 8 Mp (3280x2464), 30 frames/s or even higher (for lower resolution), CSI interface (see [7]). The second camera may be any USB connected camera and observing the requirements of the user (including available software driver).

The position receiver is an own designed module, based on the small Quectel L76, which occupies less than 1 sq cm. It can track up to 33 satellites (GPS and GLONASS) and provides a serial output, NMEA protocol. The serial interface is based on an own module, using an FTDI circuit. Fig. 4 presents the receiver module, including the antenna. Other modules are also available. The position error depends on its intrinsic error (3-5m in open space) and on the delay between camera and GPS acquisition. Having in mind that the NMEA protocol delivers the position every second, the position error is roughly around 20m, i.e. the space covered by the vehicle in this time.

Fig. 4. The positioning receiver module

The external memory device can be one of the SD card, HDD and SSD memory. However, the SD card has limited space. Supplementary, it may suffer from intermittent large writing delays (problem caused by the software driver), despite the good transfer speed (12MB/s). USB memory sticks should be avoided, as they can have even poorer write performance. Instead, HDD or SSD is the preferred solution, allowing up to 25MB/s and hundreds of GB of memory space.

The software is written in Python and is based on usual Linux functions. One software module gets the camera images, using its library, and stores them on the external memory device. The name of the file contains the sampling moment (timestamp). The second module gets the GPS position, using the Linux "gpsd" function and stores it together with a similar time stamp. The position information is not inserted in the image file, as it can delay the acquisition and synchronization process. The position information is stored in a separate file containing the similar timestamp. The matching of the image and position data will be performed offline. The third software module gets the vibration data, using the USB interface (presented in [6]).

IV. EXPERIMENTAL RESULTS AND CONCLUSIONS

Fig. 5 presents the built system, used during the tests. It contains the central board, the Pi camera and the USB camera (both on the left), the HDD connection (below), Ethernet connection (right down), the custom built GPS receiver module (right middle). The USB/RS485 interface and a vibration module are at the top.

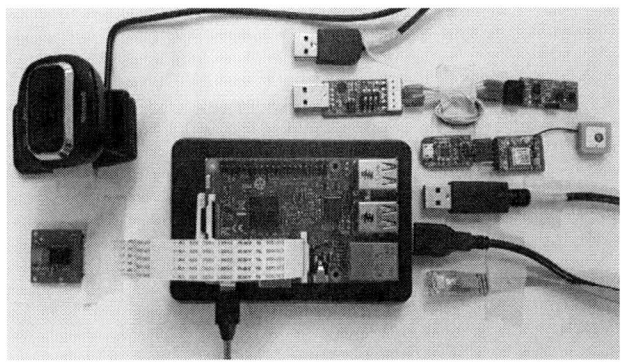

Fig. 5.   The built hardware, around the Raspberry Pi, used for tests

The write speed tests proved that recording the data on HDD or SSD observes the requirements set in section II. The speed is over 26MB/s, the native speed of Raspberry Pi, allowing more than 7 frames/s. The SD card performs the transfer at the acceptable limit, it is not a reliable solution. An ordinary memory stick can slow down to 7MB/s. If using this solution a high quality one should be chosen.

The processor load is far from its maximum, even when performing simultaneously acquisition and multiple data transfer.

The tests on the camera module proved a good behavior. The image is sampled at the right rate, then compressed and transferred at a convenient speed. An example of recorded image is presented in Fig. 6, where a pavement failure is obvious, in the middle of the road. As expected, the shadows can trick the image processing algorithms, whereas the low light degrade the quality of the image. This means that the proposed system can be used only during the day, with favorable light conditions. The tests performed with a no-IR-filter camera did not show an improved behavior.

Fig. 6.   Example of recorded image (shadow of the car and pavement distress are visible).

The power consumption does not raise problems for the Pi3 version, when using its own power supply. The board, the cameras, the attached HDD, the GPS and the vibration interface consume up to 1.5A, if image acquisition and data writing are performed simultaneously. The average consumption is lower than 1A. The power availability of the Pi2 version is at the upper limit, when connecting the HDD. Consequently a powered USB hub has to be used.

The price of the system, even produced in small quantity, is below 230 euros.

A special mention has to be made about the influence of the vibrations. As expected, large vibrations are produced when the vehicle crosses the pavement failures. However, the effect on the image is quite small. Fig. 7 presents an image taken with the 5Mp camera, when the vehicle travels smoothly (moment 0.883 seconds). Fig. 8 presents details of two similar images, taken at the moments 0.883s (low vibrations) and 2.051s (important shock, when crossing an obstacle). The details prove that there is no much difference between the quality of the images: no blur, not smeared. This is the effect of the camera quality: shutter control and fast sampling.

Fig. 7. Image of the road, low vehicle vibrations

Fig. 8. Details of the images taken in different vibration conditions

The presented experimental results prove that the image acquisition system has a good behavior, provided it works in good light conditions. The low cost and the secure data recording make it a convenient alternative to the expensive dedicated car-embedded systems, for research related to image processing and pavement evaluation. Future research will be carried on for more in-depth evaluation of the effect of the vibrations and low light conditions.

ACKNOWLEDGMENT

The work was partly supported by the Romanian Council for Research (UEFISCDI) under Grant PN-II-PT-PCCA-2013-4-1762.

REFERENCES

[1] T. M. Oguara, "Pavement Maintenance Management System: the Paradigm decision-making tools for Highway Engineers", lecture notes, Rivers State University of Science and Technology, Port Harcourt, 2007.

[2] H. Zakeri, F. Nejad, A. Fahimifar: Image Based Techniques for Crack Detection, Classification and Quantification in Asphalt Pavement: A Review, presented at CIMNE Barcelona, 2016, Archives of Computational Methods in Engineering, Springer, online Sept. 2016

[3] *** http://www.transview.org/aran/

[4] M. Borowiec, A. Sen, G. Litak, J. Hunicz, "Vibrations of a vehicle excited by real road profiles", Forschung im Ingenieurwesen, vol. 74, Springer, 2010, pp. 99-109.

[5] A. Das, V. Rosauer, J. S. Bald: Study of Road Surface Characteristics in Frequency Domain using Micro-optical 3-D Camera, KSCE Journal of Civil Engineering, vol. 19, July 2015, pp.1282-1291

[6] C. Chiculita, L. Frangu, "A Low-Cost Car Vibration Acquisition System", IEEE 21st Intl. Symp. for Design and Technology of Electronics Packaging (SIITME), Brasov, Romania, 2015, pp. 281-285

[7] *** Sony, camera IMX219 datasheet, http://www.sony-semicon.co.jp/products_en/new_pro/april_2014/imx219_e.html

FFT Based Investigations on Light Flicker in New Lighting Systems

Ciprian Ionescu

University "Politehnica" of Bucharest,

Center for Technological Electronics and Interconnection Techniques, UPB-CETTI, Bucharest, Romania

ciprian.ionescu@cetti.ro

Mihai Dima

Department of Computational Physics and IT - Institute for Nuclear Physics and Engineering, Bucharest, Romania

Detlef Bonfert

Fraunhofer Institution for Modular Solid State Technologies EMFT Munich, Germany

Abstract— **These days we live the transition in home and public lighting to Solid State Lighting (SSL) and possible others as Organic LED (OLED) and new developed gas discharge lamps as Cold Cathode Fluorescent Lamps (CCFL). Most of the new lamps are marketed to be retrofitted in existing sockets for luminaires as E14 or E27. Beside major benefits on energy consumption and environment protection, there are some concerns about the biological and psychological effects of light flicker produced by these new lamps, especially in the SSL case [1]. We mention here the effects that range from fatigue to headache and epilepsy. We propose in this paper a quantitative investigation on flicker, based on FFT of the acquired photometric data. This way, not only the flicker amplitude, but also the frequency and the human perception is taken into account. Based on the proposed processing, different lamps with different technologies and flicker level are consistently compared.**

Keywords—photometric flicker, flicker measurement, light quality, LED lamps, LED dimming.

I. INTRODUCTION

Industrial, public and home lighting was one of the first applications of electricity. Nowadays the energy consumption is an everyday problem with a great impact on the budget of every country. Hence it results the need to search for new and more economical lamps. The new "economical" lamps bring the benefit of energy savings, but there are some issues regarding the acceptance among human subjects. In Europe the penetration of the fluorescent lamp was visible during the last decade, supported by the ban in EU of tungsten lamp with power higher than 75W and the phase-out process will continue. The classical lamps manufacturers have found an "ultimate" solution. They switch the production to halogen lamps, which produce more light at up to 40% less energy, still remaining filament lamps, i.e. non-economical.

The true energy saving lamps available for home usage are Compact Fluorescent Lamps (CFLs) and LED lamps [1]. The emerging LED technology was seen in consumer environment only in the last 2-3 years. Cold Cathode Fluorescent Lamps (CCFL) is not very new in conception new, but has been recently been available as retrofit lamps. These lamps are not widespread and are limited to small and local applications. The new organic LED (OLED) based lighting will be available soon but it seams that the initial promising results are not seen so optimistic now in comparison to LED.

For proper use in the right in places the features, benefits and drawbacks of these modern lamps CFLs, CCFLs and LEDs need investigations, not only from electrical point of view, but also from the point of view of light quality and human perception.

II. LIGHT FLICKER ISSUES

Flicker of the light sources is defined as a rapid and repetitive change over time in the luminous flux or luminance of a light source. The frequency of most lamps flicker is commonly 100Hz in Europe and 120Hz in USA for very different categories of light sources. This fact results from the ac mains frequency.

Some studies have tried to identify the level at which the flicker was either detectable or induces a discomfort. The studies differentiate the visible flicker that is perceived by human viewer and invisible flicker, flicker that is not perceivable by a human viewer but can have psychological effects. Flicker can be better perceived off of a reflected surface, not directly viewing the source. Potential flicker-induced phenomena are: neurological problems, including epilepsy, headaches, fatigue, blurred vision, eyestrain, apparent slowing or stopping of motion (stroboscopic effect), distraction.

The degree of acceptance for a certain flicker value depends on the category of importance where the lighting is used. We found three categories that are defined in literature: low, medium, or high importance [2]. Of low importance for flickering concerns are outdoor lighting and stationary displays. Office activities as reading/writing are considered to be of medium importance. Flicker concern is considered to have high importance in activities that include lighting in rooms where devices or objects with high speed motion are present such as industrial machines and sporting events.

In the past, the flicker issues were made aware when in some office buildings where magnetic ballasted fluorescent lamps were used, the office employees experienced headaches and eye strains. The modern circuits that are used as replacement of magnetic ballasts operate at higher frequencies, theoretically imperceptible for human eye.

In present there are different methods that are used to estimate the flicker of a lamp. Two parameters are widely used to express the flicker: Percent Flicker (sometimes called Percent Modulation) which can be computed based on amplitudes of the modulated wave and Flicker Index computed on the corresponding areas delimited by the mean value line [3].

Definition of both parameters is based on Fig. 1.

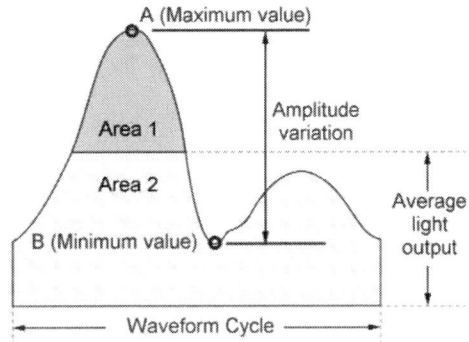

Fig. 1. Periodic waveform used to define Percent Flicker and Flicker Index, adapted from [3].

Percent Flicker can be estimated by (1)

$$PercentFlicker = \frac{A-B}{A+B} \times 100 \qquad (1)$$

with A, B, the maximum and minimum value of the periodic waveform , respectively.

The Flicker Index is calculated with (2), taking into account the areas above and below the average light line.

$$FlickerIndex = \frac{Area1}{Area1 + Area2} \qquad (2)$$

Both parameters are easy to understand and can be calculated from the acquired waveform. The obtained values and are accessible for comparison between different lamps and lamp driver circuits. Associations and correlations can be made through experiments with human subjects between the calculated values and the physiological response.

In some references the Flicker Index is preferred versus Percent Flicker mostly because it takes into account the shape of the waveform. For instance let's compare two periodical light signals with the same amplitude and a duty factor of 10% and 90% respectively. The signals will have the same percent flicker, but it is obvious will have different values for Flicker Index. It was observed actually that the human subject will perceive different the flicker from the two light sources.

Both the above mentioned parameters share the same major drawbacks, neither calculated metric takes into account the human perception. It is today clear that the calculated values of both parameters don't express a direct correlation with human eye perception. For example, two acquired flicker waveforms from different lamps having the same amplitudes and average values, but different frequencies, for instance 200Hz and 20Hz will give the same percent flicker and flicker index. The human eye will sense different the flicker of the two light sources, being more responsive to the lower frequency. There exists guides or local rules regarding about how high the Flicker Percentage or Flicker Index is acceptable, and there are no definitive conclusions and regulations in this domain regarding the acceptance of a certain lamp.

In Table 1 there are presented the accepted thresholds for percent flicker and flicker index regarding the stroboscopic effects.

TABLE I. MAXIMUM ACCEPTED PERCENT FLICKER AND FLICKER INDEX VALUES

Relative Importance of Flicker	Accepted Percent Flicker	Accepted Flicker Index
Low	100%	0.5
Medium	38%	0.32
High	25%	0.17

For most home applications the new lamps LEDs, but not only, are required to have the dimming capability. The existing dimmers are so called "phase cut dimmers" that were designed initially for incandescence bulbs, in fact for purely resistive loads. The simplest circuit to produce light dimming is the so called "phase-cut dimmer" [4]. The leading edge phase-cut dimmers or other named forward phase control can be found in many houses, due to the simple construction, low price and already proven operation lifetime and reliability. The dimmer is built around an electronic switch, a triac, which turns itself off every half-cycle when the ac voltage of the line passes through zero. The triac is turned "on" every half-cycle by an R-C circuit. By changing the time constant R·C we can shift the position of the leading edge, and hence we can control the conduction time, or the time when the lamp goes in "on" state. We can change the light level using a potentiometer that changes the time constant.

In Fig. 2 can be seen the waveform associated with the leading edge dimmer.

2016 IEEE 22ⁿᵈ International Symposium for Design and Technology in Electronic Packaging (SIITME)

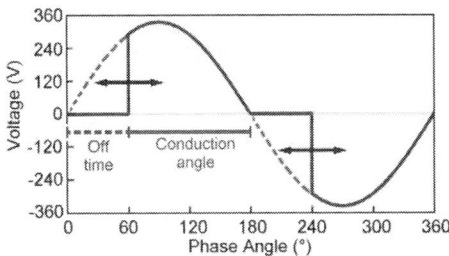

Fig. 2. Leading edge Phase-cut dimmer voltage waveform applied to the lamp.

The dashed line represents the uncut line voltage that what would be applied to the lamp if there were no dimming device between the line and the lamp. Although were initially designed to be used with filament incandescent lamps, can be used with the new retrofitted LED lamps, provided that these ones are dimmable. Dimmable LEDs have special PWM (Pulse Width Modulation) driving circuits and are more expensive than non-dimmable lamps [5],[6]. A problem that can arise is the fact that the dimmer requires the presence of a minimum load, to hold the triac in the "on" state. These values are typically in the range of 5W to 40W, and not always the LED or CCFL lamps present this minimum load.

We have acquired and we will present the results from different type of lamps: LEDs, incandescence (tungsten and halogen), compact fluorescent lamps (CFL) and CCFL. Where dimming was permitted, we will present the effect of dimming on the flicker.

III. FLICKER INVESTIGATION USING DISCRETE FOURIER TRANSFORM

As presented in previous section, both parameters used as metric for flicker don't take into account the frequency and the human eye perception.

We have developed an investigation procedure to analyze the flicker and to define a specific metric, called CFD-Compact Flicker Degree. The frequency of the flicker is taken into account inherently in the data processing based on the discrete FFT. The principle of data acquisition and processing is presented in Fig. 3.

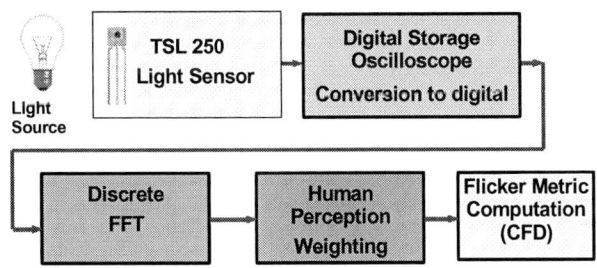

Fig. 3. Flow chart of data for CFD flicker metric calculation.

The sensor TSL 250 which is a three terminal device emulates the photopic luminous efficiency function $V(\lambda)$. It includes a photodiode and a transimpedance amplifier. The output voltage is directly proportional to light intensity (irradiance) on the sensor surface.

The analog waveform was acquired with a Tektronix Digital Storage Oscilloscope TDS3052B at a rate of 100000 samples per second.

With the obtained data a discrete Fourier transform was then performed.

$$U_k = \frac{1}{N}\sum_{n=0}^{N-1} u_n e^{-j2\pi nk/N}, k = 0,1,...,N-1 \qquad (3)$$

The data was processed using a C++ developed code, and adapted to the size of existing data sample in order to process a FFT <17>, ie with 2^{17} samples. More detail is found in [7] and [8].

The amplitude of a spectral component to a certain frequency Vk will be given by taking the modulus of the complex quantity Uk.

$$V_k = |U_k| \qquad (4)$$

The obtained amplitude values will be normalized to the dc component, or the mean value V0, resulting the modulation ratio Mk.

$$M_k = \frac{V_k}{V_0}, k = 1,2,...,N-1 \qquad (5)$$

For a signal with a single frequency the modulation ratio expressed in percent is equal to the percent flicker.

After that, the obtained normalized values of spectral components will be weighted in order to include the human perception factor.

The perceptible light flicker threshold for humans as a function of the frequency was obtained from a large number of studies. Supplemental it takes into account the unperceptible flicker or visible in the peripheral field of view, the stroboscopy effect, and the phantom array effect (Luminous flux modulation perceptible by the motion of the observer's eye, when the light source is still)

The frequency-dependent light flicker perceptibility threshold, which can be changed during following years, corresponds approximately to this graph [9].

Kelly-Henger weighting curve is named after the researcher's names, ASSIST comes from Alliance for Solid-State Illumination Systems and Technologies.

The function from the graph in Fig. 4 was digitized and then interpolated from 24 frequency points in order to obtain the weighted values Mwk (f) in every k point.

Next, each modulation ratio was multiplied with corresponding weighting coefficient and multiplied with 100 in order to express the result $MPwk$, in percentage units.

$$M_{Pwk} = M_k \cdot M_{wk}, k = 1,2,...,N-1 \qquad (6)$$

978-1-5090-4446-7/16 $31.00 © 2016 IEEE   285   20-23 Oct 2016, Oradea, Romania

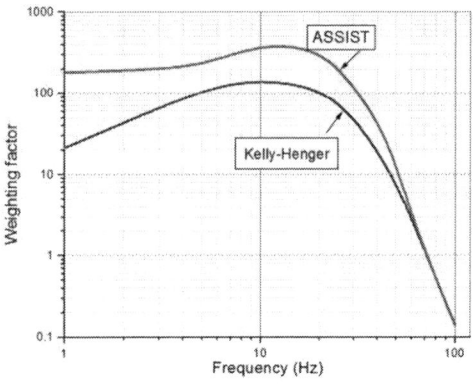

Fig.4.   Weighting factor for flicker spectral components (arranged from [9]).

The values *MPwk* are called perceptual modulation and include the human detection threshold, according to Fig. 5. We now can add quadratic these values. The resulting metric for flicker is called CFD- Compact Flicker Degree and can be used for different comparisons between lamps.

$$CFD = \sqrt{\sum_{k=1}^{N-1}\left(M_{Pwk}\right)^2} \qquad (7)$$

For practical reasons, in our examples, the frequency domain was chosen below 1000 Hz, the contribution of higher frequency components was negligible.

The CFD permits the estimation of light source quality. A CFD <10% is almost imperceptible and a good lamp "flicker free should have CFD <1%. CFD values 10% and 75% are more likely to be perceptible by more than 50% of the population. A CFD> 75% is perceived by more than 75% of the population and strong stroboscopic effects can occur, beside neurologial effects.

IV. EXPERIMENTS AND RESULTS

The measuring platform includes a dark box, which has about the JEDEC recommended one cubic feet size. The interior walls of the box are made from black and non-reflective material, as can be seen in Fig. 5(a).

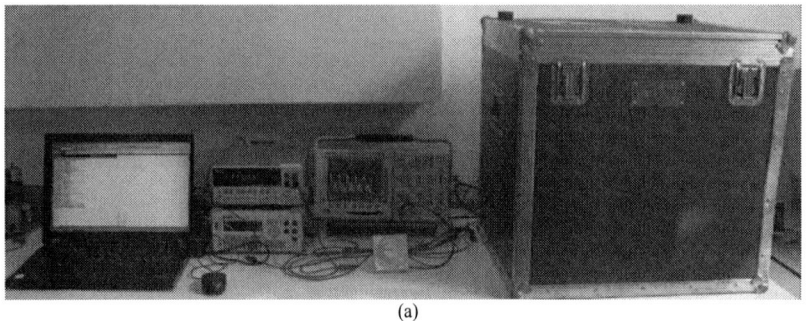
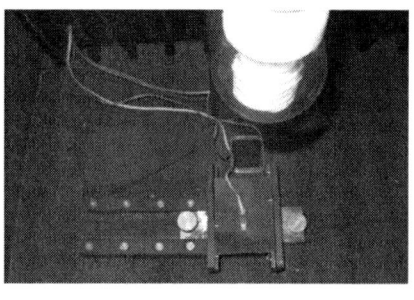

(a)   (b)

Fig. 5. Experimental setup a) equipment, b) detail with the TSL250 light sensor.

For the acquisition we have used the Digital Storage Oscilloscope at a rate of 100kSa/s connected to the light sensor, as described in Section III. An alternative acquisition was done using a DMM 2700 which has a lower sampling rate 2kSa/s but permits a higher number of samples. There were no differences in data processing between the two acquisition methods, and the oscilloscope was preferred because it offers the possibility to save the screen. In the middle of Fig. 5 (a) can be observed the rotary knob of the dimmer, that has different angular positions, for instance position 160 corresponds to 160 degrees, the zero degree position was chosen at clock hour 12 position. During the experiments, a series of lamps in socket E27were tested. The identification of each lamp can be seen in Table 2.

The idea for choosing the lamps, is to have various types and among them to have some that can be dimmable. Incandescence lamps under tests were a tungsten classical lamp and two relatively new halogen bulbs. These last ones are still classical filament lamps, but can still be used in European Union, in contrast with tungsten lamps which are accepted only for low power levels (less than 60W). The difference from classical bulb is that halogen bulbs are filled

with an iodine or bromine gas that permits a longer filament life and an operation at higher temperatures, resulting a higher color temperature. The filament lamps will serve as a reference for the dimmer used, a thyristor phase cut dimmer, being purely resistive loads. The second group is represented by the dimmable LEDs. These must incorporate a PWM type controller in order to permit the dimming process. The lamps are from the same brand, but there is no information if they came from the same manufacturer. The third group contains only one representative: the CCFL lamp. The new CCFL lamp offers moderate light level and claims to have a very long operation life expressed in on/off cycles and running hours. Supplemental, it is dimmable and is expected to have more market share in the future. The fourth group contains non-dimmable lamps: two LEDs of different power and two 23W CFL lamp. CCT in the table is the abbreviation from Correlated Color Temperature, and is not always given by manufacturers as a numeric value but usually is preferred the attribute "warm" and "cold" for lower and respectively higher values of CCT. Color Rendering Index (CRI) is a metric for the ability of the light source to make the colors visible in a manner similar to natural light [10].

TABLE II.   IDENTIFICATION AND CHARACTERISTICS OF TESTED LAMPS.

Lamp no.	Lamp Type/Name/Code	Power (W)	Luminous Flux (lumens)	CCT (K)	CRI
Group1: Filament					
1	Tungsten	40W	N/A	N/A	N/A
2	Halogen	70W	1250	N/A	N/A
3	Halogen	100W	1200	N/A	N/A
Group2: Dimmable LEDs (IKEA)					
4	LEDARE LED1477G22	22W	1800	2700	90
5	LED1480G13	13W	1000	2700	87
6	LED1466G9	8.6W	600	2700	90
7	LED1469G6	6.3W	400	2700	87
Group 3: CCFL dimmable (Aurora)					
8	CCFL	8W	320	2850	N/A
Group4: Non-dimmable LEDs and CFL					
9	Lightme LM85240 A60	10W	810	2750	N/A
10	CFL	23W	N/A	5500	N/A
11	CFL	23W	N/A	2700	N/A
12	LED 5W (No Name)	5W	200lm	5500	

In Fig. 6 we present the spectral components obtained as presented in Section III. It can be seen the dimming effect and the computed percent flicker and CFD for the same lamp (lamp 1), tungsten 40W.

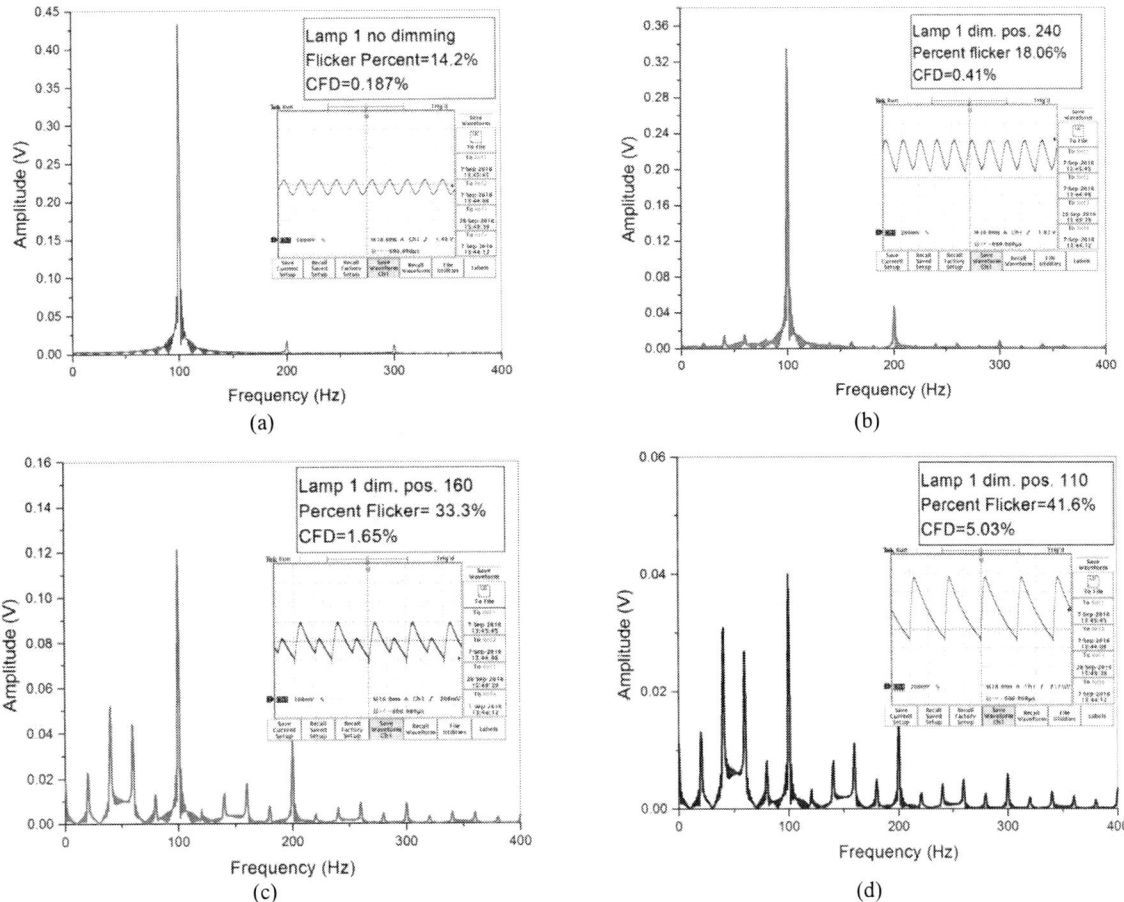

Fig. 6.   FFT graphs and corresponding waveform.

We present in Fig. 7 the flicker waveform from a tungsten lamp for four dimmer positions in absolute scale. In Fig. 8 the normalized waveform, used to compute the percent flicker is presented. The resulting percent flicker was for no dimming, positions 240, 160, 110 respectively 8.99%, 10.92%, 17.1%, 27.71%.

In Fig. 9 similar results as in Fig. 6 are presented, but for with no dimming and for different lamps.

2016 IEEE 22nd International Symposium for Design and Technology in Electronic Packaging (SIITME)

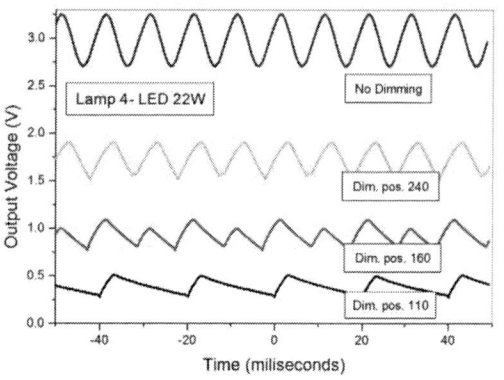

Fig. 7. Acquired light waveforms for LED lamp 4 in absolute representation.

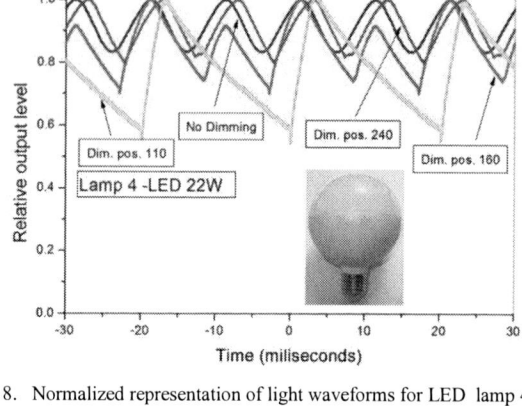

Fig. 8. Normalized representation of light waveforms for LED lamp 4.

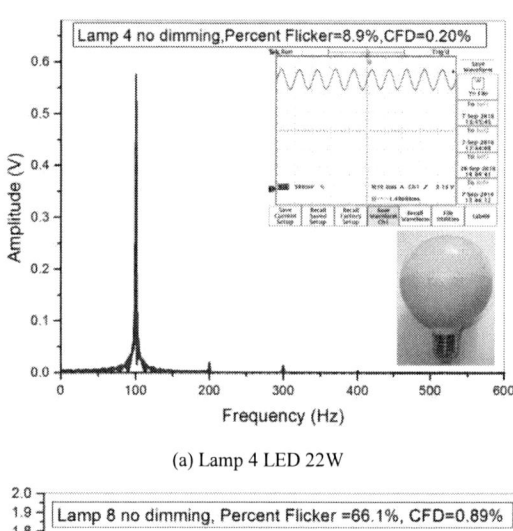

(a) Lamp 4 LED 22W

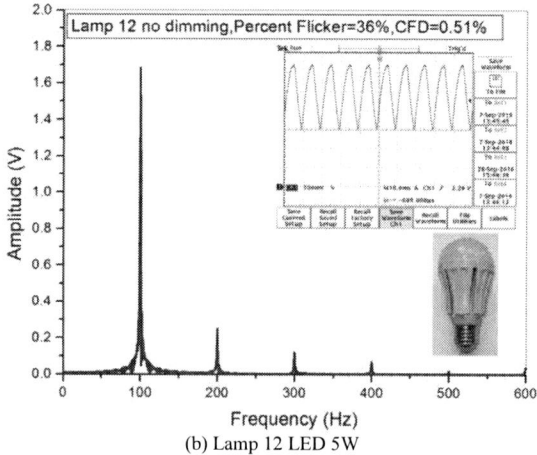

(b) Lamp 12 LED 5W

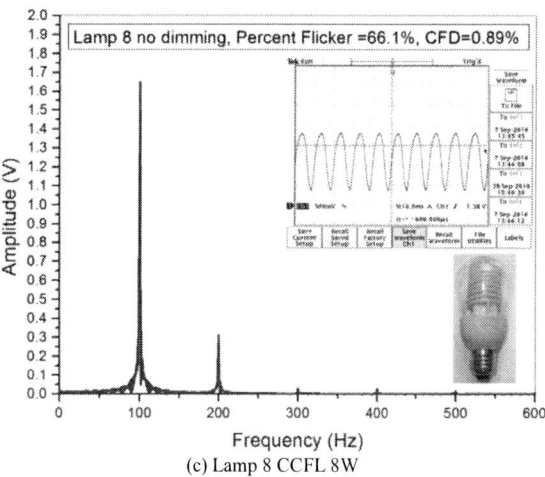

(c) Lamp 8 CCFL 8W

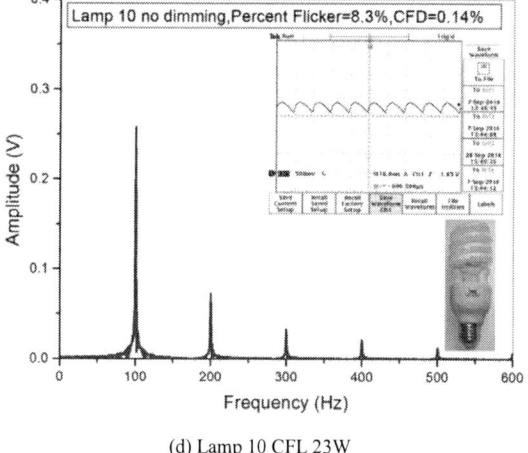

(d) Lamp 10 CFL 23W

Fig. 9. FFT graphs waveform for different lamps without dimmer.

As expected the flicker has a peak frequency of 100Hz, which originates from the double of the mains frequency, as a consequence that the luminous flux depends on the electrical power delivered to the lamp which depends quadratic on ac voltage. Some other frequency components are present, and are taken into account when comparing different lamps.

It can be seen that in Fig. 9 (c) the CCFL lamp has a very poor percent flicker but an acceptable CFD value.

978-1-5090-4446-7/16 $31.00 © 2016 IEEE   288   20-23 Oct 2016, Oradea, Romania

V. Conclusions

The method for flicker investigation using discrete Fast Fourier Transform- FFT takes into account the frequency and the human subject perception, not only the wave shape and will be more and more used, especially for new lighting systems. All the new systems is very probable to be PWM driven and hence to operate in non-sinusoidal regime at higher frequencies. Theoretically, high frequency components of light flicker are not perceptible, but studies are in development.

The dimming process as a rule increases the flicker but this could be not only an intrinsic feature of the lamp but rather a combination of lamp and dimmer. So, dimming should be used carefully in applications where flicker is of concern. The new investigated CCFL lamp, presented as "dimmable" is not fully compatible with existing phase cut dimmers, the waveform was strongly distorted and the CFD increases up to 15%, being perceptible.

References

[1] Kitsinelis, S., "Light Sources, Basics of Lighting Technologies and Applications", CRC Press, Boca Raton, Chap. 2-4 (2015).

[2] Alliance for Solid-State Illumination Systems and Technologies (ASSIST). 2015. ASSIST recommends… Application considerations related to stroboscopic effects from light source flicker. Vol. 11, Iss. 2. Troy, N.Y.: Lighting Research Center. Internet:

http://www.lrc.rpi.edu/programs/solidstate/assist/recommends/flick er.asp.

[3] Poplawski, M. E. and N. J. Miller, "Flicker in solid state lighting: Measurement techniques, and proposed reporting and application criteria", Proceedings of the Commission Internationale de l'Éclairage Centenary Conference, 188-202 (2013).

[4] NEMA Lighting Systems Division, "Document LSD 49-2010- Solid State Lighting for Incandescent Replacement, Best Practices for Dimming", 1-18 (2010).

[5] Plotog, I.; Vladescu, M.,"Power LED efficiency in relation to operating temperature", 7th International Conference on Advanced Topics in Optoelectronics, Microelectronics, and Nanotechnologies (ATOM-N 2014), Proceedings of SPIE, Volume: 9258, Article Number: 92582O, (2014).

[6] Drumea, A. and Dobre, R. , "Modelling, simulation and testing of an autonomous embedded system supplied by a photovoltaic panel", Proc. of the 20th International Symposium for Design and Technology in Electronic Packaging SIITME2014, 309-312 (2014).

[7] M. O. Dima, Y. N. Pepelyshev, L. Tayibov, "Neutron Noise Analysis with Flash-Fourier Algorithm at the IBR-2M Reactor", Advances in High Energy Physics, Volume 2013, Article ID 821709, http://dx.doi.org/10.1155/2013/821709.

[8] Nastac, D.I., and Ulmeanu A.P.: "An advanced model for electric load forecasting", 2013 IEEE Workshop on Integration of Stochastic Energy in Power Systems (ISEPS), Bucharest, 7 November 2013 (Digital Object Identifier: 10.1109/ISEPS.2013.6707942, IEEE Catalog Number: CFP1310W-ART, ISBN: 978-1-4799-1511-8).

[9] http://www.derlichtpeter.de/der-kompaktflimmergrad-cfd/ accessed September 2016.

[10] Benya, J. R., Leban, D. J., Lighting Retrofit and Relighting, A Guide to Green Lighting Solutions, Wiley, Hoboken, Chap. 2 (2011).

On Spectral Component Estimation using Neural Networks for Rolling Bearing Fault Diagnosis

V. Nicolau, M. Andrei

Electronics and Telecommunications Dep.
"Dunarea de Jos" University of Galati
Galati, Romania
viorel.nicolau@ugal.ro

Abstract—In general, the functioning state of rotating internal parts of a machine, which is inaccessible without dismantling the machine, can be obtained by indirect methods. Hence, fault diagnosis of rotating components, like rolling bearings, can be done by analyzing the external relevant information obtained by measurements, in order to evaluate their internal states. The most widely used method for detection and fault diagnosis of rolling bearings is based on vibration measurements. In signal processing, there are many applications of neural networks (NN) combined with spectral analysis, using Fast Fourier Transform (FFT) algorithm. The function approximation using neural networks is a complex task for problems with high dimension of the input space, like those based on signal spectral analysis. In this paper, some neural estimation aspects of spectral components of vibration signals for rolling bearings fault diagnosis are discussed. Inner race defect of bearings is considered in simulations. The goal is to find a feed-forward neural network (FFNN) model for estimating spectral components, with computational complexity comparable with FFT algorithm, but easier to implement in hardware. Different FFNN architectures, with different data sets and training conditions, are analyzed.

Keywords—*Neural networks, estimation, spectral components, fault diagnosis*

INTRODUCTION

In the modern world, bearings are one of the most important mechanical parts used in rotational machinery. Their malfunction almost always implies the machinery breakdown, with costly shutdowns, lapses in production, and even human casualties [1]. Hence it is imperative to establish suitable conditioning and monitoring procedures to prevent malfunction and breakage during operation. The monitoring of the operative conditions of a rotary machine provides a great economic improvement by reducing the operational and maintenance costs, as well as improving the safety level.

Bearing fault detection is a typical problem in rotating machinery fault diagnosis. In general, fault diagnosis of rotating components can be done by analyzing the external relevant information obtained by measurements, in order to evaluate the states of internal components, which are inaccessible without dismantling the machine [2]. Different methods are used for detection and fault diagnosis of rolling bearings, based on vibration measurements (the most widely used), acoustic measurements, temperature measurements and wear debris analysis [3].

The vibration signature of the damage bearing consists of an exponentially decaying sinusoid having the structure resonance frequency. The duration of the impulse is extremely short compared with the interval between impulses. Hence, its energy is distributed at a very low level over a wide range of frequency, being easily masked by noise and low frequency effects [4]. The periodicity and amplitude of the impulses are governed by the bearing operating speed, location of the defect, geometry of the bearing and the type of the bearing load.

Vibration signals are analysed in different domains: time, frequency and time-frequency domain. In time-domain, several characteristics can be analysed (RMS, peak value, probability density function, different orders of statistical moments), while in frequency-domain Fourier transforms are used [5]. Time-frequency domain techniques combine information from both time and frequency domains.

Many neural network models have been mathematically demonstrated to be universal approximators. The related results include proofs for the conventional multilayer perceptron [6], the radial basis function (RBF) NN [7], the rational function NN [8]. The feed-forward neural networks (FFNN), with a variety of activation functions, can be used as universal approximators [9]. For accurate function approximation and good generalization, many aspects must be taken into account, regarding neural network topology, training data set selection, and learning algorithm.

For problems with small dimension of the input space, the main learning concern is related to over-training, which can lead to poor generalization capability of the network. This fact must be considered when the neural network is designed and trained. If the dimension of the input space is big, the time complexity of the learning process increases, and good approximation capability is hard to be reached for the entire domain of interest in the input space. The role of input dimension on function approximation is studied in [10], for various norms, and target sets using linear combinations of adjustable computational units. It results that for good approximation, input upper bounds must decrease, as the input space dimension increases, which means that the smoothness of function being approximated would have to be increased.

In signal processing, there are many applications of neural networks (NN) combined with spectral analysis, using Fast Fourier Transform (FFT) algorithm. Frequently, the problems have small dimension of the input space, and frequency spectra are used as input data for NN. In this case, neural networks are used as feature classifiers of signal spectral components computed with FFT. If NN are used for spectral component estimation, than the dimension of the input space increases exponentially, and the estimation problem becomes very complex. Hence, NN applications focus on few spectral components, like in power industry, where FFNN are widely used for harmonics analysis and prediction [11].

In this paper, some aspects of neural estimation of signal spectral components are discussed. The goal is to find a FFNN model to estimate several spectral components, with computational complexity comparable with FFT algorithms. Different FFNN architectures, with different data sets and training conditions, are analyzed. A new FFNN was proposed, with local connections of the inputs to the hidden layers, forming input butterflies like in radix-2 FFT algorithm.

The paper is organized as follows. Elements of signal spectral analysis are presented in section 2. In section 3, aspects of neural approximation and generalization, along with neural prediction techniques are presented. Section 4 describes some simulation results based on different neural architectures, training data sets and noise conditions. Conclusions are presented in section 5.

ELEMENTS OF SIGNAL SPECTRAL ANALYSIS

Physical signals $x(t)$ are considered finite energy processes. Let $x_k(t)$ be a representative instance of an ergodic process $x(t)$ into finite time interval T. The energy of $x_k(t)$ instance is:

$$E = \int_{-\infty}^{\infty} x_k^2(t)\,dt = \frac{1}{2\pi} \int_{-\infty}^{\infty} |X_k(j\omega)|^2 \, d\omega \qquad (1)$$

where $X_k(j\omega)$ is the continuous-time Fourier transform of $x_k(t)$. It is well know that if the process energy is finite, then the integrals in (1) are bounded.

Frequency spectra are computed, according to input signal type, continuous- or discrete-time signal. Digital signal processing is based on discrete-time signals $x(n)$, as measurements of physical continuous-time signals $x(t)$, at discrete moments of time, $t = nT_S$, where T_S = sampling period, and n = sampling time index. The samples are obtained with the sampling frequency $f_S = 1/T_S$. Frequency spectra are computed with discrete transforms, like discrete-time Fourier transform (DTFT), which are complex continuous functions in the frequency domain. :

$$X(j\omega) = \sum_{n=-\infty}^{\infty} T_S \cdot x(nT_S) \cdot e^{-j\omega \cdot nT_S} = \sum_{n=-\infty}^{\infty} x[n] \cdot e^{-j\omega \cdot n} \qquad (2)$$

Discrete Fourier Transform (DFT) is a discrete transform for spectral analysis of finite-domain of discrete-time functions. It is a particular type of DTFT, applied on a time-window with a finite number (N) of input samples of a longer input data sequence.

DFT analytical expression is obtained by sampling DTFT in frequency domain, for the finite input sequence, denoted x_n, n = 0 … N-1:

$$X_k = \sum_{n=0}^{N-1} x_n \cdot e^{-j2\pi \cdot \frac{k}{N} \cdot n} \quad , \quad k = 0 \dots N-1 \qquad (3)$$

Fast Fourier Transform (FFT) is an efficient class of algorithms to compute DFT and its inverse. FFT computes more quickly the frequency components of the discrete spectrum than DFT in (3). Considering a finite input sequence of N samples, the computing complexity of DFT is $O(N^2)$, while FFT algorithms have $O(N \cdot log_2 N)$. The difference in speed is important, especially for big values of N, reducing computational time by several orders of magnitude.

There are many forms of FFT algorithms, with decimation-in-time and decimation-in-frequency. Radix-2 FFT algorithms need the length of input sequence (N) to be a power of 2. Other FFT algorithms need N to be equal to the product of mutual prime numbers (e.g. $9 \cdot 7 \cdot 5 = 315$ or $5 \cdot 16 = 80$). In addition, FFT algorithms for real input data are about 60% effort of the same sized complex data FFT.

In general, a radix-n FFT algorithm recursively breaks down the DFT of composite size $N = n \cdot m$ into n smaller DFTs of size m, where n is the radix number. By far the most commonly used radix-2 FFT is the Cooley–Tukey algorithm, which successively breaks a DFT of size N into 2 smaller DFTs of size N/2. The shape of the data-flow diagram is called butterfly diagram, with 2 inputs (x_1, x_0) and 2 outputs (y_1, y_0).

A decimation-in-time FFT algorithm with the length of input sequence $N = 2^p$, based on primitive N-th root of unity $\omega = exp(-j \cdot 2\pi / N)$ uses butterfly forms with k = 0 … N-1:

$$\begin{cases} y_0 = x_0 + x_1 \cdot \omega^k \\ y_1 = x_0 - x_1 \cdot \omega^k \end{cases} , \qquad (4)$$

FEED-FORWARD NEURAL NETWORK ESTIMATION

Neural network models are specified by the net topology, node characteristics, and training or learning rules. FFNNs are characterized by neuron equations without memory, meaning that their outputs are functions only of the current inputs.

In this paper, FFNN models are used for neural estimation of signal spectral components. Many aspects must be solved when using FFNNs. The most practical concerns are: the network size, time complexity of learning and network ability to generalize [12].

The network size is important, but in general it is not known for a given problem. If the network is too small, it will not be able to give a good model of the problem. If the network is too big, it will give poor approximations of the problem [13].

Time complexity of learning is directly connected with the complexity of the problem. Approximation of FFT function is complex due to huge dimension of the input space. One way to reduce the number of network weights is to use local connections of the inputs to the hidden layers, so that individual neurons in the hidden layers to process local regions of the input space [12]. The inputs are N consecutive signal samples of finite input sequence, denoted x_n, n = 0 … N-1, and the outputs represent estimated amplitude of the desired spectral components.

At every *k* moment of time, the time-window contains a finite number (N) of time-delayed samples, which form the input sequence of FFNN, as illustrated in Fig. 1. The input sequence x_n was denoted according with time moment considered: $x_0 = x(k-N+1)$, …, $x_{N-1} = x(k)$.

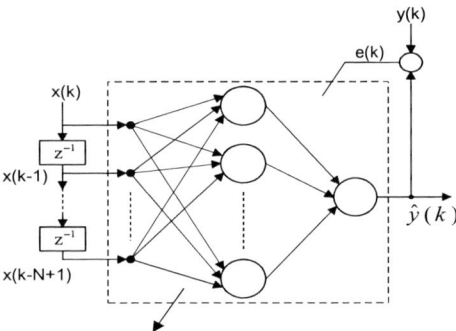

Fig. 1. FFNN estimation of signal spectral components

The estimates take into account the uncertainty in the target data or the uncertainty in the input data. Therefore, it is necessary for any learning system to be able to cope with such uncertainty. In this paper it is assumed that there is noise on the input data sets, generated by measurement system.

For every step *k*, the estimation error *e(k)* is computed based on the computed and estimated values of spectral component amplitudes. The performance goal has to be small enough to assure good estimation performance, but not too small to avoid loss of generalization. The mean squared error *mse* was chosen as performance function:

$$mse = E[e^2(k)] = \frac{\sum_{i=1}^{M} e^2(k)}{M},\qquad(5)$$

where *M* is the number of vectors in data sets.

SIMULATION RESULTS

Most of vibration-based monitoring techniques are applied in frequency-domain starting from time-domain vibration signal, which is processed by Fourier transforms, usually in the form of a FFT algorithm. The principal advantage of this approach is that the repetitive nature of the vibration signal appears as peaks in the frequency spectrum at well defined frequencies. This allows for faults to be detected in early stages and being diagnosed accurately [12].

In general, there are five basic motions that can be used to describe dynamics of bearing movements: shaft rotational speed in rpm, inner race defect (IRD), outer race defect (ORD), ball defect (BD) and cage defect (CD). Each motion generates a unique frequency (with different formulas) in frequency-domain analysis [14].

For simulations, inner race defect frequency was considered being settled at shaft rotation frequency. The shaft rotation speed was chosen in the range: [1200, 5400] rpm. This corresponds to shaft rotational and also IRD frequency range: [20, 90] Hz. In this frequency range, the spectral components should be estimated using FFNN.

Signal data sets are generated, as sequences of samples from noisy signals with different narrow frequency bandwidth within the IRD frequency range. In spectral analysis, time-windows with N = 32 samples are used, as inputs to FFNN and also to FFT algorithm, which computes the desired frequency spectra. The sampling frequency was chosen as f_S = 320 Hz, resulting a frequency resolution of deltaf = 10 Hz.

The signals are series of sinusoidal functions, with variable number of spectral components, N_{rsc} = 1, 2 or 3, with discrete frequencies randomly selected in the chosen frequency interval. Also, the amplitudes and phases are bounded and randomly selected between the limits: $A_j \in [1, 3]$ and $\varphi_j \in [-\pi/2, \pi/2]$, respectively. The signal analytical expression is:

$$x(t) = \sum_{j=1}^{N_{rsc}} A_j \cdot \sin(2\pi \cdot f_j t + \varphi_j)\qquad(6)$$

The signal parameters are: N_{rsc}, A_j, f_j, and φ_j, j = 1 … N_{rsc}. For every parameter set with chosen number of spectral components (N_{rsc} = 1, 2 or 3), a data sequence is generated. Using FFT algorithm, the computed frequency spectra have 32 spectral components, of which only 16 are used, into the useful frequency range: [0, 150] Hz. They form the desired frequency spectra in supervised learning process. To reduce the problem complexity, a smaller region of interest into the input space is selected. In this case, frequency interval is: I_I = [20, 80] Hz.

Signals are affected by different types of noise (measurement, conversion, digital processing). In general it is considered additive noise *n(t)*. As a result, the noisy signal is: *xn(t) = x(t) + n(t)*. Additive white Gaussian noise is considered in this paper. An example of data set for N_{tw} = 5 distinct time-windows, with- and without noise, is shown in Fig. 2. Time horizon includes N_{tw} = 5 distinct time-windows. Hence, each data set contains $N \cdot N_{tw}$ = 160 samples.

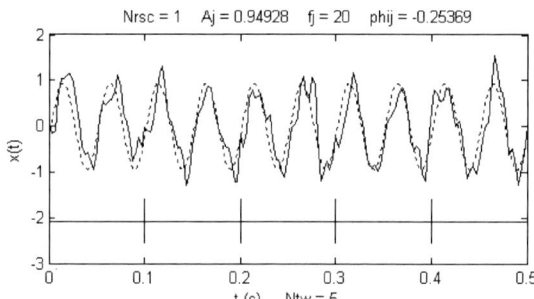

Fig. 2. Data set with 1 spectral component, f_{j1} = 20 Hz

For FFNN training and testing, each input data sequence is organized as a matrix of type *[N, N_{tw}]*, where each column contains *N* samples of one time-windows.

The samples of first time-window of data sequence presented above are illustrated on the left side in Fig. 3. The noisy samples are drawn with continuous line and the noiseless values with dotted line. On the right side, spectral components are represented with circle and diamonds marks, respectively.

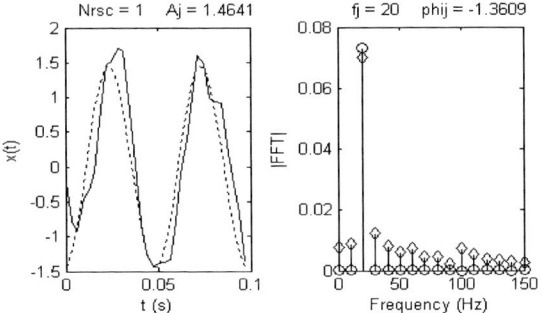

Fig. 3. Samples of first time-window, and its computed FFT magnitude

The training sets contains a number of $N_{lds} = 50$ data sequences, with $N_{tw} = 5$ time-windows each. So, each training set contains a number of learning vectors $N_{lv} = 250$. Similarly, the testing sets contains a number of $N_{tds} = 5$ data sequences, with a total number of testing vectors $N_{tv} = 25$.

Two different FFNN architectures were tested. The first type of FFNN architecture is a standard FFNN with one N-dimensional input, one hidden layer with N_{hn} neurons, and one N/2-dimensional output, as illustrated in Fig. 4. The neurons in hidden layer have *tansig* transfer functions, and output neurons have *poslin* transfer functions, so that only positive values to be generated for spectral component amplitudes. Two feed-forward neural networks were generated with different number of neurons in hidden layer, $N_{hn} = 32$, and 64, respectively.

Fig. 4. FFNN architecture, with only one N-dimensional input

The second type of FFNN architecture is a butterfly-like FFNN (BFFNN) with two N/2-dimensional inputs, two hidden layers with $N_{hn1} = N_{hn2} = N/2$ neurons, connected separately with the inputs, and one N/2-dimensional output, as illustrated in Fig. 5. The neurons in hidden layers have *linear* transfer functions, and output neurons have *poslin* transfer functions. During network learning and testing, the same performance criterion was used, the mean squared error (*mse*) of spectral component estimation related to initial noisy frequency spectrum, denoted *msel* and *mset*, respectively.

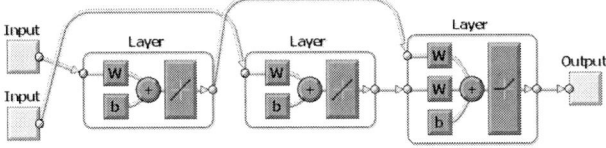

Fig.5. Butterfly-like FFNN architecture, with two N/2-dimensional inputs

The training and testing results for 3 NNs are represented in Table 1. It can be observed that butterfly-like FFNN has much better performance than FFNN.

NETWORK PERFORMANCE FOR DIFFERENT NEURAL ESTIMATORS

Comp. Freq. [Hz]	FFNN Nhn = 32		FFNN Nhn = 64		Butterfly-like FFNN	
	msel $\cdot10^{-6}$	mset $\cdot10^{-6}$	msel $\cdot10^{-6}$	mset $\cdot10^{-6}$	msel $\cdot10^{-6}$	mset $\cdot10^{-6}$
20	5.79	6.144	4.93	6.015	0.331	0.369
60	5.99	6.310	5.12	6.227	0.331	0.379

CONCLUSIONS

In this paper, some neural estimation aspects of spectral components of vibration signals for rolling bearings fault diagnosis are discussed. Inner race defect of bearings is considered in simulations. Different FFNN architectures, with different data sets and training conditions, are analyzed. A butterfly-like FFNN (BFFNN) was proposed, which has much less weight connections and better performance than FFNN.

ACKNOWLEDGMENT

The authors thank the Executive Agency for Higher Education, Research, Development and Innovation Funding (UEFISCDI) for its support under Contract PN-II-PT-PCCA-2013-4- 0044.

REFERENCES

[1] D.H. Pandya, S.H. Upadhyay, S.P. Harsha, "ANN Based Fault Diagnosis of Rolling Element Bearing using Time-Freq. Domain Feature", Int. J. of Eng. Sc. and Tech., Vol.4, No.12, pp. 2878-2886, 2012.

[2] M. Unal, M. Demetgul, M. Onat, H. Kucuk, "Fault Diagnosis of Rolling Bearing Based on Feature Extraction and NN Alg.", Recent Adv. in Telecomm, Sig. and Syst, ISBN:978-1-61804-169-2, 2013.

[3] V. Patel. and A. Patel, "Recent Advanced in AI Based Intelligent Fault Techniques for Rolling Element Bearing – A Review," Int. J. of Artificial Intelligence and Mechatronics, Vol. 2, No. 4, pp. 104-108, 2014.

[4] K.F. Al-Raheem, W. Abdul-Karem, "Rolling bearing fault diagnostics using artificial neural networks based on Laplace wavelet analysis," Int. J. of Eng. Sc. and Tech., Vol.2, No.6, pp. 278-290, 2010.

[5] N. Tandon and A. Choudhury, "A review of vibration and acoustic measurement methods for detection of defects in rolling element bearings", Tribology International, vol. 32, pp. 469-480, 1999.

[6] K. Hornik, "Approximation capabilities of multilayer feedforward networks," Neural Networks, 4: 251-257, 1991.

[7] J. Park, and I.W. Sandberg, "Universal approximation using radial-basis-function networks," Neural Comput., 3: 246-257, 1991.

[8] H. Leung, and S. Haykin, "Rational function neural network," Neural Comput., 5: 928-938, 1993

[9] R. Ilin, R. Kozma, and P.J. Wetbos, "Beyond Feedforward Models Trained by Backprop: A Practical Training Tool for a More Efficient Univ. Approx." IEEE Trans. On Neural Networks, 19(6): 929-937, 2008.

[10] P.C. Kainen, V. Kurkova, and M. Sanguineti, "Dependence of Computational Models on Input Dimension: Tractability of Approximation and Optimization Tasks," IEEE Trans. On Information Theory, 58(2): 1203-1214, 2012.

[11] Wu Xihong, He Wei, Zhang Zhanlong, Deng Jun, and Li Bing, "The harmonics analysis of power system based on Artificial Neural Network," World Automation Congress (WAC 2008), pp. 1-4, 2008.

[12] D.R. Hush, B.G. Horne, "Progress in Supervised Neural Networks," IEEE Signal Proc. Mag., 10(1): 8-39, DOI: 10.1109/79.180705, 1993.

[13] B. Baum, and D. Haussler, "What size net gives valid generalization?," Neural Computation, 1: 151-160, 1989.

[14] P.P. Kharche, and S.V. Kshirsagar, "Review of Fault Detection in Rolling Element Bearing," Int. J. of Innovative Research in Advanced Engineering, Vol. 1, No. 5, pp. 169-174, 2014.

Characterization of Tin Pest by Electrical Resistance Measurement

Agata Skwarek
Department of Microelectronics
Institute of Electron Technology,
Krakow, Poland
askwarek@ite.waw.pl

Balázs Illés,
Department of Electronics Technology
Budapest University of Technology and Economics
Budapest, Hungary
billes@ett.bme.hu

Attila Géczy,
Department of Electronics Technology
Budapest University of Technology and Economics
Budapest, Hungary
gattila@ett.bme.hu

Abstract— In this paper the allotropic transition of ß to α-Sn (so called tin pest) was characterized by electrical resistance measurements. The aim was to compare the ability and sensitivity of the usually applied optical observations by the electrical measurements. Samples were prepared from 99Sn1Cu alloy. The samples were inoculated by InSb and were stored at -18°C for 40 days. The transition was checked by optical microscopy on the surface of the samples and electrical resistance of the samples was measured by a microhmmeter. Cross-sections were also prepared to compare the resistance change with the amount of transformed tin in the samples. It was found that the electrical resistance measurement is a much more sensitive characterization method of tin pest growth than any kind of optical observation. It was also observed that in the case of InSb inoculator the nature of the process differs from the results of α-Sn inoculator.

Keywords— Tin pest, Sn allotropes, α-Sn ß-Sn, solder alloys, electrical resistance measurement..

I. INTRODUCTION

Tin (Sn) is one of the most important materials in the electronics technology, this is the base metal in most of the solder alloys and used as surface finishing of the solder pads and components leads. Tin has four allotropes: α, ß, γ and σ tin. From these β-Sn (white tin) is the metallic form and it is applied in the electronics technology. β-Sn is stable between 13.2 and 231.9°C, it has bct (body-centered tetragonal) structure crystallizes in the space-group symmetry I41/amd (No. 141) with lattice parameters a = 5.8316 Å and c = 3.1815 Å. The α-Sn (gray tin) is a nonmetallic form, which is stable below 13.2°C, it has diamond structure with cubic symmetry Fd3m (No. 227), lattice parameter a = 6.4892Å (Fig. 1). α-Sn is brittle and has semiconductor properties. The other two allotropes, γ and σ, exist only at very harsh circumstances, over 161°C and high pressure [1].

Tin pest is a spontaneous allotropic transition of ß-Sn to α-Sn below 13.2°C. In basic case the transition is slow to initiate

due to a high activation energy but the presence of materials with the same crystal structure and close lattice parameters to the α-Sn or very low temperatures (under −30°C) can speed up the process considerably. The transition causes also a large volume increase of about 27%.

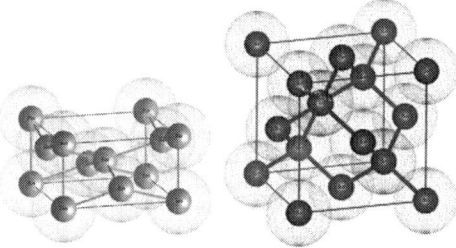

Fig. 1. : β (left) and α-Sn (right) crystal structures [2].

Firstly the transition starts from blemishes, discoloration, mottling. Later these first ones are changed into characteristic warts (Fig. 2, which originates the name "tin pest").

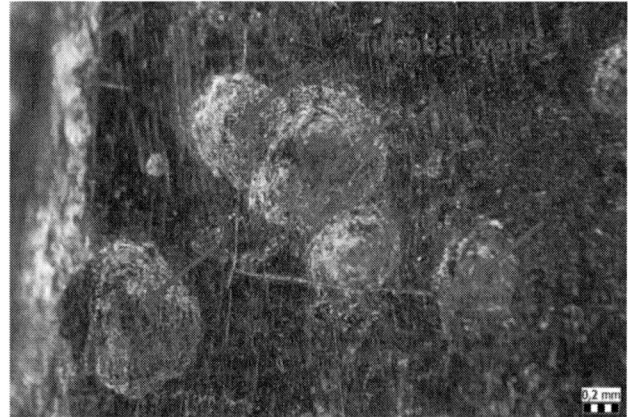

Fig. 2. SEM image of tin pest warts on the surface of tin object.

Finally the sample is changed into the form of the powder leading to the total disintegration of Sn sample which is related to the volume increase and the very low ductility of the tin results in the material blistering and cracking [3]. The tin pest is an autocatalytic reaction, which means that the appearance of α-Sn speeds up the transition leading to the sample disintegration. The identification and characterization of tin pest in Sn-rich solders and surface finishes is crucial for electronic devices working in sub-zero temperatures in aeronautical, aerospace and automobile applications [4].

The α-Sn transition can be avoided by alloying with small amount of electropositive metals which are soluble in tin's solid phase such as antimony or bismuth. In the past, another effective method was the application of high lead content alloys, since the lead blocks effectively the α-Sn transition similar as in the case of tin whisker growth [5]. Therefore, before the RoHS (Restriction of Hazardous Substances) directives of the European Commission, the tin pest phenomenon was less frequent in the electronics appliances. However since 2006, electronic manufacturers have to replace the tin-lead solder alloys with high tin content (tin content can reach 98% [6]) lead-free solder alloys, which means a real risk of a tin pest allotropic transition in solder joints as well as in surface finishes and coatings [7] and in addition there are other reliability issues of high tin content alloys [8]. There are some studies which stated that the tin pest is impossible in binary and trinary tin alloys [9, 10], but there are researches where the phenomenon is occurred in high tin content alloys too [2, 4]. Generally it is accepted that the alloying with Sn - soluble elements delays the process [11].

The tin pest phenomenon is difficult to study in natural conditions since the nucleation phase is very slow, so the first signs of transition are noticeable only after several months [12]. Application of the seeds or inoculators – such as CdTe or InSb or α-Sn itself [3] – could speed up the nucleation phase considerably. However these techniques are not enough for prediction the nucleation time, or to simulate real-life situations, but they are very useful for studying the nature of the process and the growth of the α phase.

The α-Sn growth is usually monitored or detected by XRD study [13], SEM with EDS analysis [14] optical observation [3] or Mössbauer spectroscopy [15]. In this study our aim was to characterize the α-Sn transition in 99Sn1Cu alloy with of InSb inoculator by electrical resistance measurements and to compare the ability and sensitivity of the electrical resistance measurements – as detection and monitoring method – to the optical observations.

II. MATERIALS AND METHODS

The samples were prepared with the tin pest induction method developed earlier [16]. For the investigation samples were prepared by cold-rolled sheet technology from 99Sn1Cu alloy with 45x6x1.5mm2 size. The average electrical resistance value of the samples was 0.49mΩ. The electrical resistance deviation was 4%, due to the deviation of the sample size. All together 20 pcs. of samples were inoculated by InSb inoculator powder, which was pressed into the upper layer of the samples be a mechanic laminator with 30kN force

(Fig. 3). The samples were stored at -18°C for 42 days (6 weeks) in a refrigerator.

The transition progress was checked by an OLYMPUS optical microscope on the surface of the samples and electrical resistance of the samples was measured by AGILENT 4338B µΩ meter. The electrical measurements were done at room temperature to avoid the resistance change at low temperature. The measurement accuracy of the instrument at the mΩ range was under 3%. The repeatability error of the resistance measurements was under 2%. During the different stage of the test, cross-sections were also prepared from the samples to compare the resistance changes with the amounts of transformed tin in the samples.

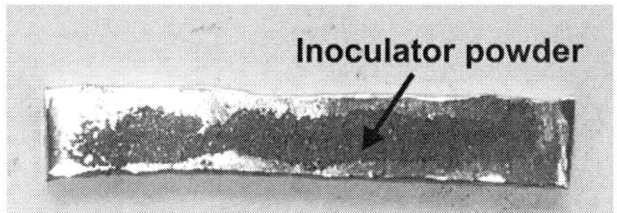

Fig. 3. Inoculated sample before the test.

From a given samples 10 pcs. of cross-sections were prepared to get more information about the uneven development of α-Sn. The even inoculation of the samples was complicated during the study because the inoculator powder has not adhered evenly on the body of the samples (as it can be seen in Fig. 3).

III. RESULTS

Fig. 4 shows the electrical resistance increases on a logarithmic scale in the function of time. During the first 7 days no detectable resistance change was observed. After the 2 weeks some resistance increases were found at some samples, but the highest increase was under 6% which value is very close to the repeatability and measurement error. However the typical of tin pest have been already appeared on the surface of certain samples (Fig. 5). After 3 weeks of temperature storage at -18°C, all of the samples showed resistance increase between 5 and 20%, therefore the nucleation phase of the α-Sn development in this study was determined in 3 weeks.

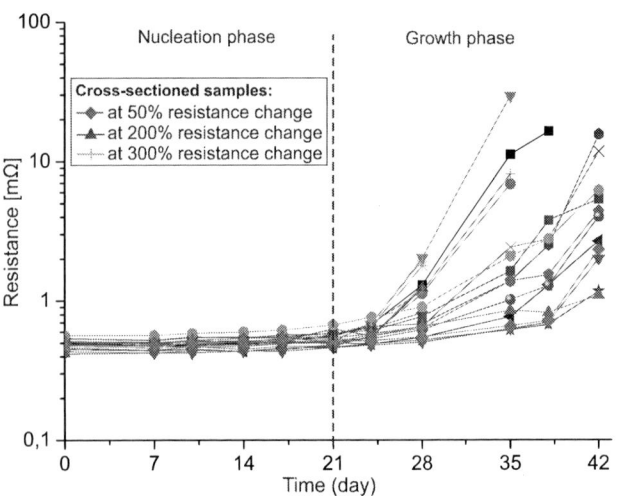

Fig. 4. Electrical resistance increase during the test.

After 3 weeks, when nucleation phase has been over, the electrical resistance increase is considerable. However the deviation of the resistance increase was also very high. After 4 weeks of temperature storage at -18°C, the changes were between 10 and 300%, meanwhile after 5 weeks they were between 25 and 2100%. This effect was caused by the not even inoculation of the samples, but much more by the stochastic development nature of the α-Sn transition.

Since the process is highly autocatalytic, it causes that only a small difference in time or in amount of the α-Sn appearance in the different samples results in big differences in transition rate. This is clearly visible in Fig. 4, as the deviation is increasing with the time-in nucleation phase.

Fig. 5. First α-Sn (tin pest) warts on the surface of a sample after 2 weeks

It was observed after 5 weeks of temperature storage at -18°C that some of the samples decomposed by the α-Sn transition. This effect is presented in Fig. 6. The disintegration of the samples usually occurs when the electrical resistance of the samples increased of 25 – 30 times.

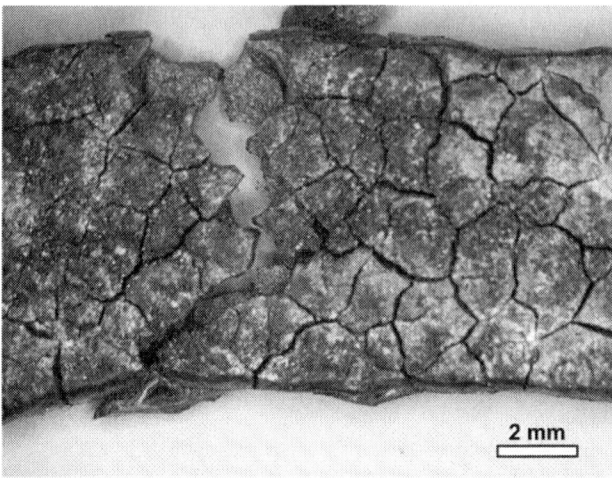

Fig. 6. Disintegrated sample by the α-Sn transition after 6 weeks of storage at -18 °C.

To this stage the tin pest (from both surfaces of the samples) growth across the samples and this effect separated the samples into powder and bigger sized morsels. The highest measurable resistance changes was 32 times after 5 weeks of storage at -18 °C (see the pink curve in Fig. 4).

After 3 weeks of storage at -18 °C, three samples with 50, 200 and 300% resistance increase respectively, were selected and pull out for the test for cross-sectioning. Samples with higher resistance changes were not suitable for cross-sectioning since they became very brittle.

Fig. 7 shows five cross-sections of the sample with 50% resistance increases. Crack like structures can be seen in Fig. 7a) –d), however these are not real cracks but holes from where the transformed α-Sn is missing. As it was discussed earlier the mechanical strength of the α-Sn is much worse than the mechanical strength ß-Sn. So these holes were formed during the gridding of the samples (as a necessary step of the cross-section preparation), when the emerging α-Sn usually fell out from the sample body. The holes always start from the surface of the sample which is also a clear evidence that they made by the tin pest. Some examples for the presence of InSb inoculator is still visible in Fig. 7e), these are the dark grey areas of the samples (a magnification can be seen in the red rectangle).

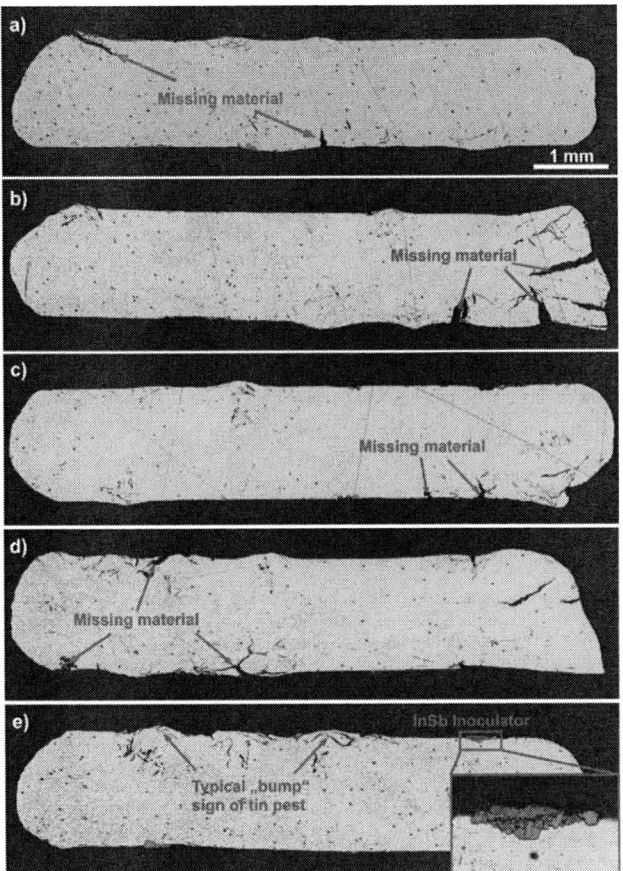

Fig. 7. Cross-sections of a sample with 50% of resistance increase.

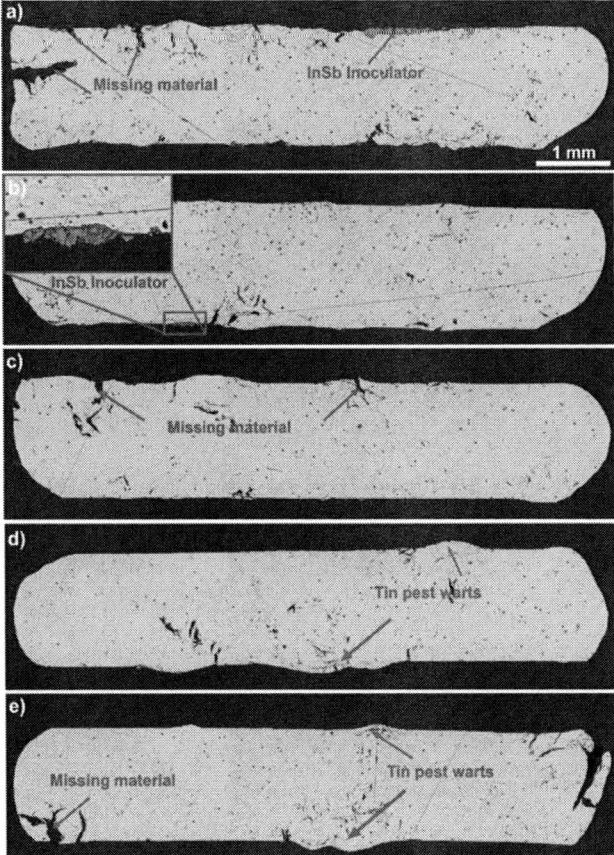

Fig. 8. Cross-sections of a sample with 200% of resistance increase.

Fig. 8 shows five cross-sections of the sample with 200% resistance increases. The holes of the fallen out α-Sn are also visible in Fig. 8, but interestingly the amount and the size of these holes are the same as in the case of the previous sample which has only 50% resistance increase. The typical tin pest wart related to the volume change during the transition, is well observable in Fig. 8d)-e), at these areas the tin pest penetrated under some tens of μm under the surface and the volume expansion of transition could push up the covering tin layer.

Fig. 9 shows five cross-sections of the sample with 300% resistance increase. In this samples more considerable holes were found also in sizes and amounts (Fig. 9b) and d)). The α-Sn transition involves greater surface across the sample (Fig. 9b)). However it is interesting to see that even in this sample there are relatively untransformed parts of β-Sn like in Fig. 9a) and c).

IV.   DISCUSSION

Only one example was found in the literature where the tin pest phenomenon was characterized by electrical measurements. Maio and Hunt studied the electrical resistivity changes of different tin alloys inoculated with CdTe at -35°C [17]. Their and our results are hardly comparable due to the different inoculators and curing temperatures, but the electrical resistance characteristics show similar shapes. They observed a relatively long nucleation phase which is followed by a fast growth of α-Sn phase where the electrical resistance increases considerably until the disintegration of the sample. The most difference between the results is the speed of the process, which was much slower in our case mainly due to the two times higher storage temperature.

It was observed that in the case of InSb inoculator the nature of the process differs from the non-inoculated case or from the case when α-Sn inoculator is used. On that cases it is usually observed that the α-Sn expands much faster horizontally than vertically in the sample body which results in the forming of "disc like" layer separations, one by one towards the middle of the sample body [3, 4]. Contrary in this study, the vertical expansion of the α-Sn was the same or sometimes faster than the horizontally. It is also assumed according to the disintegrated samples that-the α-Sn transition usually occurs at the Sn grain boundaries, however further researches are necessary to prove this hypothesis.

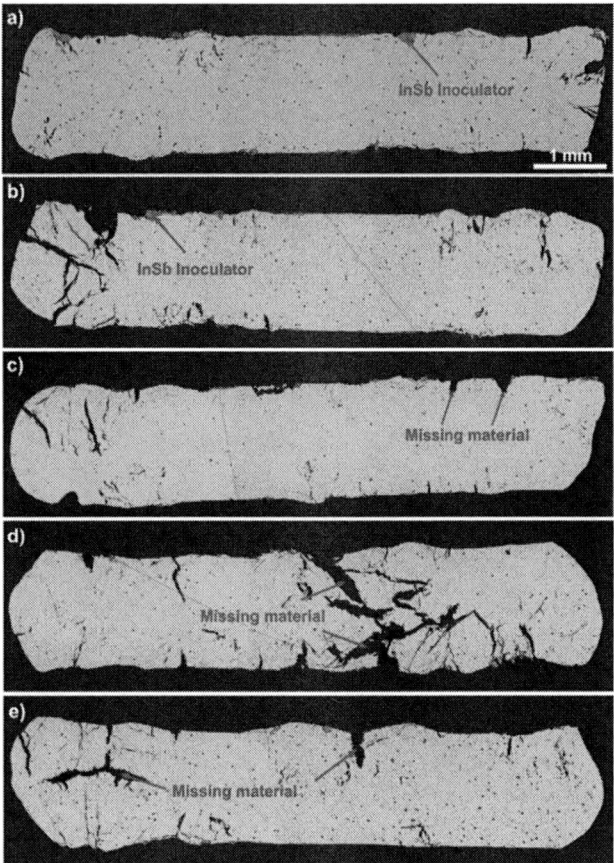

Fig. 9. Cross-sections of a sample with 300% of resistance increase.

Nevertheless the ability and sensitivity of the optical observations showed not satisfactory performance for accurate characterization of the tin pest phenomenon. The surface observations are not useful since do not give any information about the deepness of the α-Sn transition in the sample body. The observation of the cross-sections is more useful since it shows more precisely the progress of the process (the amount of the transformed tin) but unfortunately only on 2D sections which means that for enough accurate information a lot of expensive cross-sections are necessary to use. Even in the case of high number of cross-sections the possibility of the incorrect evaluation is still high, e.g. the views in Fig. 7 and 8 are almost the same, but the progress of the α-Sn transition in the given samples is very different according to the resistance values. In addition the cross-sectioning is applicable only at the beginning state of transition since the samples become brittle very fast.

The electrical resistance measurement performed the best results during the study. The sensitivity of the method is high, already 10% resistance change is surely detectable. The measurement range of the electrical measurements is also much higher than in the case of the cross-sectioning, with the resistance measurement the process can be fallowed till the disintegration of the samples. Even a 32 times resistance increase was detectable during this study but this values highly depends on the thickness of the samples since a thicker sample will decompose later.

V. CONCLUSIONS

The α-Sn transition was characterized by electrical resistance measurements. The aim of the study was to compare the ability and sensitivity of the usually applied optical observations by the electrical measurements. The main conclusions are the following:

- The electrical resistance characteristic of the tin pest phenomenon has a relatively long nucleation phase which is followed by a fast growth phase where the resistance increases considerably until the disintegration of the sample.

- The electrical resistance measurements give better results in characterization of tin pest phenomenon than the optical observations. It has much better sensitivity and measurement range than the optical observations.

- The optical observation on cross-sections is more useful than the surface imaging since it shows more precisely the progress of the process but for enough accurate information a lot of cross-sections are necessary to use, and even in this case the possibility of the incorrect evaluation is still high.

- In the case of InSb inoculator the process showed different form as it is usually observed, when the α-Sn expands much faster horizontally than vertically in the sample which results in "disc like" layer separations. Here the vertical penetration of the α-Sn was the same or sometimes faster than the horizontally.

- It is likely that α-Sn transition occurs at the Sn grain boundaries, however further researches are necessary to prove this assumption.

References

[1] A. M. Molodets, S. S. Nabatov, "Thermodynamic Potentials, Diagram of State, and Phase Transitions of Tin on Shock Compression", High Temperature vol. 38/5, 2000, 715–721.

[2] A. Skwarek, J. Kulawik, K. Witek, "Induction of Tin Pest Transformation in Solder Joints in Ceramic Packages of Sub–THz Scanner", in Proceed. of 38th IEEE-ISSE conf., Egerszalók, Hungary, 2015, pp. 47-51.

[3] A. Skwarek, P. Zachariasz, J. Kulawik, K. Witek, "Inoculator dependent induced growth of α-Sn", Materials Chemistry and Physics vol. 166, 2015, pp. 16-19.

[4] A. Skwarek, M. Sroda, M. Pluska, A. Czerwinski, J. Ratajczak, K. Witek, "Occurrence of tin pest on the surface of tin-rich lead-free alloys", Soldering & Surface Mount Technology vol. 23/3, 2011, pp. 184-190.

[5] B. Illés, B. Horváth, "Whiskering Behaviour of Immersion Tin Surface Coating", Microelectron. Reliab., Vol. 53, 2013, pp. 755-760.

[6] O. Krammer, T Garami, B Horváth, T Hurtony, B Medgyes, L Jakab, "Investigating the thermomechanical properties and intermetallic layer formation of Bi micro-alloyed low-Ag content solders", Journal of Alloys and Compounds, vol. 634, 2015, pp. 156-162.

[7] D. Di Maio, C. Hunt, "On the absence of the ß to α Sn allotropic transformation in solder joints made from paste and metal powder", Microelectronic Engineering vol. 88, 2011, pp. 117–120.

[8] O Krammer, "Comparing the Reliability and Intermetallic Layer of Solder Joints prepared with Infrared and Vapour Phase soldering", Soldering & Surface Mount Technology, vol 26/4, 2014, pp. 214-222.

[9] O. Semenova, H. Flandorfer, H. Ipser, On the non-occurrence of tin pest in tin–silver–indium solders, Scripta Materialia vol. 52, 2005, pp. 89–92.

[10] W. Peng, "An investigation of Sn pest in pure Sn and Sn-based solders", Microelectronics Reliability, vol. 49, 2009, pp. 86–91.

[11] G. Zeng, S. D. McDonald, K. Sweatman, K. Nogita, "Tin pest in lead-free solders?, Fundamental studies on the effect of impurities on phase transformation kinetics", ICEP 2014 Proceedings, pp. 135-139.

[12] W. Plumbridge, "Recent Observations on Tin Pest Formation in Solder Alloys", Journal of Electronic Materials Vol.37/2, 2008, pp 218–223.

[13] K. Nogita, C.M. Gourlay, S.D. McDonald, S. Suenaga, J. Read, G. Zeng, Q.F. Gud, "XRD study of the kinetics of β↔α transformations in tin", Philosophical Magazine vol 93/27, 2013, pp. 3627-3647.

[14] N. D. Burns, "A Tin Pest Failure", Journal of Failure Analysis and Prevention vol. 9/5, 2009, pp 461–465.

[15] A. Skwarek, P. Zachariasz, J. Zukrowski, B. Synkiewicz, K. Witek, "Early stage detection of $\beta \rightarrow \alpha$ a transition in Sn by Mössbauer spectroscopy", Materials Chemistry and Physics vol. 182 (2016), pp. 10-14

[16] A. Skwarek, J. Kulawik, K. Witek, Method of Evaluating the Susceptibility of Tin Alloys to Tin Pest, 2013 polish patent application number: P404330, patent number: 221478, filing date: 14.06.2013

[17] D. Di Maio, C.P. Hunt, "Monitoring the Growth of the α Phase in Tin Alloys by Electrical Resistance Measurements", Journal of Electronic Materials, vol. 38/9, 2009, pp. 1874 – 1880.

Adaptive User Interface for Higher Education based on webTechnology

Research and Innovation in Industry 4.0

Monica Ciolacu

Faculty of Business Informatics
Deggendorf Institute of Technology
Deggendorf, Germany
Monica.Ciolacu@th-deg.de

Rick Beer

Informatics Faculty
University Passau
Passau, Germany

Abstract—**We present an Adaptive User Interface (AUI) for online courses in higher education as a method for solving the challenges posed by the different knowledge levels in a heterogeneous group of students. The scenario described in this paper is an online beginners' course in Mathematics which is extended by an adaptive course layout to better fit the needs of every individual student. The course offers an entry-level test to check each student's prior knowledge and skills. The results are used to automatically determine which parts of the course are relevant for the student and which ones can be hidden, based on parameters set by the course teachers. Initial results are promising; the new adaptive learning platform in mathematics is leading to higher student satisfaction and better performance**

Keywords— Adaptive User Interface, Industry 4.0 in Education, blended learning, innovation, digitalisation, interactivity, mathematics, higher education, mobile learning .

I. INTRODUCTION

Digitalization is the most important step in Industrial revolution 4.0. An intelligent, digital and adaptive interface is a global challenge in educational technology for Higher Education. The impact of individual adaptive user interfaces will gain higher importance in our education system. [1]

"Industrie 4.0" is a German-government vision for advanced manufacturing. This 4.0 code has been used to mark the disruptive change which takes place. "The triumph of modern information and communication technologies (ICT) and in particular of the Internet is linked to new challenges in the area of privacy and IT security. However, the Internet does not only link people. Networked embedded systems – so called cyber-physical systems – are integrated in a growing number of everyday objects. The physical world is hooked up to the virtual world. Applications are now found in machine controls, medical devices or ABS systems in cars. The networking of objects and the use of networked embedded systems provides industry with new opportunities. Based on German industry's competences in managing industrial production processes and controlling and combining complex production and business processes, the forward-looking project Industry 4.0 will provide important prospects for technology, economy and social policy. Combining embedded systems with business application

software leads to entirely new business models and considerable potential for optimizing production and logistics. At the same time, Industry 4.0 facilitates more resource-conserving production, greater individualization and a perfect fit of products at mass-production prices." [2] Industry 4.0 describes an interwoven system of production resources that contain autonomous, intuitive machine learning and knowledge management. These machines are self-configuring, sensorially driven and dispose of their own parameters on spatial arrangement. This affects robotics, production machines, operating materials and resources, furthermore planning and control instruments.

II. DIGITAL ERA EVOLUTION

The technological foundation of Industry 4.0 combines intelligently – personalized- adaptive – flexible production and interfaces, Smart Services with self-managing production processes, Communities of Knowledge, Lean Organizations and production methods with state-of-the-art information and information technology. The digital era evolution is characterized by technology which increases the speed and breadth of knowledge turnover within the economy and society. The implications of the Digital Era are huge and will increase as technological functionality becomes more knowledge-based, our everyday lives and understanding of ourselves become more linked to it and it takes on a 'life' of its own [2]. "Smart" has become a new buzzword to describe technological, economic and social developments.

III. ACADEMIC EDUCATION 4.0

Education 4.0 is known as a higher form of blended learning. Academic Education 4.0 is going to be one that is intuitive, adaptive / personalized, knowledge-based, interactive and social (Community of Practices /Networks) – providing more value to both teachers and their students. The students will have to succeed in a working environment which is increasingly automatized, virtualized, networked and flexible. New skills and competencies such as nonlinear thinking, social and intercultural skills, meta-knowledge, self-management and self-competence will become more important. [3]

According to "The Impact of Technology on College Student Study Habits" a survey of more than 2,600 U.S. college students shows that students are embracing technology for its ability to help them learn more effectively through continual feedback:

- Almost two-thirds of students who already use such analytics report that their impact on their academic performance is "very positive" or "extremely positive."

- 75% of students using adaptive learning technology report that it is "very helpful" or "extremely helpful" in aiding their ability to retain new concepts.

- 68% of students using adaptive learning technology report that it is most helpful in making them better aware of concepts that they do not know yet.

The results show how technology—and adaptive technology in particular—can improve the entire student experience:

- 91% of all students report that adaptive capabilities in a digital study tool are "important" or "very important."

- Students reported that adaptive learning technologies are the most effective form of study technology, with 84% indicating a moderate or major improvement in grades.

- Technology increases engagement across the board: Students report that technology increases their engagement not only with course materials (77%), but with professors (64%) and fellow students (50%).

IV. THE CONCEPT OF THE BAVARIAN VIRTUAL UNIVERSITY

Set up in 2000, the Bavarian Virtual University (BVU) is a network of 31 universities and polytechnics that includes the higher education institutions of the Free State of Bavaria. BVU promotes and coordinates the development and implementation of online courses at Bavarian universities. "The BVU model based on cooperation among higher education institutions -despite of its high costs-was reported to foster the overall efficiency of the Bavarian higher education system and to reduce duplication of efforts in the area of distance and online teaching" [6].

BVU constitutes an important basic institution within Education 4.0 as it functions as a broker between course providers, students and further training providers. In contrast to the MOOC scene, this business model guarantees sustainability. BVU courses are MOOCs within Bavaria, as they are open to the entirety of Bavarian students, which means they represent a semi-open scenario.

The advantages of BVU concept:

- flexibility, independence of time and place

- high quality multimedia courses

- free of charge for students of BVU member universities

- high teaching standards with supervision and support

- complements and extends face-to-face courses

- individual choice of curriculum, selection of courses

- goal-oriented review or enrichment of specific subject areas

- acquisition of e-learning

V. ADAPTIVE USER INTERFACE FOR HIGHER EDUCATION

An intelligent, digital and adaptive interface is a global challenge in educational technology for higher education. The impact of individual adaptive user interfaces will gain higher importance in our education system [7]. The major advantage of AUIs is the possibility to respond dynamically to individual needs of a user, e.g. present only topics and course materials which are relevant for him or her and leave out everything which he or she already knows.

Since 2007, Prof. Dr. Dr. H Popp, Mathematics Teacher at Deggendorf Institute of Technology, [11] has started with the passive adaption of many online mathematics and informatics courses structured for different learning types. He created six different course contents which allow users to choose the material which fits with his/her learning behavior [10].

The active adaption by design and content of the user interface is driven by personalization and adaption of the content recommended to individual students needs [9].

A. Adaptive User Interface in Mathematics

One scenario where this is especially important is the case of beginners' courses, where students with different background knowledge and skills build a very heterogeneous group of participants. The more the users differ from each other, the more important is a flexible course design in order to be useful for every type of user. In this paper, we present an adaptive course UI for the popular learning management system "Moodle" which was developed for the beginners' course in Mathematics of BVU and hosted by Deggendorf Institute of Technology (DIT) [9].

The first step is the creation of an entry-level test, which every student is supposed to complete before he begins to work with the actual course materials.

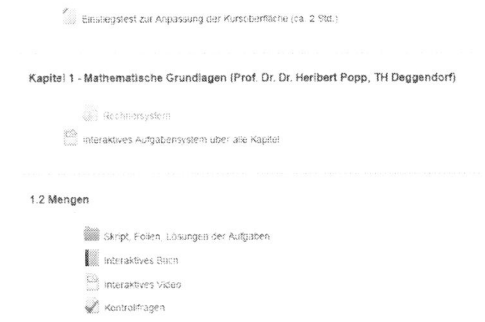

Fig. 1: Initial Mathematics course

This test should be created in a way that every part of the course (e.g. chapter) is covered by a set of questions, so that the results of these questions indicate the skills that the student already has in that topic. For the mathematics course, we chose a self-programmed, interactive Java Script framework in which users can solve mathematical exercises step-by-step by answering multiple-choice-questions and inputting missing parts of the solution.

For every chapter of the course (like fractions and equations), two appropriate exercises were chosen. The student's performance in these questions determines whether the corresponding topic in the course will later be shown or hidden, as can be seen in Fig.2.

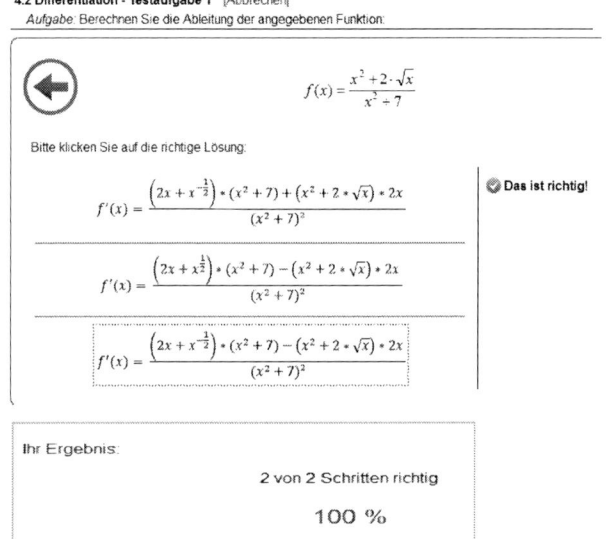

Fig 2. Solving mathematical exercises step-by-step

B. Results after completing the test

After completing all questions, the PHP and Java Script framework determines the final result for all chapters and fills out a hidden form which is sent to the Moodle course.

```
if(bearb_aufg > 0) {

    ergebnis_tblock[k] = summe_proz / bearb_aufg;

    if(ergebnis_tblock[k] >= bewertungsschluessel) {
        var p = HTMLObject({'tag':'p','class':"fb_gut",'text':"insgesamt " +
        Math.round(ergebnis_tblock[k]*100) + "% => Kapitel wird ausgeblendet"
        });
        box.appendChild(p);
    } else {
        var p = HTMLObject({'tag':'p','class':"fb_schlecht",'text':"insgesamt
        " + Math.round(ergebnis_tblock[k]*100) + "% => Kapitel wird
        eingeblendet"});
        box.appendChild(p);
    }

} else if(runs > 0) {
```

Fig 3. Extracted Source Code JavaScript & PHP

The course uses a plugin for Moodle in the form of a modified, self-programmed course layout. While in use, it requires the administrators to annotate every course chapter with a short token (e.g. "frac" for fractions). These tokens are used equivalently in the entry-level test and ensure the correct matching between questions and course topics. The plugin recognizes the arrival of the form data and saves the incoming results into a database table. When a student queries the course after completing the test, every chapter is checked for its relevance for the student during the rendering process.

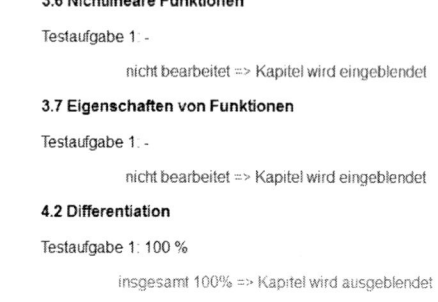

Fig 4. Correct matching between question and course topics

By comparing his or her performance to a threshold value which is defined by the admins, the chapter is then printed out or hidden.

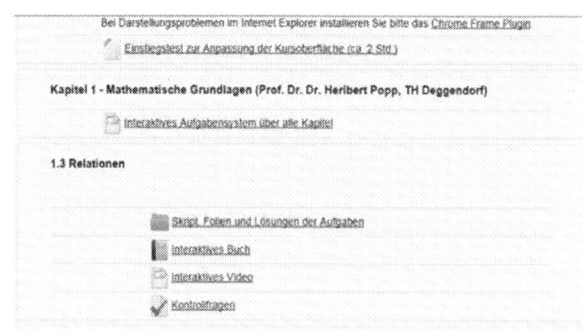

Fig 5. Presents only topics and materials which are relevant for him/her

VI. BENEFITS

Every student receives an individual course layout in which only his relevant chapters are shown. A first survey of students indicates that students are motivated and successful after using AUI.

TABLE 1 EVALUATION OF MATHEMATICS COURSE

	Evaluation of Mathematics Course DIT (69 Students)		
		Agreed	*Not*
Topics	I would visit the course again	55	14
	The course's time interval suited me well	65	4
	The course tool to long	15	54
	I'll recommand the course	60	9
	I feel familiar with the content	45	24
	I have gained subject related knowledge	65	4

VII. CONCLUSION AND FUTHER WORK

For future extension, the implemented mechanisms of course modification can be combined with additional machine learning technologies and learning analytics in order to adapt to each student's learning behavior and learning type in real time.

Baepler and Murdoch define academic analytics as an area that"...combines select institutional data, statistical analysis, and predictive modeling to create intelligence upon which learners, instructors, or administrators can change academic behavior". [12]

In "The State of Learning Analytics in 2012: A Review and Future Challenges" Rebecca Ferguson [13] tracks the "progress of analytics for learning as a development through:

- "The increasing interest in 'big data' for business intelligence

- The rise of online education focused around Virtual Learning Environments (VLEs), Content Management Systems (CMSs), and Management Information Systems (MIS) for education, which saw an increase in digital data regarding student background (often held in the MIS) and learning log data (from VLEs). This development afforded the opportunity to apply 'business intelligence' techniques to educational data."

REFERENCES

[1] NMC Horizon Report "Higher Education Edition", [Online], Available: http://cdn.nmc.org/media/2016-nmc-horizon-report-he-EN.pdf, (Accessed 08.08.2016)

[2] Bundesministerium für Bildung und Forschung (BMBF), Digital World – The Information Society, [Online], Availabe: https://www.bmbf.de/en/the-information-society-2353.html (Accessed 24.09.2016)

[3] Bundesministerium für Arbeit und Soziales, Studie „Wertewelten Arbeiten 4.0", March 2016 [Online]. Available: https://www.arbeitenviernull.de/fileadmin/Downloads/Wertestudie_Arbeiten_4.0.pdf (Accessed 01.10.2016)

[4] G. Doukidis, N. Mylonopoulos, and N. Pouloudi, "Social and Economic Transformation in the Digital Era", 2004, Idea Group Inc., pp. 1-18.

[5] McGraw-Hill Education's, "The Impact of Technology on College Student Study Habits", INDIANAPOLIS, (EDUCAUSE) – 28.20.2015, http://www.mheducation.com/news-media/press-releases/learning-analytics-new-likes-college-better-access-personalized-data-new-research.html

[6] "The concept of the Bavarian Virtual University", [Online]. Available: http://www.vhb.org/en/students/concept/ (Accessed 01.10.2016)

[7] Egham, UK, "What is Industrie 4.0" Gartner,(2015), [Online]. Available: http://www.gartner.com/newsroom/id/3054921

[8] M. Souto-Otero, A. Inamorato dos Santos, R. Shields, P. Lažetić, J. Castaño-Muñoz, A. Devaux, S. Oberheidt, Y. Punie , "OpenCases: Case Studies on Openness in Education", (2016), Institute for Prospective Technological Studies, Joint Research Centre, European Commission. EUR 27937 EN, doi:10.2791/039825

[9] H. Popp, R. Beer, "Evaluation virtueller Mathematik Kurse Lernszenarienvergleich und Learning Analytics", Tagung Graz, E-learning Tag 2014; S 98-108, (2014)

[10] H. Popp, et al."Steigerung des Humankapitals in KMUs durch virtuelle Weiterbildung, bei der sich E-Learning-Systeme an die Benutzer anpassen", In D. Krieger, A. David, A. Bellinger (Ed.). Wissensmanagement für KMU (219-236); vdf Hochschulverlag: Zürich, Schweiz 2006

[11] H. Popp, „E-Learning-System bedient die verschiedenen Lernertypen eines betriebswirtschaftlichen Fachbereichs: Didaktik, Realisierungstechnik und Evaluation", E. Seiler Schiedt, Chr. Sengstag (Ed.). E-Learning – alltagstaugliche Innovation? (141-151); Waxmann: Münster; New York; München; Berlin (2006), ISBN: 978-3830917205

[12] P. Baepler and C. J. Murdoch "Academic Analytics and Data Mining in Higher Education," International Journal for the Scholarship of Teaching and Learning: Vol. 4: No. 2, Article 17. (2010). [Online] Available: https://doi.org/10.20429/ijsotl.2010.040217

[13] R. Ferguson, "Learning analytics: drivers, developments and challenges", International Journal of Technology Enhanced Learning, 4(5/6) pp. 304–317, (2012), Available: http://oro.open.ac.uk/36374/

Experimental Module for Assistive Technologies Applications

Ioan Lita, Daniel Alexandru Visan and Alin Gheorghita Mazare

Electronics, Communications, Computers Department
University of Pitesti
Pitesti, Romania
ioan.lita@upit.ro

Abstract — This paper approaches the problem of implementing an experimental module that facilitates the teaching and learning of the Braille alphabet. The proposed system follows to improve the existing assistive technologies dedicated for helping visually impaired peoples. The main board of the experimental module is realized around the PIC 16F877A microcontroller that runs a proprietary software application for controlling the operation of the system. A set of two stepper motors and six solenoids are used for actually emboss the Braille characters on the paper or other suitable physical support. For ensuring the required level of power, the solenoids and the stepper motors are connected to the output ports of the microcontroller trough a set of drivers realized with ULN2803A circuits. The module can communicate with a PC through a serial interface. This connection is necessary for receiving the characters that mast be printed. The conversion of the plain text to the Braille format, together with other formatting steps are realized in an automate manner by the software processing application installed on the PC. The maximum printing speed achieved with the proposed module is of 50 characters/second approximately. By using a compact hardware architecture based on microcontrollers combined with a versatile software program, dedicated for this type of application, the proposed module distinguish by other existing implementations through its reliability and flexibility in operation.

Keywords — *assistive technologies, Braille alphabet, microcontroller.*

I. INTRODUCTION

For visually impaired and blind peoples the most important access methods to computers technology is by tactile and hearing senses. Voice synthesis is most efficient form the point of view of the quantity of information that can be transmitted but in the case of persons having also hearing impairment the devices based on tactile sense remain the only option [1].

Because the usual informatics systems have not been designed natively with the capabilities required by the interaction with non visual users, for this kind of persons it is necessary to develop dedicated software applications and specialized equipments that can be capable to convert visual information into an accessible format. For this reasons, in this paper is approached the problem of realizing a experimental module that allows transcription of text into Braille alphabet

and printing by actually embossing the Braille characters on the paper or other suitable physical support [2].

The hard copies of the text resulted at the output of the printing module constitute valuable tools in teaching and learning processes of the Braille alphabet because, in practice, despite of the wide spreading of this type of coding, only a small part of visual impaired peoples are capable to use it.

II. THE BRAILLE ENCODING SYSTEM

The Braille system represents a tactile method that allows visually impaired and blind people to read and even write document using dotted symbols sensed through touch.

The basic Braille characters have a uniform structure formed by six dots organized in a matrix with two columns and three lines. This organizations of the elements that compose a Braille character was imposed by the necessity that each dot matrix to fit under the user's fingerprint.

For extending the number of characters and symbols that can be encoded as much is possible, the Braille system make use of various arrangements of all 64 possible combinations that can be realized with the basic structure of points [2].

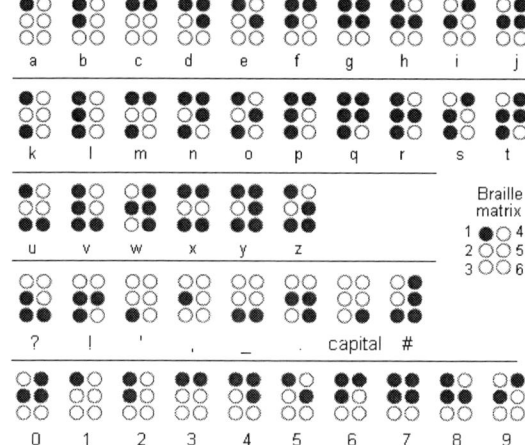

Fig. 1. Representation of usual characters and symbols using Braille encoding system.

2016 IEEE 22ⁿᵈ International Symposium for Design and Technology in Electronic Packaging (SIITME)

The available basic combinations can be used not only for representing letters, numbers, usual words and punctuation marks but also for coding mathematical and logical symbols, musical notes etc. Also, the Braille system makes use of context for conferring multiple significations to the same dots arrangement or combination of symbols. The reading of Braille characters is realized from left to right. As can be observed from Fig. 1, the first ten letters of alphabet are encoded using only for dots positioned at the top of the Braille matrix (1,2,4,5). The rest of the letters, depending also of the particularity of the language, are represented using another set of two dots (3 and 6). The punctuation marks are encoded separately, in the lower side of the Braille matrix. The hand writing in this system is possible in practice with a puncturing device. The characters are actually embossed on the paper or other suitable physical support. In this case the hand writing must be realized from right to left ensuring that when the paper is mirrored the characters appears in normal configuration for the reader. The operation is difficult and very slow due to the mirroring step. This represents another reason why it is necessity to implement a reliable module that allows transcription and printing of text into Braille alphabet.

III. THE STRUCTURE OF THE EXPERIMENTAL MODULE

The simplified block diagram of the module is illustrated in the Fig. 2. The experimental module presented in this paper ensures the access of visual impaired people to the information and open a way for efficiently use of a wide range of characteristic offered by computers and communication equipments [3].

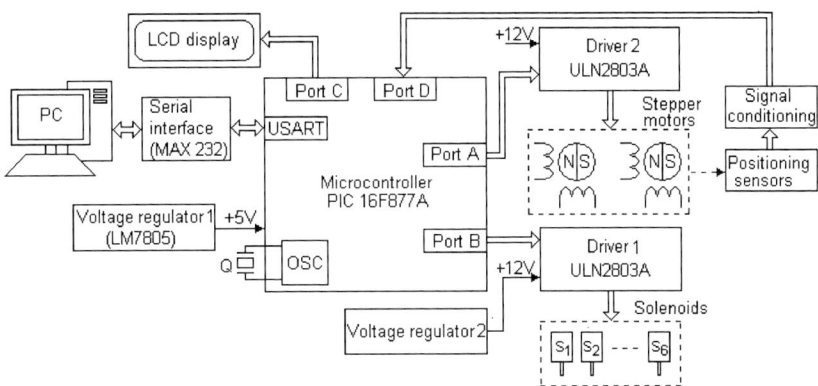

Fig. 2. The simplified block diagram of the experimental module used for applications in the field of teaching and learning of the Braille alphabet.

The proposed system use a microcontroller based hardware architecture combined with two software applications, one written in assembler language, for controlling the operation of the system, and the second written in LabVIEW and dedicated for conversion of text information into the Braille format [3]. The experimental module is realized around the PIC16F877A microcontroller that implements the control functions necessary into the system. For serial communication between experimental module and the PC it was used a MAX232 circuit that ensures the bidirectional conversion of the voltage levels that are specific to RS232 standard and TTL logic levels that are generated by the USART module localized inside the microcontroller. The PIC16F877A microcontroller is clocked at a frequency of 20 MHz stabilized by a external quartz. As can be seen from Fig. 2, the block diagram of the experimental module contains two voltage regulators for separately supply the proper voltage to the control module and the execution elements. The microcontroller, the LCD and the serial communication circuit, together with the signal conditioning module that is responsible with the reading of the positioning sensor's outputs are together supplied from a separate voltage regulator having a +5V output voltage. A second voltage regulator having a output voltage of +12V was used in the schematic for eliminating the possibility that the control unit to be influenced by spurious signals generated by the execution elements represented by drivers, electrical motors and printing solenoids [4].

The software application for the PIC16F877A microcontroller was written in the PIC C programming environment. A general view for the state diagram of this application is presented in Fig. 3.

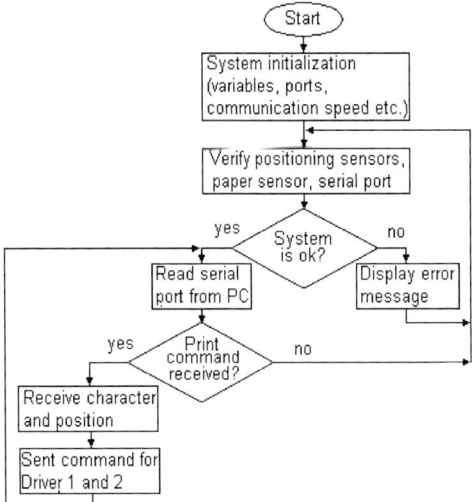

Fig. 3. The simplified state diagram of the software application for the microcontroller that was used in the implementation of experimental module.

IV. IMPLEMENTATION AND RESULTS

As been mentioned in the previous paragraph, the experimental module for Braille printing was implemented around the versatile microcontroller PIC16F877A. The device operates with a reduced instruction set (35 single word instructions only) and has relatively large flash memory that can be programmed easily for a very large number of times. This is an important aspect in designing and prototyping new experimental equipments that usually requires frequently program modification and debugging.

The maximum time period of 200ns, necessary for every instruction cycle execution of PIC16F877A, ensure a higher enough operating speed for the proposed application.

The 10 bits, internal analog-to-digital converter with 8 multiplexed input channels represents another feature that was used in this project for reading the signals from the sensors that monitors the temperatures and the current consumption of the printing solenoids. Also, two positioning sensors were integrated in the system for proper control of the printing head that contains the matrix with six solenoids.

The ULN2803A driver circuit is used for connecting the solenoids and the stepper motors to the output ports of the microcontroller. The circuit contains eight Darlington transistors in common emitter configurations that ensures the required level of power for the solenoids and the stepper motors.

The internal transistors are protected by a set of suppression diodes that cancel the voltage overshoots specific to inductive loads. Also, the ULN2803A driver contains an internal $2,7k\Omega$ bias resistor that configures the circuit for proper operation with TTL and CMOS that are supplied at voltage of +5V for compatibility with typical microcontrollers.

In the Fig. 3 is presented the detailed structure of the electronic circuit used for hardware implementation of the proposed experimental module.

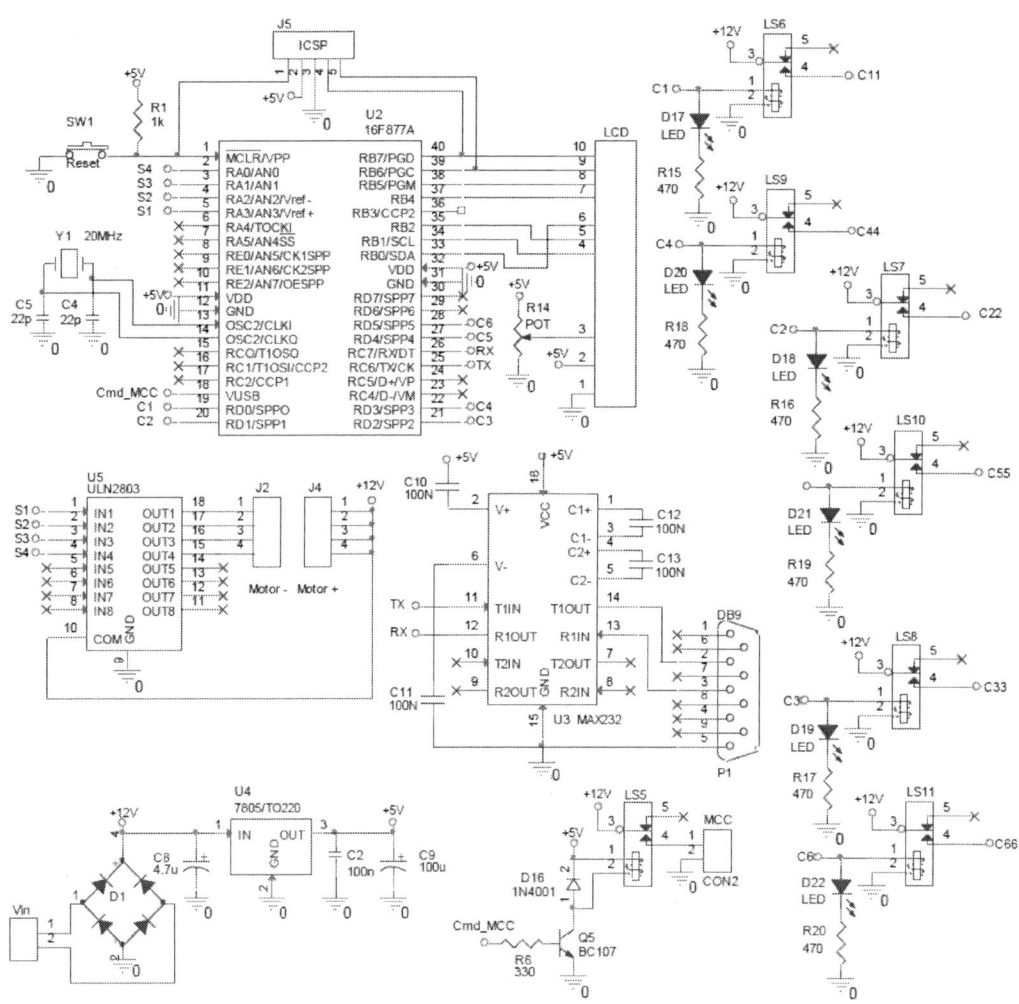

Fig. 4. The electronic diagram of the experimental module used for applications in the field of teaching and learning of the Braille alphabet.

The maximum level of current that can be sustained by these groups of transistor is of 600mA. Because this limitation in our experiments we also used a driver based on relays for controlling the matrix of six solenoids. This approach proved to be efficient but the reliability of the system for long operation time can be affected. Another solution is to operate with pairs of transistors connected in parallel in order to be capable to sustain a higher level of current required by more powerful solenoids.

In Fig. 5 can be observed the general aspect of the hardware implementation of the proposed module. The six solenoids, organized in the typical matrix that is used for printing Braille characters are depicted in Fig. 6.

In the practical experiments was used EMA-0420-L-06 open frame solenoids characterized by an internal resistance of 30Ω and an actuating inductor with 1095 turns that can be operated at 25% duty cycle when the supply voltage is of 12Vdc. The force generated at the activation of the solenoid depends on the stroke of the embossing had. For allowing the system to work with thicker printing supports, in the proposed design was established a maximum stroke of 1 mm. Another method for increasing the force of the printing had is to supply the solenoids from a higher voltage source, having a output level of up to 20V, but in this case the duty cycle must be reduced at 10% or less.

The performances of the module were evaluated through various tests that proved improved efficiency, reliability and flexibility in operation, compared with other similar equipments.

It was found that the maximum printing speed achieved with the proposed module is around of 50 characters/second considering A4 size normal or thicker copying paper sheets having a weight of maximum $160 g / m^2$.

Fig. 5. The hardware implementation of the module used for teaching and learning the Braille alphabet.   Fig. 6. The solenoids used for printing Braille characters.

V. CONCLUSIONS

The informatics systems have not been designed natively with the capabilities required by the interaction with non visual users so it is necessary to develop specialized equipments for this kind of persons.

The experimental module presented in this paper allows transcription of text into Braille alphabet and embossing the Braille characters on the paper or other suitable physical support at a maximum printing speed of 50 characters/second.

Because only a small part of visual impaired peoples are capable to use Braille alphabet the designed system can become a valuable tool in teaching and learning processes of this type of encoding.

REFERENCES

[1] Mohammed Kaleemur Rahman, Saurabh Sanghvi, Noura El-Moughny, "Enhancing an automated Braille Writing Tutor", 2009 IEEE/RSJ International Conference on Intelligent Robots and Systems, pp. 2327 - 2333, 2009.

[2] Hélène Pigot, Sylvain Giroux, "Living labs for designing assistive technologies", 17th International Conference on E-health Networking, Application & Services (HealthCom), pp. 170 – 176, 2015.

[3] Andrei Drumea, Robert Dobre, "Clicks counting methods for a scope knob," Hidraulica, vol. 4, pp. 79-84, October 2013.

[4] Dianne T. V. Pawluk, Richard J. Adams, Ryo Kitada, "Designing Haptic Assistive Technology for Individuals Who Are Blind or Visually Impaired", IEEE Transactions on Haptics, Vol. 8, Issue 3, pp. 258 - 278, 2015.